# Microbiology of Waterborne Diseases

**Steven Percival**
*Department of Microbiology, Leeds General Infirmary, Leeds, UK*

**Rachel Chalmers**
*Cryptosporidium Reference Unit, Singleton Hospital, Swansea, UK*

**Martha Embrey**
*The George Washington School of Public Health and Health Services,
Washington DC, USA*

**Paul Hunter**
*Chester Public Health Laboratory, Chester, UK*

**Jane Sellwood**
*Environmental Virology Unit, UK Health Protection Agency,
Reading, UK*

**Peter Wyn-Jones**
*Institute of Pharmacy, Chemistry and Biomedical Sciences,
University of Sunderland, UK*

ELSEVIER
ACADEMIC
PRESS

AMSTERDAM • BOSTON • HEIDELBERG • LONDON • NEW YORK • OXFORD
PARIS • SAN DIEGO • SAN FRANCISCO • SINGAPORE • SYDNEY • TOKYO

Permissions may be sought directly from Elsevier's Science & Technology Rights Department in Oxford, UK: phone: (+44) 1865 843830, fax: (+44) 1865 853333, e-mail: permissions@elsevier.co.uk. You may also complete your request on-line via the Elsevier homepage (http://www.elsevier.com), by selecting 'Customer Support' and then 'Obtaining Permissions'

Elsevier Academic Press
525 B Street, Suite 1900, San Diego, California 92101-4495, USA
http://www.elsevier.com

Elsevier Academic Press
84 Theobald's Road, London WC1X 8RR, UK
http://www.elsevier.com

**British Library Cataloguing in Publication Data**
A catalogue record for this book is available from the British Library

Library of Congress Catalog Number: 2003115959

ISBN     0-12-551570-7

Typeset by Charon Tec Pvt. Ltd, Chennai, India
Printed and bound in Great Britain
04 05 06 07 08 09 9 8 7 6 5 4 3 2 1

# Contents

# Preface

The microbiological quality of drinking water varies widely and constitutes a constantly fluctuating ecosystem composed of a dynamic complexity of niches. With these multifacted ecological niches in water comes man's ambition to detect and ultimately control microorganisms that represent a concern to public health. The health consequences of exposures to waterborne pathogens are significant and constitute a grave concern to both developed and developing countries. However, due to the deficiencies of current knowledge in both identifying and understanding the ecology of waterborne pathogens, the determination of safe exposure levels of these pathogens to man still represents an area of intensive investigation. With poor methods available for the detection and monitoring of pathogens, it is presently impractical and costly to try to determine all the types of pathogens found in drinking water.

Water is a chaotic ecosystem and through every drinking water outlet comes a multitude of problems when considering health concerns due to waterborne pathogens. Surely, therefore, in water when it comes to knowing and understanding the microbes of human significance we are only touching the 'tip of the iceberg' when it comes to highlighting microbes that have an impact on public health. However, the emphasis in drinking water research is placed upon understanding conditions likely to ensure the safety of drinking-water supplies and monitoring their fulfilment more directly.

Historically we still adhere to the concept that with the absence of *E. coli* from drinking water comes the reassurance that pathogens of human significance are also absent. However, what we must consider in water is not just those microbes that produce short-term suffering, but the ones that may have long-term health implications not yet discovered. Have we therefore to leave well alone or to strive to understand pathogens and their behaviour in water and try to determine other microbes that may well compound other human diseases?

This book looks at the pathogens that are known to be associated with drinking water and acknowledged to constitute a concern to 'the health of the nation'. The book brings together the current knowledge of drinking water pathogens highlighting their basic microbiology, clinical features, survival in the environment and, in particular, risk assessment. We hope you enjoy reading this book.

*Steven Percival*

# Part 1

# Introduction

# 1

# Risk assessment and drinking water

## What is risk?

The provision of safe drinking water was one of the triumphs of 20th century public health. The use of chlorine disinfection, especially, has resulted in a drastic decrease in exposure to waterborne pathogens, and therefore, waterborne disease. In the 1970s, water pollution from chemical waste received special attention following incidents such as the Cuyahoga River in the USA catching fire. This prominence of chemical risk incidents accelerated the practice of environmental risk assessment. Consequently, early risk assessment paradigms focused exclusively on chemical contamination. When the 1993 outbreak of cryptosporidiosis from drinking water made some 400 000 Milwaukee residents ill and killed over 100, the focus of risk in drinking water shifted to the microbial. The influences of agricultural and municipal wastewater, as well as the ageing of water treatment and distribution systems, continue to raise concerns about the microbiological quality of drinking water.

In 1994, the US Congress directed the formation of a Commission to investigate the appropriate uses of risk assessment and risk management in federally mandated programmes. Their Presidential/Congressional Commission on Risk Assessment and Risk Management – Final Report (1997) defines risk as '... the probability that a substance or situation will produce harm under specified conditions. Risk is a combination of two factors: the probability that an adverse event will occur (such as a specific disease or type of injury) and the consequences of the adverse event. Risk encompasses impacts on public health and on the environment, and arises from exposure and hazard. Risk does not exist if exposure to a harmful substance or situation does not or will not occur. Hazard is determined by whether a particular substance or situation has the potential to cause harmful effects.'

## Risk assessment

Risk assessment is carried out by international agencies (e.g. International Agency for Research on Cancer and the World Health Organization), national governments, state and local governments and private industry. This process uses the broadest sort of regulatory decision making on a national and international level – from a national government regulating drinking water treatment protocols to the local public health department making a decision about issuing a boil-water advisory.

The practice of quantitative risk assessment gathered momentum in the 1960s with scientists' attempts to estimate risks of small exposures to carcinogens. Quantitative risk assessment would improve decision makers' ability to set both research and regulatory priorities. During the 1970s and 1980s, methods of risk assessment began to evolve, and better data were available to support the assessments and their conclusions. At this point in history, risk assessment was used primarily to estimate chemical risks by extrapolating toxicological data from animals to humans.

In 1983, the process of risk assessment was formalized in a report from the US National Research Council (NRC) called *Risk Assessment in the Federal Government: Managing the Process*. This pivotal publication still serves as the basis for most risk assessment frameworks. The NRC formula comprises four steps that lead to the evaluation of data on the hazards of the agent in question, on the extent of human exposure to it, and on the characterization of the possible risk. This framework was the first to present a systematic approach to analysing scientific information about substances that may pose health risks under given conditions.

This four-step framework is now a universally recognized method to characterize the likelihood of adverse health effects from (mainly chemical) exposures:

1. Hazard identification
   - determines the types and quantities of the contaminants under study
   - identifies the nature of hazards they pose to health

- determines if exposure to the agents can cause an increased incidence of adverse health effects and
- characterizes the nature and strength of the evidence of causation.

2. Dose–response assessment
   - determines the relationship between concentrations (dose) and adverse effects from exposure (response)
   - characterizes the relationship between exposure or dose and the incidence and severity of the adverse health effect
   - considers factors that influence dose–response relationships, such as intensity and pattern of exposure and age and lifestyle variables that could affect susceptibility
   - involves extrapolation of high-dose responses to low-dose responses and from animal responses to human responses.

3. Exposure assessment
   - describes the conditions under which people could be exposed to contaminants
   - determines the intensity, frequency and duration of exposures of humans to the agents in question
   - estimates concentrations of the substances at various points from their sources through the environment.

4. Risk characterization
   - describes the nature of adverse effects that can be attributed to contaminants, estimates the likelihood of exposures and evaluates the strength of evidence and uncertainty
   - combines the assessments of exposure and response under various exposure conditions to estimate the probability of specific harm to an exposed individual or population
   - should include the distribution of risk in the population.

Risk assessment can be controversial because risk evaluation is based on both scientific data and judgement. So much uncertainty exists because our current knowledge regarding exposure and effects of agents, and their relationship to humans and the environment, is unclear. Within this framework, there continues to be active debate about the most appropriate risk assessment approaches, the impact of various kinds of data on risk projections, and the level and appropriateness of conservatism to use in estimates. The most frequently debated issues within risk assessment include the use of default options, which are mostly conservative and more likely to overestimate risk; the validation of methods and models; and the variability in those exposed – within individuals, among individuals, and among populations. The question of how to accommodate susceptible subpopulations in the framework is crucial, and the uncertainty in data and models cannot always be quantified.

Risk assessment uses research and data from a variety of disciplines – epidemiology, toxicology, statistics, molecular biology, clinical medicine, exposure modelling, dosimetry and others. Determining what constitutes sound science requires time to gather information and consider opposing views. The risk assessor must evaluate the context, source, presumptions, and biases of scientific studies that will be used as evidence. The policies and procedures for conducting risk assessment are not static; they have evolved over time in the face of new demands and the availability of new and better kinds of information. A spectrum of beliefs and data typically must be deciphered and evaluated to extract those scientific facts, assumptions and beliefs that will stand the test of time.

## Microbial risk assessment in drinking water

### Introduction

While techniques for assessing chemical risks have been in widespread use for over 40 years, we have less experience performing risk assessments for microbial pathogens. Differences between chemicals and pathogens in environmental fate and transport and pathogenesis require new approaches to quantitative risk assessment (Gibson *et al.*, 1998). Quantitative risk assessment methodology has not been thoroughly applied to environmentally-transmitted infectious diseases. The use of risk assessment to evaluate infectious microbes in drinking water, shellfish, recreational water, and foods is now being reported, but not in a systematic way.

### Regulatory history

Risk assessment in general is closely tied to governmental policies focusing on the control of contaminants. In the US Safe Drinking Water Act (SDWA), risk assessment was used to set maximum contaminant levels (MCLs) for chemicals that may have the potential to cause health problems in humans. Risk assessment has not been used to develop microbial standards for drinking water in the USA, and microbial contaminants in drinking water have not been individually prioritized for regulation (Rose and Gerba, 1991). The maximum contaminant level goal (MCLG) for pathogens is zero, but this has been accomplished through specifying treatment methods for various types of source water and monitoring for faecal coliform bacteria, which may indicate contamination. The US Environmental Protection Agency's (EPA) Surface Water Treatment Rule requires a safety goal of 99.9% reduction in *Giardia* and viruses (Environmental Protection Agency, 1998). EPA believes that this level of removal would result in an annual risk of no more than one infection/10 000 people exposed to drinking water (Rose and Gerba, 1991). The European Union standard also does not permit any pathogen level above zero in drinking water (Barrell *et al.*, 2000). Their regulatory guidance relies on total coliforms, faecal coliforms and total colony counts as pollution indicators. Although this combination of pollution indicators and treatment controls has worked fairly well over the years, major deficiencies have become known because of emerging pathogens such as *Cryptosporidium* and *Mycobacterium avium* complex (MAC); in fact, the UK has introduced precedent-setting legislation that singles out *Cryptosporidium* oocysts as a detection parameter for drinking water quality (Barrell *et al.*, 2000). Problems have been identified with indicator organisms (e.g. members of the *Enterobacteriaceae*), such as the fact that viruses and protozoa can be present and viable when indicator organisms are inactive. Also, coliforms and other indicator bacteria may be more sensitive to chlorine than some pathogenic organisms, so the resulting treated water quality assessment can be inadequate. Many communities have experienced waterborne disease outbreaks even though their water supplies have met mandated coliform standards (Craun *et al.*, 1997).

## Application of data

The role of toxicology is large in chemical risk assessment. For many reasons – the greatest of which are ethical considerations – animal studies are relied on extensively in chemical evaluation. Usually, the only available type of data on human exposure to chemicals is occupational in origin or the result of an accidental spill or release. In the absence of human data, toxicological evaluation usually begins with the simplest, fastest, and most economical tests and proceeds to complexity only as warranted by the initial results (National Research Council, 1994). In the case of microbial infection, it is difficult to develop animal models for most pathogens because of their specificity to humans, but some do exist. On the other hand, the high incidence of many waterborne diseases gives us data on human exposure and health effects in the form of surveillance and outbreak reports that would not typically be available for chemicals.

Although toxicology, epidemiology, clinical medicine and exposure assessment all contribute data to risk assessment, epidemiology is important because it provides evidence on human subjects in real-world situations. This is especially important in microbial risk assessment. The WHO formed a working group to look at evaluating and using epidemiological data for health risk assessment. Their guidelines on using epidemiological information have been published (World Health Organization Working Group, 2000). Because most epidemiology is observational, drawing conclusions about causation is problematic. People engage in many different behaviours that affect their exposures: choice of food, drinking water, recreational activities, workplace and contact with others. All of these affect a person's susceptibility to becoming infected and to developing illness once infected. Epidemiologists must try to sort through these many factors in order to draw inferences about their hypotheses. Because epidemiology is observational and inductive, no single study can provide a definite answer about cause and effect, even if bias is minimal (Pontius, 2000).

*Use of surveillance*

In the USA and the UK, formal surveillance data are collected through collaboration between federal, state and local government entities. The UK system also includes a central automated screening process that looks for unusual patterns of microbial infection in weekly monitoring reports (Tillett *et al.*, 1998). The goals of a surveillance programme consist of characterizing the epidemiology of outbreaks, identifying the agents causing outbreaks, and identifying how an outbreak occurred (Barwick *et al.*, 2000). Information that can be gleaned from surveillance includes magnitude of community impact, attack rates, hospitalization and mortality, demographics, sensitive populations, level of contamination, duration, medical costs, community costs and secondary transmission.

The US Centers for Disease Control and Prevention (CDC) defines a waterborne outbreak as two people experiencing a similar illness after exposure to drinking or recreational water, and evidence must indicate that water is the probable source of the outbreak (Barwick *et al.*, 2000). However, the role of waterborne exposure is difficult to estimate for most pathogens. Those that are transmitted via the faecal-oral route can contaminate food as well as water or be spread by person-to-person contact. Waterborne outbreaks of caliciviruses and microsporidia, for example, have been documented, but although other viruses, such as echovirus and adenovirus, are present in sewage, they have not been associated with drinking-water transmission.

Outbreak surveillance uses both epidemiological data and water-quality data, when available. Epidemiology, which is the study of occurrence and causes of diseases in a population, identifies associations between exposures and health outcomes using statistical methods, and has been the main science used to study the relationship between drinking water and infective disease. Because infectious diseases are caused by many exposures besides just drinking water, epidemiology can help define the contribution of drinking water. Water quality data are helpful, but frequently unavailable because of the transient nature of contamination and the analytical difficulties in detection (Tillett *et al.*, 1998). However, the UK has begun regular monitoring for the presence of cryptosporidial oocysts in municipal water supplies (Hunter, 2000).

Waterborne disease surveillance data are useful in evaluating our treatment and supply of safe drinking and recreational water; however, quality and completeness of investigations vary from place to place, and risks are not fully identified. The extent of that underestimation varies by locale and is unknown overall. The ability to recognize an outbreak relies on a number of factors including public awareness, ill people seeking health care, the extent of laboratory testing, and local infrastructure available for investigating and reporting outbreaks. Interventions to increase waterborne illness surveillance programmes have included the use of 'sentinel' populations, like nursing home residents, pharmacy sales of anti-diarrhoeal drugs, and school absentee logs (Proctor *et al.*, 1998).

Standardizing the way data are collected, managed, transmitted, analysed, accessed and disseminated is critical in evaluating the adequacy of current regulations for water treatment and monitoring, as well as producing important information on the pathology and epidemiology of waterborne diseases. Timely and accurate surveillance data are crucial to identifying an outbreak's aetiology and therefore methods for mitigation. Surveillance studies are perhaps the most important source of information we have on waterborne disease, and improving the infrastructure for surveillance programmes would enhance our knowledge on waterborne disease (Gostin *et al.*, 2000).

## Microbial risk assessment frameworks

A number of frameworks to address some of the issues unique to microbial risk assessment have been developed. Some have taken the traditional four-step chemical assessment paradigm and altered it; others have used a different basis for analysis. Rose and Gerba (1991) propose the following alteration of the traditional four steps for use in microbial risk assessment:

1 Hazard identification
   - identifies the microorganisms of concern with food or water through scientific literature such as clinical studies, epidemiological studies and surveillance, animal toxicology studies, and *in vitro* toxicology studies.
2. Dose response
   - describes the severity and duration of adverse effects that may result from the ingestion of a microorganism in water through qualitative or quantitative methods
   - relates to the microbe and to the human host: virulence and infectivity, secondary transmission, asymptomatic infection, low doses of some organisms resulting in severe effects.
3. Exposure assessment
   - assesses the extent of human exposure, such as the occurrence and concentration of microbial contaminants in raw and treated drinking water or recreational water, and the actual consumption of drinking water
   - relates to socioeconomic and cultural backgrounds, ethnicity, seasonality, population demographics, geographical differences, and consumer preferences.
4. Risk characterization
   - depends on the variability, uncertainty and assumptions used in the previous steps. Data uncertainties include those that might arise in the evaluation and interpretation of epidemiological, microbiological and animal studies. Biological variation includes the differences in virulence that exist within a microbiological population and variability in the susceptibility within the human population.

The Codex Committee on Food Hygiene (1999) has published guidelines regarding microbial risk assessment. Some of the general principles the committee has defined include:

- separation between risk assessment and risk management
- inclusion of hazard identification, hazard characterization, exposure assessment and risk characterization in the approach
- use of a transparent and iterative process
- use of precise, best-available data
- consideration of the dynamics unique to microbes, such as growth, survival, death and the interaction between host and pathogen.

An extensive risk framework has been developed that attempts to address the issues unique to waterborne microbial risk assessment (Neumann and Foran, 1997). US EPA has adopted this approach to evaluate future contaminants for regulation (Risk Policy Report, 2001). This three-part framework comprises:

1. Problem formulation
2. Analysis phase
3. Risk characterization.

*1) Problem formulation.* The purpose of problem formulation is to identify the purpose, goals and extent of the risk assessment and consider the major factors that will come into play. An initial risk model will be formulated in this step. This conceptual model describes the interactions of a particular pathogen and a defined population within a defined exposure scenario. An initial characterization of exposure and health effects is conducted. This step considers pathogen characteristics, host characteristics and host–pathogen interactions. The exposure scenarios and health effects are defined, including the media, exposure routes, endpoints and essential assessment variables.

*2) Analysis phase.* The analysis phase of the microbial risk assessment consists of the technical evaluation of data concerning the potential exposure and associated health effects and is based on the conceptual model developed during problem formulation. This phase consists of two elements: characterization of exposure and characterization of human health effects. The quality and quantity of data available affect analyses – many uncertainties exist. This analysis characterizes the relationship between dose, infectivity, and the manifestation and magnitude of health effects in an exposed population. The relationship is complex and, in many cases, a complete understanding will not be possible. Data obtained from animal studies, human clinical studies and outbreaks are used to generate a curve or model for the quantitative relationship between dose and response. Animal models may be useful for determining these relationships, but should be interpreted carefully because of the host specificity of most pathogens. Another difficulty that may be encountered in a dose–response analysis is the availability of data regarding infection. In many cases, infection data will not be available, so the analysis may only be able to describe the relationship between dose and clinical illness, rather than dose, infection and clinical illness.

Characterizing exposure involves the evaluation of the interaction between the pathogen, the environment and the human population and results in the development of an exposure profile that quantitatively or qualitatively evaluates the magnitude, frequency and pattern of human exposure for the scenarios developed during problem formulation. Some of the questions asked during this process have to do with the nature of the pathogen, such as its ability to be transmitted and cause disease in the host, its occurrence and distribution in the environment and its ability to survive and multiply. Human factors to be considered include the demographic characteristics of the exposed population, how the population is exposed and how long the population is exposed.

Characterization of human health effects involves the interactive analysis of three critical components: host characterization, evaluation of human health effects and quantification of the dose–response relationship. Elements include individual susceptibilities like age, immune status, nutritional status, etc.; clinical manifestations regarding infection, illness and sequelae and characterizing the nature of the relationship between exposure, symptomatic and asymptomatic infection and the duration and severity of illness.

*3) Risk characterization.* Risk characterization is the final phase of the microbial risk assessment and is the product of combining the information from the exposure profile and the host–pathogen profile. Risk characterization consists of two major steps: risk estimation and risk description. Risk estimation describes the types and magnitude of effects anticipated from exposure to the microbe and the likelihood of those effects occurring. It can be qualitative or quantitative depending on the data and methods used. The second component of risk characterization – risk description – involves describing the event according to its nature, severity and consequences. Uncertainties associated with all phases – problem formulation, analysis and risk characterization – are identified and quantified when possible in this section.

Risk management is the process by which the results of risk assessment are integrated with other variables, such as political, social, economic and engineering factors, to arrive at decisions about the needs and methods for risk reduction. Although not included as part of the original four-part paradigm of risk assessment designed by the NRC, risk management entails the important process of creating actions and strategies to reduce risks. Examples of how microbial risk assessment data can be applied to practical risk management approaches include:

- predicting endemic rates of drinking-water-related infections
- estimating pathogen densities that drive standard-setting and treatment protocols
- determining the effectiveness of water treatment
- forecasting the risk increase during water-treatment failure

- balancing microbial risks with risks of disinfection by-products
- identifying the most cost-effective options to reduce microbial health risks (Gale, 1996).

### Susceptible subpopulations

Risk assessments are generally used to guide regulatory decisions regarding a level of exposure resulting in 'acceptable' risks. When the exposure exceeds the defined risk level, action is taken to reduce the concentration of pollutants to an acceptable level. Defining what kind of vulnerability merits exposure protection and where to draw the regulatory line on whom to protect and how best to protect them are the issues with which policy makers struggle. Even the definition of 'susceptible' in this context has been a moving target, with scientists and policy makers approaching it from different directions (Parkin and Balbus, 2000).

A greater emphasis on susceptible subpopulations has resulted in a more complex risk assessment process. People are concerned about vulnerable members of the population, but characterizing who is most sensitive to which particular risk is an impossible task. Therefore, concepts of susceptibility have focused on:

1. the probability that an individual will be exposed to a questionable agent and then react to it
2. the comparison of an individual's susceptibility to that of the majority of the population
3. the variation of individual states of vulnerability within a population. Important factors in overall susceptibility include immune status, pregnancy, underlying illness and lifestyle (e.g. smoking and drinking habits). Due to environmental effects on expression, genetic factors may either result in lifelong or periodic susceptibility.

Host characterization contains susceptible subpopulation factors, for example age, immune status, genetic background, pregnancy, etc. The elderly, children and the immunocompromised are perceived as most susceptible to infection and illness from waterborne pathogens, however, that assumption is not always founded in fact. Certain illnesses are closely associated with the immunocompromised, such as MAC and microsporidia in AIDS patients. Other pathogens, such as *Helicobacter pylori* do not appear to affect the immunocompromised any differently than the immunocompetent (Edwards *et al.*, 1991) and, though *H. pylori* infection occurs primarily in childhood, its sequelae do not manifest themselves until adulthood. Viruses have a higher incidence in childhood – probably because of a child's naive immune system – but it is common for adults to have a higher degree of morbidity once infected. MAC manifests itself completely differently in the host. In children, it causes lymphadenitis. In adults with underlying lung damage, it causes pulmonary MAC, which is a chronic disorder, like tuberculosis. Pulmonary MAC has also been showing up in elderly women who are without predisposing factors (Prince *et al.*, 1989; Kennedy and Weber, 1994). In AIDS patients, MAC manifests itself as a disseminated, end-stage disease. Up to 40% of AIDS patients develop disseminated MAC in their lifetimes, though the incidence has dramatically decreased with the advent of retroviral drug therapy (Horsburgh, 1991; Havlir *et al.*, 2000).

There are different ways a population can be categorized as susceptible in terms of the frequency and severity of infection. A pathogen may affect a certain population by:

1. having a higher rate of infection than the general population, but the same disease response
2. having a higher rate of infection than the general population, but a lessened disease response
3. having the same incidence of infection as the general population, but a greater level of morbidity and/or mortality
4. both – a higher rate of infection and a more serious outcome. For example, children can be more susceptible to infection from certain pathogens than adults because of their inherently reduced immunity, which is a natural state. Although they are more likely to become infected, many children's outcomes are mild or asymptomatic and infection usually confers some level of protection from later exposures. In this sense, children fall into the category of having a higher rate of infection, but decreased morbidity and/or mortality compared with adults. This is also a transient category; children are not automatically in this state at birth, though they may acquire temporary immunity from their mothers from breastfeeding. Also, once a child has been infected by a particular pathogen and develops immunity, he or she leaves that category and enters either the 'general' population or another category based on some other susceptibility. This cycling in and out of categories of susceptibility varies for each pathogen (see Table 1.1).

## Risk communication

Initially, risk communication was added to the risk assessment process as an afterthought, along with risk management. Over time, risk managers began to experience the need for help in communicating environmental risks to the public and there has been recognition that public input into the goals and mechanics of risk assessment helps create trust in the overall process. In their 1994 publication, the NRC said, 'Public confidence that risk managers are addressing real concerns, as opposed to going through a process perfunctorily, is critical to the future of risk assessment as an activity capable of improving the quality of life.'

Current risk communication research efforts aim to help risk analysis by providing a basis for understanding and anticipating public responses to hazards and improving the communication of risk information among lay people, technical experts and decision makers. Risk managers who uphold

**Table 1.1** Variation in infection rate and degree of morbidity and mortality among susceptible subpopulations

| Pathogen | Susceptible population | Incidence | Morbidity/ mortality | References |
| --- | --- | --- | --- | --- |
| MAC | AIDS/elderly | Decreasing in AIDS; increasing in elderly | One of many end-stage infections in AIDS; chronic pulmonary disease in elderly | Prince *et al.*, 1989; Reich and Johnson, 1991; Kennedy and Weber, 1994; Palella *et al.*, 1998; Havlir *et al.*, 2000 |
| Adenovirus | Immuno-compromised (e.g. transplant) | Same as general population | Greater risk of severe/fatal outcome | Hierholzer, 1992; Saad *et al.*, 1997 |
| *Helicobacter pylori* | Immuno-compromised (e.g. AIDS) | Same as general population | Same as general population | Edwards *et al.*, 1991; Battan *et al.*, 1990 |
| Calicivirus | Unidentified genetic factor | Higher than general population | Higher rate of illness/ reinfection than general population | Moe *et al.*, 1999 |

and regulate health and safety need to understand how people think about and respond to risk. Experience shows that merely disseminating information without reliance on communication principles can lead to ineffective health messages and public health actions (Angulo *et al.*, 1997; Owen *et al.*, 1999; Harding and Anadu, 2000).

When the focus of risk communication to the public is to warn against a particular environmental health risk, the task is to notify as well as motivate people to act to mitigate the risk. However, people tend to underestimate these sorts of risks and fail to take any protective actions (Wiedemann and Schutz, 1999). Also, targeting the messages to the right audience is challenging: one must both identify the populations at risk and design appropriate outreach campaigns (usually through brochures or media outlets); however, these methods may reach a very small percentage of the intended audience. Even if the right audience attains the risk information, they may see themselves as personally not at risk, or they may not understand what action to take. A California study (in Harding and Anadu, 2000) of 900 consumers found that 80% of the respondents did not take any action in response to public notification regarding drinking water hazards; the author speculated that this resulted from the lack of preventive measures detailed in the notification. In addition, resources are often not available to those responsible for public outreach to construct and carry out a carefully researched risk communication strategy.

There are many ways that formal risk communication is used as a part of water use risk management, including fish consumption advisories, boil-water notices and other regulatory provisions, which require public notification when treatment violations occur. In the case of a boil-water advisory or a fish consumption warning, there may be little opportunity for officials to interact with the intended audience, and especially with a boil-water advisory, the timing of the message can be essential to the public's perception of trust in the source. In Milwaukee, health officials delayed the release of information, including the issuance of a boil-water order, even though the outbreak appeared to be waterborne. This delay resulted in the public's outrage and long-lasting distrust of government officials (Griffin *et al.*, 1998; Sly, 2000). Although quick action to alert communities may enhance trust in the source, it still may not have any effect on people's willingness to adopt the risk-reduction behaviour (O'Donnell *et al.*, 2000); 25% of people failed to boil their water after they had learned about an alert because they did not believe it to be true (Angulo *et al.*, 1997).

Most monitoring for specific pathogens in the water is done *after* there is evidence suggesting an outbreak. The UK has instituted standard monitoring for cryptosporidial oocysts in the drinking water supply, which has raised the question of what to do when oocysts are detected, but no cases of illness are indicated. The relationship between pathogens found in the course of regular water monitoring and actual health risks is still very unclear, but health officials had to develop guidelines on what sort of actions to take under those circumstances. Depending on the follow-up information available, under these guidelines, health officials have a spectrum of options available for public notification ranging from taking no action to issuing a boil-water advisory (Hunter, 2000).

People given the same environmental risk information will come to different conclusions based on their perceptions and values. Risk communication, then, is challenging because of everyone's different interpretations. Risk mitigation usually depends on voluntary compliance and the ability to communicate to all stakeholders is important to obtain compliance. Two challenges of environmental risk communication are to identify specific groups that may be at greater risk and to understand the information needs of all stakeholders. Delivering the details in a way that people understand is important, but not absolute. To be able to communicate the complexity of the different issues effectively is more than challenging for the risk communicator and the temptation to disseminate information without stakeholder interaction and assume that people are satisfied still lingers. However, the risk assessment community recognizes that the more interaction there is among the interested parties, the better chance for success in the entire risk process.

# References

Angulo, F.J., Tippen, S., Sharp, D.J. *et al.* (1997). A community waterborne outbreak of salmonellosis and the effectiveness of a boil water order. *Am J Public Hlth*, 87(4): 580–584.

Barrell, R.A.E., Hunter, P.R. and Nichols, G. (2000). Microbiological standards for water and their relationship to health risk. *Commun Dis Public Hlth*, 3(1): 8–13.

Barwick, R.S., Levy, D., Craun, G.F. *et al.* (2000). Surveillance for waterborne-disease outbreaks – United States, 1997–1998. *MMWR*, **49**(4): 1–21.

Battan, R., Raviglione, M.C., Palagiano, A. *et al.* (1990). *Helicobacter pylori* infection in patients with acquired immune deficiency syndrome. *Am J Gastroenterol*, **85**: 1576–1579.

Codex Committee on Food Hygiene. (1999). Principles and guidelines for the conduct of microbiological risk assessment. Rome: Codex Alimentarius Commission. CAC/GL-30 (1999).

Craun, G.F., Berger, P.S. and Calderon, R.L. (1997). Coliform bacteria and waterborne disease outbreaks. *JAWWA*, **89**(3): 96–100.

Edwards, P.D., Carrick, J., Turner, J. *et al.* (1991). *Helicobacter pylori*-associated gastritis is rare in AIDS: antibiotic effect or a consequence of immunodeficiency? *Am J Gastroenterol*, **86**: 1761–1764.

Environmental Protection Agency. (1998). National Primary Drinking Water Regulations: Interim Enhanced Surface Water Treatment; Final Rule. *Federal Register*, **63**(241): 69477–69521.

Gale, P. (1996). Developments in microbiological risk assessment models for drinking water – a short review. *J Appl Bacteriol*, **81**(4): 403–410.

Gibson, C.J., Haas, C.N. and Rose, J.B. (1998). Risk assessment of waterborne protozoa: current status and future trends. *Parasitology*, **117**(Suppl.): S205–S212.

Gostin, L.O., Lazzarini, Z., Neslund, V.S. *et al.* (2000). Water quality laws and waterborne diseases: *Cryptosporidium* and other emerging pathogens. *Am J Public Hlth*, **90**: 847–853.

Griffin, R.J., Dunwoody, S. and Zabala, F. (1998). Public reliance on risk communication channels in the wake of a cryptosporidium outbreak. *Risk Anal*, **18**(4): 367–375.

Harding, A.K. and Anadu, E.C. (2000). Consumer response to public notification. *JAWWA*, **92**(8): 32–41.

Havlir, D.V., Schrier, R.D., Torriani, F.J. *et al.* (2000). Effect of potent antiretroviral therapy on immune responses to *Mycobacterium avium* in human immunodeficiency virus-infected subjects. *J Infect Dis*, **182**(6): 1658–1663.

Hierholzer, J.C. (1992). Adenovirus in the immunocompromised host. *Clin Microbiol Rev*, **5**: 262–274.

Horsburgh, C.R. (1991). *Mycobacterium avium* complex infection in the acquired immunodeficiency syndrome. *New Engl J Med*, **324**: 1332–1338.

Hunter, P.R. (2000). Advice on the response from public and environmental health to the detection of cryptosporidial oocysts in treated drinking water. *Commun Dis Public Hlth*, **3**(1): 24–27.

Kennedy, T.P. and Weber, D.J. (1994). Nontuberculous mycobacteria. An underappreciated cause of geriatric lung disease. *Am J Respir Crit Care Med*, **149**: 1654–1658.

Moe, C., Rhodes, D., Pusek, S. *et al.* (1999). Determination of Norwalk virus dose–response in human volunteers. Presented at health Effects Stakeholder Meeting for the Stage 2 DBPR and LT2ESWTR, February 12, 1999. Washington, DC.

National Research Council (1983). *Risk Assessment in the Federal Government: Managing the Process*. Washington, DC: National Academy Press.

National Research Council (1994). *Science and Judgement in Risk Assessment*. Washington, DC: National Academy Press.

Neumann, D.A. and Foran, J. (1997). Assessing the risks associated with exposure to waterborne pathogens: An expert panel's report on risk assessment. *J Food Prot*, **60**(11): 1426–1431.

O'Donnell, M., Platt, C. and Aston, R. (2000). Effect of a boil water notice on behaviour in the management of a water contamination incident. *Commun Dis Public Hlth*, **3**(1): 56–59.

Owen, A.J., Colbourne, J.S., Clayton, C.R.I. *et al.* (1999). Risk communication of hazardous processes associated with drinking water quality – a mental models approach to customer perception, Part 1 – a methodology. *Water Sci Technol*, **39**(10–11): 183–188.

Palella, F.J. Jr, Delaney, K.M., Moorman, A.C. *et al.* (1998). Declining morbidity and mortality among patients with advanced human immunodeficiency virus infection. HIV Outpatient Study Investigators. *New Engl J Med*, **338**: 853–860.

Parkin, R.T. and Balbus, J.M. (2000). Variations in concepts of 'susceptibility' in risk assessment. *Risk Anal*, **20**(5): 603–611.

Pontius, F.W. (2000). Defining sound science. *JAWWA*, **92**(10): 16–20, 92.

Presidential/Congressional Commission on Risk Assessment and Management. (1997). The Presidential/Congressional Commission on Risk Assessment and Risk Management. Vol. 1. Washington, DC.

Prince, D.S., Peterson, D.D., Steiner, R.M. *et al.* (1989). Infection with *Mycobacterium avium* complex in patients without predisposing conditions. *New Engl J Med*, **321**: 863–868.

Proctor, M.E., Blair, K.A. and Davis, J.P. (1998). Surveillance data for waterborne illness detection: an assessment following a massive waterborne outbreak of *Cryptosporidium* infection. *Epidemiol Infect*, **120**: 43–54.

Reich, J.M. and Johnson, R. (1991). *Mycobacterium avium* complex pulmonary disease. Incidence, presentation, and response to therapy in a community setting. *Am Rev Respir Dis*, **143**: 1381–1385.

Risk Policy Report. (2001). EPA Adopts Risk Approach to Potential Drinking Water Contaminants. January 22.

Rose, J.B. and Gerba, C.P. (1991). Use of risk assessment for development of microbial standards. *Water Sci Technol*, **24**(2): 29–34.

Saad, R.S., Demetris, A.J., Lee, R.G. *et al.* (1997). Adenovirus hepatitis in the adult allograft liver. *Transplantation*, **64**: 1483–1485.

Sly, T. (2000). The perception and communication of risk: a guide for the local health agency. *Can J Public Health*, **91**(2): 153–156.

Tillett, H.E., deLouvois, J. and Wall, P.G. (1998). Surveillance of outbreaks of waterborne infectious disease: categorizing levels of evidence. *Epidemiol Infect*, **120**: 27–42.

Wiedmann, P.M. and Schutz, H. (1999). Risk communication for environmental health hazards. *Zentralbl Hyg Umweltmed*, **202**: 345–359.

World Health Organization Working Group. (2000). Evaluation and use of epidemiological evidence for environmental health risk assessment: WHO guideline document. *Environ Hlth Perspect*, **108**(10): 997–1002.

# Part 2

# Bacteriology

# 2

# *Acinetobacter*

## Basic microbiology

Acinetobacters are strictly aerobic, short and plump rod-shaped bacteria, often capsulated and classified as Gram-negative but often typified as being 'Gram-variable' when present in a pure culture. Morphologically, *Acinetobacter* rods are 1–1.5 μm in diameter and 1.5–2.5 μm in length, becoming coccoid (0.6–0.8 μm × 1.0–1.5 μm) in shape during the stationary phase of growth. *Acinetobacter* are catalase-positive, oxidase-negative and non-spore forming. While *Acinetobacter* exhibit so called 'twitching motility' as a result of the presence of fimbriae (polar), as a group they are non-motile. Regardless of many medical strains of *Acinetobacter* having optimum growth conditions at a temperature of 35°C, environmental strains are able to grow over a wide temperature range.

Acinetobacter occur frequently as part of the commensal flora of animals and humans and as such are regularly contaminating patients in hospitals, particularly those with bronchopneumonia and septicaemia.

## Origin and taxonomy

The genus *Acinetobacter*, first identified in 1911 by Beijerinck, was originally classified as *Micrococcus calcoaceticus* following its initial isolation from soil

(Baumann *et al.*, 1968). After 1911 at least 15 other 'generic' names had been used to describe the organisms now classified as members of the genus. The most documented ones have included *Bacterium anitratum*, *Herellea vaginicola*, *Mima polymorpha*, *Achromobacter*, *Alcaligenes*, 'B5W', *Moraxella glucidolytica* and *Moraxella lwoffii* (Henriksen, 1973). It was a group of French microbiologists who first proposed the genus *Acinetobacter*, which comprised a collection of non-motile, Gram-negative, oxidase-positive (*Moraxella*) and oxidase-negative saprophytes which could be distinguished from other bacteria by their lack of pigmentation when grown on agar plates (Brisou and Prévot, 1954). It was not until 1971 that the Subcommittee on the Taxonomy of Moraxella and Allied Bacteria recommended that the genus *Acinetobacter* should include only the oxidase-negative strains (Lessel, 1971).

Up until recently the genus *Acinetobacter* was included in the family Neisseriaceae (Juni, 1984). After extensive taxonomic developments it was proposed that members of the genus *Acinetobacter* should be classified in the new family Moraxellaceae. To date the family includes *Moraxella*, *Acinetobacter*, *Psychrobacter* and related organisms (Rossau *et al.*, 1991) which constitutes a discrete phylometric branch in superfamily II of the Proteobacteria on the basis of 16S rRNA studies and rRNA-DNA hybridization assays (Rossau *et al.*, 1989).

Historically, the genus *Acinetobacter* has undergone an extensive amount of taxonomic reclassification. This reclassification of *Acinetobacter* has been based on DNA-DNA homology studies. Presently, based on this method of classification, there is evidence of at least 19 DNA-DNA homology groups. These have been accepted as *A. baumannii*, A. *calcoaceticus*, A. *haemolyticus*, A. *lwoffii*, A. *radioresistens*, A. *johnsonii* and at present 13 unnamed genomic species.

In taxonomical terms the DNA G+C content of *Acinetobacter* has been calculated at between 39 and 47 mol%.


## Metabolism and physiology

The majority of the strains of *Acinetobacter* have no major growth factor, being able to use a large number of organic carbon and energy sources. However, *Acinetobacter* spp. fail to utilize carbohydrates, with most documented strains being unable to use glucose as a carbon source (Juni, 1978). However, acidification of certain sugars, including glucose, arabinose, cellobiose, galactose, lactose, maltose, mannose, ribose and xylose, via an aldose dehydrogenase, has been extensively documented. Most acinetobacters are unable to reduce nitrate to nitrite but some strains can use both nitrate and nitrite as nitrogen sources by means of an assimilatory nitrate reductase. Numerous *Acinetobacter* strains are also documented as being able to metabolize many diverse compounds, including aliphatic alcohols, some amino acids, decarboxylic and fatty acids, unbranched hydrocarbons and many relatively recalcitrant aromatic compounds such as benzoate, mandelate, n-hexadecane, cyclohexanol and 2,3-butanediol (Towner *et al.*, 1991).

*Acinetobacter calcoaceticus* are oxidase-negative, catalase-positive and indole-negative with some strains able to produce urease.

# Clinical features

*Acinetobacter* are ubiquitous but there are significant population differences between the genomic species found as part of the normal human flora in clinical and other environments. In the general population, *Acinetobacter* appears to be characterized by predominant groups of genomic species. These include *A. baumannii* and *Acinetobacter* spp. 3, which can be isolated from the skin and numerous body sites of infected or colonized patients, particularly in hospital. Rarely they have been isolated from the non-hospitalized population. In the non-clinical environments *A. lwoffii*, *A. johnsonii* and *Acinetobacter* spp. 12 seem to be the predominant natural inhabitants of human skin.

Acinetobacters are more increasingly being associated with nosocomial infections. These include septicaemia, urinary tract infections, eye infections, meningitis, skin and wound infections, brain abscesses, lung abscesses, pneumonia and endocarditis. A survey of nosocomial infections indicated that this organism might be responsible for over 0.5% of endemic nosocomial infections, particularly in critically ill patients, with 3–24% of pneumonia patients using mechanical ventilation devices becoming infected with at least one *Acinetobacter* spp. Mortality rates associated with nosocomial *Acinetobacter* infections are higher than those for other bacterial species, apart from *Pseudomonas aeruginosa*.

In recent years, antibiotic-resistant strains of *Acinetobacter* have been recognized as important pathogens involved in outbreaks of hospital infection, particularly in high-dependency or intensive care units. As they are being isolated from a wide range of clinical specimens, including tracheal aspirates, blood cultures, cerebrospinal fluid and pus, this constitutes a major public health concern. As with all nosocomial-acquired organisms the control of antibiotic usage, to minimize development of resistant strains, and good housekeeping practices and effective isolation procedures of infected patients, are important control factors for acinetobacters in hospitals. The most documented species which is associated with nosocomial outbreaks is *A. baumannii*, although other genomic species, particularly *Acinetobacter* spp. 3, *A. johnsonii* and *A. lwoffii*, have also been reported.

# Pathogenicity and virulence

*Acinetobacter* are classified as low-grade primary pathogens stereotyped into the broad expression 'opportunistic pathogen'. Comprising a number of virulence mechanisms, *Acinetobacter* spp. have become involved in a vast array of clinically acquired infections. The majority of these can be located above.

The virulence mechanisms inherent to *Acinetobacter* include: the presence of a polysaccharide capsule; adhesions (fimbriae and capsular polysaccharide), used to adhere to human epithelial cells; the production of cytotoxic enzymes; and the lipopolysaccharide (LPS) component of the cell wall and the presence of lipid A. The production of an endotoxin *in vivo* has been acknowledged and this may be responsible for the disease symptoms observed particularly during acinetobacter septicaemia. *Acinetobacter* have been shown to obtain iron from the human body, an important virulence determinant, aided by siderophores such as aerobactin, and iron-repressible outer-membrane receptor proteins (Smith *et al.*, 1990; Actis *et al.*, 1993).

## Treatment

Historical documented data have shown that all strains of *Acinetobacter* are resistant to penicillin, ampicillin, first-generation cephalosporins and chloramphenicol, whereas carbenicillin, tetracyclines (particularly minocycline, oxytetracycline and chlortetracycline) and aminoglycosides seemed to be effective at killing this organism. This still generally remains the situation to date with environmental isolates. However, with clinical isolates, it is now evident that *Acinetobacter* is able to develop or acquire resistance to any new antibiotics it may be challenged with including broad-spectrum cephalosporins and 4-fluoroquinolones. Because of this fact it has become essential with *Acinetobacter* infections to guide antimicrobial management by antimicrobial sensitivity tests.

## Survival in the environment

Acinetobacters are ubiquitous, free-living saprophytes. They have been isolated in soil, seawater, freshwater, estuaries, sewage, contaminated food and mucosal and outer surfaces of animals and humans (Towner *et al.*, 1991). Also included in this list is the clinical environment where *Acinetobacter* have been isolated in hospital sink traps, hospital floor swab cultures and in air samples in wards (Towner *et al.*, 1991).

Acinetobacters have frequently been isolated on granular activated carbon (GAC) and sand filters, in biofilms and point-of-use devices suggesting water as an important mode of transmission. In groundwater *Acinetobacter* has been detected in large numbers, approximating to a mean of 8 colony-forming units (cfu)/100 ml (range, <1–178). In one study, however, it was found that in groundwater *Acinetobacter* constituted less than 0.1% of the heterotrophic plate count (HPC) population (AWWA, 1999). Some studies, though have reported that 54% of HPC isolates obtained from groundwater have been acinetobacters. Despite this, of particular concern in water, particularly drinking water, is the fact that acinetobacters have been isolated in a large number of sites containing no coliforms, which suggests the inadequacy of coliforms as an indicator of these organisms. Conventional treatment of water by coagulation and filtration is known to remove between 80 and 99% (0.6–2 logs) and 50 and 99.5% (0.3–2.3 logs), respectively, of bacteria. The effectiveness of disinfection processes during the water-treatment process can vary appreciably depending on the disinfectant used and the relative resistance of the organism; however, typical removals range from 99 to 99.99% (AWWA, 1999). In a number of studies of a chlorinated distribution system, *Acinetobacter* has been found to be the most commonly isolated organism often comprising over 5% of the total organisms identified. In studies conducted in Canada, approximately 1–2% of organisms isolated from the distribution system were identified as *Acinetobacter*.

Naturally-occurring *Acinetobacter* spp. have been observed to have inactivation rates similar to other heterotrophic bacteria, such as *Moraxella*, *Aeromonas*, *Pseudomonas*, and *Alcaligenes*, when exposed to chloramines. However, some studies have indicated that acinetobacters can develop increased resistance to chlorine, chloramines and chlorine dioxide when grown under conditions favouring cell aggregation (AWWA, 1999). Despite these problems to date there are no regulatory guidelines for acinetobacters in drinking water.

## Methods of detection

Acinetobacters can be readily isolated and cultivated on conventional ordinary laboratory media without necessary growth factor requirements. Commonly used laboratory media used for the isolation of *Acinetobacter* have included, organic medium 79, mineral medium with crude oil, peptone yeast extract medium, Trypticase soy agar with glycerol and trypticase phytone medium. The selective and differential media, such as Sellers agar, Herella agar (contains bile salts, sugars and bromocresol purple) (Mandel *et al.*, 1964) and MacConkey agar, have also been used for the isolation of *Acinetobacter* (AWWA, 1999). A novel antibiotic-containing selective medium, Leeds Acinetobacter medium, that combines selectivity with differential characteristics has also been described for the improved isolation of *Acinetobacter* spp. (Jawad *et al.*, 1994) from both clinical and environmental sources.

In potable water environments, in order to differentiate *Acinetobacter* from other normal heterotrophic organisms, Eosin-Methylene Blue Agar and mAC agar has been used (AWWA, 1999). For the improved recovery of *Acinetobacter* spp. from the environment, samples can be enriched with the addition of 20 ml of an acetate-mineral medium with 5 ml of a water sample or a filtered 10% soil suspension followed by vigorous aeration at 30°C or at room temperature (AWWA, 1999).

Clinical isolates of *Acinetobacter* grow at 37°C with some being documented as growing often up to 42°C. However, a temperature of 30°C has often been recommended for the growth of *Acinetobacter*. On media such as nutrient agar and trypticase soya agar, *Acinetobacter* are known to form smooth, sometimes mucoid, pale yellow to greyish white colonies, about 1–2 mm in diameter.

Much research is now being focused upon the use of molecular fingerprinting techniques, including analysis of plasmid profiles, restriction endonuclease digestion and pulsed-field gel electrophoresis of total chromosomal DNA, random amplified polymorphic DNA profiles, ribotyping, cell envelope and outer-membrane protein profiles or multilocus enzyme electrophoretic typing (Thurm and Ritter, 1993) for the profiling of *Acinetobacter*. In terms of epidemiology these methods have all been used successfully to investigate outbreaks of infection associated with *Acinetobacter*, but no single system has so far gained overall acceptance for typing *Acinetobacter* spp.

## Epidemiology of waterborne outbreaks

There have been numerous hospital outbreaks of *Acinetobacter*-related infections reported in the literature. These have tended to occur on certain wards such as neurosurgery, burns units and intensive therapy units. There have been a small number of outbreaks where the strain isolated from clinical specimens was also isolated from taps, though it was not quite clear whether the tap water was the source of the outbreak or represented environmental contamination from colonized patients (Debast *et al.*, 1996; Pina *et al.*, 1998). There have also been outbreaks that have been associated with contamination and overgrowth of humidifiers and aerators (McDonald *et al.*, 1998; Kappstein *et al.*, 2000).

In treated drinking water no outbreaks due to acinetobacters have been acknowledged. However, outbreaks in hospital settings are well documented and do constitute a cause for concern (Ritter *et al.*, 1993; Pina *et al.*, 1998).

## Risk assessment

*Health effects*: occurrence of illness, degree of morbidity and mortality, probability of illness based on infection:

- *Acinetobacter* spp. can cause infection in virtually every organ system: septicaemia, urinary tract infections, eye infections, meningitis, skin infections, pneumonia and endocarditis.
- Although *Acinetobacter* causes mainly opportunist infections in hospitalized patients, community-acquired infections have been reported. It has been estimated that 1% of nosocomial infections are known to be caused by acinetobacters.

- Without disruption of normal host defence mechanisms, the role of *Acinetobacter* in human infection remains limited.

*Exposure assessment*: occurrence in source water, environmental fate, routes of exposure and transmission:

- The acinetobacters are found ubiquitously in the environment and have been isolated from fresh water, estuaries, sewage, seawater, drinking water and biofilms in drinking-water distribution systems. It is a very hardy organism – especially on fomites and surfaces.
- *Acinetobacter* is primarily spread from person to person and from fomites, such as medical equipment. It is possible that aerosolization is a route of exposure.
- Drinking water has not been shown to be a transmission pathway for *Acinetobacter*, though it has been isolated from drinking water.

*Risk mitigation*: drinking-water treatment, medical treatment:

- Data indicate that *Acinetobacter* spp. are inactivated with chlorine disinfectant at similar rates to other heterotrophic bacteria, such as *Moraxella*, *Aeromonas* and *Pseudomonas*. However, some studies have indicated that acinetobacters can develop increased resistance to chlorine disinfectants.
- The impact of the presence of *Acinetobacter* in disinfection system biofilm is unknown.
- Antibiotic resistance has hindered the medical management of *Acinetobacter* infections, and the overall trend is one of increasing resistance. However, mild to moderate infections can usually be treated successfully with monotherapy and severe infections with multiple-antibiotic therapy.

# References

Actis, J.A., Tomasky, M.E. *et al.* (1993). Effect of iron-limiting conditions on growth of clinical isolates of *Acinetobacter baumannii*. *J Clin Microbiol*, **31**: 2812–2815.

AWWA (1999). *Manual of Water Supply Practices: Waterborne Pathogens*. Washington, DC: Amercican Waterworks Association.

Baumann, P., Doudoroff, M. and Stanier, R.Y. (1968). A study of the Moraxella group. II. Oxidase-negative species (genus Acinetobacter). *J Bacteriol*, **95**: 1520–1541.

Brisou, J. and Prévot, A.R. (1954). Études de systématique bacterienne. X. Révision des espèces réunies dans le genre Achromobacter. *Ann Inst Pasteur (Paris)*, **86**: 722–728.

Debast, S.B., Meis, J.F., Melchers, W.J. *et al.* (1996). Use of interrepeat PCR fingerprinting to investigate an *Acinetobacter baumannii* outbreak in an intensive care unit. *Scand J Infect Dis*, **28**: 577–581.

Henriksen, S.D. (1973). Moraxella, Acinetobacter, and the Mimeae. *Bacteriol Rev*, **37**: 522–561.

Jawad, A., Hawkey, P.M. *et al.* (1994). Description of Leeds Acinetobacter Medium, a new selective and differential medium for isolation of clinically important *Acinetobacter* spp., and comparison with Herellea agar and Holton's agar. *J Clin Microbiol*, **32**: 2353–2358.

Juni, E. (1978). Genetics and physiology of Acinetobacter. *Annu Rev Microbiol*, **32**: 349–371.

Juni, E. (1984). Genus III. Acinetobacter Brisou et Prévot 1954. In *Bergey's Manual of Systematic Bacteriology*, vol. 1, 9th edn, Krieg, N.R. and Holt, J.G. (eds). Baltimore: Williams and Wilkins, pp. 303–307.

Kappstein, I., Grundmann, H., Hauer, T. *et al.* (2000). Aerators as a reservoir of *Acinetobacter junii*: an outbreak of bacteraemia in paediatric oncology patients. *J Hosp Infect*, **44**: 27–30.

Lessel, E.F. (1971). Minutes of the Subcommittee on the Taxonomy of Moraxella and Allied Bacteria. *Int J Syst Bacteriol*, **21**: 213–214.

McDonald, L.C., Walker, M., Carson, L. *et al.* (1998). Outbreak of *Acinetobacter* spp. bloodstream infections in a nursery associated with contaminated aerosols and air conditioners. *Pediatr Infect Dis J*, **17**: 716–722.

Mandel, A.D., Wright, K. and McKinnon, J.M. (1964). Selective medium for isolation of Mima and Herellea organisms. *J Bacteriol*, **88**: 1524–1525.

Pina, P., Guezenec, P., Grosbuis, S. *et al.* (1998). An *Acinetobactr baumanii* outbreak at the Versailles Hospital Center. *Pathol Biol (Paris)*, **46**: 385–394.

Ritter, E., Thurm, V., Becker-Boost, E. *et al.* (1993). Epidemic occurrences of multiresistant *Acinetobacter baumannii* strains in a neonatal intensive care unit. *Zentralbl Hyg Umweltmed*, **193**: 461–470.

Rossau, R., van den Bussche, G. *et al.* (1989). Ribosomal ribonucleic acid cistron similarities and deoxyribonucleic acid homologies of Neisseria, Kingella, Eikenella, Alysiella, and Centers for Disease Control Groups EF-4 and M-5 in the amended family Neisseriaceae. *Int J Syst Bacteriol*, **39**: 185–198.

Rossau, R., van Landschoot, A. *et al.* (1991). Taxonomy of *Moraxellaceae famnov.*, a new bacterial family to accommodate the genera Moraxella, Acinetobacter and Psychrobacter and related organisms. *Int J Syst Bacteriol*, **41**: 310–319.

Smith, A.W., Freeman, S. *et al.* (1990). Characterization of a siderophore from *Acinetobacter calcoaceticus*. *FEMS Microbiol Lett*, **70**: 29–32.

Thurm, V. and Ritter, E. (1993). Genetic diversity and clonal relationships of *Acinetobacter baumannii* strains isolated in a neonatal ward: epidemiological investigations by allozyme, whole-cell protein and antibiotic resistance analysis. *Epidemiol Infect*, **111**: 491–498.

Towner, K.J., Bergogne-Bérézin, E. and Fewson, C.A. (1991). *The Biology of Acinetobacter: Taxonomy, Clinical Importance, Molecular Biology, Physiology, Industrial Relevance.* New York: Plenum Press.

# 3

# *Aeromonas*

## Basic microbiology

*Aeromonas* spp. are rod-shaped (0.3–1.0 μm by 1.0–3.5 μm), Gram-negative, non-spore forming, oxidase-positive, and facultatively anaerobic bacteria. While commonly found in freshwater reservoirs, soil and agricultural produce, *Aeromonas* have also been isolated from the gastrointestinal contents of fish, reptiles, amphibia and higher vertebrates. Occasionally *Aeromonas*, which are pathogenic to humans, are found in the marine environment.

The genus is divided into two groups, namely the non-motile psychrophilic aeromonads, which are pathogenic to fish, and the motile mesophiles, which grow at a temperature range of 15–38°C. It is the mesophilic aeromonads, namely *Aeromonas hydrophila*, *Aeromonas caviae* and *Aeromonas sobria* which are of human significance and thus important to the water industry and public health. Despite the mesophilc *Aeromonas* spp. being highlighted as primary agents involved in gastroenteritis, controversy still presides over their pathogenicity and epidemiology. However, in light of the available literature, it seems more likely that a number of the species within the genus, specifically *Aeromonas hydrophila*, deserve recognition as a candidate involved in gastroenteritis and related diarroheal infections.

## Origin and taxonomy

The earliest documented evidence of the existence of the bacterium *Aeromonas* occurred in 1891. Initially it was classified as *Bacillus hydrophilus fuscus* following its isolation from the blood and lymph of an infected frog (Sanarelli, 1891). Numerous reports of the presence of *Bacillus hydrophilus fuscus* occurred in 1897 (Chester, 1897) until its reclassification to *Bacterium hydrophila* in 1901 (Chester, 1901). As its name implies this was a bacterium that loved water. During the decades that passed following its first acknowledgment *Bacterium hydrophila* was isolated from a wide array of different animals ranging from birds to reptiles. However, due to the lack of consensus on the name, *Bacterium hydrophila*, many researchers misclassified it into different genera. This is apparent when you consider the collection of groups to which *Bacterium hydrophila* has been assigned during the many years of reclassification. These groups have included *Aerobacter*, *Proteus*, *Pseudomonas*, *Escherichia*, *Achromobacter*, *Flavobacterium* and *Vibrio* (Table 3.1).

It was not until 1936 that Kluyver and van Niel (1936) proposed the genus *Aeromonas*, which literally meant 'gas-producing unit'. *Aeromonas* was endorsed in the seventh edition of *Bergey's Manual of Determinative Bacteriology*. Incorporation of *Aeromonas* into the family Vibrionaceae in the eighth edition of the manual occurred later (Schubert, 1974).

The acceptance of the genus *Aeromonas*, during its early developments as a human pathogen, was generally slow despite small pockets of evidence emerging in the 1930s suggesting a plausible link to disease. In spite of this, it took over 20 years to conclude that some infections in humans were due to the colonization and subsequent pathology associated with infection by *Aeromonas* species. It was not until 1954 that a relationship between *Aeromonas* and human disease was established. This evolved following a report compiled by

**Table 3.1** Historical names given to the genus *Aeromonas* (adapted from Carnahan and Altwegg, 1996)

| Name | Year of first description |
| --- | --- |
| *Bacillus punctatus* | |
| *Bacillus ranicida* | |
| *Bacillus hydrophilus fuscus* | 1891 |
| *Bacterium punctatum* | 1891 |
| *Aerobacter liquefaciens* | 1900 |
| *Bacillus hydrophilus* Sanarelli | 1901 |
| *Bacillus (Proteus, Pseudomonas, Escherichia) ichthyosmius* | 1917 |
| *Achromobacter punctatum* | 1923 |
| *Pseudomonas (Flavobacterium) fermentans* | 1930 |
| *Pseudomonas punctata* | 1930 |
| *Proteus melanovogenes* | 1936 |
| *Pseudomonas caviae* | 1936 |
| *Pseudomonas formicans* | 1954 |
| *Vibrio jamaicensis* | 1955 |

Hill *et al.* (1954) suggesting that *Aeromonas*, isolated from a biopsy of a women who had died, had possibly caused fulminant septicaemia with metastatic myositis. Greater evidence of *Aeromonas* causing disease did not transpire until about 1968 when infections caused by *Aeromonas hydrophila* were apparent. This occurred because of an outbreak of infections in a hospital in New Haven, Connecticut, USA where 27 cases of infection due to *Aeromonas hydrophila* were documented. This paper appeared in the *New England Journal of Medicine* and suggested the greatest evidence to date of the positive relationship between aeromonads and diarrhoea (von Graevenitz and Mensch, 1968).

Within the three main groups of mesophilic *Aeromonas*, namely the species *caviae*, *hydrophila* and *salmonicida*, multiple hybridization groups (HGs) are known to exist, indicating the existence of phenospecies. Studies by Popoff *et al.* (1981) and the Centers for Disease Control have shown 12 HGs are evident in the group *Aeromonas*. The type strains of named species are assigned to specific HGs, as in the case of *A. hydrophila* (HG 1), *A. salmonicida* (HG 3), *A. caviae* (HG 4), *A. media* (HG 5) and *A. sobria* (HG 7). Investigations using DNA-DNA reassociation have identified a number of new species. These included *A. eucrenophila* and HG 6, *A. jandaei* and HG 9, *A. veronii* and HG 10 and *A. schubertii* and HG 12. HGs 2 and 11 are currently unnamed with HG 8 now suggested as a biotype of *A. veronii*. Two new *Aeromonas* species, which do not correspond to any of the original 12 HGs, have also been proposed. These are *A. trota* (Carnahan *et al.*, 1991) and *A. allosaccharophila*.

To date there are 14 species in the genus *Aeromonas* with only nine of these implicated in human disease (Janda, 1991) with 17 DNA hybridization groups (Table 3.2). A further new species, *A. popoffi* (unassigned DNA hybridization group), has also been proposed.

## Metabolism and physiology

Aeromonads are chemo-organotrophs and have the ability to utilize a wide collection of different sugars and carbon as a source of energy. Glucose is metabolized both aerobically and fermentatively with or without the production of gas ($CO_2$, $H_2$). Salt at high concentrations (6–7%) has been found to have an inhibitory effect on the growth of *Aeromonas*, a useful characteristic involved in bacterial isolation. Despite this, it is well documented that *Aeromonas* spp. are able to tolerate pH values in the range 5–10 suggesting these organisms are very resilient to the demands inflicted on them in unfavourable environments, particularly those found in fresh waters. The mesophilic aeromonads, in particular *Aeromonas hydrophila*, are capable of utilizing a vast array of compounds including amino acids, carbohydrates and carboxylic acids, peptides and long-chain fatty acids. As a result of this ability *Aeromonas* are ideally suited for growth in biofilms in distribution systems.

**Table 3.2** Genospecies and phenospecies of the genus *Aeromonas* (from WHO)

| DNA hybridization | Reference strain (T = type strain) | Genospecies | Phenospecies |
|---|---|---|---|
| 1 | ATCC 7966T | *A. hydrophila* | *A. hydrophila* |
| 2 | ATCC 51108T | *A. bestiarum* | *A. hydrophila* |
| 3 | ATCC 33658T | *A. salmonicida* | *A. salmonicida* |
| 3 | CDC 0434-84 | *A. salmonicida* | *A. hydrophila* |
| 4 | ATCC 15468T | *A. caviae* | *A. caviae* |
| 5A | CDC 0862-83 | *A. media* | *A. caviae* |
| 5B | CDC 0435-84 | *A. media* | *A. media* |
| 6 | ATCC 23309T | *A. eucrenophila* | *A. eucrenophila* |
| 7 | CIP 7433T | *A. sobria* | *A. sobria* |
| 8 | ATCC 9071 | *A. veronii* | *A. veronii* biovar sobria |
| 9 | ATCC 49568T | *A. jandaei* | *A. jandaei* |
| 10 | ATCC 35624T | *A. veronii* | *A. veronii* |
| 11 | ATCC 35941 | Un-named | *Aeromonas* spp. (ornithine-positive) |
| 12 | ATCC 43700T | *A. schubertii* | *A. schubertii* |
| 13 | ATCC 43946 | Un-named | *Aeromonas* Group 501 |
| 14 | ATCC 49657T | *A. trota* | *A. trota* |
| 15 | CECT 4199T | *A. allosaccharophila*[a] | *A. allosaccharophila*[a] |
| 16 | CECT 4342T | *A. encheleia*[a] | *A. encheleia*[a] |

[a] The taxonomic status of *A. allosaccharophila* and *A. encheleia* remains to be confirmed.

## Clinical features

Aeromonads have been documented as being involved in both intestinal and extraintestinal human infections. The principal species of *Aeromonas* associated with gastroenteritis are *A. caviae*, *A. hydrophila* and *A. veronii* biovar *sobria* (Joseph, 1996), with the incidence of *A. caviae* more specifically evident in young children under the age of three. Clinical data have suggested that strains of *A. hydrophila* and *A. sobria* are inherently more pathogenic than *A. caviae* (Janda *et al.*, 1994).

While certain strains of *Aeromonas* have been proposed as aetiological agents of diarrhoeal disease, their role in specific disease causation still remains inconclusive. Gastrointestinal conditions associated with *Aeromonas* are usually self-limiting, but cases have been documented which show that *Aeromonas* can produce a severe life-threatening cholera-like disease (Joseph, 1996). Despite a possible link between *Aeromonas* and gastrointestinal infections a large number of investigations have suggested that mesophilic aeromonads are not primary enteropathogens.

*Aeromonas* spp. have also been associated with human wounds that are often localized and mild, and known to cause septicaemia, respiratory tract infections and a large array of systemic infections (Janda and Duffey, 1988; Janda and Abbott, 1996; Nichols *et al.*, 1996). The range of infections associated with mesophilic *Aeromonas* can be seen in Table 3.3.

**Table 3.3** Occurrence of human infections associated with mesophilic *Aeromonas*[a] (from WHO)

| Type of infection | Characteristics | Relative frequency[b] |
|---|---|---|
| **Diarrhoea** | | |
| Secretory | Acute watery diarrhoea, vomiting | Very common |
| Dysenteric | Acute diarrhoea with blood and mucus | Common |
| Chronic | Diarrhoea lasting more than 10 days | Common |
| Choleraic | 'Rice water' stools | Rare |
| **Systemic** | | |
| Cellulitis | Inflammation of connective tissue | Common |
| Myonecrosis | Haemorrhage, necrosis with/without gas gangrene | Rare |
| Erythema gangrenosum | Skin lesions with necrotic centre, sepsis | Uncommon |
| Septicaemia | Fever, chills, hypotension, high mortality | Fairly common |
| Peritonitis | Inflammation of peritoneum | Uncommon |
| Pneumonia | Pneumonia with septicaemia, sometimes necrosis | Rare |
| Osteomyelitis | Bone infection following soft-tissue infection | Rare |
| Cholecystitis | Acute infection of gallbladder | Rare |
| Eye infections | Conjunctivitis, corneal ulcer, endophthalmitis | Rare |

[a] Modified from Janda and Duffey, 1988, and Nichols *et al.*, 1996.
[b] Frequency of occurrence relative to all cases of *Aeromonas* infection.

It is highly probable that people are generally unaffected by enteric *Aeromonas* and that aeromonads form a natural part of the gut microbiota. However, we cannot rule out the fact that certain conditions including age, immunocompetence, infection dose, and sufficient virulence factors of *Aeromonas* spp. may aid in its ability to cause disease (Nichols *et al.*, 1996).

## Pathogenicity and virulence

While a number of virulence mechanisms for *Aeromonas* have been documented the pathogenesis of *Aeromonas* is still not fully understood (Gosling, 1996). This is probably due to the virulence mechanisms being multifactorial. Many extracellular enzymes are produced by *Aeromonas* that may be important pathogenicity mechanisms. These include nucleases, cytolytic toxins, stapholysin, lipases, sulphatases, lecithinase and amylase (Gosling, 1996; Howard *et al.*, 1996).

In gastroenteritis, the chief *Aeromonas* virulence factor in the disease process seems to be the enterotoxin (Janda, 1991). Cytolytic beta-haemolysin, or aerolysin, produced by several different *Aeromonas* species, including *A. hydrophila* and *A. veronii*, seems to be the most studied enterotoxin. Aerolysins are part of a family of *Aeromonas* toxins that have similar structural properties with dissimilar genetic, immunological and biological characteristics (Kozaki *et al.*, 1989). One enterotoxin, proaerolysin, has been shown to have a major impact on cell integrity by producing a transmembrane channel that ultimately destroys the host cell (Tucker *et al.*, 1990; Parker *et al.*, 1994). A second non-haemolytic toxin, identified in *Aeromonas*, produces a cytotonic response in adrenal cells. The weak haemolysin produced by *A. hydrophila* and *A. salmonicida* is called glycerophospholipid cholesterol acyltransferase (GCAT) (Howard *et al.*, 1996). One cytotonic enterotoxin with similar activity to cholera toxin has also been demonstrated (Ljungh *et al.*, 1982; Gosling *et al.*, 1992; Gosling, 1996). There is also documented evidence for plasmid-encoded expression by *A. hydrophila* and *A. caviae* of another cytotoxin. This seems to be very similar to the Shiga-like toxin 1.

*Aeromonas* also possess an array of other virulence-mechanisms possibly playing a role in gastrointestinal disease. These include, but are no means exhaustive, adhesins, mucinases and mechanisms aiding in their ability to penetrate specific eukaryotic cells (Janda, 1991). Also documented are the S layers (providing protection in the harsh environment and aiding possible resistance to phagocytosis), beta-haemolysin activity (Chakraborty *et al.*, 1987), resistance to complement-mediated lysis (Janda *et al.*, 1994), lipopolysaccharides and proteases (Gosling, 1996).

All the virulence mechanisms mentioned above have been demonstrated in *A. hydrophila* and *A.veronii* but not at present in *A. caviae*.

## Treatment

Aeromonads are known to be ampicillin-resistant but susceptible to the third generation cephalosporins, aminoglycosides, tetracycline, trimethoprim-sulphamethoxazole, chloramphenicol and quinolones, carbapenems and monobactams (Burgos *et al.*, 1990). Resistance to beta-lactam antibiotics in *Aeromonas* species is often documented.

## Presence in the environment

Aeromonads can be isolated from virtually any freshwater source and poikilothermic animals during the cold months of the year, when numbers are relatively low. It is during the warmer months of the year that *Aeromonas* numbers are often high and can be isolated from potable water sources. Levels of *Aeromonas* are also found at very high densities, all year round, within both domestic and industrial wastewater. This is due predominantly as a result of faecal contamination and also high temperatures evident in these situations.

The density of *Aeromonas* numbers varies substantially within different environmental situations. In certain environments it has been suggested that *Aeromonas* can exist at levels of $>10^8$ cfu/ml in sewage sludge, $1–10^2$ cfu/ml in lakes and reservoirs, $10^2$–$10^7$ cfu/ml in wastewater, $10^4$ cfu/ml in rivers, 1–$10^2$ cfu/ml in drinking water and 1–10 cfu/ml in groundwater (Carnahan and Joseph, 1991).

In raw sewage and many effluents the mesophilic *Aeromonas* species that shows dominance is *A. caviae* (Ramteke *et al.*, 1993). Evidence of *A. caviae* in sewage treatment and ultimately the receiving waters has often led to proposals for its use as an indicator of faecal pollutions. However, it has been noted that a higher percentage of *A. hydrophila* and *A. sobria* are toxigenic compared with *A. caviae*. Aeromonads in sewage effluents are of concern where such effluents are used for irrigation of crops or are discharged into recreational waters.

Aeromonads are frequently isolated from surface waters with the most predominant species being *Aeromonas hydrophila* closely followed by *Aeromonas caviae*. Low numbers of *Aeromonas* have been documented as occurring in groundwater with maximum counts of 35 cfu/100 ml being reported in deep aerobic and anaerobic groundwaters, even in the absence of coliforms (Havelaar *et al.*, 1990). Other evidence of the existence of *Aeromonas* in groundwater has been documented (Huys *et al.*, 1995). In this study two strains of aeromonads isolated from two groundwaters in Belgium consisted solely of strains belonging to hybridization groups 2 and 3 of the *A. hydrophila* complex.

Despite water treatment plants reporting mean reductions of 99.7% in aeromonad numbers following flocculation-decantation and chlorination (Huys *et al.*, 1995; Kersters *et al.*, 1995), low numbers of mesophilic *Aeromonas* are often found in drinking waters (Havelaar *et al.*, 1990). Surveys in London and Essex, UK have shown isolation rates for *A. hydrophila* from chlorinated drinking water of 25% in summer and 7% in the winter, compared with rates of 82% in the summer and 75% in the winter for untreated water. Researchers have found that 90% of domestic water supplies in areas of Cairo contain *Aeromonas* (Ghanem *et al.*, 1993), while from a survey of three distribution systems in Sweden (Krovacek *et al.*, 1992), it has been reported that 85% of samples were positive for *Aeromonas* (860 cfu/100 ml). A number of studies have also been carried out looking at the occurrence of *Aeromonas* species in the metropolitan water supplies of Perth, Western Australia (Burke *et al.*, 1984). From these studies a relationship between *Aeromonas* in chlorinated water with water temperature and residual chlorine was demonstrated. Despite *Aeromonas*, like the majority of organisms, entering the water-distribution system in very low numbers, development in biofilms results in bacterial aftergrowth and possible public health concerns (van der Kooij *et al.*, 1995).

When it comes to controlling the numbers of *Aeromonas* spp. it has been found, when using clinical and environmental strains of *A. hydrophila*, *A. sobria*, *A. caviae* and *A. veronii* ( Knochel, 1991), the mesophilic aeromonads are generally more susceptible to chlorine and monochloramine than the Enterobacteriaceae (Medema *et al.*, 1991). However, *A. hydrophila*, within a mixed heterotrophic bacterial biofilm, are unaffected by the addition of 0.3 mg/l monochloramine (Mackerness *et al.*, 1991). From this study it was also evident that the biofilm-associated *A. hydrophila* was able to also survive 0.6 mg/l monochloramine.

## Methods of isolation and detection

At present there are no regulatory standards for the acceptable number of aeromonads in drinking water both in Europe and the USA. However, it must be assumed that if a water treatment process is functioning properly the heterotrophic plate counts (HPCs) should not increase substantially. On the whole heterotrophic plate counts should remain at levels below 10 cfu/100 ml in drinking water. Within Europe, guidelines as to the numbers of *Aeromonas* in drinking water have been established. Here it is generally accepted that levels of *Aeromonas* in drinking water should be no higher than 20 cfu/100 ml when leaving the water treatment plant and no more than 200 cfu/100 ml in distribution water.

Membrane filtration is commonly used for the enumeration of *Aeromonas* from treated water. The most widely used medium for the recovery of *Aeromonas* from drinking water is ampicillin–dextrin agar (ADA) (Havelaar *et al.*, 1987; Havelaar and Vonk, 1988). Ryan's *Aeromonas* medium (Holmes and Sartory, 1993) has also been employed. As with many media employed for the recovery of drinking-water bacteria these media contain selective agents that are nutrient-rich, which may result in the recovery of low numbers of some aeromonads (Gavriel and Lamb, 1995; Holmes *et al.*, 1996). As *Aeromonas* species are sensitive to the presence of copper at concentrations as low as 10 mg/l, a complexing agent (50 mg/l sodium ethylenediamine tetra-acetate, $Na_2EDTA$) should therefore be added to samples (Versteegh *et al.*, 1989; Schets and Medema, 1993). This is particularly relevant during the sampling of domestic households and municipal buildings that use copper pipes to transport water. There is also evidence that pre-enrichment with alkaline peptone water before subculturing to selective media has proved successful for recovery of *Aeromonas* from water (e.g. well water) (Moyer *et al.*, 1992).

There have been a large number of different media used for the recovery of *Aeromonas* from environmental waters. These have included m-aeromonas agar (Rippey and Cabelli, 1979), ADA, starch–ampicillin agar (Palumbo *et al.*, 1985), pril–xylose–ampicillin agar (Rogol *et al.*, 1979) and SGAP-10C agar (Huguet and Ribas, 1991).

Because growth requirements are simple, aeromonads can be routinely cultured on many non-selective, differential and selective agars. Since most strains ferment sucrose, lactose or both sucrose and lactose, they can be easily missed on media like MacConkey, Hektoen enteric and xylose-lysine-desoxycholate (XLD) agars where they resemble non-pathogenic coliforms. The best differential/selective media for the isolation of pathogenic aeromonads is blood agar and cefsulodin-Irgasan-novobiocin (CIN) agar plates. Blood agar can be used as a screening medium for *Aeromonas*. Many *Aeromonas* strains also produce beta-haemolysis on sheep blood agar, a phenotypic characteristic useful in separating aeromonads from other enteric bacilli.

On non-selective agar media *Aeromonas* isolates typically appear as buff-coloured, smooth, convex colonies after 24 hours incubation at 37°C. It has been documented that most aeromonads are able to grow in media containing up to 4% NaCl and over a pH range of 5–9 (Austin *et al.*, 1989).

Primary identification of the genus *Aeromonas* is straightforward. Identification to phenospecies or genospecies level is, however, more complex (Millership, 1996). Molecular based technology using polymerase chain reaction (PCR) (Khan and Cerniglia, 1997) has enabled the detection of *A. caviae* and *A. trota* in water and environmental samples (Dorsch *et al.*, 1994).

## Epidemiology

The health significance of detecting mesophilic aeromonads in public water supplies is poorly understood. Early studies have correlated a possible link between *Aeromonas* in drinking water and gastroenteritis (Burke *et al.*, 1984; Picard and Goullet, 1987). However, these studies did not adequately type strains. A number of studies have now reported subtyping of strains of *Aeromonas* isolated from water and human cases (Havelaar *et al.*, 1992; Hanninen and Siitonen, 1995; Borchardt *et al.*, 2003). All the typing studies reported found little similarity between human and environmental strains suggesting that *Aeromonas* in drinking water does not constitute a risk to human health. This problem is complex when we consider that *Aeromonas* may in fact be part of the normal microbiota of the human host. It is possible that *Aeromonas* may not even be a true enteric pathogen.

Some epidemiological studies have been carried out that suggest people have become colonized by *Aeromonas* from drinking untreated water. However, the possible role of potable water implicated in the transmission of infections from *Aeromonas* is still under discussion.

To date even though aeromonads are frequently isolated from drinking water systems, and some strains may exhibit enterotoxic properties, there is a need for further epidemiological studies. This will help to ascertain a relationship between cases of *Aeromonas*-associated diarrhoea and presence of these organisms in drinking water.

## Risk assessment

*Health effects*: occurrence of illness, degree of morbidity and mortality, probability of illness based on infection:

- *Aeromonas*-related gastroenteritis is generally self-limited, acute, watery diarrhoea of a few days' to a few weeks' duration. Chronic diarrhoea of more than a few weeks and more serious sequelae, such as sepsis, usually occur only in immunocompromised people. Disseminated, systemic disease in the immunocompromised (especially those with underlying liver disease or cancer) causes a high fatality rate. Soft-tissue and skin infection is related to fresh water exposure.

- The probability of illness given infection is unknown for *Aeromonas*-related gastroenteritis. Symptomatic and asymptomatic people shed *Aeromonas* in their faeces.

*Exposure assessment*: routes of exposure and transmission, occurrence in source water, environmental fate:

- Routes of exposure may include ingestion of contaminated water and food and dermal contact with water or soil. Most infections are community-acquired; some are nosocomial (especially in the immunocompromised). The role of drinking water in transmission is under debate.
- *Aeromonas* spp. are found at high densities in all types of water: lakes, rivers, marine environments, wastewater and chlorinated drinking water. Because *Aeromonas* is also found frequently in all types of food, animals and is ubiquitous in the environment, the role of water ingestion in the development of gastroenteritis is unknown. Exposure to fresh water is the primary risk factor associated with wound infection.
- The aeromonads occur at generally high concentrations in all types of source water. River and marine water concentrations have been reported up to $10^4$ cfu/ml; lake concentrations have been reported up to $1–10^2$ cfu/ml; and $>10^8$ cfu/ml in sewage sludge. Concentrations of $1–10^2$ cfu/ml have been found in chlorinated drinking water.
- *Aeromonas* spp. are able to tolerate pH values in the range of 5–10, which suggests that they are hardy in the environment, and particularly, in natural waters.
- Survival and amplification in drinking water distribution is significant. *Aeromonas* can colonize wells and distribution systems for months and years. Substantial regrowth can occur after disinfection, and the aeromonads colonize biofilm, which makes them resistant to disinfectant residuals.

*Risk mitigation*: drinking-water treatment, medical treatment:

- Water treatment with filtration and chlorine disinfectant decreases concentrations significantly, but not necessarily completely. Water temperature, contact time, and level of organic material in the source water are all related to disinfection efficacy. Biofilm-associated *A. hydrophila* was shown to be resistant to 0.6 mg/l of monochloramine.
- *Aeromonas* is resistant to many antibiotics, including ampicillins; however, many other antibiotics, such as third generation cephalosporins and tetracycline, are efficacious in treating all kinds of infections, including chronic gastroenteritis.

# References

Austin, D.A., McIntosh, D. and Austin, B. (1989). Taxonomy of fish associated *Aeromonas* spp., with the description of *Aeromonas salmonicida subsp. smithia* subsp. nov. *Syst Appl Microbiol*, **11**: 277–290.

Burke, V., Robinson, R., Gracey, M. *et al.* (1984). Isolation of *Aeromonas hydrophila* from a metropolitan water supply: seasonal correlation with clinical isolation. *Appl Environ Microbiol*, **48**: 361–366.

Burgos, A., Quindós, G. *et al.* (1990). *In vitro* susceptibility of *Aeromonas caviae, Aeromonas hydrophila* and *Aeromonas sobria* to fifteen antibacterial agents. *Eur J Clin Microbiol Infect Dis*, **9**: 413–417.

Carnahan, A.M. and Altwegg, M. (1996). Taxonomy. In *The Genus* Aeromonas, Austin, B. *et al.* (eds). London: Wiley, pp. 1–38.

Carnahan, A.M. and Joseph, S.W. (1991). *Aeromonas* update: new species and global distribution. 3rd International Workshop on *Aeromonas and Plesiomonas. Experienta*, **47**: 402–403.

Chakraborty, T., Huhle, B. *et al.* (1987). Marker exchange mutagenesis of the aerolysin determinant in Aeromonas hydrophila. *Infect Immun*, **55**: 2274–2280.

Chester, F.D. (1897). A preliminary arrangement of the species of the genus *Bacterium* Contribution to determinative bacteriology. Part 1. *9th Annu Rep Delaware College Agric Exp Sta*, 92.

Chester, F.D. (1901). *A Manual of Determinative Bacteriology*. New York: Macmillan.

Dorsch, M. *et al.* (1994). Rapid identification of *Aeromonas* species using 16S rDNA targeted oligonucleotide primers: a molecular approach based on screening of environmental isolates. *J Appl Bacteriol*, **77**: 722–726.

Gavriel, A. and Lamb, A.J. (1995). Assessment of media used for selective isolation of *Aeromonas* spp. *Lett Appl Microbiol*, **21**: 313–315.

Ghanem, E.H., Mussa, M.E. and Eraki, H.M. (1993). Aeromonas-associated gastro-enteritis in Egypt. *Zentralbl Mikrobiol*, **148**: 441–447.

Gosling, P.J. (1996). Pathogenic mechanisms. In *The Genus* Aeromonas, Austin, B. *et al.* (eds). London: Wiley, pp. 245–265.

Gosling, P.J. *et al.* (1992). Isolation of Aeromonas sobria cytotonic enterotoxin and beta-haemolysin. *J Med Microbiol*, **38**: 227–234.

Hanninen, M.L., Saimi, S. and Siitonen, A. (1995). Maximum growth temperature ranges of *Aeromonas* spp. isolated from clinical or environmental sources. *Microb Ecol*, **29**: 259–267.

Havelaar, A.H. and Vonk, M. (1988). The preparation of ampicillin dextrin agar for the enumeration of *Aeromonas* in water. *Lett Appl Microbiol*, **7**: 169–171.

Havelaar, A.H., During, M. and Versteegh, J.F.M. (1987). Ampicillin-dextrin agar medium for the enumeration of *Aeromonas* species in water by membrane filtration. *J Appl Bacteriol*, **62**: 279–287.

Havelaar, A.H., Versteegh, J.F.M. and During, M. (1990). The presence of Aeromonas in drinking water supplies in the Netherlands. *Zentralbl Hyg*, **190**: 236–256.

Havelaar, A.H., Schets, F.M., van Silfhout, A. *et al.* (1992). Typing of *Aeromonas* strains from patients with diarrhoea and from drinking water. *J Appl Bacteriol*, **72**: 435–444.

Hill, K.R., Caselitz, F.H. and Moody, L.M. (1954). A case of acute metastatic myosistis caused by a new organism of the family *Pseudomonodaceae:* a preliminary report. *W Ind Med*, **3**: 9–11.

Holmes, P. and Sartory, D.P. (1993). An evaluation of media for the membrane filtration enumeration of *Aeromonas* from drinking-water. *Lett Appl Microbiol*, **17**: 58–60.

Holmes, P., Niccolls, L.M. and Sartory, D.P. (1996). The ecology of mesophilic *Aeromonas* in the aquatic environment. In *The Genus* Aeromonas, Austin, B. *et al.* (eds). London: Wiley, pp. 127–150.

Howard, S.P., MacIntyre, S. and Buckley, J.T. (1996). Toxins. In *The Genus* Aeromonas, Austin, B. *et al.* (eds). London: Wiley, pp. 267–286.

Huguet, J.M. and Ribas, F. (1991). SGAP-10C agar for the isolation and quantification of *Aeromonas* from water. *J Appl Bacteriol*, **70**: 81–88.

Huys, G., Gersters, l., Vancanneyt, M. *et al.* (1995). Chemotaxonomic analysis and genomic fingerprinting of *Aeromonas* sp. isolated from Flemish drinking water production plants. *Abstracts of the 5th International* Aeromonas–Plesiomonas *Symposium*, Edinburgh.

Huys, G., Gersters, L., Vancanneyt, M. *et al.* (1995). Diversity of *Aeromonas* sp. in Flemish drinking water production plants as determined by gas-liquid chromatographic analysis of cellular fatty acid methyl esters (FAMEs). *J Appl Bacteriol*, **78**: 445–455.

Janda, J.M. (1991). Recent advances in the study of the taxonomy, pathogenicity, and infectious syndromes associated with the genus *Aeromonas*. *Clin Microbial*, **4**: 397–410.

Janda, J.M. and Abbott, S.L. (1996). Human pathogens. In *The Genus* Aeromonas, Austin, B. *et al.* (eds). London: Wiley, pp. 151–173.

Janda, J.M. and Duffey, P.S. (1988). Mesophilic aeromonads in human disease: current taxonomy, laboratory identification, and infectious disease spectrum. *Rev Infect Dis*, **10**: 980–987.

Janda, J.M., Kokka, R.P. and Guthertz, L.S. (1994). The susceptibility of S-layer-positive and S-layer-negative Aeromonas strains to complement-mediated lysis. *Microbiology*, **140**: 2899–2905.

Joseph, S.W. (1996). *Aeromonas* gastrointestinal disease: a case study in causation? In *The Genus* Aeromonas, Austin, B. *et al.* (eds). London: Wiley, pp. 311–335.

Kersters, I., Huys, G., Janssen, P. *et al.* (1995). Influence of temperature and process technology on the occurrence of *Aeromonas* sp. and hygienic indicator organisms in drinking water production plants. Presented at the Fifth International *Aeromonas–Plesiomonas* Symposium, Edinburgh.

Khan, A.A. and Cerniglia, C.E. (1997). Rapid and sensitive method for the detection of *Aeromonas caviae* and *Aeromonas trota* by polymerase chain reaction. *Lett Appl Microbiol*, **24**: 233–239.

Kluyer, A.J. and Niel, C.B. (1936). Prospects for a natural system of classification of bacteria. *Zentralbl Bakteriol Parasitenk Intekionskr Hyg Abt II*, **94**: 369–403.

Knochel, S. (1991). Chlorine resistance of motile *Aeromonas* spp. *Water Sci Technol*, **24**: 327–330.

Kozaki, S., Asao, T. *et al.* (1989). Characterization of Aeromonas sobria hemolysin by use of monoclonal antibodies against *Aeromonas hydrophila* hemolysins. *J Clin Microbiol*, **27**: 1782–1786.

Krovacek, K., Faris, A., Baloda, S.B. *et al.* (1992). Isolation virulence profiles of *Aeromonas* spp. from different municipal drinking water supplies in Sweden. *Food Microbiol*, **9**: 215–222.

Ljungh, A., Eneroth, P. and Wadstrsm, T. (1982). Cytotonic enterotoxin from *Aeromonas hydrophila*. *Toxicon*, **20**: 787–794.

Mackerness, C.W., Colbourne, J.S. and Keevil, C.W. (1991). Growth of *Aeromonas hydrophila* and *Escherichia coli* in a distribution system biofilm model. *Proc UK Symp Health-Related Water Microbiology*. London: IAWPRC, pp. 131–138.

Medema, G.J., Wondergem, E., van Dijk-Looyaard, A.M. *et al.* (1991). Effectivity of chlorine dioxide to control *Aeromonas* in drinking water distribution systems. *Water Sci Technol*, **24**: 325–326.

Millership, S.E. (1996). Identification. In *The Genus* Aeromonas, Austin, B. *et al.* (eds). London: Wiley, pp. 85–107.

Millership, S.E., Barer, M.R. and Tabaqchali, S. (1986). Toxin production by *Aeromonas spp.* from different sources. *Med Microbiol*, **22**: 311–314.

Moyer, N.P. *et al.* (1992). Application of ribotyping for differentiating aeromonads isolated from clinical and environmental sources. *Appl Environ Microbiol*, **58**: 1940–1944.

Nichols, G.L. *et al.* (1996). Health significance of bacteria in distribution systems – review of *Aeromonas*. London: UK Water Industry Research Ltd (Report DW-02/A).

Palumbo, S.A. *et al.* (1985). Starch-ampicillin agar for the quantitative detection of *Aeromonas hydrophila*. *Appl Environ Microbiol*, **50**: 1027–1030.

Parker, M.W., Buckley, J.T. *et al.* (1994). Structure of the Aeromonas toxin proaerolysin in its water-soluble and membrane-channel states. *Nature*, **367**: 292–295.

Picard, B. and Goullet, P. (1987). Epidemiological complexity of hospital Aeromonas infections revealed by electrophoretic typing of esterases. *Epidemiol Infect*, **98**: 5–14.

Popoff, M.Y., Coynault, C., Kiredjian, M. *et al.* (1981). Polynucleotide sequence relatedness among motile *Aeromonas* species. *Curr Microbiol*, **5**: 109–114.

Ramteke, P.W., Pathak, S.P., Gautam, A.R. *et al.* (1993). Association of *Aeromonas caviae* with sewage pollution. *J Environ Sci Health*, **A28**: 859–870.

Rippey, S.R. and Cabelli, V.J. (1979). Membrane filter procedure for enumeration of *Aeromonas hydrophila* in fresh waters. *Appl Environ Microbiol*, **38**: 108–113.

Rogol, M. *et al.* (1979). Pril-xylose-ampicillin agar, a new selective medium for the isolation of *Aeromonas hydrophila*. *J Med Microbiol*, **12**: 229–231.

Sanarelli, G. (1891). Ober einen neuen Mikroorganismus des Wassers, welcher für Thiere mit veriinderlicher und konstanter Temperatur pathogen ist. *Zentralbl Bakt Parasitenk*, **9**: 222–228.

Schets, F.M. and Medema, G.J. (1993). Prevention of toxicity of metal ions to *Aeromonas* and other bacteria in drinking-water samples using nitrilotriaceticacid (NTA) instead of ethylenediaminetetraaceticacid (EDTA). *Lett Appl Microbiol*, **16**: 75–76.

Schubert, R.H.W. (1974). Genus 11. *Aeromonas* Kluyver and Van Niel, 1936. In *Bergey's Manual of Determinative Bacteriology*, 8th edn, Buchanan, R.E. and Gibbons, N.E. (eds). Baltimore: Williams and Wilkins, pp. 345–348.

Tucker, A.D., Parker, M.W. *et al.* (1990). Crystallization of a proform of aerolysin, a hole-forming toxin from *Aeromonas hydrophila*. *J Mol Biol*, **212**: 561–562.

van der Kooij, D., Veenendaal, H.R., Baars-Lorist, C. *et al.* (1995). Biofilm formation on surfaces of glass and Teflon exposed to treated water. *Wat Res*, **29**: 1655–1662.

van der Goot, F.G., Pattus, F. *et al.* (1993). Oligomerization of the channel-forming toxin aerolysin precedes insertion into lipid bilayers. *Biochemistry*, **32**: 2636–2642.

Versteegh, J.F.M. *et al.* (1989). Complexing of copper in drinking-water samples to enhance recovery of *Aeromonas* and other bacteria. *J Appl Bacteriol*, **67**: 561–566.

Von Graevenitz, A. and Mensch, A.H. (1968). The genus *Aeromonas* in human bacteriology. *N Engl J Med*, **278**: 245–249.

# 4

# *Arcobacter*

## Basic microbiology

*Arcobacter* are Gram-negative, non-spore forming rods, generally 0.2–0.9 μm wide by 1–3 μm long. One example of this is *Arcobacter butzleri* which are curved to S-shaped rods, 0.2–0.4 μm wide by 1–3 μm long, catalase-positive or weakly positive. Under some conditions the cells of *Arcobacter* become very long. These cells are usually motile with a darting, corkscrew-like movement aided by an unsheathed single polar flagellum. *Arcobacter* are able to grow over a diverse temperature range with evidence of growth documented at 15°C, with viability also evident at 42°C. They are micro-aerophilic but do not require hydrogen for growth and are aerotolerant at 30°C.

## Origin of the organism

The genus *Arcobacter* was first described by Vandamme and others in 1991 (Vandamme *et al.*, 1991). It was a name given to a bacterium which was aerotolerant and able to grow at low temperatures. Originally called *Campylobacter cryaerophila* (Neill *et al.*, 1985), it has since been renamed *Arcobacter cryaerophilus*. It is an organism primarily associated with the abortion of pigs and cattle but strains have now been associated with diarrhoea, mainly in children in developing countries (Tee *et al.*, 1988; Taylor *et al.*, 1991; Kiehlbauch *et al.*, 1991). *Arcobacter cryaerophilus* at present contains two subgroups. Sometime after the first isolation of *Arcobacter cryaerophilus*, *Arcobacter butzleri* (16 biotypes and 65 serotypes) was proposed as a species by Kiehlbauch and coworkers, originally it was only a strain of *Arcobacter cryaerophilus* (Kiehlbauch *et al.*, 1991). *Arcobacter nitrofigilis* was isolated later from the roots of plants, it was found not to be associated with any human disease or infection.

Initially all three species of *Arcobacter* were classified as *Campylobacter* spp. To date *Arcobacter* are integrated into the family Campylobacteraceae and there are currently four species, the final one being *A. skirrowii*.

## Metabolism and biochemistry

Arcobacters are metabolically inactive with d-glucose and other carbohydrates neither fermented nor oxidized. They are, however, able to hydrolyse indoxyl acetate but not hippurate and able to reduce nitrite and produce indole and hydrogen sulphide. All *Arcobacter* strains are oxidase-positive, catalase-positive and urease-negative (except *A. cryaerophila*).

There are slight differences in the biochemistry of *Arcobacter* that aid in their identification. In the case of *A. butzleri*, these have weak catalase activity, are able to grow on MacConkey agar and in 8% glucose and reduce nitrate. *A. skirrowii*, on the other hand, differs from *A. butzleri*, as it is unable to grow on MacConkey agar and generally does not grow in the presence of 1.5% NaCl. *A. cryaerophila* are not able to reduce nitrate. *A. nitrofigilis* can be differentiated from other arcobacters by its nitrogenase activity and from *A. cryaerophilus* and *A. skirrowii* in particular by its ability to grow in the presence of 1.5% NaCl.

## Clinical features

*A. butzleri* is associated with diarrhoea in humans (Lerner *et al.*, 1994) and animals, occasional systemic infection in humans, and abortion in cattle and pigs. The prevalence of such organisms in abortion and enteritis in other species of livestock is unknown and their pathogenic role has yet to be defined. *A. cryaerophilus* is isolated from normal and aborted ovine and equine fetuses and has been the cause of an outbreak of bovine mastitis. *A. skirrow* has been isolated from the preputial fluid of bulls, aborted bovine, porcine and ovine fetuses, and the faeces of animals with diarrhoea.

Occasionally *Arcobacter* species can be isolated from blood and peritoneal fluid of patients following acute appendicitis.

The role of *A. butzleri* as an enteric pathogen is not yet clearly established.

## Treatment

At present all species of *Arcobacter* are susceptible to nalidixic acid, with variable susceptibility to cephalothin.

## Environment

*A. cryaerophilus* has been found in the faeces of domestic animals and occasionally from humans, although some strains originally identified as this species have since been identified as *A. butzleri*. *Arcobacter butzleri* is found in water, sewage, poultry and other meats. Recent findings have shown that several biotypes and serogroups of *A. butzleri* have been associated with human disease as well as fresh poultry carcasses. Food may therefore be a possible source of exposure to *Arcobacter*. *A. nitrofigilis* is a nitrogen-fixing bacterium found on the roots of a small marsh plant but is not found in animals or humans.

There is no available evidence on the removal of Arcobacters by water-treatment processes, though they are likely to be removed to a similar extent as other bacteria. However, during a two-investigation study, while six strains of *C. jejuni* and *C. coli* were isolated 100 strains of *Arcobacter butzleri* were isolated from drinking water treatment plants in Germany (Jacob *et al.*, 1998).

If we consider the similarity of the genus *Arcobacter* with that of *Campylobacter* then we could conclude that *Arcobacter* should be susceptible to inactivation by disinfectants such as chlorine and ozone. However, this may turn out not to be the case.

## Isolation and detection

The primary isolation of *Arcobacter* involves a two-step motility enrichment in a semisolid medium at 30°C, followed by culture on blood agar containing a selective agent such as carbenicillin. Direct filtration onto enrichment media as well as non-selective media has also been used.

A primary isolation procedure based on swarming in a semisolid blood-free selective medium at 24°C has been developed for the examination of meat products. The selective agents used were piperacillin, cefoperazone, trimethoprim and cycloheximide, and these were incorporated in a basal medium comprising Mueller Hinton broth and 0.25% agarose (*Arcobacter* Selective Medium). Pure cultures of *Arcobacter* are then isolated from the swarming zones which may extend 30–40 mm. Optimal conditions for the isolation of *Arcobacter* spp. from clinical specimens have not been determined. *Arcobacter* spp. are aerotolerant and have been recovered on certain selective media used for *Campylobacter*, such as Campy-CVA.

*Arcobacter* strains grown on blood agar for 24 hours under microaerobic conditions at 37°C form small, flat, watery, translucent beige to yellow, irregular colonies with a campylobacter-like appearance. *A. nitrofigilis* has a slightly different appearance with typical colonies that are white and round. Growth occurs at 15°C and 30°C but only slightly at 42°C. *A. cryaerophilus* colonies on blood agar are domed, entire and yellowish in the case of subgroup 2 strains. *A. skirrowii* colonies are flat, greyish and alpha-haemolytic on blood agar. It has been observed that most aerotolerant strains, except *A. butzleri*, grow weakly on the common blood agar bases, although recently brain heart infusion agar containing 0.6% yeast extract and 10% blood agar has been used successfully for routine culturing.

PCR based identification assays (targetted at 16S ribosomal RNA sequences) have been developed to differentiate *Arcobacter* species from other closely related campylobacters, including *Campylobacter* and *Helicobacter*. Biotyping of *A. butzleri* can be performed using four biochemical tests and serotyping by slide agglutination based on heat labile antigenic factors of live bacteria using absorbed and unabsorbed antisera.

## Epidemiology

The exact route as to how humans become infected with *Arcobacter* is at present unknown. It possibly could be due to the consumption of or contact with contaminated water. However, there is documented evidence which suggests that *Arcobacter* may be transmitted by food. One example of this occurred when *A. butzleri*-like organisms had been isolated from retail meat products, in particular poultry meat.

The epidemiology of *Arcobacter* infections is not well understood. However, *A. butzleri* has been isolated from urban drinking water, river water, faeces and sewage. Isolation from a drinking water reservoir in Germany has also been documented. It is likely that consumption of contaminated water is a source of exposure to arcobacters, but there is no direct evidence of this to date. This seems probable if we consider its pathogenic potential (Vandamme *et al.*, 1991, 1992).

## Risk assessment

*Health effects*: occurrence of illness, degree of morbidity and mortality, probability of illness based on infection:

- Three *Arcobacter* species have been recovered from humans: *A. butzleri*, *A. cryaerophilus*, and *A. skirrowii*.
- *A. butzleri* has most often been associated with human illness – diarrhoea and occasional systemic infection. *A. butzleri* has been isolated from patients with chronic diarrhoea and stomach cramps (Lerner *et al.*, 1994).
- Occasionally *Arcobacter* species can be isolated from blood and peritoneal fluid of patients following acute appendicitis.
- The role of *A. butzleri* as an enteric pathogen is not yet clearly established.

*Exposure assessment*: routes of exposure and transmission, occurrence in source water, environmental fate:

- The routes of exposure and transmission in humans are unknown. Because it has been found in sewage, surface water, drinking water treatment plants and groundwater, contaminated water is a possible route; however, the prevalence of *Arcobacters* in meat, especially poultry, makes food transmission probable. There is no direct evidence of either route of exposure.
- *A. butzleri* was found to attach easily to water-distribution pipe surfaces made of stainless steel, copper and plastic, making regrowth in the distribution system a possibility (Assanta *et al.*, 2002).
- *A. butzleri* is found in water, sewage, poultry and other meats. Recent findings have shown that several biotypes and serogroups of *A. butzleri* have been associated with human disease. *Arcobacter* are able to grow over a diverse temperature range with evidence of growth documented at 15°C, with viability also evident at 42°C.
- *A. cryaerophilus* has been found in the faeces of domestic animals and occasionally from humans.
- *A. skirrowii* is slow growing, so may be missed in sample analysis and therefore overlooked as a human pathogen.

*Risk mitigation*: drinking-water treatment, medical treatment:

- We know little about the effects of drinking-water treatment on *Arcobacter*. One study found *A. butzleri* to be sensitive to chlorine disinfection (Rice *et al.*, 1999).
- At present all species of *Arcobacter* are susceptible to nalidixic acid, with variable susceptibility to cephalothin. A majority of *A. butzleri* isolates were found resistant to antibiotics commonly used for the treatment of infectious bacterial diseases in humans (Atabay and Aydin, 2002).

# References

Assanta, M.A., Roy, D., Lemay, M.J. *et al.* (2002). Attachment of *Arcobacter butzleri*, a new waterborne pathogen, to water distribution pipe surfaces. *J Food Prot*, **65**(8): 1240–1247.

Atabay, H.I. and Aydin, F. (2002). Susceptibility of *Arcobacter butzleri* isolates to 23 antimicrobial agents. *Lett Appl Microbiol*, **33**(6): 430–433.

Jacob, J., Woodward, D., Feuerpfeil, I. *et al.* (1998). Isolation of *Arcobacter butzleri* in raw water and drinking water treatment plants in Germany. *Zentralbl Hyg Umweltmed*, **201**: 189–198.

Kiehlbauch, K.A., Brenner, D.J., Nicholson, M.A. *et al.* (1991). *Campylobacter butzleri* sp. nov. isolated from humans and animals with diarrhoeal illness. *J Clin Microbiol*, **29**: 376.

Lerner, J., Brumberger, V. and Preac-Mursic, V. (1994). Severe diarrhea associated with *Arcobacter butzleri*. *Eur J Clin Microbiol Infect Dis*, **3**(8): 660–662.

Neill, S.D., Campbell, J.N., O'Brien, J.J. *et al.* (1985). Taxonomic position of *Campylobacter cryaerophila* sp. nov. *Inst J Syst Bact*, **35**: 342.

Rice, E.W., Rodgers, M.R., Wesley, I.V. *et al.* (1999). Isolation of *Arcobacter butzleri* from ground water. *Lett Appl Microbiol*, **28**(1): 31–35.

Taylor, D.N., Kiehlbauch, J.A., Tee, W. *et al.* (1991). Isolation of Group 2 aerotolerant *Campylobacter* species from Thai children with diarrhoea. *J Inf Dis*, **163**: 1062.

Tee, W., Baird, R., Dyall-Smith, M. *et al.* (1988). Campylobacter cryaerophila isolated from a human. *J Clin Microbiol*, **26**: 2469.

Vandamme, P., Falsen, E., Rossau, R. *et al.* (1991). Revision of Campylobacter, Helicobacter and Wolinella taxonomy: emendation of generic descriptions and proposal of Arcobacter gen. Nov. *Inst J Syst Bacteriol*, **41**: 88–103.

Vandamme, P., Pugina, P., Benzi, G. *et al.* (1992). Outbreak of recurrent abdominal cramps associated with *Arcobacter butzleri* in an Italian school. *J Clin Microbiol*, **30**: 2335–2337.

# 5

# *Campylobacter*

## Basic microbiology

*Campylobacter* are Gram-negative, curved or spiral rods 0.2–0.4 μm wide by 0.5–5 μm long and non-spore forming. All campylobacters are oxidase-positive, catalase-positive, urease-negative and motile, by means of a single polar unsheathed flagellum at one or both ends of the cell, except *Campylobacter gracilis* (oxidase-negative and aflagellate). Campylobacters are 'microaerophilic'. Generally *Campylobacter* require a 3–5% concentration of carbon dioxide, a 3–15% concentration of oxygen and a temperature of 42°C for optimum growth. The ability of campylobacters to tolerate oxygen is thought to result from the vulnerability of their strongly electronegative dehydrogenases to super-oxides and free radicals, especially when they are in a resting state. The coccoid form of *Campylobacter*, often formed in old cultures or cultures that have been exposed to air, has been suggested as the 'viable but non-culturable state' (VBNC) of the organism (Cappelier and Federighi, 1998; Holler *et al.*, 1998). However, a large amount of controversy still presides over the VBNC state of *Campylobacter*, particularly when we consider its colonization, pathogenicity and transmission. The significance of this VBNC state for infection of animals, and as the cause of disease for humans, is still very tentative. In a review by Thomas *et al.* (1999) the statement was made, 'The virulence of VBNC forms should be considered to be equivalent to that of the culturable forms, with the added risk that they are not detectable by conventional culturable methods.' This statement has been fully backed up by research conducted in 1999 suggesting that the VBNC state of *Campylobacter* has an important role to play in the transmission of infection for a number of strains (Tholozan *et al.*, 1999; Lazaro *et al.*, 1999; Cappelier *et al.*, 1999).

*Campylobacter* are commensals in most animals and are only a major concern in humans, predominantly *C. jejuni* and *C. coli*, which are known to cause both food-borne diarrhoea and enterocolitis (Blaser, 2000).

Campylobacters that are of relevance to the water industry are ascribed to the class of campylobacters known as the 'thermophilic' group. This group is generally composed of *C. jejuni*, *C. coli* and *C. upsaliensis*.

## Origin and taxonomy

Over 100 years ago Theodor Escherich showed evidence of campylobacter enteritis (Escherich, 1886). He visualized 'campylobacter' cultures in smears made from the colonic contents of babies who had died of 'cholera infantum'. In 1906, campylobacter was first isolated from the uterine exudate of aborting sheep (McFadyean and Stockman, 1913). In 1919, Smith isolated a similar organism from fetuses of aborting cows. The isolated organism was named *Vibrio fetus* (Smith and Taylor, 1919). In 1959, Florent showed that a form of the infection known as bovine infectious infertility was due to a variety of *V. fetus* transmitted from carrier bulls to cows during coitus. He named the organism *Vibrio foetus* var. *venerialis* (now *Campylobacter fetus* subsp. *venerealis*).

In 1927, another microaerophilic vibrio was isolated from the jejunum of calves with diarrhoea and later named *Vibrio jejuni* (Jones *et al.*, 1931). In 1944, Doyle isolated a similar vibrio from pigs suffering from swine dysentery; he named it *Vibrio coli* (Doyle, 1948). These became *C. jejuni* and *C. coli*, respectively, with the formation of the genus *Campylobacter* proposed by Sebald and Véron in 1963, although no examples of the original *V. jejuni* or *V. coli* strains had at that time survived.

The true first isolations of the organism from the faecal samples of patients with diarrhoea occurred in Australia in 1971. Other evidence emerged in 1972 (Dekeyser *et al.*, 1972). In Europe, however, the extent and prevalence of *Campylobacter* and its relevance to gastroenteritis did not occur until 1977. This was primarily due to the problems of isolating this organism.

## Metabolism and physiology

Most species of *Campylobacter* produce catalase and, apart from *C. jejuni* subsp. *doylei*, all reduce nitrate to nitrite, with *C. jejuni* the only species able to hydrolyse sodium hippurate. Campylobacters obtain their energy from amino acids or tricarboxylic acid cycle intermediaries and are known not to utilize sugars or produce indole, except *C. gracilis*. The possession of a complete citric acid cycle and a complex highly branched respiratory chain together with a large number of regulatory functions enables *Campylobacter* to survive in a large number of different niches. These will be discussed in more detail below. Some species of *Campylobacter* are able to grow anaerobically in the presence of fumarate, aspartate or nitrate, which act as electron acceptors (Véron *et al.*, 1981).

## Clinical features

The principal symptom of infection by *Campylobacter* in humans is acute diarrhoea. The incubation period, following infection, ranges from 1 to 8 days with 2–3 days the more common time period. The infectious dose has been shown to vary considerably, although infection has been caused by ingestion of a few hundred organisms. Following infection, the onset of diarrhoea is usually sudden, often preceded by a prodromal flu-like illness, acute abdominal pain, or both. These symptoms often mimic the symptoms of appendicitis, resulting in incorrect diagnosis. The diarrhoea, often just self-limiting to the patient, may be profuse and watery, probably due to the production of a cholera-like enterotoxin, or may be dysenteric and contain blood and mucus. In young children, mild symptoms of watery, non-inflammatory diarrhoea are often seen (Ketley, 1997). With cases of *Campylobacter* diarrhoea stools are often culture negative after 3 weeks. Asymptomatic carriage of *Campylobacter* is, however, a common feature following infection and a patient may disseminate *Campylobacter*, in the faeces, for over 4 months.

*Campylobacter* enteritis, specifically *C. jejuni*, can be associated with complications, although these are relatively rare. Such complications are reactive arthritis and Guillain-Barré syndrome. These are known to occur in about 1–2% of individuals who have been infected with *Campylobacter*. Guillian-Barré syndrome is an autoimmune disorder of the peripheral nervous system leading to acute flaccid paralysis. The characteristics of this syndrome are rapidly progressing weakness of the limbs and respiratory muscles.

## Pathogenicity and virulence

*C. jejuni* and *C. coli* are by far the most important members of the Campylobacteraceae of human significance, expressing a number of different virulence factors (van Vliet and Ketley, 2001). However, studying *Campylobacter* spp. at the molecular level has posed many problems to investigators, resulting in limited success in the identification of virulence factors. To date, the only identified virulence factors that have been studied in any great detail are the flagellar genes and antibiotic resistance genes (Taylor, 1992).

For *Campylobacter* to induce gastrointestinal problems, like most enteric organisms, it must be able to colonize the intestines. For it to do this it must be able to penetrate the mucus layer of the gut. Motility, aided by the flagella, combined with its spiral shape allows *Campylobacter* to penetrate effectively the mucus layer of the intestinal wall (Guerry *et al.*, 1992). Motility and flagella are very important determinants for attachment and invasion of *Campylobacter* (Wassenaar *et al.*, 1991). The specific components of intestinal mucin known to be chemotactic to *Campylobacter* include L-fucose and L-serine. Movement towards these components may be important in campylobacters' pathogenesis of infection (Hugdahl *et al.*, 1988; Takata *et al.*, 1992). Once *Campylobacter* has crossed the mucus layer it has to adhere to the host's epithelial cells and then invade them. The adhesion of *Campylobacter* to these cells is mediated by fimbriae and other adhesions, such as the proteins PEB1 and CadF (fibronectin binding protein). Once *Campylobacter* is adhered to the host cell it then has to internalize itself. The mechanism for this is not known at present. However, the uptake of *Campylobacter* into the host cell appears to be dependent upon bacterial protein synthesis. It is probable that once invasion of the host cell is successful an inflammatory response is initiated (Konkel *et al.*, 1992). *C. jejuni* appear to produce proteins which may be important in internalization of the organism (Konkel *et al.*, 1993). *C. jejuni* has also been reported to produce cytotoxins and a cholera-like enterotoxin (Ruiz-Palacios *et al.*, 1992; Gillespie *et al.*, 1993).

Several investigators have examined the ability of *C. jejuni* to acquire iron, that is essential for bacterial pathogenesis, from exogenous sources. As levels of iron in host tissues are low, an organism that has the ability to complex iron is at a selective advantage. *C. jejuni* does not appear to produce its own siderophores to utilize iron. However, it is able to utilize exogenous siderophores (ferrichrome and enterochelin) from other bacteria as iron carriers (Baig *et al.*, 1986).

Another important virulence factor, particularly in *C. fetus*, is its microcapsule, or S layer, which protects the bacteria from serum killing and phagocytosis (Blaser and Pei, 1993).

During the past decade, evidence has appeared showing an association of *Campylobacter* infection with Guillain-Barré syndrome (GBS) (Mishu and Blaser, 1993). Several studies have shown that the disease is particularly associated with serogroup O:19 (Penner) strains. Surface polysaccharides, as well as flagella, following *sialylation* may be responsible for GBS as a result of molecular mimicry.

*Campylobacter*, specifically *C. jejuni* and *C. coli*, also have oxidative stress defence systems. The main superoxide stress defence in *Campylobacter* is superoxide dismutase (sodB) and is involved in converting superoxides into hydrogen peroxide. It is probable that this plays some role in the intracellular survival of *Campylobacter* in the host (Pesci *et al.*, 1994).

Heat shock proteins are also evident as a virulence factor in *Campylobacter*. These enable *Campylobacter* to survive the extremes of temperature it may be associated with.

*Campylobacter* are also associated with the oral cavity and may play a role in the pathogenesis of periodontal disease (Ogura *et al.*, 1995).

## Treatment

*Campylobacter* enteritis is generally self-limiting and requires no more than fluid and electrolyte replacement. All species of *Campylobacter* show sensitivity to erythromycin, ciprofloxacin, metronidazole and fluoroquinones at present but resistance to these agents is developing rapidly. To date *C. jejuni* and *C. coli* are showing good resistance to penicillins and cephalosporins.

## Survival in the environment and water

Campylobacters are widespread in the environment. They have been isolated and identified in fresh and marine waters. Campylobacters have also been found in high numbers in domestic sewage and undisinfected treated sewage effluents. Numbers of campylobacters in surface waters are usually low as opposed to the high numbers detected in sewage effluent. In groundwater (4°C) campylobacters are able to survive for several weeks (Gondrosen, 1986). In aquatic environments *C. jejuni* appears to be the predominant species, more so than *C. coli* or *C. lari*. In developed countries domestic animals, birds (caged), pigs, sheep and cows have been suggested as sources of *Campylobacter* infections. The risk to human health of the presence of *Campylobacter* in these environments is unknown and remains so to date. This seems to be due to the fact that there is still no clear route for the transfer of *Campylobacter* from the environment to the consumer, apart from food of course. Little is known about the survival of *Campylobacter* in the environment apart from a number of laboratory-based studies used to simulate the environment (Koenraad *et al.*, 1997; Beutling, 1998; Buswell *et al.*, 1998; Holler *et al.*, 1998). These studies have shown that *Campylobacter* is only able to survive a few hours in adverse conditions indicative of the environment with temperature being a major factor in campylobacter's survival. However, from these studies it was found that the survival of *Campylobacter* was greater with decreasing temperature, reaching several days at 4°C. In fact the survival of *Campylobacter* was substantially enhanced when it was present with other organisms within a biofilm (Buswell *et al.*, 1998). Research has shown that campylobacters can survive in water for many weeks, even months, at temperatures below 15°C.

On the available evidence so far it is consensually accepted that fully treated water, which is subjected to correct disinfection procedures, is regarded as being free from *Campylobacter*. If, however, campylobacters are found in chlorinated drinking water, they are present usually as a result of post-treatment contamination. Saying this, however, much more research is needed on the survival potential and the role of the VBNC state in *Campylobacter* in the environment and in drinking water, particularly biofilms, that provide appropriate microniches for their proliferation.

In the environment, specifically water, campylobacters are only found in the presence of faecal streptococci and faecal coliforms (Carter *et al.*, 1987; Arvanitidou *et al.*, 1995). When compared to the coliforms, it is generally accepted that methodologies used to inactivate these coliforms, specifically in drinking water, are fully effective against campylobacters. This is further accepted when we consider the numerous studies that have shown *Campylobacter* to be very vulnerable to chlorine.

## Methods of detection

A wide spectrum of media has been used in the isolation of *Campylobacter*, although most have not been designed for water. In the 1980s most work on the detection of campylobacters in water and the environment used Preston broth (a nutrient broth base containing lysed blood, trimethoprim, rifampicin, polymyxin B and amphotericin B) together with a supplement known as FBP (containing ferrous sulphate, sodium metabisulphite, and sodium pyruvate) followed by plating on Preston agar.

As a general procedure, *C. jejuni* and *C. coli* detection in water samples involves concentration using a 0.22-$\mu$m filter. Following filtration the filter is placed in non-selective broth containing FBP, supplemented with numerous antibiotics, and incubated at 42°C for 4 hours. Some studies have shown that the period of incubation should then be continued for another 24 hours. Following this the broth is then plated onto Preston agar, in a microaerophilic environment and incubated for 48 hours. Following the incubation stage appropriate tests can then be performed.

In clinical terms, faecal specimens from patients who present with a gastro-intestinal infection are analysed for the presence of *Campylobacter*. Faecal samples are usually transported in Cary-Blair media. Maximum recovery of *Campylobacter* spp. is achieved in a microaerobic atmosphere containing approximately 5% $O_2$, 10% $CO_2$ and 85% $N_2$. *Campylobacter* spp. may then be detected by direct Gram-stain examination of faecal samples. In areas where species other than *C. jejuni* and *C. coli* are common, a filtration method using non-selective medium should also be used (Mishu Allos *et al.*, 1995). In the clinical laboratory most campylobacters produce visible growth after 24 hours at 37°C. However, a further incubation period of 24 hours enables the development of appropriate-sized colonies which are generally circular and convex.

# Epidemiology and waterborne outbreaks

Faecal material, transported from farms, is the probable cause of ground-water contaminated with *Campylobacter* (Pearson *et al.*, 1993; Stanley *et al.*, 1998; Jones, 2001). It is possible that the conditions found within subsurface aquifers favour the survival of *Campylobacter* in these environments aiding in their survival (Jones, 2001).

Waterborne outbreaks due to *Campylobacter* have arisen when individuals have consumed untreated or contaminated water (Taylor *et al.*, 1983). In fact campylobacters have been the main cause of private water outbreaks in the UK in water that had not been adequately disinfected, a prerequiste for human consumption (Duke *et al.*, 1996; Furtado *et al.*, 1998). Campylobacter outbreaks from private water supplies have been principally due to *Campylobacter jejuni* HS50 PT35 (Anon, 2000), which is an organism rarely cultured.

Outbreaks due to campylobacters have been associated with drinking water (Sobsey, 1989; Skirrow and Blaser, 1992; Jones and Roworth, 1996; Pebody *et al.*, 1997). In Canada, *Campylobacter* has been indirectly incriminated as a cause of a waterborne outbreak of enteritis (Borczyk *et al.*, 1987) and is considered the most important bacterial agent in waterborne diseases in many European countries (Strenstrom *et al.*, 1994; Furtado *et al.*, 1998). *Campylobacter* has caused a large number of outbreaks in Sweden involving over 6000 individuals (Furtado *et al.*, 1998).

*Campylobacter* enteritis incidences on average are 60–100 cases per 100 000 population per year. However, laboratory diagnosed cases of *Campylobacter* do not represent a true picture of the true incidence of *Campylobacter* infections (Kendall and Tanner, 1982). In the UK over 550 000 and in the USA 2.4 million *Campylobacter* infections are reported annually (Sibbald and Sharp, 1985).

A large number of waterborne outbreaks of *Campylobacter* have been reported in the literature, often affecting hundreds or even thousands of individuals (Pebody *et al.*, 1997) (Table 5.1). Despite a large number of outbreaks

**Table 5.1**  Campylobacter water (and unknown) outbreaks in England and Wales (1992–1994) (Pebody *et al.*, 1997; Frost, 2001)

| Vehicle | Setting | Evidence |
|---------|---------|----------|
| Water | College | Microbiological |
| Water | College | Microbiological |
| Water | Community | Microbiological |
| Water | Adventure camp | Cohort |
| Water | Function | Descriptive |
| Water | College | Descriptive |
| Unknown | Home of elderly | Descriptive |
| Unknown | Residential school | Descriptive |
| Unknown | Dining centre | Descriptive |
| Unknown | Home for elderly | Descriptive |
| Unknown | Function | Descriptive |

due to *Campylobacter* the organism causing the outbreaks has seldom been isolated. This lack of isolation is possibly due to sporadic occurrence or the presence of viable but non-culturable *Campylobacter*. Based on these findings, it seems logical to suggest that while drinking water may be a major risk factor of *Campylobacter* enteritis, there is no evidence to date to indicate that campylobacters can survive in water-distribution systems.

Many typing systems including phage typing, biotyping (Bolton *et al.*, 1984), bacteriocin sensitivity, detection of preformed enzymes, auxotyping, lectin binding, serotyping, multilocus enzyme electrophoresis, and genotypic methods such as restriction endonuclease analysis, ribotyping and restriction analysis of polymerase chain reaction (PCR) products have been devised to study the epidemiology of *Campylobacter* infections (Patton and Wachsmuth, 1992). However, for routine epidemiological investigations combinations of phenotypic and genotypic tests are being used to investigate the presence of *Campylobacter* infections in the population (Salama *et al.*, 1990; Khakhria and Lior, 1992; Nachamkin *et al.*, 1993, 1996).

## Risk assessment

*Health effects*: occurrence of illness, degree of morbidity and mortality, probability of illness based on infection:

- In most industrialized countries, *Campylobacter* enteritis is the most frequent form of acute infective diarrhoea. Laboratory reports give incidences in the order of 50–100 cases per 100 000 population per year.
- The principal symptom of infection by *Campylobacter* is acute diarrhoea. The diarrhoea may be profuse and watery or may be dysenteric and contain blood and mucus. The diarrhoea produced is usually self-limiting.
- *Campylobacter* enteritis can be associated with complications, although these are relatively rare. One such complication is that of reactive arthritis and Guillain-Barré syndrome which occurs in about 1–2% of individuals who have been infected.
- Most patients who become infected with *C. jejuni* were previously healthy and recover rapidly from infection. However, patients with *C. fetus* infections are usually immunocompromised with conditions such as chronic alcoholism, liver disease, old age, diabetes mellitus and malignancies.
- *C. fetus* infections may cause intermittent diarrhoea or non-specific abdominal pain without localizing signs.
- Not all *Campylobacter* infections produce illness. Two of the most important factors related to infection appear to be the dose of organisms reaching the small intestine and the specific immunity of the host to the pathogen.
- Volunteers rechallenged with the *C. jejuni* organism developed infection, but were protected from illness. Also, in developing countries, where

*C. jejuni* infection is hyperendemic, the decreasing case-to-infection ratio with age suggests the acquisition of immunity.

*Exposure assessment*: routes of exposure and transmission, occurrence in source water, environmental fate:

- Campylobacters are transmitted by the faecal-oral route; person-to-person transmission is relatively uncommon; direct transmission from animals to humans is relatively common; indirect transmission, through consumption of contaminated food or water, is by far the most common route of infection.
- Many waterborne outbreaks of *Campylobacter* have been reported. Despite these large outbreaks, the organism has seldom been isolated from the drinking water supply. Sources have included surface water, unchlorinated water storage tanks contaminated with bird faeces, groundwater contaminated by surface run-off, and water mains contaminated by cross-connection.
- The infectious dose has been shown to vary considerably, although infection has been caused by ingestion of a few hundred organisms.
- Though they cannot grow in water, campylobacters have been isolated and identified in fresh and marine waters and are also found in high numbers in domestic sewage and undisinfected treated sewage effluents.
- Campylobacters have been shown to survive in water at 4°C for many weeks, but for only a few days at temperatures above 15°C.
- In water, campylobacters are generally sensitive to adverse conditions, such as heat, disinfectants and gamma-irradiation.
- While drinking water may be a risk factor, there is no evidence that campylobacters can colonize or survive in water-distribution systems, thus, consumption of properly treated water is unlikely to result in infection.

*Risk mitigation*: drinking-water treatment, medical treatment:

- Treatment methodologies used to inactivate coliforms are fully effective against campylobacters. *Campylobacter* has been shown to be very vulnerable to chlorine, more so than *E. coli*.
- Most infections are self-limited. However, antibiotic treatment is advised in patients with high fever, bloody diarrhoea, or more than eight stools per day; whose symptoms have not lessened or are worsening at the time the diagnosis is made; or whose symptoms have lasted more than one week (Blaser, 2000).

# References

Anon. (2000). Surveillance of waterborne disease and water quality, July to December 1999. *Commun Dis Rep Wkly*, **10**: 65–67.

Arvanitidou, M., Stathopoulos, G.A., Constantinidis, T.C. *et al.* (1995). The occurrence of *Salmonella*, *Campylobacter* and *Yersinia* spp in river and lake waters. *Microbiol Res*, **150**: 153–158.

Baig, B.H., Wachsmuth, I.K. and Morris, G.K. (1986). Utilization of exogenous siderophores by Campylobacter species. *J Clin Microbiol*, **23**: 431–433.

Beutling, D. (1998). Incidence and survival of *Campylobacter* in foods. *Arch Lebensm, Hyg*, **39**: 53–62.

Blaser, M.J. (2000). *Campylobacter jejuni* and related species. In Mandell, Douglas, and Bennett's *Principles and Practice of Infectious Diseases*, vol. 1, 5th edn. Philadelphia: Churchill Livingstone, pp. 2276–2285.

Blaser, M.J. and Pei, Z. (1993). Pathogenesis of *Campylobacter fetus* infections: critical role of high-molecular weight S-layer proteins in virulence. *J Infect Dis*, **167**: 372–377.

Bolton, F.J., Holt, A.V. and Hutchinson, D.N. (1984). Campylobacter biotyping scheme of epidemiological value. *J Clin Pathol*, **37**: 677–681.

Borczyk, A., Thompson, S. *et al.* (1987). Water-borne outbreak of Campylobacter laridis-associated gastroenteritis. *Lancet*, **1**: 164–165.

Buswell, C.M., Herlihy, Y.M., Lawrence, L.M. *et al.* (1998). Extended survival and persistence of *Campylobacter* spp in water and aquatic biofilms and their detection by immuno-fluorescent-antibody and rRNA staining. *Appl Environ Microbiol*, **64**: 733–741.

Cappelier, J.M. and Federighi, M. (1998). Demonstration of viable but non culturable state *Campylobacter jejuni*. *Rev Med Vet*, **149**: 319–326.

Cappelier, J.M., Minet, J., Magras, C. *et al.* (1999). Recovery in embryonated eggs of viable but nonculturable *C.jejuni* cells and maintenance of ability to adhere to HeLa cells after resuscitation. *Appl Environ Microbiol*, **65**: 5154–5157.

Carter, A.M., Pacha, R.E., Clarke, G.W. *et al.* (1987). Seasonal occurrence of *Campylobacter* spp in surface waters and their correlation with standard indicator bacteria. *Appl Environ Microbiol*, **53**: 523–526.

Dekeyser, P., Gossuin-Detrain, M. *et al.* (1972). Acute enteritis due to related vibrio: first positive stool cultures. *J Infect Dis*, **125**: 390–392.

Doyle, L.P. (1948). The etiology of swine dysentery. *Am J Vet Res*, **9**: 50–51.

Duke, L.A., Breathnach, A.S., Jenkins, D.R. *et al.* (1996). A mixed outbreak of cryptosporidium and campylobacter infection associated with a private water supply. *Epidemiol Infect*, **116**: 303–308.

Escherich, T. (1886). Beiträge zur Kenntniss der Darmbacterien. III. Ueber das Vorkommen von Vibrionen im Darmcanal und den Stuhlgängen der säuglinge. (Articles adding to the knowledge of intestinal bacteria. III. On the existence of vibrios in the intestines and faeces of babies.) *Münch Med Wochenschr*, **33**: 815–817.

Florent A. (1959). Les deaux vibrioses génitales de la bête bovine: la vibriose vénérienne, due à V. foetus venerialis, et la vibriose d'origine intestinale due à V. foetus intestinalis, *Proceedings of the 16th International Veterinary Congress*, Madrid, **2**: 953–957.

Frost, J.A. (2001). Current epidemiological issues in human campylobacteriosis. *J Appl Microbiol*, **90**: 85S–95S.

Furtado, C., Adak, G.K., Sturt, M. *et al.* (1998). Outbreaks of waterborne infections intestinal disease in England and Wales, 1992–1995. *Epidemiol Infect*, **121**: 109–119.

Gillespie, M.J., Haraszthy, G.G. and Zambon, J.J. (1993). Isolation and partial characterization of the Campylobacter rectus cytotoxin. *Microb Pathog*, **14**: 203–215.

Gondrosen, B. (1986). Survival of thermotolerant campylobacters in water. *Acta Vet Scand*, **79**: 1–47.

Guerry, P., Alm, R.A. *et al.* (1992). Molecular and structural analysis of Campylobacter flagella. In Campylobacter jejuni: *Current Status and Future Trends*, Nachamkin, I., Blaser, M.J. and Tompkins, L.S. (eds). Washington, DC: American Society for Microbiology, pp. 267–281.

Holler, C., Witthuhn, D. and Janzen-Blunck, B. (1998). Effect of low temperatures on growth, structure and metabolism of *Campylobacter coli* SP10. *Appl Environ Microbiol*, **64**: 581–587.

Hugdahl, M.B., Beery, J.T. and Doyle, M.P. (1988). Chemotactic behavior of *Campylobacter jejuni*. *Infect Immun*, **56**: 1560–1566.

Jones, F.S., Orcutt, M. and Little, R.B. (1931). Vibrios (*Vibrio jejuni*) associated with intestinal disorders of cows and calves. *J Exp Med*, **53**: 853–864.

Jones, I.G. and Roworth, M. (1996). An outbreak of *Escherichia coli* 0157 and campylobacterosis associated with contamination of a drinking water supply. *Public Hlth*, **110**: 277–282.

Jones, K. (2001). Campylobacters in water, sewage and the environment. *J Appl Microbiol*, Symposium Supplement, **90**: 68S–79S.

Kendall, E.J.C. and Tanner, E.I. (1982). Campylobacter enteritis in general practice. *J Hyg*, **88**: 155–163.

Ketley, J.M. (1997). Pathogenesis of enteric infection by Campylobacter. *Microbiology*, **143**: 5–21.

Khakhria, R. and Lior, H. (1992). Extended phage-typing scheme for *Campylobacter jejuni* and *Campylobacter coli*. *Epidemiol Infect*, **108**: 403–414.

Koenraad, P.M.F.J., Rombouts, F.M. and Notermans, S.H.W. (1997). Epidemiological aspects of thermophilic Campylobacter in water-related environments: a review. *Water Environ Res*, **69**: 52–63.

Konkel, M.E., Mead, D.J. and Cieplak, W. (1993). Kinetic and antigenic characterization of altered protein synthesis by *Campylobacter jejuni* during cultivation with human epithelial cells. *J Infect Dis*, **168**: 948–954.

Konkel, M.E., Hayes, S.F. *et al.* (1992). Characteristics of the internalization and intracellular survival of *Campylobacter jejuni* in human epithelial cell cultures. *Microb Pathog*, **13**: 357–370.

Lazaro, B., Carcamo, J., Audicana, A. *et al.* (1999). Viability and DNA maintenance in nonculturable spiral *C.jejuni* cells after long-term exposure to low temperatures. *Appl Environ Microbiol*, **65**: 4677–4681.

McFadyean, J. and Stockman, S. (1913). Report of the Departmental Committee appointed by the Board of Agriculture and Fisheries to inquire into Epizootic Abortion. Part III. Abortion in Sheep. London: HMSO.

Mishu Allos, B., Blaser, M.J. and Lastovica, A.J. (1995). Atypical campylobacters and related microorganisms. In *Infections of the Gastrointestinal Tract*, Blasé, M.J., Smith, P.D. *et al.* (eds). New York: Raven Press, pp. 849–865.

Mishu, B. and Blaser, M.J. (1993). Role of infection due to *Campylobacter jejuni* in the initiation of Guillain-Barré syndrome. *Clin Infect Dis*, **17**: 104–108.

Nachamkin, I., Bohachick, K. and Patton, C.M. (1993). Flagellin gene typing of *Campylobacter jejuni* by restriction fragment length polymorphism analysis. *J Clin Microbiol*, **31**: 1531–1536.

Nachamkin, I., Ung, H. and Patton, C.M. (1996). Analysis of O and HL serotypes of Campylobacter by the flagellin gene typing system. *J Clin Microbiol*, **34**: 277–281.

Ogura, N., Shibata, Y. *et al.* (1995). Effect of Campylobacter rectus LPS on plasminogen activator-plasmin system in human gingival fibroblast cells. *J Periodont Res*, **30**: 132–140.

Patton, C.M. and Wachsmuth, I.K. (1992). Typing schemes: are current methods useful? In Campylobacter jejuni: *Current Status and Future Trends*, Nachamkin, I., Blaser, M.J. and Tompkins, L.S. (eds). Washington, DC: American Society for Microbiology, pp. 110–128.

Pearson, A.D., Greenwood, M. *et al.* (1993). Colonization of broiler chickens by waterborne *Campylobacter jejuni*. *Appl Environ Microbiol*, **59**: 987–996.

Pebody, R.G., Ryn, M.J. and Wall, P.G. (1997). Outbreaks of campylobacter infection: rare events for a common pathogen. Communicable disease report. *CDR Rev*, **7**: 33–37.

Pesci, E.C., Cottle, D.L. and Pickett, C.L. (1994). Genetic, enzymatic and pathogenic studies of the iron superoxide dismutase of *Campylobacter jejuni*. *Infect Immun*, **62**: 2687–2694.

Ruiz-Palacios, G.M., Cervantes, L.E. *et al.* (1992). *In vitro* models for studying Campylobacter infections. In Campylobacter jejuni: *Current Status and Future Trends*, Nachamkin, I., Blaser, M.J. and Tompkins, L.S. (eds). Washington, DC: American Society for Microbiology, pp. 176–183.

Salama, S.M., Bolton, F.J. and Hutchinson, D.N. (1990). Application of a new phagetyping scheme to campylobacters isolated during outbreaks. *Epidemiol Infect*, **104**: 405–411.

Sebald, M. and Véron, M. (1963). Teneur en bases de l'ADN et classification des vibrions. *Ann Inst Pasteur (Paris)*, **105**: 897–910.

Sibbald, C.J. and Sharp, J.C.M. (1985). Campylobacter infections in urban and rural populations in Scotland. *J Hyg*, **95**: 87–93.

Skirrow, M.B. and Blaser, M.J. (1992). Clinical and epidemiological considerations. In Campylobacter jejuni: *Current Status and Future Trends*, Nachamkin, I., Blaser, M.J. and Tompkins, L.S. (eds). Washington, DC: American Society for Microbiology, pp. 3–9.

Smith, T. and Taylor, M. (1919). Some morphological and biological characters of the spirilla (*Vibrio fetus*, n. sp.) associated with disease of the fetal membranes in cattle. *J Exp Med*, **30**: 299–311.

Sobsey, M.D. (1989). Inactivation of health related microorganisms in water by disinfection processes. *Water Sci Technol*, **21**: 179–196.

Stanley, K.N., Cunningham, R. and Jones, K. (1998). Thermophilic campylobacters in groundwater. *J Appl Microbiol*, **85**: 187–191.

Stenstrom, T.A., Boisen, F., Georgsen, F. *et al.* (1994). *Vattenburna Infektioner I Norden* (Waterborne Outbreaks in Northern Europe). Copenhagen, Den: TemaNord, Nordisk Minist.

Takata, T., Fujimoto, S. and Amako, K. (1992). Isolation of nonchemotactic mutants of *Campylobacter jejuni* and their colonization of the mouse intestinal tract. *Infect Immun*, **60**: 3596–3600.

Taylor, D.N. (1992). Campylobacter infections in developing countries. In Campylobacter jejuni: *Current Status and Future Trends*, Nachamkin, I., Blaser, M.J. and Tompkins, L.S. (eds). Washington, DC: American Society for Microbiology.

Taylor, D.N., McDermott, K.T. *et al.* (1983). Campylobacter enteritis from untreated water in the Rocky Mountains. *Ann Intern Med*, **99**: 38–40.

Tholozan, J.L., Cappelier, J.M., Tissier, J.P. *et al.* (1999). Physiological characterization of viable but nonculturable *C. jejuni* cells. *Appl Environ Microbiol*, **65**: 1110–1116.

Thomas, C., Gibson, H., Hill, D.J. *et al.* (1999). Campylobacter epidemiology: an aquatic perspective. *J Appl Microbiol*, Symposium Supplement, **85**: 168S–177S.

van Vliet, A.H.M. and Ketley, J.M. (2001). Pathogenesis of enteric Campylobacter infection. *J Appl Microbiol*, **90**: 45S–56S.

Véron, M., Lenvoisé-Furet, A. and Beaune, P. (1981). Anaerobic respiration of fumarate as a differential test between *Campylobacter fetus* and *Campylobacter jejuni*. *Curr Microbiol*, **6**: 349–354.

Wassenaar, T.M., Bleumink-Pluym, N.M. and van der Zeijst, B.A. (1991). Inactivation of *Campylobacter jejuni* flagellin genes by homologous recombination demonstrates that flaA but not flaB is required for invasion. *EMBO J*, **10**: 2055–2061.

# 6

# Cyanobacteria

## Basic microbiology

Cyanobacteria are Gram-negative bacteria, formerly known as blue-green algae, that are able to fix nitrogen in the dark. They range in size from 1 μm for unicellular cyanobacteria to over 30 μm for multicellular species. They perform oxygenic photosynthesis, deriving electrons from water to reduce carbon dioxide to cellular material. During the warmer months of the year cyanobacteria produce algal blooms, specifically in freshwater lakes and reservoirs. The abundance and concentration of these blooms have been increasing over the last decade, more specifically in the summer months. These toxic blooms have been reported in many parts of Europe, the USA, Australia, Africa, Asia and New Zealand. A survey in the UK has found that 75% of cyanobacterial blooms contain toxins (Baxter, 1991; Carmichael, 1992).

## Origin and taxonomy

In the USA, five genera of cyanobacteria have been identified as toxin producers, including two strains of *Anabaena flosaquae*, *Aphanizomenon flosaquae*, *Microcystis aeruginosa* and *Nodularia* spp. World-wide, six genera and at least 13 species, have been identified. Cyanobacteria were originally classified into groups on the basis of morphology, namely the unicellular cyanobacteria and the filamentous forms. This taxonomic principle was not upheld by rDNA studies. One classically defined group that has withstood rDNA analysis was the filamentous cyanobacteria that contained heterocysts, specialized cells that fix nitrogen, referred to as *Anabaena cylindrica*. Heterocysts protect the oxygen-labile enzyme nitrogenase, since, unlike vegetative cells, heterocysts do not carry out oxygen-evolving photosynthesis and further protect the nitrogenase by having an oxygen-impermeable glycolipid layer and a suite of respiratory enzymes.

There are about 25 species of cyanobacteria that have been associated with adverse health effects (Gold *et al.*, 1989).

## Clinical features

While high densities of cyanobacteria may appear in the faeces of infected animals, cyanobacterial illness is not infectious, but due to contact with or consumption of various toxins (Carmichael, 1994). The exact symptoms experienced following cyanobacterial poisoning depend on the route of exposure and the nature and concentration of the particular toxin (Hunter, 1998). Following recreational contact the most common reported features are allergic in nature, acute dermatitis or a hay-fever-like syndrome characterized by rhinitis, conjunctivitis and asthma. Gastroenteritis following recreation exposure is generally mild and short lived. Atypical pneumonia has been described in one outbreak (Turner *et al.*, 1990).

Acute illness following consumption of drinking water contaminated by cyanobacteria is more commonly gastroenteritis (Hunter, 1998). A subclinical hepatitis has been reported in one study (Falconer *et al.*, 1983) and there has been one outbreak of severe systemic illness including hepatitis and nephropathy that has been linked to cyanobacterial contamination of drinking water (Byth, 1980; Bourke *et al.*, 1983). An outbreak of hepatitis linked to cyanobacterial poisoning occurred in a dialysis unit; patients presented with visual disturbances, nausea and vomiting and almost 50% of those affected died (Jochimsen *et al.*, 1998).

Endotoxic reaction in dialysis patients, characterized by chills, fever, myalgia, nausea and vomiting, and hypotension, have also been described (Hindman *et al.*, 1975).

Probably the most concerning outcome is primary liver cell (PLC) carcinoma. Most of the epidemiological evidence of an association comes from the southeast coastal area of China where the mortality due to PLC is particularly high (>30/100 000 person years) (Yu, 1989, 1995; Yeh, 1989; Zhu *et al.*, 1989). The most significant risk factor for PLC is infection with either hepatitis B or C virus.

However, in people who are infected with these viruses, drinking water from ponds and ditches that are at increased risk of cyanobacterial contamination substantially increases the risk. Furthermore, the incidence of PLC in a community declined when the source of water was changed from ponds to other wells.

## Pathogenicity and virulence

The cyanobacteria do not invade the animal/human body but, as mentioned above, some produce potent toxins. Cyanobacterial toxins are of three main types: hepatotoxins, neurotoxins and lipopolysaccharide (LPS) endotoxins.

Several hepatotoxins have been described, though the most important are microcystin and nodularin, low molecular weight cyclic peptide toxins. Microcystin is a seven amino acid ring, two amino acids of which are unique to microcystin, N-methyl-dehydroalanine (Mdha) and 3-amino-9-methoxy-2,6,8-trimethyl-10-phenyldeca-4,6-dienoic acid (ADDA). Nodularin is a five amino acid ring hepatotoxin. Following absorption across the ileum, hepatotoxins are transported to the liver and then taken up by hepatocytes. These toxins are highly potent inhibitors of protein phosphatases types 1 and 2A, being active at nanomolar concentrations (MacKintosh *et al.*, 1990; Matsushima *et al.*, 1990; Yoshizawa *et al.*, 1990; Runnegar *et al.*, 1995). Protein phosphatases play an essential role in various cellular processes and their inhibition can lead to cell death and to mutagenic change, including cell growth and tumour suppression. Following acute ingestion, experimental animals develop weakness, anorexia, pallor of the mucous membranes, vomiting, cold extremities and diarrhoea. When death occurs it may be within a few hours due to intrahepatic haemorrhage and shock or a after a few days due to hepatic insufficiency (Hunter *et al.*, 1999). The $LD_{50}$ by intraperitoneal injection in mice for most hepatotoxins is in the range 60–70 µg/kg body weight. Microcystin has also been shown to be an extremely active tumour promoter for primary liver cell cancer in both *in vivo* and *in vitro* models (Zhou and Yu, 1990; Nishiwaki-Matsushima *et al.*, 1992; Yu, 1995). In a two-stage hepatocarcinogenesis model in rats, the number of gamma-glutamyl-transferase (GGT) altered foci after partial hepatectomy was increased in rats given pond/ditch water to drink. This is strong evidence that a component of the pond water was a cancer promoter.

The two main neurotoxins are the anatoxins and neosaxitoxin. Anatoxin-a (antx-a), is a secondary amine, 2-acetyl-9-azabicyclo[4.2.1]non-2-ene, an alkaloid neurotoxin and is reported to be the most potent nicotinic agonist so far described. Antx-a will cause a depolarizing neuromuscular blockade (Carmichael, 1992). In experimental animals, acute poisoning causes muscle fasciculations, decreased movement, collapse, exaggerated abdominal breathing, cyanosis, convulsions and death, due to respiratory failure (Hunter *et al.*, 1999). Death will occur within minutes of administration. The $LD_{50}$ (intraperitoneal in mice) is about 200 µg/kg body weight.

Neosaxitoxin is related to saxitoxin, the causative toxin in paralytic shell-fish poisoning. These two toxins act by inhibiting nerve conduction by blocking sodium, but not potassium, transport across the axon membrane. Features of acute administration include loss of coordination, twitching, irregular breathing and death by respiratory failure (Hunter *et al.*, 1999). The $LD_{50}$ (intraperitoneal in mice) is about 10 µg/kg.

Cyanobacterial lipopolysaccharide (LPS) endotoxin differs from that found in other Gram-negative bacteria in that it lacks phosphate in the lipid A core (Keleti and Sykora, 1982). Cyanobacterial LPS is also somewhat less toxic than enterobacteriaceal LPS, at least in animal studies. Although there has been relatively little research into the pathological effects of cyanobacterial LPS, it is possible that it may play a role in the allergic reactions and the gastroenteritis seen in people after contact with cyanobacterial material.

It is still not known to any large extent what factors are responsible for toxin production by cyanobacteria. It is known that toxin production varies substantially between strains and in the same strain over time (Hunter *et al.*, 1999). There remains a considerable amount of uncertainty about the factors promoting the synthesis and release of cyanobacterial toxins. A number of environmental factors have been proposed as affecting toxin production such as temperature, light intensity and phosphate levels, though different studies have often yielded conflicting results.

## Survival in the environment and water

Cyanobacteria are not dependent on a fixed source of carbon and, as such, are widely distributed throughout aquatic environments. These include freshwater and marine environments and in some soils. In fact, cyanobacteria are found in the early stages of soil formation being associated with converting bare rock or decomposing debris. Stagnant water, sediments and soil appear to be the significant reservoirs for these organisms. In the environment, cyanobacteria are at a selective advantage over eukaryotes under adverse conditions, as they can fix nitrogen.

No species in the cyanobacteria group is classed as a true pathogen, but certain freshwater strains, such as *Anabaena* and *Microcystis*, produce saxitoxin-like neurotoxins in the waters in which they grow. This is usually not a problem for humans, but farm animals can be poisoned by drinking water from ponds containing dense cyanobacterial blooms. It has been hypothesized that these toxins are produced by cyanobacteria to kill fish, thereby releasing nutrients.

Cyanobacteria naturally occur in stream sediments, slow-moving streams, receiving waters for a variety of waste discharges and treatment effluents, rural storm runoff, drainage canals, and marine waters. Massive growth of these bacteria often occurs during the summer in surface waters. Densities of 500 cells or more per millilitre have been recorded at this time of the year.

Polluted surface waters that become stagnant due to slow flows under summer drought conditions often support persisting populations of cyanobacteria. Growth of cyanobacteria is stimulated by high water temperatures and high concentrations of inorganic nitrogen (N) and phosphorus (P). Their capability to be a source of biological nitrogen fixation in soils and water is also a significant contributor to long-term survival and an important role in the ecological succession of microorganisms in the environment.

Control of cyanobacteria is a problem as research has shown that the toxins can remain potent for days even after the organisms have been destroyed by chlorination and copper sulphate (El Saadi *et al.*, 1995). While there are toxicity data obtained from mouse models, further research is needed on the acute and chronic toxicity of cyanobacterial toxins and suitable methods need to be developed for monitoring the types and concentrations of cyanobacterial toxins in natural as well as treated drinking water.

## Methods of detection

Direct microscopic examination of bloom material will allow identification of the cyanobacterial species present. In many cases this will enable distinction between those species that are potentially toxic and those that are not (Chorus and Bartram, 1999). A semiquantitative estimate of cyanobacterial biomass can also be made from a simple microscopic analysis. A more accurate quantitative method has been recommended. To culture cyanobacteria, water samples are first blended with glass beads or treated by ultrasound to break filamentous forms prior to streaking on agar plates (AWWA, 1999). There are a variety of selected mineral media available (D-medium, ASM-1, BG-11, and WC) with incubation at 25°C under cool white fluorescent light. Some recalcitrant cyanobacteria may not be freed easily of contaminants, thus, physical and chemical separation schemes may be necessary.

There are also several methods for detecting cyanobacterial toxins in water (Chorus and Bartram, 1999). Such methods include bioassays in mice, or the brine shrimp (*Artemia salina*). There are also a range of analytical methods including high pressure liquid chromatography (HPLC), gas chromatography (GC) and liquid chromatography/mass spectroscopy (Chorus and Bartram, 1999).

## Epidemiology of waterborne outbreaks

There have been numerous reports of poisonings of livestock, pets and wildlife with waters containing cyanobacteria blooms (Hunter *et al.*, 1999). Compared to animal poisoning, reports of outbreaks of human illness have been less common. Most reports of human illness have been associated with recreational contact (Hunter, 1998). Nevertheless, there have still been several reports of illness associated with consumption of drinking water.

Most of the reported outbreaks associated with drinking water have been of gastroenteritis. Zilberg (1966) reported a sharp annual increase in admissions during winter from an area supplied by a reservoir that regularly suffered from algal blooms in Salisbury, Rhodesia (now Harare, Zimbabwe) but not in neighbouring populations. The peak incidence in cases corresponded with the death of the algae in the affected reservoir. Lippy and Erb (1976) reported an outbreak of gastroenteritis that affected an estimated 5000 people (62% of the population) supplied by a single reservoir in Sewickley, Pennsylvania, USA. On examination of the water reservoir, the remains of a recently dead bloom of *Schizothrix calicola* were found. These two outbreaks suggest that the main risk of gastroenteritis is associated with death of a cyanobacterial bloom. Cell death is probably associated with the release of various toxins into the water.

An outbreak of gastroenteritis affecting about 2000 people, of whom 88 died, in Brazil was found to be associated with drinking water taken from a new dam (Teixeira *et al.*, 1993). The only abnormal results were high levels of *Anabaena* and *Microcystis* in untreated dam water and the outbreak declined rapidly after copper sulphate treatment of the dam water to reduce cyanobacterial counts.

In a prospective case-control study of people taking their drinking water from the Murray River, cases of gastroenteritis were more likely to have drunk chlorinated river water rather than rain or spring water compared to controls (El Saadi *et al.*, 1995). There was also a correlation between the weekly mean log cyanobacterial cell counts in the river and the number of patients presenting to medical practitioners with gastroenteritis.

There is also evidence that drinking water can be associated with hepatitis. An outbreak of hepatitis illness affected 139 children and 10 adults of aboriginal descent in the Palm Island Community, Queensland, Australia (Byth, 1980; Bourke *et al.*, 1983). Illness was strongly correlated with consumption of the main supply, which came from a dam that had suffered from a heavy algal bloom. The outbreaks started 5 days after the dam had been treated with copper sulphate to kill the algae. Evidence of subclinical hepatitis was found to be related to an algal bloom on a drinking water reservoir when increased liver enzymes were detected in blood samples taken for other purposes (Falconer *et al.*, 1983).

There have been at least two outbreaks linked to contamination of water used for renal dialysis. The first outbreak was of pyogenic reactions in a dialysis centre in Washington DC, USA (Hindman *et al.*, 1975). This was found to be due to high levels of endotoxin in the potable water supply to the clinic, which correlated in turn with a period of high blue-green algal bloom counts in the supply reservoir. The second was an outbreak of hepatitis in a dialysis centre in Brazil during February 1996 (Jochimsen *et al.*, 1998). Of 130 patients attending the centre, 116 (89%) became ill and 50 patients died with liver failure. Microcystin was detected in the water used for dialysis and in samples from affected patients. The water had been taken from a local reservoir and not treated before delivery to the hospital.

For utilities using surface water supplies, cyanobacteria are well known for their association with taste-and-odour problems, often regarded as a matter of aesthetics. In light of recent information on cyanobacteria, granular activated carbon (GAC) may be very important to toxin removal. Furthermore, for those water systems using disinfection as the only surface water treatment, there is always the threat of a seasonal passage of cyanobacteria and deposition of their dead cells in the distribution pipe network. Such an occurrence provides a source of assimilable organic carbon (AOC), which is a potential nutrient for bacterial regrowth.

The US Environmental Protection Agency (USEPA) and European governments regulate neither cyanobacteria nor their metabolites, except under guidelines stating that drinking water must be potable. The World Health Organization has established guideline values for the tolerable concentration of microcystin in drinking water – a value of 1 μg/l was obtained (Fawell, 1993; Falconer, 1994). Contrary to this a value of <0.5 μg/l was determined by work in Canada (Kuiper-Goodman *et al.*, 1994). The engineering and water supply department of South Australia has also developed interim guidelines for acceptable numbers of cyanobcteria in water supplies (El Saadi *et al.*, 1995).

## Risk assessment

*Health effects*: occurrence of illness, degree of morbidity and mortality, probability of illness based on infection:

- There are about 25 species of cyanobacteria that have been associated with adverse health effects.
- Cyanobacteria act primarily as hepatotoxins and neurotoxins, but can also cause skin irritation. They are extremely potent toxins and, therefore, have the potential to be fatal to humans. However, acute oral or dermal exposures have not resulted in any known human deaths. Reported illnesses in humans exposed to cyanobacterial toxins range from dermatitis and gastroenteritis to hepatitis and allergic reactions. Illness is self-limited.
- Outcomes from cyanobacterial toxins are mostly acute; however, the microcystins and other toxins have been shown to be tumour promoters in animal studies, and epidemiological evidence in humans suggests that chronic exposure to microcystins in drinking water is associated with an increase in hepatocellular cancer.

*Exposure assessment*: routes of exposure and transmission, occurrence in source water, environmental fate:

- World-wide, the number of humans acutely affected by cyanobacterial toxins is low compared with other waterborne contaminants. However, because of decreasing water quality, the potential for an increase in incidents is high.

- Routes of exposure are primarily ingestion through drinking water or recreational water contact, also, dermal exposure and possibly aerosolization.
- Most acute exposures result from recreational water use; low levels in drinking water are associated with an increase in hepatocellular cancer in certain exposed populations.
- Cyanobacteria are found in all types of water: lakes, rivers, marine environments, and drinking-water reservoirs. Surface waters that receive waste effluents are at special risk for contamination. High water temperatures and high concentrations of inorganic nitrogen and phosphorus stimulate growth.
- Surface water systems that use only disinfectant may get deposition of dead cells in the distribution system with potential for regrowth.
- Cyanobacterial toxins are ubiquitous, though their occurrence is dependent on the conditions that contribute to algal bloom formation. Concentrations vary widely depending on the species of bloom and the stage of the bloom's formation and deterioration. Toxin concentrations have been reported as ranging from 0.2 μg/l to 8.5 mg/l.
- The toxic dose is unknown for humans. The no observed adverse effect level (NOAEL) for mice dosed orally with microcystin-LR has been reported to be 40 μg/kg/day for 13 weeks. The NOAEL reported for mice dosed orally with anatoxin-a has been reported to be 0.1 mg/kg/day. Intraperitoneal and intranasal exposure is more potent than oral ingestion for both toxins.

*Risk mitigation*: drinking-water treatment, medical treatment:

- There is some question as to the efficacy of standard drinking water treatment (e.g. coagulation, sedimentation, disinfection and filtration) for removing all but large concentrations of cyanobacterial toxins, though current methods are effective enough to prevent any acute effects.
- Evidence on the efficacy of chlorine on the microcystins is equivocal; chlorine is ineffective on anatoxin-a. Activated carbon treatment appears to be the best removal method for treated water.
- Preventing the formation of blooms in the source water is the best way to assure cyanobacteria-free drinking water.
- Membrane filtration technology has the potential to remove virtually any cyanobacteria or their toxins from drinking water.
- Efficacious medical treatment is unknown in an acute exposure; however, antihistamines and steroids may be helpful for allergic reactions. If given in a timely manner, activated charcoal or an emetic could have a positive effect on the toxic response.

# References

AWWA. (1999). *Manual of Water Supply Practices: Waterborne Pathogens*. Washington, DC: American Water Works Association.

Baxter, P.J. (1991). Toxic marine and freshwater algae: an occupational hazard? *Br J Ind Med*, **49**: 505–506.

Bourke, A.T.C., Hawes, R.B., Neilson, A. *et al.* (1983). An outbreak of hepato-enteritis (the Palm Island mystery disease) possibly caused by algal intoxication. *Toxicon*, 3(Suppl.): 45–48.

Byth, S. (1980). Palm Island mystery disease. *Med J Aust*, 2: 40–42.

Carmichael, W.W. (1992). Cyanobacteria secondary metabolites – the cyanotoxins. *J Appl Bacteriol*, 72: 445–459.

Carmicheal, W.W. (1994). The toxins of cyanobacteria. *Sci Am*, 270: 78–86.

Chorus, I. and Bartram, J. (1999). *Toxic Cyanobacteria in Water*. London: E & FN Spon.

El Saadi, O., Easterman, A.J., Camerson, S. *et al.* (1995). Murray River water, raised cyanobacterial cell counts and gastrointestinal and dermatological symptoms. *Med J Aust*, 162: 122–125.

Fawell, J.K. (1993). Toxins from blue-green algae: *Toxicological Assessment of Microcystin-LR*. Vol. 4. Microcystin-LR:13 week oral (gavage) toxicity study in the mouse. Final Report. Medmenham, UK: Water Res. Cent., pp. 1–259.

Falconer, I.R. (1994). Health problems from exposure to cyanobacteria and proposed safety guidelines for drinking water and recreational water. In *Detection Methods for Cyanobacterial Toxins*, Codd, G.A., Jefferies, T.M., Keevil, C.W. *et al.* (eds). Cambridge, UK: R. Soc. Chem., pp. 3–10.

Falconer, I.R., Beresford, A.M. and Runnegar, M.T.C. (1983). Evidence of liver damage by toxin from a bloom of the blue-green alga, Microcystis aeruginosa. *Med J Aust*, I: 511–514.

Gold, G.A., Bell, S.G. and Brooks, W.P. (1989). Cyanobacterial toxins in water. *Water Sci Technol*, 21: 1–13.

Hindman, S.H., Favero, M.S., Carson, L.A. *et al.* (1975). Pyogenic reactions during haemodialysis caused by extramural endotoxin. *Lancet*, ii: 732–734.

Hunter, P.R. (1998). Cyanobacterial toxins and human health. *J Appl Bacteriol*, 84(Suppl.): 35S–40S.

Hunter, P.R., Petersen, A., Merrett, H. *et al.* (1999). Investigations of toxins produced by cyanobacteria: Literature review of the hazards and risks posed by freshwater cyanobacteria to human and animal health. *R&D Report Series No. 4*. Johnstown Castle Estate, County Wexford, Ireland: Environmental Protection Agency.

Jochimsen, E.M., Carmichael, W.W., An, J.S. *et al.* (1998). Liver failure and death after exposure to microcystins at a haemodyalisis center in Brazil. *New Engl J Med*, 338: 873–878.

Keleti, G. and Sykora, J.L. (1982). Production and properties of cyanobacterial endotoxins. *Appl Environ Microbiol*, 43: 104–109.

Kuiper-Goodman, T., Gupta, S., Combley, H. *et al.* (1994). Microcystins in drinking water: risk assessment and derivation of a possible guidance value for drinking water. In *Toxic Cyanobacteria: A Global Perspective*, Steffensen, D.A. and Nicholson, B.C. (eds). Salisbury, S. Australia: Aust. Cent. Water Qual. Res., pp. 17–23.

Lippy, E.C. and Erb, J. (1976). Gastrointestinal illness at Sewickley, Pa. *JAWWA*, 76: 60–70.

MacKintosh, C., Beattie, K.A., Klumpp, S. *et al.* (1990). Cyanobacterial microcystin-LR is a potent and specific inhibitor of protein phosphatases 1 and 2A from both mammals and higher plants. *FEBS Lett*, 264: 187–192.

Matsushima, R., Yoshizawa, S., Watanabe, M.F. *et al.* (1990). *In vitro* and *in vivo* effects of protein phosphatase inhibitors, microcystin and nodularin, on mouse skin and fibroblasts. *Biochem Biophys Res Comm*, 171: 867–874.

Nishiwaki-Matsushima, R., Ohta, T., Nishiwaki, S. *et al.* (1992). Liver cancer promoted by the cyanobacterial cyclic peptide toxin microcystin LR. *J Cancer Res Clin Oncol*, 118: 420–424.

Runnegar, M., Berndt, N. and Kaplowitz, N. (1995). Microcystin uptake and inhibition of protein phosphatases: effects of chemoprotectants and self-inhibition in relation to known hepatic transporters. *Toxicol Appl Pharmacol*, 134: 264–272.

Teixeira, M.G.L.C., Costa, M.C.N., de Carvalho, V.L.P. *et al.* (1993). Gastroenteritis epidemic in the area of the Itaparica dam, Bahia, Brazil. *Bull PAHO*, 27: 244–253.

Turner, P.C., Gammie, A.J., Hollinrake, K. *et al.* (1990). Pneumonia associated with contact with cyanobacteria. *Br Med J*, 300: 1440–1441.

Yeh, F.-S. (1989). Primary liver cancer in Guangxi. In *Primary Liver Cancer*, Tang, Z.-Y., Wu, M.-C. and Xia, S.-S. (eds). Beijing: China Academic Publishers, pp. 223–236.

Yoshizawa, S., Matsushi, R., Watanabe, M.F. *et al.* (1990). Inhibition of protein phosphatases by microcystis and nodularin associated with hepatotoxicity. *J Cancer Res*, **116**: 609–614.

Yu, S.-Z. (1989). Drinking water and primary liver cancer. In *Primary Liver Cancer*, Tang, Z.-Y., Wu, M.-C. and Xia, S.-S. (eds). Beijing: China Academic Publishers, pp. 30–37.

Yu, S.-Z. (1995). Primary prevention of hepatocellular carcinoma. *J Gastroenterol Hepatol*, **10**: 674–682.

Zhou, T.L. and Yu, S.Z. (1990). Laboratory study on the relationship between drinking water and hepatoma: Quantitative evaluation using GGT method. *Chin J Prevent Med*, **24**: 203–205.

Zhu, Y.-R., Chen, J.-G. and Huang, X.-Y. (1989). Hepatocellular carcinoma in Qidong County. In *Primary Liver Cancer*, Tang, Z.-Y., Wu, M.-C. and Xia, S.-S. (eds). Beijing: China Academic Publishers, pp. 204–222.

Zilberg, B. (1966). Gastroenteritis in Salisbury European children – a five year study. *Cent Afr J Med*, **12**: 164–168.

# 7

# *Escherichia coli*

## Basic microbiology

*Escherichia coli* are non-spore-forming, Gram-negative bacteria, usually motile by peritrichous flagella. They are facultative anaerobes with gas usually produced from fermentable carbohydrates. *E. coli* form rod-shaped cells 2.0–6.0 μm in length and 1.1–1.5 μm in width, however, cells may vary from coccal to long filamentous rods. Capsules or microcapsules, made of acidic polysaccharides, are common in *E. coli*. Mucoid strains sometimes produce extracellular polymers, generally referred to as K antigens and acid poly-saccharides, composed of colanic acid, known as M antigens. *E. coli* produce different kinds of fimbriae which are important during the adhesion of host cells. Fimbriae vary both structurally and antigenically in different strains of *E. coli*.

Some less commonly encountered *E. coli* strains found both within the environment and potable water systems are capable of giving rise to diseases usually in the form of diarrhoea. The process by which *E. coli* causes diarrhoea varies between strains. These strains can be grouped depending upon which mechanism is used by a particular strain.

## Origins of the organism

*E. coli* was first identified by the German paediatrician Theodor Escherich during his studies of the intestinal flora of infants. He described the organism in 1885 as *Bacterium coli commune* and established its pathogenic properties in extraintestinal infections. The name *Bacterium coli* was widely used until 1919. The genus *Escherichia* and the type species *E. coli* was used thereafter. It was not until 1964 that the genus *Escherichia* was defined in Wilson and Miles, *Topley and Wilson's Principles of Bacteriology and Immunity* as a motile or non-motile organism that produced characteristics conforming to the definition Enterobacteriaceae. Work conducted in the 1920s concluded that *Bacterium coli* was a very antigenically heterogeneous species. However, it was not until the 1940s that a classification scheme dividing *E. coli* into more than 70 different serogroups, primarily based on the O (somatic) antigens (Kauffmann, 1947) was developed. Today over 50 H (flagella) antigens and over 100 K (capsular antigens) are now recognized which enable *E. coli* to be further subdivided into serotypes.

In 1987, the type genus *Escherichia* contained four new species in addition to *E. coli*: *E. blattae* (isolated from the hind gut of the cockroach and has not been reported in clinical material) was first described in 1973 (Burgess *et al.*, 1973), *E. fergusoni* (isolated from clinical material) first described by Farmer and coworkers in 1985, *E. hermannii* (isolated from wounds) first described by Brenner and others in 1982 (Brenner *et al.*, 1982a) and *E. vulneris* (isolated from human wounds) first described by Brenner and coworkers also in 1982 (Brenner *et al.*, 1982b).

## Metabolism and physiology

The biochemical characteristics of the genus *Escherichia* are shown in Table 7.1 and the characteristics of *E. coli* are shown in Table 7.2. Most strains of *E. coli* ferment lactose. They produce indole, fail to hydrolyse urea and to grow in Møller's KCN broth. $H_2S$ production is not detectable on triple sugar iron (TSI) agar, phenylalanine is not deaminated, and gelatine is not liquefied. Most strains decarboxylate lysine and utilize sodium acetate, but they do not grow on Simmons' citrate agar.

## Clinical features

In 1921, Muir and Ritchie first described the pathogenic properties of *B. coli* being associated with infections of the intestine and urinary tract, some cases

**Table 7.1** Biochemical characteristics of *Escherichia* (adapted from Wilson and Miles, 1964; Barrow and Feltham, 1995)

| Characteristics | Reaction |
| --- | --- |
| Motility | + |
| MacConkey growth | + |
| Mannitol fermentation | +, usually gas |
| Lactose, 37°C | Acid +, gas + |
| Lactose, 44°C | Acid +, gas + |
| Adonitol | Seldom fermented |
| Inositol | Seldom fermented |
| Indole at 37°C | Usually produced |
| Indole at 44°C | Usually produced |
| Methyl red reaction | + |
| Voges-Proskauer reaction | − |
| Urea | No hydrolysis |
| Phenylalanine deamination | − |
| Kligler's $H_2S$ (hydrogen sulphide) medium | No blackening |
| Møller's KCN (potassium cyanide) medium | No growth |
| Gluconate oxidation | − |
| Gelatine liquefaction | − |
| Glutamine acid decarboxylase | + |
| Lysine decarboxylase | + |

**Table 7.2** Characteristics of *Escherichia coli* (adapted from Wilson and Miles, 1964; Barrow and Feltham, 1995)

| Characteristics | Reaction |
| --- | --- |
| Gram stain | Negative |
| Morphology | Straight rods |
| Motility | + (peritrichous) some non-motile |
| Aerobic and anaerobic growth | + |
| Oxidase | − |
| Catalase | + |
| MacConkey growth | + |
| D-mannitol fermentation | +, usually gas (over 90% of strains) |
| Lactose, 37°C | Acid+, gas + (over 90% of strains) |
| Lactose, 44°C | Acid +, gas + (over 90% of strains) |
| D-adonitol | Seldom fermented (over 90% of strains) |
| Inositol | Seldom fermented |
| D-glucose | Acid |
| Indole at 37°C | Usually produced |
| Indole at 44°C | Usually produced |
| Methyl red reaction | + (over 90% of strains) |
| Voges-Proskauer reaction | − (over 90% of strains) |
| Urea | No hydrolysis |
| Phenylalanine deamination | − (over 90% of strains) |
| $H_2S$ (triple sugar iron) medium | No blackening (over 90% of strains) |
| KCN (potassium cyanide) medium | No growth |
| Gelatine liquefaction | − (over 90% of strains) |
| Glutamine acid decarboxylase | + |
| Lysine decarboxylase | + (75–89% of strains) |

of summer diarrhoea (cholera nostras) and some cases of infantile diarrhoea and food poisoning. From Topley and Wilson's first addition (1929) the pathogenicity of *Bact. coli* was summarized as: '*Bact. Coli* is a normal inhabitant of the intestine of man and other animals. In certain circumstances it acquires pathogenicity, and may cause local or general infection. It is a frequent cause of acute and chronic infection of the urinary tract, and may give rise to an acute or chronic cholecystitis.'

To date, *E. coli* is classed as a harmless member of the normal microbiota of the human located at the distal end of the intestinal tract. The organism is generally acquired at birth or by the faecal-oral route from the mother and also from the environment. Most strains of *E. coli* are not pathogenic. A list of some of the strains of *E. coli* that can cause a number of illnesses is given in Table 7.3. The illnesses related to *E. coli* are shown in Table 7.4.

*E. coli* is the most common cause of acute urinary tract infections as well as urinary tract sepsis. It has also been known to cause neonatal meningitis and sepsis and also abscesses in a number of organ systems. *E. coli* may also cause acute enteritis in humans as well as animals and is a general cause of 'traveller's diarrhoea', a dysentery-like disease affecting humans, and haemorrhagic colitis often referred to as 'bloody diarrhoea'. Many oral challenge studies have been done with a number of *E. coli* types to determine necessary doses required to cause infection. The results of these studies suggest that levels of $10^5$–$10^{10}$ enteropathogenic (EPEC) organisms, $10^8$–$10^{10}$ enterotoxigenic (ETEC) and $10^8$ cells of enteroinvasive (EIEC) need to be ingested to produce diarrhoea and infection. These numbers will of course vary depending on age, sex and acidity of the stomach. In the case of Vero cytotoxigenic *E. coli* (VTEC) the infective dose that is capable of causing infection is <100 cells (Advisory committee on the Microbiology Safety of Food, 1995; Bolton *et al.*, 1996).

VTEC is the major *E. coli* of concern. It is known to cause haemolytic uraemic syndrome (HUS) that is characterized by acute renal failure, haemolytic anaemia and thrombocytopaenia that usually occurs in children under the age of 5. On average 10% of patients infected with VTEC O157 develop HUS and some go onto develop thrombotic thrombocytopaenic purpura (TTP). Approximately 5% of cases of VTEC develop haemorrhagic colitis which then develop into HUS, in which case fatality can be as high as 10%. Diarrhoea caused by *E. coli* O157 is sometimes self-limiting. Strains of *E. coli* fall into at least five groups with different pathogenic mechanisms. These will be discussed in turn.

## Enterotoxigenic *E. coli* (ETEC)

Enterotoxigenic *E. coli* (ETEC) cause gastroenteritis with profuse watery diarrhoea with abdominal cramps, vomiting and fever evident in a small percentage of patients. The severity of illness as a result of infection with ETEC varies from relatively mild and short-lived to a severe life-threatening illness. They are one of the major causes of death in children below the age of 5 years in

**Table 7.3**  Serogroups and disease associations of *E. coli* (Rowe, 1983; Salyers and Whitt, 1994; Beutin *et al.*, 1997; Sussman, 1997; Willshaw *et al.*, 1997; Bell and Kyriakides, 1998).

| Virulence type | Serogroup | Disease | Summary of host cell interaction |
|---|---|---|---|
| Enteropathogenic (EPEC) | 055 H6, NM<br>086 H34, NM<br>0111 H2, H12, NM<br>0119 H6, NM<br>0125ac H21<br>0126 H27, NM<br>0128 H2, H12<br>0142 H6 | Enteritis in infants<br>Traveller's diarrhoea | EPEC attach to intestinal mucosal cells causing cell structure alterations (attaching and effacing), EPEC cells invade the mucosal cells |
| Enterotoxigenic (ETEC) | 06 H16<br>08 H9<br>011 H27<br>015 H11<br>020 NM<br>025 H42, NM<br>027 H7<br>078 H11, H12<br>0128 H7<br>0148 H28<br>0149 H10<br>0159 H20<br>0173 NM | Diarrhoea, vomiting and fever<br>Traveller's diarrhoea | ETEC adhere to the small intestinal mucosa and produce toxins that act on the mucosal cells |
| Vero cytotoxigenic (VTEC) (including enterohaemor-rhagic, EHEC) | 026 H11, H32, NM<br>055 H7<br>0111ab H8, NM<br>0113 H21<br>0117 H14<br>0157 H7 | Shigella-like dysentery (stools contain blood and mucus)<br>Haemolytic uraemic syndrome | EHEC attach to and efface mucosal cells and produce toxin |
| Enteroinvasive (EIEC) | 028ab NM<br>029 NM<br>0112ac NM<br>0124 H30, NM<br>0136 NM<br>0143 NM<br>0144 NM<br>0152 NM<br>0159 H2, NM<br>0164 NM<br>0167 H4, H5, NM | Shigella-like dysentery | EIEC invade cells in the colon and spread laterally, cell to cell |
| Enteroaggregative (EAEC) | 03 H2<br>015 H18<br>044 H18<br>086 NM<br>077 H18<br>0111 H21<br>0127 H2<br>0? H10 | Persistent diarrhoea in children | EAEC bind in clumps (aggregate to cells of the small intestine) and produce toxins |
| Diffusely adherent (DAEC) | Not yet established | Childhood diarrhoea | Fimbrial and non-fimbrial adhesions identified |

**Table 7.4** *E. coli* related illnesses (adapted from Bell and Kyriakides, 1998)

| Pathogenic *E. coli* | Time to onset of illness | Duration of illness | Symptoms |
| --- | --- | --- | --- |
| EPEC | 17–72 h (average 36 h) | 6 h to 3 days (average 24 h) | Severe diarrhoea in infants which may persist for more than 14 days. Also fever, vomiting and abdominal pain. In adults, severe watery diarrhoea, with prominent amounts of mucus without blood, and nausea, vomiting, abdominal cramps, headache, fever and chills. |
| ETEC | 8–44 h (average 26 h) | 3–19 days | Watery diarrhoea, low-grade fever, abdominal cramps, malaise, nausea. When severe, causes cholera-like extreme diarrhoea with rice water-like stools, leading to dehydration. |
| VTEC | 3–9 days (average 4 days) | 2–9 days (average 4 days) | Haemorrhagic colitis (HC): sudden onset of severe crampy abdominal pain, grossly bloody diarrhoea, vomiting, no fever. Haemolytic uraemic syndrome (HUS): prodrome of bloody diarrhoea, acute renal failure in children, thrombocytopaenia, acute nephropathy, seizures, coma, death. Thrombotic thrombocytopaenic purpura (TTP): similar to HUS but also fever, central nervous system disorders, abdominal pain, gastrointestinal haemorrhage, blood clots in the brain, death. |
| EIEC | 8–24 h (average 11 h) | Days to weeks | Profuse diarrhoea or dysentery, chills, fever, headache, muscular pain, abdominal cramps. |

developing countries. Adults and older children in tropical countries can also become infected by ETEC but, generally, these become asymptomatic carriers as a result of mucosal immunity. Individuals who do not acquire this sort of immunity develop a condition known as traveller's diarrhoea. Traveller's diarrhoea is usually brief episodes of loose stools sometimes accompanied with nausea, vomiting and abdominal pains.

ETEC produce two enterotoxins, a heat stable toxin and a heat labile oligopeptide toxin. Both toxins enter the cell and increase the concentration of cyclic guanine monophosphate (cGMP) or cyclic adenine monophosphate (cAMP) known to affect electrolyte transport causing excessive fluid loss. The incubation period associated with infection of this group of *E. coli* is 12–72 hours, with duration of illness often 3–5 days.

Approximately 2–8% of the *E. coli* found in water are enteropathogenic *E. coli* which cause traveller's diarrhoea. While water and also foods are instrumental in the transmission and spread of *E. coli* the required dose for this pathogen to cause infection is high, typically in the range of $10^6$–$10^9$ organisms.

## Enteropathogenic *E. coli* (EPEC)

EPEC is a major cause of infant (less than 2 years of age) watery diarrhoea infections. It has, however, been documented as a major cause of watery diarrhoea in children less than 6 months of age specifically in developing countries. Mortality can be as high as 30%. The infectious dose of EPEC is very high, about $10^8$–$10^{10}$ organisms. The transmission of infection is directly from person to person with no evidence to date that it is transmitted in water.

## Enterohaemorrhagic *E. coli* (EHEC) or Vero cytotoxigenic *E. coli* (VTEC)

Enterohaemorrhagic *E. coli* (EHEC) produces a shiga-like toxin that is cytotoxic to Vero cells. The commonest EHEC is O157. The incubation period following ingestion is 3–8 days and duration of infection is 1–12 days. Following ingestion of the required dose of less than 100 organisms symptoms develop which include watery and bloody diarrhoea associated with vomiting. EHEC can also cause two distinct conditions, haemorrhagic colitis and haemolytic uraemic syndrome (HUS). HUS is characterized by thrombocytopaenia, microangiopathic, haemolytic anaemia and renal failure.

Symptoms resulting from EHEC include a colicky abdominal pain, followed by diarrhoea, vomiting in some cases, that becomes bloody. This loss of blood is often severe after only a few days following infection. EHEC has also been associated with sporadic outbreaks of haemorrhagic colitis which is a grossly bloody diarrhoea. Fatality rates are very high in the elderly and the very young with 10% of children under 10 years developing a combination of haemolytic anaemia, thrombocytopaenic purpura and acute renal failure. In the UK, HUS is now the commonest cause of acute renal failure in children.

### Enteroinvasive *E. coli* (EIEC)

Patients who become infected with EIEC present with watery diarrhoea, with a small proportion developing bloody diarrhoea. The infectious dose is high, between $10^6$ and $10^{10}$ organisms. EIEC are closely linked to shigellae in that they both cause the bacillary dysentery using similar processes. The incubation period following ingestion is usually 1–3 days and the duration of infection is 1–2 weeks.

### Enteroaggregative *E. coli* (EAEC)

EAEC causes a watery mucoid diarrhoea, which usually contains no blood and there is no fever. The infectious dose is generally high and it is primarily a disease of developing countries.

## Virulence and pathogenicity

### ETEC

In order for ETEC to initiate effects on the host it must be able to adhere to the mucosal surface of the epithelial cells in the small intestine. Many colonization factors have been discovered in ETEC, suggesting the process of adhesion is more complex than originally thought and it still remains an area of great debate. However, the process of adhesion seems to be achieved by fimbriae, known to have specific binding receptors. Once ETEC is adhered the role of toxins in the infection process comes into play. ETEC strains have been found to produce two toxins, a short polypeptide chained toxin called the heat-stable (ST) toxin, and/or a heat-labile (LT), large oligometric enterotoxin. Both the ST and LT toxins have two antigenic types, namely STa and STb and LT-I and LT-II respectively. LT-I is closely related to the toxin produced by *Vibrio cholerae*. It is an oligomeric toxin composed of one polypeptide A subunit and five identical polypeptide B subunits. The molecular weights of subunits A and B are 25 000 and 115 000 respectively. The five B subunits form a very stable doughnut shape with a central aqueous channel and are important in binding the toxin to the epithelial cells. The A subunit sits above and is partly inserted into this central channel.

After the toxin molecule has bound to the enterocyte cell surface through its five B subunits the toxin molecule is taken into the cell by the process of endocytosis. Within the cell the A subunit is split into $A_1$ and $A_2$ fragments. The activity lies in the $A_1$ fragment which is an ADP-ribosyltransferase. It is subunit A which catalases the nicotinamide adenine dinucleotide (NAD) dependent activation of adenylate cyclase, causing an increase in the concentration of intracellular cyclic adenosine 5′-monophosphate (cAMP). There is a marked net excretion of chloride from the enterocyte into the gut lumen. This is because cAMP inhibits the absorption of sodium in the intestinal villus cells and therefore chloride ions and water. The net result of these reactions is that the intraluminal osmolality increases and water is drawn into the gut. In the crypt cells cAMP causes an increase in sodium secretion, with a loss of chloride ions and water. This then leads to the development of a watery diarrhoea. Other more complex mechanisms have also been suggested as alternative to this possible mechanism (Sears and Kaper, 1996).

The A subunit in LT-II causes a similar effect to that of LT-I. LT-II also causes a net excretion of chloride through an increase in intracellular cAMP, though the initial intracellular binding site is different.

The heat stable enterotoxins (STa) of *E. coli* are low molecular weight toxins, generally poorly immunogenic. As mentioned there are two major classes, STa and STb. The STa (18 amino acids) toxin activates guanylate cyclase C (GC-C), an enzyme present in the luminal membrane of enterocytes. The binding of STa to GC-C causes an increase in intracellular cGMP. The activity of STa is quick and the end result is increased chloride excretion in the same way as for LT. LT, as opposed to STa, does have a lag period before it acts. STb differs from STa in being larger, 48 amino acids, and not causing an increase in cAMP or cGMP. Also chloride transport is not directly affected, instead there is a net increase in bicarbonate excretion. Furthermore, STb is associated with histological evidence of cellular damage and STa is plasmid encoded and can be distinguished from STb as it is soluble in methanol.

## EPEC

EPEC produce characteristic histological features in the intestinal tract of patients. These are known as attaching and effacing lesions which show the adherence of the bacteria to the epithelial cell membrane with effacement of the microvilli. The lesions develop in three stages. At stage one *E. coli* attaches, aided by fimbriae, to the intestinal enterocyte cells. Following attachment to the enterocyte the secretion of various extracellular proteins causes dissolution of the microvilli (Bain *et al.*, 1998). Following on from this the bacterium binds closely to the enterocyte membrane. There has been much recent interest in EPEC (DeVinney *et al.*, 1999; Vallance and Finlay, 2000). Recent work has shown that EPEC inserts its own receptor for intimate adherence, Tir (translocated intimin receptor) into the host cell membrane. Once attached, translocated EPEC proteins activate signalling pathways within the underlying cell. These cause the reorganization of the host actin cytoskeleton. EPEC once attached become partially surrounded by cup-like projections (pedestal-like structures) beneath the adherent bacteria.

The most prevalent serogroups of EPEC are O6, O8, O25, O111, O119, O125–O128.

## EHEC

EHEC has the same virulence mechanisms as EPEC but produces potent toxins known as Shiga (Stx) toxins. Stx1 has been shown to be identical to the Shiga toxin of *S. dysenteriae* type 1 with Stx 2 only some 55–57% homologous. Both Stx1 and 2 are made up of an A subunit and five B subunits. The A subunit possesses the biological activity while the B subunits are thought to mediate binding to the cell surface and thus aid uptake of the toxin.

Shiga toxins bind to a glycolipid receptor, globotriaosylceramide. The toxin can then be taken into the cell by a process of endocytosis and transported to the endoplasmic reticulum. Once the toxin is attached subunit A is released into the cytoplasm where it disrupts protein synthesis, leading to cell death. The toxin removes a single adenine residue from the 28S rRNA of eukaryotic ribosomes which brings about this effect.

Associated with EHEC is the production of diarrhoea possibly caused by the death of the intestinal absorptive cells, leaving the secretory cells intact. Haemolytic uraemic syndrome (HUS) is thought occur because the toxin is transported via the blood to the kidneys where it can cause kidney failure. However, some research has shown that while strains of *E. coli* O157:H7 had been isolated from cases of HUS they in fact lacked the ability to produce the Shiga toxin (Schmidt *et al.*, 1999).

## EIEC

Enteroinvasive *E. coli* bind to enterocysts of the large intestine where they destroy cells resulting in tissue damage which results in an inflammatory response. The invasive property of EIEC is controlled by a 140 MD plasmid. As the name suggests there is invasion of the epithelial cell itself with subsequent intracellular multiplication and lateral movement through the cell to allow subsequent penetration of adjacent cells. Although not yet fully understood, there also seems to be production of an enterotoxin which is thought to be responsible for the initial watery diarrhoea.

EIEC causes dysentery-like symptoms by invading colonic epithelium (cf. Shigella). These strains belong to serogroups O28, O52, O112, O115, O124, O136, O143, O145, O147. These strains account for <0.5% infection in the West but about 0.5% of dysentery in Asia.

## EAEC

EAEC are named after their characteristic aggregative adherence to Hep-2 cells in tissue culture. The mechanism of the pathogenicity of EAEC is poorly understood, however, a model of pathogenicity has been suggested by Nataro and Kaper (1998). This model is composed of three stages with the first two stages involving the adherence of EAEC to the mucosal membrane and the stimulation of mucus secretion leading to the formation of a thick mucus biofilm. In the biofilm the bacteria tend to clump forming an enteroaggregative. The final stage of the pathogenicity model involves the secretion of a cytotoxin responsible for causing damage to the mucosal lining.

## DAEC

Infection with DAEC causes a watery diarrhoea, mostly in older children. DAEC like EAEC adhere to Hep-2 cells. The pattern by which DAEC attaches is, however, more diffuse than EAEC, a distinguishing feature. However, the pathogenic mechanisms for DAEC are to date still inconclusive.

## Treatment

With *E. coli* diarrhoeal infections fluid and electrolyte correction is obligatory. Extraintestinal *E. coli* infections, however, are treated with antibiotics. If this method of treatment is to be employed it must be borne in mind that *E. coli* does possess intrinsic resistance to benzylpenicillin. As a general rule *E. coli* are still sensitive to the antibiotics ampicillin, tetracycline, aminoglycosides, trimethoprim and the cephalosporins. However, because of the widespread evidence of antibiotic resistance being acquired by plasmid transfer, mounting numbers of *E. coli* are becoming resistant to streptomycin and tetracycline. For this reason antibiograms should be performed, especially for epidemiological purposes. The use of antibiotics in treating *E. coli* O157 is questionable due to the problem with the growing numbers of *E. coli* O157 developing antibiotic resistance.

## Survival in the environment

*E. coli* has a reservoir in the intestines of man and other warm-blooded animals, and is excreted in the faeces. It is known to survive in the environment but not reproduce (Feachem *et al.*, 1983), however, in the tropical environment there is evidence that *E. coli* can survive and multiply (Rivera *et al.*, 1988). Subsequently, if *E. coli* is detected in the environment it is indicative of faecal pollution but in the tropics this may not be the case which warrants further investigation. For this reason this research suggests that in fact *E. coli* may not be a good indicator of faecal pollution. *Escherichia coli* was introduced into water bacteriology because it was a useful marker of faecal pollution and thus became an important marker in food and water hygiene. The theory was that if *E. coli* was present then so could be pathogenic enteric bacteria such as *Shigella* and *Salmonella* spp. (Gleeson and Gray, 1997).

The natural reservoirs of enteropathogenic strains are the intestines of humans (EPEC, ETEC, EIEC, EAEC) or domestic animals (ETEC, EHEC). The organisms are transmitted by direct contact or via contaminated food and water.

## Water

All enterovirulent *E. coli* are acquired directly or indirectly from a human or animal carrier. Risk from drinking water, therefore, only follows from faecal contamination of the supply. Given the sensitivity of *E. coli* to chlorine and other disinfectants, even if the organisms did contaminate the supply adequate chlorination would effectively remove any risk (Hunter, 2003).

However, standards do not exist for *E. coli* O157 (EHEC) within the 1980 European Drinking Water Inspectorate Directive. In 1997 the Environment Group of the former Scottish Office of Agriculture, Environment and Fisheries Department (SOAEFD) commissioned the Water Research Council (WRC) to carry out a study to examine the existing evidence for waterborne transmission of *E. coli* O157. From this study they found no evidence to indicate that *E. coli* O157 was more persistent in the environment or more resistant to water-treatment processes than the non-pathogenic *E. coli* found in the human gastrointestinal tract. Currently the Scottish Agricultural College and the University of Aberdeen are doing research into 'The survival and dispersal of *E. coli* O157 in Scottish soils and potential for contamination of private water supplies'. It is expected results from this study will be available in 2004. Findings to date suggest that *E. coli* O157 can survive for up to 21 days in water but that *E. coli* O157 is as susceptible to chlorination as any other *E. coli* strain.

Concern exists, however, about the potential role of biofilm in protecting enterovirulent *E. coli*. The very high infectious doses required for all enterovirulent *E. coli*, other than EHEC, suggests that this possible route of transmission is unlikely as a risk. The lower infectious dose of EHEC does potentially increase the risk of infection from biofilms in water but there have been no outbreaks or studies of sporadic cases of EHEC implicating inadequately disinfected water supply.

## Detection

*Escherichia* spp. may be detected as part of the normal microbiota (gastro-intestinal tract) of both man and animals. *E. coli* can also be recovered from patients suffering from urinary tract and wound infections, meningitis and septicaemia. Samples, which have been recovered from sterile sites on the human body, can be plated out on non-selective media such as blood, chocolate or nutrient agar. If contamination of samples is probable then MacConkey or eosin-methylene blue (EMB) agar should also be incorporated into any form of microbiological analysis.

For isolating enterovirulent *E. coli* from fresh stool samples MacConkey or EMB agar is used. For convalescents or patients with previous antibiotic treatment a number of samples should be taken for analysis. As *E. coli* is part of the normal intestinal flora, at least 10, but preferably more, colonies should be picked from the isolation media and tested for the presence of virulence markers.

Polymerase chain reaction (PCR) targeting virulence-associated gene sequences can be performed on colony sweeps from MacConkey. *Escherichia* are identified by biochemical reactions and slide agglutination. For the identification of different enterovirulent *E. coli* biological, immunological and molecular methods can be employed.

## EPEC

For the detection of EPEC, stool specimens are plated out on MacConkey agar. Lactose-fermenting colonies of *E. coli* groups can be tested by use of agglutinating antisera for the EPEC serogroups. For EPEC outbreaks ten or more lactose-positive colonies are subcultured on blood or nutrient agar and agglutinated with a set of polyvalent and monovalent OK antisera. After subculture a clear agglutination with a monovalent OK antiserum is heated at 100°C for 30–60 minutes and titrated in parallel with a reference culture of the same O group. After overnight incubation at 50°C the agglutination is read. Presence of the particular O group is confirmed only if both the test and the reference O antigens are agglutinated to nearly the same titre. EPEC strains can be identified by fluorescence actin staining method and cell culture tests with HEp-2 or HeLa cells to demonstrate localized or diffuse adherence. Molecular diagnosis of EPEC using PCR or colony hybridization of genes encoding for intimin (eaeA), bundle-forming pili (bfpA) or the EPEC adherence factor (EAF) plasmid have also been used for the diagnosis of EPEC.

## EAEC

EAEC have the ability to clump and are characterized by a unique, 'stacked brick-like' adherence which can be demonstrated in HEp-2 cells. Based on these characteristics and a particular heat-stable enterotoxin, EAEC can be confirmed.

## ETEC

Cell culture techniques (heat-labile enterotoxin, LT) or animal models (the suckling mouse assay for the heat-stable enterotoxin, ST, and the rabbit ileal loop for LT and ST) are used for the detection of LT. Exposure of cells, e.g. Vero monkey kidney cells to culture supernatant containing LT leads to morphological changes in cell structure which can be observed using microscopic techniques.

Many immunological tests are available for the detection of LT. These are available commercially and include enzyme-linked immunoassay (ELISA) and solid phase radio-immunoassay (RIA).

PCR or DNA probes can detect genes encoding for LT and ST production. Many of these probes have also been developed for use in food and water.

## EIEC

EIEC do not decarboxylate lysine and are mostly non-motile and usually non- or late lactose fermenters. After an appropriate infection period in Hep-2 and HeLa cells the presence of EIEC cells can be observed by microscopy. EIEC strains are confirmed by proof of invasivenes (Serény or cell culture tests) or by demonstration of the ipaH gene present on the chromosome and on the virulence plasmid. When invasion by EIEC has been demonstrated biochemical tests will enable identification.

## VTEC

Serovars O157:H7 and O157:H−, plus a few others, are the most important VTEC strains of *E. coli*. One of the most important problem *E. coli* is the VT-producing strains of O group 157. While a very high percentage of *E. coli* ferment sorbitol, the strain O157 does not ferment sorbitol in 24 hours. This characteristic is then used as a basic test used to detect them by using MacConkey (SMAC) agar. However, it is now documented that some strains of *E. coli* O157:H – isolated in Europe – are able to ferment sorbitol rapidly. Therefore SMAC alone should not be used for their detection. By supplementing SMAC with cefixime or cefixime-tellurite the selectivity for sorbitol-negative EHEC O157 can be improved.

Agglutinating sera and a latex coagglutination tests are available for VTEC detection.

Most strains of *E. coli* O157:H7 produce a plasmid-encoded haemolysin. This can now be detected by PCR.

VT production in *E. coli* can be detected by the cytotoxin effects observed in the Vero cell culture test. Commercially available enzyme immunoassay (Meridian Diagnostics, Cincinnati, USA and Milan, Italy) can be used for cytotoxin detection in both stool specimens and *E. coli* isolates.

Because of the low number of excreted VTEC evident in stool specimens, especially in patients with the haemolytic uraemic syndrome (HUS), testing of 20–100 isolated colonies of *E. coli* is necessary to achieve a sufficient level of diagnostic sensitivity. PCR can be performed on colony sweeps from

MacConkey plates giving an excellent sensitivity for the detection of VTEC strains.

Following the identification of a VT-producing colony biochemical determination of the species should follow. To prevent the loss of encoding genes, *E. coli* cultures should be suspended in Luria-Bertani or tryptic soy broth supplemented with 20–50% glycerol and stored at −70°C.

# Epidemiology

The epidemiology of waterborne *E. coli* infections are reviewed elsewhere (Hunter, 2003).

## EPEC

Few epidemics of EPEC have been documented in Europe and the USA. However, the incidence of sporadic cases of infantile enteritis peaks in summer months. Cases seem to be more common in areas that have very poor hygiene (Regua *et al.*, 1990). While water may be implicated in its transmission, the evidence to date suggests that EPEC is not transmitted in water.

## ETEC

ETEC have occasionally been reported in regions of the world that have good hygiene but outbreaks of diarrhoea, due to ETEC, are one of the major causes of deaths in children under 5 in developing countries. ETEC are probably the commenest cause of traveller's diarrhoea. The most probable cause of ETEC transmission seems to be via a water source with person-to-person transmission uncommon.

Waterborne outbreaks due to ETEC have been documented. A large outbreak, affecting more than 2000 staff and visitors to an American National Park in Oregon, occurred in the summer of 1975 (Rosenberg *et al.*, 1977). From this study enterotoxigenic *E. coli* were isolated from 20 (16.7%) of 120 rectal swabs that were examined. A strong correlation between illness and drinking park water in park staff and visitors was also found ($P < 0.00001$). However, no association with drinking water occurred in visitors on 7–9 July when chlorination of the water supply was being more closely monitored. Another outbreak affecting 251 passengers and 51 crew on a Mediterranean cruise (O'Mahony *et al.*, 1986) has also suggested a route of transmission of ETEC. From this study faecal coliforms were isolated from tap water suggesting the only plausible cause of this outbreak of ETEC. From an investigation of this outbreak faulty chlorination and faulty covers, possibly allowing bilge water into the drinking water tanks, was a probable cause of the outbreak. Other studies of ETEC transmission in water have been documented by Huerta *et al.* (2000). In a further outbreak, 175 Israeli military personnel and at least 54 civilians in the Golan Heights were infected by ETEC. Samples of water from several points along the distribution system which supplied the community showed inadequate chlorination and high concentrations of *E. coli*.

Outbreaks on ships are quite common which have been due to consuming drinks with ice cubes and drinking unbottled water (Daniels *et al.*, 2000).

Children aged 7–10 months in Ecuador have been shown to have antibodies to ETEC which has been linked to the consumption of low quality drinking water (Brussow *et al.*, 1992).

## EHEC

EHEC was recognized in the early 1980s as a severe disease of the very young and the elderly (Riley *et al.*, 1983). As with most *E. coli* EHEC is found in the intestines of several animal species and as such a faecal-oral spread from infected animals or other humans is evident. Also faecal contamination of food or water provides a means of EHEC transmission. Serotype O157:H7 is linked to outbreaks in drinking water in Europe and North America. Other non-O157 EHEC are now becoming recognized as causes of foodborne outbreaks, principally beef products, and person-to-person transmission.

Evidence of *E. coli* O157:H7 was found to be strongly linked to the consumption of drinking water in Missouri in 1990 (Swerdlow *et al.*, 1992). From this study it was found that, of a population of 3126, 243 people developed illness of whom 86 developed bloody diarrhoea, 36 were hospitalized and four died. Based on a case-control study of 53 cases, the only significant factor was that those infected drank more cups of municipal water per day (7.9) than did controls (6.1) ($P = 0.04$). Following an investigation of the water supply to the city it was noted that two mains water breaks had occurred after the start of the outbreak but before its main peak. The town's drinking water came from two deep-ground water sources and it was found that the sewage system was inadequate and sewage overflow crossing drinking water mains, provided the means for contamination of water supplies.

Outbreaks of *E. coli* O157:H7 have occurred in Scotland. In this study, during the hot summer of 1990, four people developed haemorrhagic colitis (Dev *et al.*, 1991). Because of the hot weather water levels in the water supply extraction points were low. As a result of this water from two subsidiary reservoirs was used. However, one of the reservoirs was fed from a source which may have been contaminated by cattle slurry.

In South Africa and Swaziland, in 1992, an outbreak of bloody diarrhoea affecting thousands of individuals was documented (Isaacson *et al.*, 1993). There were fatalities and cases of renal failure. Most cases were from men who drank surface water in the fields and women and children who drank borehole water. From the microbiological analysis it was found that *E. coli*

O157:H7 was isolated from 14.3% of 42 samples of cattle dung and 18.4% of 76 randomly collected water samples. The conclusion drawn from this study was that cattle carcasses and dung washed into rivers and dams by heavy rains after a period of drought contaminating the water.

In Europe a large outbreak in Fuerteventura, Canary Islands occurred in March 1997 (Pebody *et al.*, 1999). From this study 14 confirmed and one probable case were identified. The cases occurred in four different hotels. It was established following investigations that three of the four hotels were supplied with water from a private well.

The largest and most disreputable outbreak of EHEC associated with drinking water occurred during May and June 2000 among residents of Walkerton, Ontario (Anon, 2000). It was found in this study that 1346 cases of illness were identified. However, many people who presented with symptoms and were then analysed were infected with *Campylobacter* instead of EHEC. However, in addition to the 1346 cases a further 65 people were admitted to the local hospital. Twenty-seven of these patients developed haemolytic uraemic syndrome with six fatalities. An association between drinking water consumption was identified. The drinking water came from a number of wells and there was strong evidence suggesting that one of the wells had become contaminated with cattle faeces following heavy rains and flooding.

## EIEC

EIEC transmission is common from person to person, although it is principally thought to be acquired by a food or waterborne route. A water route is uncommon with evidence of only one outbreak of EIEC due to water, reported in 1959 (Lanyi *et al.*, 1959).

## EAEC

One outbreak of EAEC linking it to drinking water has been documented. This occurred in a small Indian village where people who drank water from a borehole were found to be much less likely to have diarrhoea than people drinking from shallow wells. From this study it was found that *E. coli* were isolated from the shallow wells but not the borehole (Pai *et al.*, 1997).

## Risk assessment

*Health effects*: occurrence of illness, degree of morbidity and mortality:

- The pathogenic *E. coli* serotypes are grouped based on their mechanism of causing symptoms. The six are enteropathogenic, enterotoxigenic, verocyto-genic (which included enterohaemorrhagic), enteroinvasive, enteroaggregative

and diffusely adherent. All pathogenic *E. coli* cause diarrhoea to various degrees of severity.
  – Enteropathogenic: traveller's diarrhoea, enteritis in infants
  – Enterotoxigenic: traveller's diarrhoea, diarrhoea, vomiting and fever
  – Verocytotoxigenic: Shigella-like dysentery (stools with blood and mucus), haemolytic uraemic syndrome (especially in children)
  – Enteroaggregative: chronic diarrhoea in children
  – Diffusely adherent: diarrhoea in children
* *E. coli* also causes urinary tract infections and can cause sepsis and meningitis in neonates.
* Pathogenic *E. coli* strains cause the majority of childhood diarrhoea in the world.
* The degree of morbidity and mortality varies according to the strain and the host's characteristics. In developing countries, diarrhoeal disease is much more likely to result in serious illness and death. In developed countries, though childhood diarrhoea is less of a problem, infection with verocytotoxigenic *E. coli* can result in haemolytic uraemic syndrome – especially in children under five – and thrombotic thrombocytopaenia purpura. These conditions can cause acute kidney failure and death.

*Exposure assessment*: infectious dose, routes of exposure and transmission, occurrence in source water, environmental fate:

* The infectious doses of most pathogenic *E. coli* are high: ranging from $10^5$ to $10^{10}$ organisms. This varies based on the host characteristics, such as age and stomach acidity. The exception is the group of verocytotoxigenic *E. coli* serogroups, which includes *E. coli* O157:H7. The infectious dose for this group appears to be less than 100 organisms.
* The routes of exposure and transmission in humans are faecal-oral; therefore, food, water, and person to person. Water contaminated with sewage has caused water-related outbreaks. The verocytotoxigenic types are most associated with waterborne outbreaks, probably because of their low infectious doses.
* The major source of pathogenic strains is human sewage, or in the case of verocytotoxigenic strains, faecal contamination from cattle.
* Humans and domestic animals are the reservoirs for enteropathogenic strains of *E. coli*; the connection between undercooked beef and *E. coli* O157:H7 infection is well known. Though *E. coli* strains can survive in the environment, they do not reproduce. Survival times in water vary based on the water characteristics (e.g. source, temperature, etc.).

*Risk mitigation*: drinking-water treatment, medical treatment:

* *E. coli* is very sensitive to chlorine and other disinfectants. Adequate residual disinfection should take care of any contamination in the distribution system. Waterborne outbreaks have resulted from treatment failures or from untreated water sources contaminated with faecal matter.

- All diarrhoeas should be treated with fluids and electrolytes, if appropriate. Though most *E. coli* strains have been sensitive to many antibiotics, including ampicillin, cephalosporins, and tetracycline, strains are becoming more and more antibiotic-resistant. Antibiotic treatment is generally not recommended for infection with *E. coli* O157:H7.

# References

Advisory Committee on the Microbiology Safety of Food. (1995). *Report on Vero Cytotoxin-Producing* Escherichia coli. London: HMSO.

Anon. (2000). Waterborne outbreak of gastroenteritis associated with a contaminated municipal water supply, Walkerton, Ontario, May–June 2000. *Canada Commun Dis Rep*, **26**: 170–173.

Bain, C., Keller, R., Collington, G.K. *et al.* (1998). Increased levels of intracellular calcium are not required for the formation of attaching and effacing lesions by enteropathogenic and enterohemorrhagic *Escherichia coli*. *Infect Immun*, **66**: 3900–3908.

Barrow, G.I. and Feltham, R.K.A. (eds) (1995). *Cowan and Steel's Manual for the Identification of Medical Bacteria*, 3rd edn. Cambridge University Press.

Bell, C. and Kyriakides, A. (1998). E. coli: *A Practical Approach to the Organism and Its Control in Foods*. London: Blackie Academic and Professional.

Beutin, L., Gleier, K., Kontny, I. *et al.* (1997). Origin and characteristics of enteroinvasive strains of *Escherichia coli* (EIEC) isolated in Germany. *Epidemiol Infect*, **118**: 199–205.

Bolton, F.J., Crozier, L. and Williamson, J.K. (1996). Isolation of *Escherichia coli* O157 from raw meat products. *Lett Appl Microbiol*, **23**: 317–312.

Brenner, D.J., Davis, B.R., Steigerwalt, A.G. *et al.* (1982a). Atypical biogroups of Escherichia coli found in clinical specimens and description of *Eschericha hermannii* sp. nov. *J Clin Microbiol*, **15**: 703.

Brenner, D.J., McWhorter, A.C., Knutson, J.K.L. *et al.* (1982b). *Escherichia vulneris*: a new species of Enterobacteriaceae associated with human wounds. *J Clin Microbiol*, **15**: 1133.

Brussow, H., Rahim, H. and Freire, W. (1992). Epidemiological analysis of serologically determined rotavirus and enterotoxigenic *Escherichia coli* infections in Ecuadorian children. *J Clin Microbiol*, **30**: 1585–1587.

Burgess, N.R.H., McDermott, S.N. and Whiting, J. (1973). Laboratory transmission of Enterobacteriaceae by the oriental cockroach, *Blatta orientalis*. *J Hyg Camb*, **71**: 9.

Daniels, N.A., Neimann, J., Karpati, A. *et al.* (2000). Traveler's diarrhea at sea: three outbreaks of waterborne enterotoxigenic Escherichia coli on cruise ships. *J Infect Dis*, **181**: 1491–1495.

DeVinney, R., Gauthier, A., Abe, A. *et al.* (1999). Enteropathogenic *Escherichia coli*: a pathogen that inserts its own receptor into host cells. *Cell Mol Life Sci*, **55**: 961–976.

Dev, V.J., Main, M. and Gould, I. (1991). Waterborne outbreak of *Escherichia coli* O157. *Lancet*, **337**: 1412.

Farmer, J.J. III, Fanning, G.R., Davis, B.R. *et al.* (1985). *Escherichia fergusonii* and *Enterobacter taylorae*, two new species of Enterobacteriaceae isolated from clinical specimens. *J Clin Microbiol*, **21**: 77.

Feachem, R.G., Bradley, D.J., Garelick, H. *et al.* (1983). *Sanitation and Disease: Health Aspects of Excreta and Wastewater Management*. Chichester: John Wiley & Sons.

Gleeson, C. and Gray, N. (1997). *The Coliforms Index and Waterborne Disease*. London: E and FN Spon.

Huerta, M., Grotto, I., Gdalevich, M. *et al.* (2000). A waterborne outbreak of gastro-enteritis in the Golan Heights due to enterotoxigenic *Escherichia coli*. *Infection*, **28**: 267–271.

Hunter, P.R. (1997). *Waterborne Disease: Epidemiology and Ecology*. Chichester: Wiley.

Hunter, P.R. (2003). Drinking water and diarrhoeal disease due to *Escherichia coli. J Water Hlth*, **1**: 65–72.

Isaacson, M., Canter, P.H., Effler, P. *et al.* (1993). Haemorrhagic colitis epidemic in Africa. *Lancet*, **341**: 961.

Kauffmann, F. (1947). Review, the serology of the coli group. *J Immunol*, **57**: 71–100.

Lanyi, B., Szita, J., Ringelhann, A. *et al.* (1959). A waterborne outbreak of enteritis associated with *Escherichia coli* serotype 124:72:32. *Acta Microbiol Hung*, **6**: 77–78.

Muir, R. and Ritchie, J. (1921). *Manual of Bacteriology*. London: Oxford University Press, pp. 353–359.

Nataro, J.P. and Kaper, J.B. (1998). Diarrheagenic *E. coli. Clin Rev Microbiol*, **11**: 142–201.

O'Mahony, M.C., Noah, N.D., Evans, B. *et al.* (1986). An outbreak of gastroenteritis on a passenger cruise ship. *J Hyg*, **97**: 229–236.

Pai, M., Kang, G., Ramakrishna, B.S. *et al.* (1997). An epidemic of diarrhoea in south India caused by enteroaggregative *Escherichia coli. Ind J Med Res*, **106**: 7–12.

Pebody, R.G., Furtado, C., Rojas, A. *et al.* (1999). An international outbreak of Vero cytotoxin-producing *Escherichia coli* O157 infection amongst tourists; a challenge for the European infectious disease surveillance network. *Epidemiol Infect*, **123**: 217–223.

Regua, A.H., Bravo, V.L.R., Leal, M.C. *et al.* (1990). Epidemiological survey of the enteropathogenic *Escherichia coli* isolated from children with diarrhoea. *J Trop Paediatr*, **36**: 176–179.

Riley, L.W., Remis, R.S., Helgerson, S.D. *et al.* (1983). Hemorrhagic colitis associated with a rare *Escherichia coli* serotype. *New Engl J Med*, **308**: 681–685.

Rivera, S.C., Hazen, T.C. and Toranzos, G.A. (1988). Isolation of fecal coliforms from pristine sites in a tropical rain forest. *Appl Environ Microbiol*, **54**: 513–517.

Rosenberg, M.L., Koplan, J.P., Wachsmuth, I.K. *et al.* (1977). Epidemic diarrhea at Crater Lake from Enterotoxigenic *Escherichia coli*. A large waterborne outbreak. *Ann Intern Med*, **86**: 714–718.

Rowe, B. (1983). *Escherichia coli* diarrhoea. *Culture*, **4**(1): 1–3.

Salyers, A.A. and Whitt, D.D. (1994). *Bacterial Pathogenesis – A Molecular Approach*. Washington, DC: ASM Press, pp. 190–204.

Schmidt, H., Scheef, J., Huppertz, H.I. *et al.* (1999). *Escherichia coli* O157:H7 and O157:H(−) strains that do not produce Shiga toxin: phenotypic and genetic characterization of isolates associated with diarrhea and hemolytic-uremic syndrome. *J Clin Microbiol*, **37**: 3491–3496.

Sears, C.L. and Kaper, J.B. (1996). Enteric bacterial toxins: mechanisms of action and linkage to intestinal secretion. *Microbiol Rev*, **60**: 167–215.

Sussman, M. (1997). *Escherichia coli* and human disease. In Escherichia coli: *Mechanisms of Virulence*, Sussman, M. (ed.). Cambridge: Cambridge University Press, pp. 3–48.

Swerdlow, D.L., Woodruff, B.A., Brady, R.C. *et al.* (1992). A waterborne outbreak in Missouri of *Escherichia coli* O157:H7 associated with bloody diarrhoea and death. *Ann Intern Med*, **117**: 812–819.

Vallance, B.A. and Finlay, B.B. (2000). Exploitation of host cells by enteropathogenic *Escherichia coli. Proc Natl Acad Sci USA*, **97**: 8799–8806.

Willshaw, F.A., Scotland, S.M. and Rowe, B. (1997). Vero cytotoxin-producing *Escherichia coli*. In Escherichia coli: *Mechanisms of Virulence*, Sussman, M. (ed.). Cambridge: Cambridge University Press, pp. 421–448.

Wilson, G.S. and Miles, A.A. (1964). *Topley and Wilson's Principles of Bacteriology and Immunity*, vol. 1, 5th edn. London: Edward Arnold, pp. 806–826.

# 8

# *Helicobacter pylori*

## Basic microbiology

*Helicobacter pylori* is an important pathogen which colonizes the mucus layer and epithelial mucus of the stomach in approximately 50% of humans worldwide. *H. pylori* is a curved (0.6 μm width, 2–5 μm length), Gram-negative bacterium which is typically motile with five to six sheathed unipolar flagella. Morphologies include spiral, curved, rod (bacillary), gull-winged, U-shaped and circular (coccoid) forms. *H. pylori* is urease- and oxidase-positive and negative for indoxyl acetate and hippurate hydrolysis (Foliguet *et al.*, 1989; Popovic-Uroic *et al.*, 1990; On and Holmes, 1992). Genomes of two strains of *H. pylori* have been completely sequenced (Tomb *et al.*, 1997; Alm *et al.*, 1999). The genomes contain nearly 1.7 million base pairs encoding approximately 1600 genes, 17% of which are unique to *H. pylori* (Doig *et al.*, 1999; Pawlowski *et al.*, 1999).

Under conditions of stress, *H. pylori* can transform from a spiral, bacillary morphology to a condensed, coccoid morphology. While it has been speculated that it may be a dormant form and play a significant role in transmission, considerable controversy surrounds the function and viability of *H. pylori* coccoid forms (Cellini *et al.*, 1994; Kusters *et al.*, 1997; Mizoguchi *et al.*, 1998; Enroth *et al.*, 1999; Kurokawa *et al.*, 1999; Ren *et al.*, 1999). Certain *H. pylori* coccoid forms may represent a viable but non-culturable (VBNC) state similar to those of *Vibrio* and *Campylobacter* spp. in the natural environment (Roszak and Colwell, 1987; Cappelier *et al.*, 1999). The putative VBNC state has been considered to represent a specific form of dormancy occurring in non-sporulating organisms (Oliver, 1995). Cells in this state cannot be grown by the culture method normally used for the organism concerned but are believed to have the capacity to return to the vegetative state when exposed to appropriate stimuli (Gribbon and Barer, 1995). Therefore, by definition, VBNC bacteria are unable to form colonies on traditional media, while potentially maintaining their pathogenic capabilities (McFeters, 1990). It is probable that VBNC indicator bacteria can pass undetected from a water-treatment system into the distribution network and result in the underestimation of indicator and pathogenic bacterial populations (Bucklin *et al.*, 1991). A distinction between VBNC *H. pylori* and coccoid *H. pylori* should be emphasized, as the VBNC state of *H. pylori* is morphology-independent. *H. pylori* enter a VBNC state without change from vibrioid morphology (Cellini *et al.*, 1994). Coccoid *H. pylori* exhibit diversity in ultrastructure following exposure to different stresses and it is likely that coccoids exhibit variation in viability (Mizoguchi *et al.*, 1998). Some coccoid forms are electron-dense with intact cellular membranes and flagella, indicating that they are likely to be viable (Zheng *et al.*, 1999). These forms remain capable of reducing tetrazolium (INT) for extended periods in water (Gribbon and Barer, 1995). SDS-PAGE (sodium dodecyl sulphate polyacrylamide gel electrophoresis) demonstrated that most protein bands appeared to be similar in both the spiral and coccoid forms, thus supporting the notion that some of the coccoid forms of *H. pylori* are likely to be viable (Zheng *et al.*, 1999). *H. pylori* undergoes substantial modification of its unique muropeptide composition during morphological transition to coccoid forms. The accumulation of dipeptide monomers and a concomitant reduction in tri- and tetrapeptide monomers constitutes the most dramatic modification (Costa *et al.*, 1999). This suggests that activation of a γ-glutamyl-diaminopimelate endopeptidase leads to massive conversion of tri- and tetrapeptide monomers into dipeptide monomers, as previously observed in sporulating *Bacillus sphaericus* (Vacheron *et al.*, 1979). This accumulation of disaccharide-dipeptides appears to be a result of convergent evolution between the distantly related bacteria *H. pylori* and *B. sphaericus* in the genesis of resistant forms, i.e. coccoid cells and endospores, respectively. Muropeptide composition remains essentially stable from the time coccoid cells become predominant and remains so for at least 2 weeks (Costa *et al.*, 1999).

The role of the VBNC *H. pylori* and coccoid forms in infection and transmission remains unclear. Attempts at successful infection of animals with coccoid *H. pylori* are conflicting (Cellini *et al.*, 1994; Eaton *et al.*, 1995; Wang *et al.*, 1997). *In vitro* reversion of the coccoid morphology to the bacillary form has been reported (Anderson *et al.*, 1997; Kurokawa *et al.*, 1999). Deprivation of nutrients and a non-permissive temperature act as a powerful trigger for *H. pylori* programmed cell death, apparently as a means of species preservation (Cellini *et al.*, 2001). Pre-incubation in non-nutrient solution and high density of bacterial concentration appear to be important for recovery of *H. pylori* cultured for a prolonged time under anaerobic conditions (Yamaguchi *et al.*, 1999).

## Origin and taxonomy

The discovery of *H. pylori* by Warren and Marshall in Western Australia in 1983 (Warren and Marshall, 1983) not only introduced a whole new group of bacteria to science, but also revolutionized our concept of gastroduodenal pathology, in particular peptic ulcer disease. Although the presence of spiral bacteria in the human stomach had been reported in the literature (Freedberg and Barron, 1940; Steer and Colin-Jones, 1975) *H. pylori* was not successfully cultured until the 1980s (Marshall and Warren, 1984). *H. pylori* was subsequently found to cause peptic ulcer disease and gastric cancer (Marshall and Warren, 1984; Blaser, 1987). This spiral organism isolated from the stomach was originally called *Campylobacter pyloridis*, but this was later changed to the more grammatically correct *C. pylori*. Further studies showed that the organism differed sufficiently from true campylobacters justifying the need of a new genus, *Helicobacter*. It soon became apparent that similar organisms colonized the stomach of a wide variety of animals other than humans, and those spiral bacteria colonizing the intestines of rodents and other animals also belonged to *Helicobacter*.

*Helicobacter* are phylogenetically closely related to *Campylobacter*, *Wolinella* and *Arcobacter*, thus belonging to the delta-epsilon group of Proteobacteria (Wesley *et al.*, 1995; Bunn *et al.*, 1997). Initially classified as a species in the genus *Campylobacter*, its unique 16S rRNA sequence data, presence of sheathed flagella, and a distinct fatty acid and outer membrane protein profile led to the establishment of the genus *Helicobacter* (Goodwin *et al.*, 1989). Approximately 20 *Helicobacter* species have been described in the literature, and some are now regarded as human pathogens. Figure 8.1 summarizes the taxonomy of the *Helicobacter* genus as determined by sequencing of the 16S ribosomal RNA.

## Metabolism and physiology

*H. pylori* is a fascinating organism. It has a mixture of aerobic and anaerobic physiologies that combine to produce a microaerophilic physiology, the molecular basis for which is not fully understood. Part of the microaerophilic nature is probably due to oxygen-sensitive enzymes within its central metabolic pathways. The biochemical basis for the requirement for $CO_2$ has not been completely explained and there is a surprising lack of anaplerotic carboxylation ($CO_2$ fixation) enzymes (Kelly, 1998).

The organism has solute transport systems and an incomplete tricarboxylic acid (TCA) cycle, all contributing to complex nutritional requirements. *H. pylori* has an absolute growth requirement for arginine, histidine, leucine, isoleucine, valine, methionine and phenylalanine (Doig *et al.*, 1999). The respiratory chain of *H. pylori* is remarkably simple, apparently with a single

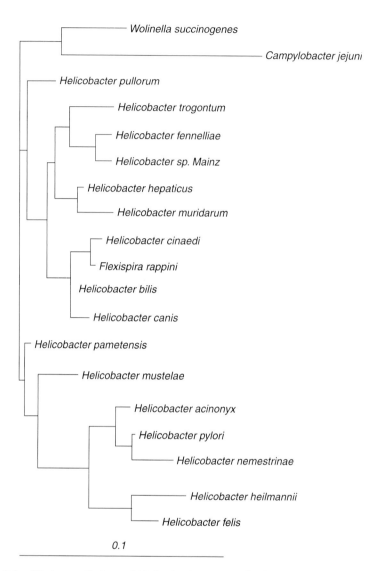

**Figure 8.1** Phylogenetic tree of *Helicobacter* spp. and related organisms (adapted from Cantet *et al.*, 1999).

terminal oxidase and with fumarate reductase as the only reductase for anaerobic respiration. NADPH appears to be the preferred electron donor *in vivo*, rather than NADH as in most other bacteria (Kelly, 1998). The non-cyclic TCA cycle, characteristic of anaerobic metabolism, is directed towards the production of succinate in the reductive dicarboxylic acid branch and alpha-keto-glutarate in the oxidative tricarboxylic acid branch. Both branches are metabolically linked by the presence of alpha-ketoglutarate oxidase activity. *H. pylori* does not possess a gamma-aminobutyrate shunt, owing to the absence of both gamma-aminobutyrate transaminase and succinic semialdehyde dehydrogenase activities (Pitson *et al.*, 1999). Genomic analyses suggest that glucose or malate is not the primary source for production of pyruvate in *H. pylori* but, rather, that lactate, L-alanine, L-serine, and D-amino acids are the primary sources. The Entner-Doudoroff and pentose phosphate pathways have been shown to be active rather than glycolytic pathways (Doig *et al.*, 1999).

While the urease of most bacterial species has been found to possess three distinct subunits, urease-producing *Helicobacter* produce a unique two-subunit enzyme which exhibits remarkable submillimolar K values (Mobley *et al.*, 1995). The small subunit (urea) of *H. pylori* probably represents a structural gene fusion. A complex urease-independent chemotaxis signal transduction system suggests that chemotaxis (toward urea, $Na^+$, and bicarbonate) is an important feature of *H. pylori* physiology (Yoshiyama *et al.*, 1998). Since urea and sodium bicarbonate are secreted through the gastric epithelial surface and hydrolysis of urea by urease on the bacterial surface is essential for colonization, the chemotactic response of *H. pylori* may be crucial for its colonization and persistence in the stomach (Mizote *et al.*, 1997). Urease enzyme activity is the best-characterized colonization factor, protecting the bacteria from gastric acid exposure (Dunn, 1993). Urease catalyses the hydrolysis of urea to yield carbonic acid and two molecules of ammonia, which in solution are in equilibrium with their protonated and deprotonated forms, respectively. The net effect of this reaction is an increase in pH. In addition to buffering acidic environments, *Helicobacter* urease activity plays a critical role in nitrogen assimilation (Chin *et al.*, 2001). Ammonia can be assimilated in *H. pylori* by conversion of glutamate to glutamine. Lack of a glutamate synthase suggests that $\alpha$-ketoglutarate is transformed into glutamate by glutamate dehydrogenase and that *H. pylori* is adapted to an ammonia rich environment (Marais *et al.*, 1999).

## Clinical features

Once colonized by *Helicobacter pylori*, the human host can remain infected for life unless intensive antimicrobial therapy is administered. It is the principal cause of chronic active gastritis, stomach and peptic ulceration, and classified as a Class I carcinogen for gastric cancer and gastric mucosa-associated lymphoid tissue (MALT) lymphoma (Tytgat and Rauws, 1990; Aruin, 1997; Nedrud and Czinn, 1999). Acquisition of *H. pylori* infection early in life appears to be associated with early-onset gastric corpus atrophy and metaplasia, and a higher risk of cancer. Peptic ulcer disease (PUD) occurs at the rate of 1% per annum in infected individuals (Figure 8.2).

Gastric carcinogenesis is a multifactorial, multistep process, in which chronic inflammation plays a major role. *H. pylori* infection induces physiological changes and DNA adduct formation in the gastric microenvironment (Farinati *et al.*, 1988). Reduction in antioxidant (ascorbate) levels increases the risk of carcinogenesis and damage to DNA from intragastric release of

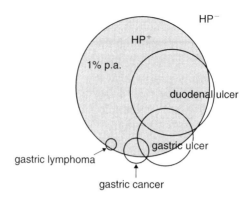

**Figure 8.2** Relation of *H. pylori* infection to clinical pathology (source: The *Helicobacter* Foundation, www.pylori.com).

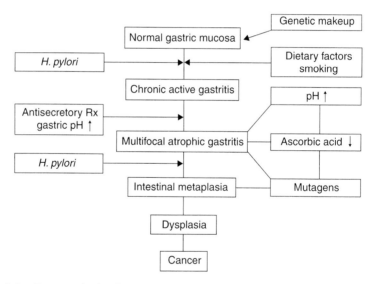

**Figure 8.3** Proposed role of *H. pylori* in gastric cancer.

free radicals. Ingestion of dietary carcinogens, deficiencies in dietary antioxidants, smoking, and anti-secretory (proton pump inhibitor) medications are thought to be important cofactors in the genesis of *Helicobacter*-related cancer (Figure 8.3).

In persons who do not develop peptic ulcer, long-term *H. pylori* gastritis can lead to intestinal metaplasia, associated with gastric adenocarcinoma. Chronic gastritis is also associated with the presence of lymphoid follicles in the lamina propria. This mucosa-associated lymphoid tissue (MALT) can develop into lymphoma. In cases of gastric MALT lymphoma, eradication of

*H. pylori* infection has been shown to cure over 50% of patients (Wotherspoon *et al.*, 1993).

According to observations from human and animal *Helicobacter* infections, the genus has the capacity to colonize and cause inflammation in the stomach (*H. pylori*, *H. heilmannii*, *H. mustelae*), colon (*H. fennelliae*) and liver (*H. bilis*). Many people who are infected with *H. pylori* are asymptomatic. For those who have gastritis or ulcers, the most common symptoms are nausea, abdominal pain, heartburn, or bleeding. Gastritis is an irritation in the lining of the stomach; an ulcer is a sore in the lining of the stomach or duodenum. If left untreated, ulcers can become life threatening. While some medications or too much stomach acid can also cause gastritis and ulcers, the most common cause is *H. pylori* infection. Acute *H. pylori* infection manifests as nausea and abdominal pain, lasting between 3 and 14 days. During this time hypochlor-hydria develops and may persist for up to a year. Following infection of the gastric epithelium, a marked inflammatory response in the mucosa occurs within days. This is induced initially by secretion of interleukin-8 from surface epithelial cells, accompanied by release of the intercellular adhesion molecule ICAM-1. The initial superficial gastritis can progress to one of three forms. The first is chronic active non-atrophic gastritis, predominantly antral, referred to as diffuse antral gastritis. This form increases the risk of duodenal ulcer. The second is chronic active gastritis with some atrophy, which places the person at risk from gastric ulcers. The third is chronic atrophic pangastritis with significant atrophy, which increases the risk of gastric cancer (Blaser, 1987). There are more than 2 million physician visits per year for duodenal ulcers, 90% of which are attributable to *H. pylori* (other causes include non-steroidal anti-inflammatory drug (NSAID) usage and Zollinger-Ellison syndrome), and more than 3 million physician visits per year for gastric ulcers, in which 80% of patients with non-NSAID-induced gastric ulcers are infected (Blaser, 1997). In a prospective study, the risk of developing duodenal ulcer disease in *H. pylori*-infected patients followed for 10 years exceeded 10%; in contrast, it was less than 1% in uninfected patients (Monath *et al.*, 1998). Development of atrophy and metaplasia of the gastric mucosa is strongly associated with *H. pylori* infection. Approximately 40–50% of infected subjects develop these conditions, which are rare in non-infected subjects (Kuipers, 1999). The risk of developing gastric cancer is estimated to be three- to sixfold higher in infected than in uninfected individuals (Antonioli, 1994). *H. pylori* is found in close contact with the mucosa in the antrum of the stomach, often between cells under a stable layer of mucus that shields the organisms from gastric acid. Evidence suggests that *H. pylori* infection of humans is ancient, and often the interaction is not overtly destructive. *H. pylori* is almost always present when inflammation is present, and is rare when inflammation is absent, indicating *H. pylori* are not simple commensals. An equilibrium between virulence of *H. pylori* and the mucosal defence mechanism has been proposed (Loffeld and Arends, 1993). In this case, the bacterium is neither commensal nor pathogen, being able to change its role depending on the local microenvironment.

## Virulence

*H. pylori* is a remarkably well-adapted organism which can persist indefinitely in the hostile environment of the stomach, including the vigorous humoral and cellular immune response mounted against it. Despite the vigour of this response, the infection in many cases is not eradicated. *H. pylori* must possess several physiological adaptations to allow it to persist in the hostile environment of the human stomach. The mechanisms by which *H. pylori* can colonize and persist at $10^4$–$10^7$ *H. pylori* per gram of gastric mucus are probably complex (Nowak *et al.*, 1997). Motility, urease activity and association with gastric mucosal cells are important colonization factors (Nakamura *et al.*, 1998; Beier *et al.*, 1997). *H. pylori*, with optimal growth in the neutral pH range, is not an acidophile and requires mechanisms to survive stomach acid. By breaking down urea present in the gastric juice and extracellular fluid, the organism is able to generate bicarbonate and ammonia in its pericellular environment, so that hydrogen ions are effectively neutralized before damaging the cell. *H. pylori* is thereby able to survive in gastric acid long enough for it to colonize the gastric mucosa.

Once attached to the gastric mucosa, *H. pylori* can damage host tissue by causing vacuolation in the epithelial cells. This vacuolation is caused by the production of a cytotoxin called vacuolating cytotoxin A (VacA), a protein which is endocytosed by epithelial cells where it causes endosome-lysosome fusion (Ge and Taylor, 1999). Variations of cytotoxin occur such that more aggressive forms are likely to be associated with peptic ulcer, whereas more benign forms of cytotoxin may be associated with gastritis without ulcers (Atherton *et al.*, 1995). Although individuals infected with *H. pylori* are often asymptomatic, they always have histological changes of chronic gastritis. Colonization of gastric mucosa by *H. pylori* induces gastritis with resultant neutrophil infiltration (Fujioka, 1995). Chronic gastritis is characterized by the accumulation of oxidative DNA damage with mutagenic and carcinogenic potential (Farinati *et al.*, 1988). *H. pylori* seems to be particularly resistant to the oxidative inflammatory response of neutrophils, which can in turn damage the host gastric mucosa (Basso *et al.*, 1999). The nature of this resistance is not well understood (Odenbreit *et al.*, 1996). Most *H. pylori* bacteria found in the stomach are in the layer of mucus overlying the epithelium. Some penetrate the mucous layer and adhere to the gastric epithelial cells with formation of a pedestal. A few *H. pylori* are seen between adjacent cells in proximity to tight junctions.

Several *H. pylori* adhesins have been described that bind to Lewis b antigens with terminal fucose residues (found in humans with blood group O), to sialic acid-lactose residues and to phosphatidylethanolamine, a glycolipid receptor on the gastric antral mucosa. The adhesin binding to blood group O-specific antigens may help explain the predispostion of people with blood group O to peptic ulcer disease and gastric adenocarcinoma.

Apart from urease, which may cause damage to host cells, a haemolysin and vacA have been described. CagA, a high molecular weight protein product of a cytotoxin associated gene (cagA), is produced by about 60% of *H. pylori* strains and is associated with the expression of vacA. CagA-positive strains are also associated with the presence of duodenal ulceration, although it is unclear whether CagA has an independent pathogenic role. CagA is an island of approximately 30 genes which have apparently been acquired by *H. pylori* from another organism since the guanine and cytosine content of this island differs from that seen in the rest of the *H. pylori* genome. The CagA pathogenicity island contains genes which function as a Type IV secretory system. Subsequent induction of interleukin-8 production attracts neutrophils through the lamina propria, and emerges between the epithelial cells. Rapid recruitment of inflammatory cells into the mucosa results, and these cells then express a repertoire of cytokines. The intense release of inflammatory mediators, along with substances that *H. pylori* itself elaborates, results in damage to the gastric epithelial barrier. In addition, *H. pylori* lipopolysaccharide (LPS) is suspected of driving production of auto-antibodies through molecular mimicry of human Lewis blood group antigens present on parietal cell $H^+/K^+$-ATPase (Namavar *et al.*, 1998).

The lipopolysaccharide of *H. pylori* appears to have an unusual structure related to the composition of the fatty acids that form the hydrophobic region of lipid A with the presence of 3-hydroxyoctadecanoic acid. The lipid A portion of *Helicobacter* LPS appears to have lower biological activity than LPS from other enteric bacteria. Some strains of *H. pylori* possess LPS with a ladder-like side chain like those of 'smooth' strains of Enterobacteriaceae; others give a 'rough'-type profile. Strains may change from smooth to rough LPS when grown on conventional solid media and may be reversible when grown in liquid medium. Although LPS core antigens are shared, side chain antigens are strain-specific. The distribution of specific LPS antigens among strains is different from that of protein antigens. Antigenic differences in LPS from different strains has been detected by immunoblotting and haemagglutination assays.

# Treatment

*Helicobacter pylori* is sensitive to penicillins (including benzylpenicillin), cephalosporins, tetracycline, erythromycin, rifampicin, aminoglycosides and nitrofurans, but resistant to nalidixic acid, though sensitive to the more active quinolones such as ciprofloxacin. It is highly resistant to trimethoprim and moderately resistant to polymyxins. *H. pylori* is usually susceptible to metronidazole but resistance rates are variable and may reach 50% in some areas. *H. pylori* is sensitive to colloidal bismuth compounds commonly prescribed for gastric disease in concentrations easily attainable in the stomach. The proton pump inhibitor omeprazole has mild *in vitro* activity against *H. pylori*; 1% bile salts are also inhibitory. Eradication of *H. pylori* has been shown to be a definitive cure for duodenal ulcer and most gastric ulcers. In the 1980s, treatment for

*H. pylori* was difficult since combinations of bismuth, tetracycline and metronidazole were required for adequate eradication of the organism (Borody *et al.*, 1989). The action of amoxicillin was found to be greatly enhanced when gastric acid was suppressed with a proton pump inhibitor, notably omeprazole (Unge *et al.*, 1989). As a result, *H. pylori* is treated with a 7-day therapy of omeprazole (to render the gastric pH neutral) in combination with two antibiotics, usually amoxicillin and clarithromycin. Omeprazole, clarithromycin, and metronidazole combinations have achieved similar high cure rates. For difficult to eradicate infections, bismuth, tetracycline, metronidazole and omeprazole are usually successful (Kung *et al.*, 1997).

No current treatment regimen for *H. pylori* is universally effective, even with triple and quadruple therapies (Vyas and Sihorkar, 1999). Three of the six most widely prescribed medications in the USA are for the treatment of ulcers. PUD treatment costs exceed $4 billion each year, not including indirect costs due to work and productivity loss (Vakil, 1997). Recent efforts toward therapeutic immunization by oral recombinant *H. heilmannii* or *H. pylori* urease given with cholera toxin or *E. coli* heat-labile enterotoxin, respectively, induced gastric corpus atrophy in mice and failed to eradicate *H. pylori* in infected individuals (Dieterich *et al.*, 1999; Michetti *et al.*, 1999). Expression of stable immunogenic antigens in an attenuated *Salmonella typhi* vector, an effective vaccine approach for typhoid fever, failed to induce detectable mucosal or systemic antibody responses for *H. pylori* in human volunteers (DiPetrillo *et al.*, 1999).

Once the exact mode of transmission is understood in different communities then effective public health measures can be started. In areas where *H. pylori* exists in the environment, humans do not seem to be able to mount a protective immune response following natural infection. Therefore, treatment is useless, and the only effective way of eliminating *H. pylori* from the population would be via public health measures, i.e. improved sanitation and standard of living, or vaccination.

## Survival in the environment

Helicobacters can be broadly divided into those that colonize the stomach mucosa and those that colonize the intestines of humans and animals, although some intestinal species are capable of colonizing the stomach when acid secretion is defective. Some gastric species only colonize the non-acid secreting mucosa, whereas others, such as *Helicobacter felis*, can colonize the canaliculi of acid-secreting oxyntic cells.

The surface of the human stomach mucosa is the major habitat of *H. pylori*. Almost all isolations are from gastric biopsy specimens, but the organism has occasionally been detected in gastric juices, saliva, dental plaque, bile and faeces. In developed countries prevalence rates increase with age, from about 20% in young adults to about 50% in people over 50 years old, but in developing countries rates are much higher and children are commonly infected. *H. pylori* has been isolated from domestic cats which raises the question about the role of this domestic animal in transmission of disease to humans. There is some suggestion that water may be a source for *Helicobacter* infection, but whether the organisms merely exist or have some ecological niche in natural waters is unknown.

The precise mechanism of transmission of *H. pylori* is unknown, but any mode that introduces the organism into the stomach of a susceptible person may lead to infection. Several routes of transmission have been proposed: gastric-oral, oral-oral, faecal-oral, zoonotic and water/food-borne. Data supporting all routes have been published and it is likely that infection occurs through multiple transmission pathways (Goodman *et al.*, 1996; Velazquez and Feirtag, 1999). Recent observations in persons infected with *H. pylori* caused to vomit or have diarrhoea showed that an actively unwell person with these symptoms could spread *H. pylori* in the immediate vicinity by aerosol, splashing of vomitus, infected vomitus and infected diarrhoea (Parsonnet *et al.*, 1999). In developed countries, *H. pylori* is more difficult to acquire and usually is transmitted from one family member to another, possibly by the faecal-oral route, or by the oral-oral route.

Faecal-oral transmission of *H. pylori* has been suggested by several researchers being implicated with crowding, socioecomonic status and consumption of raw, sewage-contaminated vegetables as risk factors for infection. Studies in Peru have identified the type of water supply as a risk factor for infection and have found that water source appeared to be more important as a risk factor than socioecomonic factors. The number of people in the household, sharing a bed, and a lack of proper sanitation and a permanent hot water supply all increase the risk (Mendall *et al.*, 1992; Whitaker *et al.*, 1993; Malaty and Graham, 1994). The gastric-oral mode of transmission of *H. pylori* would favour spread in small units of interacting people, such as in families. Interfamilial transmission is likely to occur, however, diverse *H. pylori* species have been detected in infected couples, suggesting other possible reservoirs (Kuo *et al.*, 1999). In developing countries, many children are already infected by the age of 10 years, and relapse can be a serious problem (Dooley *et al.*, 1989; Ramirez-Ramos *et al.*, 1997; Gurel *et al.*, 1999). Data regarding acquisition and loss of *H. pylori* infection are critical to understanding the epidemiology of the infection and to developing treatment and vaccination strategies. In a Canadian prospective 3-year cohort study of 316 randomly selected subjects, a continuous risk of acquisition of 1% per year rather than a cohort effect was concluded.

The prevalence of *H. pylori* infection varies according to age and country of origin. In developed countries prevalence of infection with *H. pylori* is clearly age related, but varies according to ethnic group and socioeconomic status with only about 0.5% of adults developing new infections each year (Graham *et al.*, 1991; Vaira *et al.*, 1998). *H. pylori* seropositivity was found among 1161 (45%) of 2598 healthy Italian adults in which age (67% seropositivity in subjects aged 50 or older) and sociocultural class were identified as persisting determinant features of *H. pylori* infection (Russo *et al.*, 1999). In countries whose economy has gone from developing to affluent over the past 50 years, marked decline in *H. pylori* prevalence has been noted.

Although the natural niche for *H. pylori* is the human stomach, for widespread infection the organism may need to survive in the external environment (Brown, 2000). PCR detection of *H. pylori* in oral and fingernail samples in a rural population in Guatemala suggests that oral carriage of *H. pylori* may play a role in the transmission of infection and that the hand may be instrumental in transmission (Dowsett *et al.*, 1999). However, the low frequency (<1%) of *H. pylori* in dental plaque suggests that oral transmission is not a significant reservoir in adult populations (Luman *et al.*, 1996; Cave, 1997; Kamat *et al.*, 1998; Oshowo *et al.*, 1998a, b). A recent study found that 20 of 21 Japanese couples (both members were positive for *H. pylori*) showed restriction enzyme patterns that differed between spouses, indicating that in Japan, interspousal transmission of *H. pylori* occurs rarely (Suzuki *et al.*, 1999).

Evidence seems to favour a faecal-oral route by which the bacterium, excreted with faeces, might colonize water sources and subsequently be transmitted to humans (Xia and Talley, 1997). *H. pylori* can be cultured from the faeces of infected individuals (Thomas *et al.*, 1992; Kelly *et al.*, 1994) and specific DNA sequences have been amplified from raw sewage (Forrest *et al.*, 1998). Experimental murine models support a faecal-oral, rather than oral-oral route as the mode of transmission of *H. pylori* infection (Cellini *et al.*, 1999; Yoshimatsu *et al.*, 2000).

*H. pylori* infection in Italian shepherds approaches 100%, who are about 80 times as likely to be infected with *H. pylori* as are their siblings who are not shepherds (Dore *et al.*, 1999a). Raw milk samples from 60% of sheep on Italian farms contained traces of *H. pylori* DNA, and two *H. pylori* isolates from sheep milk and tissue were isolated, indicating that *H. pylori* infection includes phases in the environment (Dore *et al.*, 1999b). The authors suggest that sheep may be the source of infection in humans, however, they did not rule out the possibility that humans might contaminate the sheep samples. Domestic cats and old-world macaques have been found to be colonized with *H. pylori*, however, it is doubtful whether this species provides an important reservoir for human infection (Bode *et al.*, 1997; Osata *et al.*, 1997). A potential role of flies in vectoral spread of *H. pylori* from human faeces to food was supported by the amplification of *H. pylori* DNA from flies in developing countries (Grubel *et al.*, 1998). *H. pylori* was rarely detected by PCR on chopsticks following eating and hence, the risk of infection via the use of chopsticks was concluded to be low (Leung *et al.*, 1999).

While the principal route of transmission remains inconclusive, geographic prevalence of *H. pylori* infection (Figure 8.4) shows a strong correlation with access to clean water (Figure 8.5).

Children in Peru whose homes were supplied with municipal water were 12 times more likely to be infected than those whose water supply came from community wells (Klein *et al.*, 1991). Similar seroprevalence patterns in a community

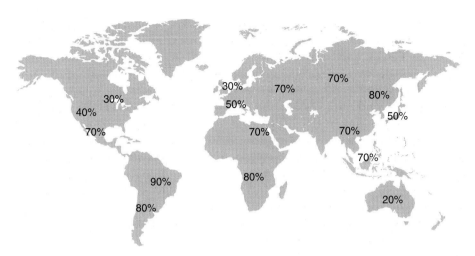

**Figure 8.4**   Geographic prevalence of *H. pylori* infection (data: *Helicobacter* Foundation, www.pylori.com).

PERCENT OF POPULATION IN URBAN AREAS HAVING ACCESS TO CLEAN WATER

▮ LESS THAN 75    ▮ 75 TO 94.9    ▯ 95 OR MORE    ▯ NO DATA

**Figure 8.5**   Geographic access to clean water (1980–90) (source: The Water Resources Institute, www.sciam.com/1197issue/1197scicit5.html).

study between *H. pylori* and hepatitis A virus (HAV) indicate *H. pylori* may have spread in a manner similar to organisms transmitted by a faecal-oral route (Pretolani *et al.*, 1997). A survey of 3289 residents (416 families) in northern Italy for prevalence of *H. pylori* IgG antibodies suggested a common source of exposure for infection (Dominici *et al.*, 1999). In the Colombian Andes, swimming in streams, using streams as a drinking water source, and frequent consumption of raw vegetables increased the odds of infection (Goodman *et al.*, 1996). Other epidemiological reports do support an association between *H. pylori* infection and consumption of untreated well or spring water (Carballo *et al.*, 1997; Bunn *et al.*, 1997). A longitudinal seroconversion study of children from Bolivia found a striking increase in the incidence of infection, which rose from less than 10% in 2 year olds or younger to over 35% in children over 2 years (Friedman, 1998). Interestingly, this study was set up on the back of a social programme to improve the hygiene by giving families plastic canisters of chlorinated drinking water. Factors associated with seroconversion included *H. pylori* infection in another household member as a positive risk factor and the use of chlorinated drinking water as a strong protective factor.

Water and water biofilm reservoirs of infection could account for many of the published conclusions, which have been attributed to other factors: e.g. socioeconomic factors such as overcrowding, which will also bring associated sharing of common water supplies. This idea is supported by a study of *H. pylori* infection rates in French submarine crewmembers with a common tanked water supply (Hammermeister *et al.*, 1992). The argument for a waterborne route of *H. pylori* transmission is supported by maintenance of viability in spiked natural water (West *et al.*, 1992; Shahamat *et al.*, 1993; Hunter, 1997; Fan *et al.*, 1998; Jiang and Doyle, 1998; Sato *et al.*, 1999).

*H. pylori* conserve the capability to produce acid-inducible proteins for at least 100 days when stored at 4°C in either phosphate-buffered saline (PBS) or distilled water (Mizoguchi *et al.*, 1999). While attempts to culture *H. pylori* from environmental water samples have been unsuccessful, closely related microaerophilic organisms, *Campylobacter jejuni* and *Arcobacter butzleri*, have been cultured from ground and surface waters (Arvanitidou *et al.*, 1994; Stanley *et al.*, 1998; Rice *et al.*, 1999) and associated with waterborne outbreaks (MMWR, 1999).

Initial molecular probe evidence of *H. pylori* from environmental water samples came from PCR amplification of samples from Columbia, where infection rates are over 90% (Schauer *et al.*, 1995; Handwerker *et al.*, 1995). A two-stage *in vitro* method for detection of *H. pylori* in spiked water and faecal samples using an initial concentration by immunomagnetic separation followed by PCR detection (IMS/PCR) has been described (Enroth and Engstand, 1995). IMS/PCR was used to detect *H. pylori* from water in Peru and Sweden (Hulten *et al.*, 1996, 1998). Water samples from a water delivery truck and two lakes in the Canadian Arctic were PCR positive for *H. pylori* (McKeown *et al.*, 1999). In a region of Japan with a high infection rate, *H. pylori*-specific DNA was detected in water, field soil, flies and cow faeces by nested PCR, and the aligned urease partial sequences of the PCR products were highly homologous (96–100%) with the *H. pylori* sequence in the GenBank database (Sasaki *et al.*, 1999). PCR analysis of river water in Japan detected *H. pylori*-specific DNA in 8 of 62 (13%) of samples (Fukuyama *et al.*, 1999). Actively respiring *H. pylori* from surface and well water in the USA were detected using fluorescent antibody-tetrazolium reduction (FACTC) microscopy (Hegarty and Baker, 1999) and confirmed using species-specific PCR (Baker and Hegarty, 2001). All of these findings suggest the presence of *H. pylori* in the natural environment and a possible waterborne route of transmission.

The challenge remains to determine the importance of waterborne, as opposed to other modes, of transmission. It is possible that a specifically adapted form of *H. pylori*, or association with a biofilm community, may be required for environmental persistence and transmission (Vincent, 1995). Waterborne bacteria can attach to surfaces by aggregating in a hydrated exo-polymer glycocalyx matrix of their own synthesis to form biofilms (Costerton *et al.*, 1999). Association with biofilm communities within a water distribution system can offer bacteria protection from disinfection and protozoan predation (Sibille *et al.*, 1998).

## Water

The geographic prevalence of *H. pylori* infection shows a strong correlation with access to clean water. In 1998, the US Environmental Protection Agency (US EPA) Office of Ground Water and Drinking Water included *H. pylori* on its Contaminant Candidate List (CCL) (62 Federal Register 52193). This list designates contaminants which are not regulated under current EPA drinking water regulations, but which are anticipated to occur in water systems. Inadequate, interrupted or intermittent drinking water treatment has repeatedly been associated with waterborne outbreaks (Moore *et al.*, 1994). Among waterborne disease outbreaks between 1971 and 1994, the distribution system has been considered the cause of 30% of the 272 outbreaks in community systems. At least two of these outbreaks resulted in hospitalizations and six deaths (Shaw and Regli, 1999). Regulations to reduce microbial risk have focused primarily on reducing risk from contaminants in the source water by providing some means of control at a treatment plant. Risks from growth of bacteria in the distribution system or from intrusion of pathogens have not been comprehensively addressed.

Distribution-system failures have been attributed to intrusion of pathogens through leaks, surges, backflow, cross-connections or as the result of unsanitary maintenance practices. Transient pressure gradients routinely occur in distribution pipes and even well-run systems can experience leakage of 10–20% of the produced water. Therefore ample opportunities exist for intrusion of contaminated water. Sources of pathogenic microorganisms come primarily from leaking sewer lines located within 18 inches of the drinking water pipeline. Bacteria can enter water via either point or non-point sources of contamination. Point sources are those that are readily identifiable and typically discharge water through a system of pipes. Sewered communities may not have enough capacity to treat the extremely large volume of water sometimes experienced after heavy rainfalls, and treatment facilities may need to bypass some of the wastewater. During bypass or other overflow events, bacteria-laden water is discharged directly into the surface water.

The existing literature on oxidative disinfection of *H. pylori* is not well developed. *H. pylori* in brain-heart infusion with 10% horse serum survived up to 24 hours following addition of 0.1 mM $H_2O_2$, a potent source of reactive oxygen metabolites (ROM), whereas *E. coli* had limited (<5 hours) survival under identical conditions (Barton *et al.*, 1997). Chlorine and ozonation studies indicate that *H. pylori* may be capable of surviving disinfection practices adequate to remove *E. coli*, thereby allowing entry into public and private drinking water distribution systems (Johnson *et al.*, 1997; Hegarty *et al.*, 1999). The mechanism by which *H. pylori* effectively persists within an environment of chronic oxidative stress (Hazen *et al.*, 1996; McGowan *et al.*, 1997; Suzuki *et al.*, 1997; Henderson *et al.*, 1999; Santra *et al.*, 2000) is intriguing in regard to transmission implications.

The dissemination of viable *H. pylori* cells through faecal material (Thomas *et al.*, 1992; Mapstone *et al.*, 1993; Kelly *et al.*, 1994) may provide a route for the contamination of drinking water and *H. pylori* may survive for prolonged periods in water over a range of physical variables (West *et al.*, 1992). *H. pylori* strains were found to survive for longer periods in physiological saline concentrations, low temperatures and a pH ranging from 5.8 to 6.9. The addition of urea was found to result in a reduction in survival times, whereas the addition of bovine serum albumin led to variable survival times. Using autoradiography, temperature was found to be an important factor in survival of the bacterium in water. Indeed *H. pylori* was found to remain viable for periods ranging from 48 hours to between 20 and 30 days depending on the conditions under which the organism was studied (Shahamat *et al.*, 1993) indicating strong evidence for support of a waterborne for route for *H. pylori*. However, *H. pylori* cells are readily inactivated by chlorine suggesting that the organism would be controlled by disinfection regimes normally employed in the treatment of drinking water (Johnson *et al.*, 1997).

The absence of a proven analytical method or a standard protocol for assessing cell viability in water has resulted in the publication of a variety of methods and survival times for *H. pylori* in the aquatic environment. Although evidence has been published that viable but non-culturable *H. pylori* cells are still alive (Barer *et al.*, 1993; Oliver, 1993; Shahamat *et al.*, 1993), no evidence has been published on resuscitation of these cells or on their ability to cause infection.

As mentioned previously, in drinking-water systems microorganisms are predominantly associated with surfaces as biofilms rather than in the water itself (Costerton *et al.*, 1999). Waterborne enteric pathogens such as *E. coli* (Mackerness *et al.*, 1993) and *Campylobacter* spp. make use of biofilms as a vector and reservoir, but do not survive for prolonged periods in drinking water itself. MacKay's work suggests that biofilms in water distribution systems may also harbour *H. pylori* (MacKay *et al.*, 1999; Park *et al.*, 2001). However, viability and culturability was not measured in these studies.

*Disinfection study evidence for water transmission of* Helicobacter pylori
There is currently a paucity of data concerning the effectiveness of standard drinking water disinfection processes on *Helicobacter pylori*. In a paper by Baker *et al.* (2002) *Helicobacter pylori* was found to be more resistant to low doses of free chlorine than either *Campylobacter jejuni* or *E. coli*. From this study it was found that exposure to 0.2 mg/l free chlorine for 1 minute resulted in a 4 log reduction in the number of *E. coli* and *C. jejuni* cells but less than a two log reduction of *Helicobacter pylori* cells. Calculated free chlorine $CT_{99}$ values, in this study, for *C. jejuni*, *E. coli* and *H. pylori* were 0.03, 0.13 and 0.25 mg/l-min respectively. A similar result was observed in the ozonation studies carried out by the same group.

Based on the study by Baker *et al.* (2002) it seems feasible to suggest that under conditions of inadequate disinfection, *Helicobacter pylori* may persist in certain aquatic environments in the absence of *E. coli*. It may therefore be reasonable to suggest that *H. pylori* may be present in water when enumeration of coliforms on standard selective media indicates that the water is potable. Therefore current monitoring methods for drinking water may indicate that water is safe while *Helicobcter pylori* may actually be present. In addition to this, it may be conceivable to suggest that *H. pylori* may be able to persist for extended periods in drinking-water distribution systems containing low levels of residual chlorine and/or chloramines. This is reinforced when we consider the $CT_{99}$ values (at pH 6.0) of 0.045 and 0.12 mg/l-min for *E. coli* and *Helicobacter pylori*, respectively (Johnson *et al.*, 1997) which suggest that *H. pylori* may be more resistant to free chlorine than *E. coli*. The data generated by Baker (2002) do not support the speculation by Johnson *et al.* (1997) that the difference in $CT_{99}$ values between *E. coli* and *H. pylori* in their disinfection experiments can be solely attributed to the presence of agar debris or *H. pylori* cell aggregates. Differences between the Baker *et al.* (2002) and Johnson *et al.* (1997) studies might have accounted for differences in observed $CT_{99}$ values. These differences include the presence and absence of inorganic salts, absence and presence of phosphate buffer and temperature (20°C and 5°C) respectively. In addition, while disinfection $CT_{99}$ values for *E. coli* were determined at pH 7.0 in Baker's study, Johnson *et al.* referred to pH 6.0 data determined by an unrelated study. The somewhat higher *H. pylori*:*E. coli* $CT_{99}$ ratio (0.12:0.045) of Johnson *et al.* compared with Baker's study (0.25:0.13) can be attributed to the reported presence of *H. pylori* cell aggregates and debris in the former study. Considering the tendency of *H. pylori* to form aggregates and biofilms, the actual $CT_{99}$ of *H. pylori* in a natural aquatic environment may be considerably larger.

Chlorine residuals in distribution water systems often range from 0.1 to 0.3 mg/l (Geldreich, 1990). Inadequate, interrupted, or intermittent treatment has repeatedly been associated with waterborne disease outbreaks. Reduced chlorine residuals may not provide adequate inactivation of *H. pylori* to prevent entry into and persistence within the water distribution systems. Assuming the mean chlorine residual (free and combined chlorine) of treated water is 1.1 mg/l (WQDC, 1992), then approximately half of all treatment systems have a residual less than that, which may leave the distribution systems exposed to surface infiltration, susceptible to *H. pylori* contamination. This would be a particular concern if infiltration occurs prior to point of use, resulting in limited disinfection residual contact time. Short-term survival in the presence of low residual disinfectants would present a risk of infection of individuals consuming marginally treated drinking water.

Association with biofilm communities (i.e. water-distribution pipe) can offer bacteria a source of nutrients and protection from disinfection. *H. pylori* is known to produce a water-soluble biofilm when grown under high carbon:nitrogen ratio conditions (Stark *et al.*, 1999). *H. pylori* incorporated into a laboratory-scale biofilm, created by continuous 28-day flow of natural water through a modified Robbins device (MacKay *et al.*, 1999) persisted >196 hours (as determined by PCR).

### Recovery of water adapted Helicobacter pylori

The crux of developing a culturable method for *Helicobacter* in water is dependent upon effective enumeration of all viable *H. pylori* cells including 'stressed' cells which, while not-immediately culturable, can maintain their infectivity. This is illustrated by the rapid decline of colonies observed on standard plate media, irrespective of changes in morphology (i.e. formation of coccoid cells), following introduction of *H. pylori* into an oligotrophic water microcosm. Understanding the mechanisms affecting persistence and reversible loss of colony formation is critical to the development of an effective culture method for *H. pylori* isolated from environmental water. Low nutrient and hyperosmotic conditions can induce a rapid VBNC state when *H. pylori* are exposed to defined natural water systems, similar to that observed with *Campylobacter jejuni* (Tholozan *et al.*, 1999). *H. pylori* conserves the capability to produce acid-inducible proteins for at least 100 days when stored at 4°C in either phosphate-buffered saline or distilled water (Mizoguchi *et al.*, 1999). This induction of protein expression may determine the potential for morphological reversion during incubation.

Spent culture supernatant from early stationary phase *M. tuberculosis* cultures increased the viability of bacilli from aged cultures and allowed small inocula to initiate growth in liquid culture. The resuscitation factor was associated with an acid-labile, heat stable component. Density dependent growth of *H. pylori* has been observed in a number of laboratories and the use of *H. pylori* spent culture could be investigated for critical recovery from environmental sources.

Human gastrin was recently reported to be a specific dose-dependent growth factor for *H. pylori*. Human gastrin shortened the lag time, increased growth rate in the log phase and increased final bacterial concentration at the stationary phase. Controls consisting of cholecystokinin, pentagastrin, somatostatin and epidermal growth factor did not stimulate growth (Chowers *et al.*, 1999). A structurally restricted receptor-mediated gastrin-specific effect, which may have a role in the adaptation of *H. pylori* to its unique gastric habitat, was suggested by the authors.

The use of water-adapted bacteria in a defined natural water provides conditions more representative of real-world water disinfection compared to common disinfection models using log phase cultures in water. Water adaptation (hyperosmotic shock, nutrient limitation, temperature downshift) of bacterial cultures grown in rich media can cause bacteria to assume significantly altered physiologies (Kolter *et al.*, 1993). Nutrient limitation can induce polyphosphate (poly P) accumulation (Zago *et al.*, 1999) and temperature downshift has a significant effect on the activity of metabolic enzymes and can induce cold-shock proteins in some bacteria.

*Possible mechanisms of persistence and culturability of* Helicobacter pylori *in water*

Physiological adaptations (including economy of global response regulators) intrinsic to *H. pylori* may allow this organism better to survive oxidative stress. Compared to *E. coli* and *Campylobacter jejuni*, *Helicobacter pylori* has far fewer regulatory and DNA binding proteins that coordinate gene expression as a bacterium enters a new environment (Tomb *et al.*, 1997). Notably lacking are two component 'sensor-regulator' systems. A two component regulatory system designated the RacR-RacS (reduced ability to colonize) system that is involved in a temperature-dependent signalling pathway has been described in *Campylobacter jejuni* (Bras *et al.*, 1999). Flagellar biosynthesis is not as highly regulated in *H. pylori* as in other bacteria and appears to be linked with urease activity (Doig *et al.*, 1999; McGee *et al.*, 1999). Unpublished observations indicate that the *H. pylori* urease operon is regulated by decay of mRNA (Akada *et al.*, 1999). This indicates a minimal degree of environmentally responsive gene expression suggesting that in the human stomach the enzymatic pathway of *H. pylori* (at least those required for gastric survival) are continually switched on. A noteworthy feature of *H. pylori* is the absence of a global regulator such as OxyR, which controls the expression of important oxidative stress-related genes in enteric bacteria. This characteristic is shared by *Mycobacterium tuberculosis*, an organism with high resistance to killing by hydrogen peroxide and organic peroxides, compared to related *Mycobacterium* spp. with functional OxyR genes. Resistance may be mediated by unregulated peroxidase or alkyl hydroperoxidase reductase. It is also interesting to note that OxyR regulatory protein is not induced during exposure of *E. coli* to free chlorine.

Although iron repression in Gram-negative bacteria is usually carried out by the Fur protein, the iron repressed ahpC gene of *C. jejuni* is Fur independent. Recent work (van Vliet *et al.*, 1999) has demonstrated that a *C. jejuni* fur homologue PerR acts as a 'peroxide stress regulator', whereby PerR insertion mutants showed depressed expression of both AhpC and KatA, but not other iron regulated genes. Interestingly, the PerR mutation made by *C. jejuni* was hyper-resistant to oxidative stress caused by hydrogen peroxide and cumene hydroperoxide, a finding consistent with the high levels of KatA and AhpC expression. *C. jejuni* is the first Gram-negative bacterium where non-OxyR regulation of peroxide stress genes has been described. The differences in the regulation of AhpC gene expression in the closely related organism, *C. jejuni* (regulated, iron repressed) and *H. pylori* (unregulated, constitutive) may cause differing sensitivities to oxidative stresses. To support this idea, depression of OxyR control of AhpC, KatG (catalase) and Dps expression in *Salmonella typhimurium* has been shown to allow survival of oxidative stress within macrophages (Farr *et al.*, 1991). In addition, a mechanism of increased oxidative resistance (100 mM $H_2O_2$) caused by upregulated, constitutive expression of AhpC, KatA and Dps homologues in a *Bacillus fragilis* mutant has been demonstrated (Rocha and Smith, 1998).

*H. pylori* may be protected from chlorination by the production of a constitutive gene product(s) that may rapidly neutralize this oxidative disinfectant effect. *H. pylori* has a lack of transcriptional and post-transcriptional regulation (i.e. multiple sigma factors, positive activators and post-translational adenylation) of metabolic gene expression (i.e. urease, glutamate synthetase, AhpC, NapA (DPs), gamma glutamyltranspeptidase) that results in constitutive gene expression under all physiological conditions (Garner *et al.*, 1998; Chevalier *et al.*, 1999). Purified disinfection by-products have recently been shown to form extremely stable crystals with DNA almost instantaneously within which DNA is sequestered, providing effective protection against varied assaults (Wolf *et al.*, 1999). Constitutive enzyme expression can provide rapid response to environmental stress, while storage of large quantities of preformed products (i.e. urease, polyposhpate), characteristic for *H. pylori* (McNulty and Dent, 1987; Bode *et al.*, 1993; Xia *et al.*, 1994; Bauerfeind *et al.*, 1997), could assist response to sudden oxidative stress by bypassing the inherent lag of regulated gene expression.

In adaptation to stress, cells must coordinate major changes in the rates of transcription, translation and replication as well as make choices in the genes expressed (Kolter *et al.*, 1993). An *E. coli* mutant lacking the enzyme polyphosphate kinase (ppK) that makes long chains of inorganic polyphospahte (polyp) was deficient in functions expressed in the stationary phase of growth. After 2 days of growth in a medium-limited carbon source, only 7% of the mutant survived compared with nearly 100% of the wild type; loss in viability of the mutant was even more pronounced in a rich medium (Rao and Kornberg, 1999). Likewise a ppk insertion mutant of *H. pylori* exhibited a dramatic decrease in survival during stationary phase.

Polyphospahte (polyP) could provide activated phosphates or coordinate an adaptive response by binding metals and/or specific proteins (Rao and Kornberg, 1999). Many distinctive functions appear likely for polyP depending on its abundance, chain length, biological source and subcellular location: an energy supply and ATP substitute, a reservoir for inorganic phosphate, a chelator of metals, a buffer against alkali, a channel for DNA entry, a cell capsule and/or of major interest a regulator of responses to stresses and adjustments for survival in the stationary phase of culture growth (Kornberg, 1999).

Due to the lack of a stringent response in *H. pylori* (Scoarughi *et al.*, 1999) low exogenous phosphate levels could promote a decline of intracellular polyphosphate levels and may result in rapid loss of colony formation (Crooke *et al.*, 1994). Hypo-osmotic shock causes rapid reversible hydrolysis up to 95% of intracellular polyphosphate with concomitant increase in cytoplasmic pH in *Neurospora crassa* and exogenous pyrophosphate (10m) has been shown significantly to increase growth yield of *E. coli* cells in glucose minimal media and enhanced stationary phase survival (Biville *et al.*, 1996). Pre-incubation in a non-nutrient phosphate buffer, at a high bacterial density (possible quorum effect) might be important for recovery of *H. pylori* (Yamaguchi *et al.*, 1999). Bacteria within biofilms encounter higher osmolarity conditions, oxygen limitations, and higher cell density than in the liquid phase. It is possible that persistence of *H. pylori* in aquatic environments may be associated with biofilms.

Bacteria exposed to hyperosmotic stress conditions efflux large amounts of intracellular potassium. Potassium has a critical role in the maintenance of cell turgor pressure, maintenance of intracellular energy, internal pH homeostasis and enzyme activation (Bakker, 1993), all of which could contribute to rapid loss (reversible or irreversible) of colony formation. Potassium/putrescine (diamine) has been readily absorbed by *E. coli* cells and has been accompanied with a proportional loss of intracellular putrescine with enhanced DNA supercoiling. Interestingly, heat shock results in a decrease in the extent of DNA supercoiling due to the dissociation of the putrescine-DNA complexes. The global level of DNA supercoiling influences the topology of oriC (origin of replication) and thereby the sequence of events leading to initiation of DNA replication in *E. coli* (von Freiesleben and Rasmussen, 1992). Changes in DNA topology may therefore serve as important global regulatory and replication (colony formation) factors.

## Methods of detection

*H. pylori* is most commonly detected in infected patients by its potent urease enzyme. Biopsies of gastric mucosa are placed in a gel containing urea and the subsequent ammonia production causes a pH change which is a highly accurate indicator of the infection. Similarly, patients can swallow urea labelled with an isotope, either $^{14}C$ or $^{13}C$, and isotope-labelled $CO_2$ in the breath is indicative of urease activity. Infected persons carry antibodies, usually IgG, directed against *H. pylori*. Serological tests exist for detecting antibodies against the organism.

While *H. pylori* accounts for the vast majority of *Helicobacter* infections of humans, evidence for the presence of *H. heilmannii*, *H. felis*, *H. rappini*, *H. cinaedi*, *H. fennelliae* and *H. pullorum* has been associated with gastroenteritis in humans (Kusters and Kuipers, 1998; Gibson *et al.*, 1999). *H. heilmannii* (initially named *Gastrospirillum hominis*, but reclassified as a *Helicobacter* (based on 16S rRNA analysis) is the most commonly described non-pylori *Helicobacter* in humans, with colonization usually associated with mild gastritis (Holck *et al.*, 1997). Differentiation between the closely-related *H. heilmannii* and *H. pylori* requires molecular characterization as they can assume identical morphologies and *H. heilmannii* is rarely culturable *in vitro* (Fawcett *et al.*, 1999). PCR amplification and sequencing of more than 80% of the 16S rDNA is believed to be the only method to reach a precise identification (Chen *et al.*, 1997; Cantet *et al.*, 1999).

Gastric biopsy specimens are the only ones likely to be used for the primary isolation of *H. pylori*. They should be transported in a moist state and cultured within 2 hours of collection. Storage beyond this time should be at 4°C, or at $-20$°C if the period is more than 2 days. Various transport media have been described for transporting biopsy samples, including cysteine brucella broth, normal saline, glucose, milk, Stuart's medium, semi-solid agar, brain-heart infusion broth, Cary-Blair medium and, more recently, cysteine-Albimi medium containing 20% glycerol. Various non-selective and selective culture media have been used for the isolation of *H. pylori* and other *Helicobacter* spp. Chocolate agar, or more complex media such as brain-heart infusion or brucella agar supplemented with 5–7% horse blood, are suitable non-selective media. Examples of selective media are Dent's, Glupczynski's Brussels charcoal medium and Skirrow's campylobacter medium. Samples should be prepared for direct examination and plating on non-selective and selective media and plates should be incubated in a microaerobic environment at 35–37°C. Colonies may be visible after 3–5 days of incubation but may take longer to appear on primary isolation. Colonies are small, domed, translucent and sometimes weakly haemolytic.

Because of the unique characteristics of *H. pylori*, only a few simple tests are needed for identification, including Gram-stain appearance, catalase, oxidase and urease tests. Intestinal helicobacters, such as *H. cinaedi* and *H. fennelliae*, can be isolated on some campylobacter-selective agars, or by filtration on non-selective agar. Like campylobacters, *H. pylori* is strictly microaerophilic and $CO_2$ (5–20%) and high humidity are required for growth. *H. pylori* requires media containing supplements similar to those used for campylobacters: blood, haemin, serum, starch or charcoal. However, *H. pylori* is inhibited by the bisulphite in the FBP campylobacter 'aerotolerance' supplement. Growth is best on media such as moist freshly prepared heated (chocolated) blood agar, or brain-heart infusion agar with 5% horse blood and 1% IsoVitaleX. Strains grow in various liquid media supplemented with fetal calf or horse serum. Some strains grow in serum-free media, notably bisulphite-free brucella broth. Sheep blood, either whole or laked, when used as a broth supplement, has been shown to inhibit the growth of *H. pylori*. All strains grow at 37°C, some grow poorly at 30° and 42°C but none grows at 25°C. Colonies from primary cultures at 37°C usually take 3–5 days to appear and are circular, convex and translucent like those of *C. fetus*. They seldom grow bigger than 2 mm in diameter even if incubation is extended beyond 1 week. They are weakly haemolytic on 5% horse blood agar. Motility is weak or absent when grown on agar.

Like campylobacters, *H. pylori* is inactive in most conventional biochemical tests. Notable exceptions are the strong production of urease, catalase and alkaline phosphatase. All strains produce DNAase, leucine aminopeptidase, and gamma-glutamylaminopeptidase. The urease is a nickel containing high molecular weight protein (c. 600 kDa) with maximum activity at 45°C and pH 8.2.

## Epidemiology of waterborne outbreaks

There are no recorded waterborne outbreaks associated with this organism. This may be due to a latent period of many years prior to phenotypic pathology and difficulty in culturing the organism from water.

## Risk assessment

While epidemiological evidence suggests transmission by multiple pathways, *H. pylori* infection has been associated with consumption of contaminated drinking water (Klein *et al.*, 1991; Baker and Hegarty, 2001). To assess accurately risks from waterborne disease, it is necessary to understand pathogen distribution and survival within water-distribution systems and to apply methodologies that can detect not only the presence, but also the viability and infectivity of the pathogen. *H. pylori* ($10^4$–$10^6$ cells) have been demonstrated to colonize the stomach mucosa of mice. Interestingly, coccoid forms of *H. pylori* induced colonization with only $10^2$ cells (Aleljung *et al.*, 1996). Compared with illness to infection ratios of other infectious diseases, that of *H. pylori*-associated peptic ulcer, 1:5, is high (Cullen *et al.*, 1993). In comparison, such ratios are about 1:25 for hepatitis B-associated chronic liver disease and about 1:10 for *Mycobacterium tuberculosis*. Although the ratio for *H. pylori*-associated gastric adenocarcinoma is lower (1:200), the morbidity and mortality associated with this disease are substantial (Monath *et al.*, 1998).

Although *H. pylori* colonizes half of the world's human population, no large reservoir outside the human stomach has been identified. Current understanding of the microbial ecology of this organism outside of its gastric niche is extremely limited. The 13 000 research papers on *H. pylori* in the past decade have focused primarily upon clinical treatment issues. There are few data concerning the occurrence, persistence or ecology of *H. pylori* organisms in environmental waters or the efficacy of water-treatment and disinfection practices for controlling this organism. The potential presence of a microbial Class I carcinogen in source water necessitates the study of procedures to prevent human infection. Elucidation of the primary transmission routes of *H. pylori* may allow for preventative control of this human pathogen. Experimental and epidemiological evidence suggests that *H. pylori* transmission may involve, under certain circumstances, the consumption of contaminated drinking water. However, no evidence yet exists for waterborne transmission outside the developing world. *H. pylori* survives poorly in drinking water compared to other pathogens such as *Escherichia coli*, but survival for up to 12 hours under environmental conditions would, in some parts of the developing world, be long enough for transmission to occur. Overall, the transmission of *H. pylori* within water is of particular concern within developing countries where contaminated drinking water and poor sanitation and hygiene conditions enable the organism to be transmitted by the faecal-oral routes. Studies using mature heterotrophic mixed-species biofilms, using continuous chemostat systems, have shown that when challenged with *H. pylori*, the bacteria can be associated with the biofilm 8 days post-challenge. This suggests that biofilms in potable water systems could be a possible and unrecognized reservoir of *H. pylori*. Water and water biofilm reservoirs of infection could account for many of the published conclusions which have been attributed to others factors: e.g. socioeconomic factors such as overcrowding, which will also bring associated sharing of common water supplies. This idea is supported by a study of *H. pylori* infection rates in French submarine crew members with a common tanked water supply (Hammermeister *et al.*, 1992). Overall, the argument for a waterborne route of *H. pylori* transmission is supported primarily by maintenance of viability in spiked natural water (West *et al.*, 1992; Shahamat *et al.*, 1993; Hunter, 1997; Fan *et al.*, 1998; Jiang and Doyle, 1998; Sato *et al.*, 1999).

### Overall risk assessment

*Health effects*: occurrence of illness, degree of morbidity and mortality:

- Approximately 50% of the world's population is infected with *H. pylori* (Parsonnet, 1998). *H. pylori* infection is more prevalent in developing countries; the rate of infection has been decreasing steadily in developed countries over the last few decades.
- The majority of people infected with *H. pylori* are asymptomatic, though they may live their whole lives with the organism. Some will develop peptic ulcer disease, and a very small fraction will develop gastric cancer. Though only a small percentage of *H. pylori*-infected people develop gastric cancer, it is the second leading cause of cancer deaths in the world – and is directly related to the prevalence level of *H. pylori* in the population.
- Compared with illness-to-infection ratios of other infectious diseases, 1:5, the ratio of *H. pylori*-associated peptic ulcer is high (Cullen *et al.*, 1993). In comparison, other such ratios are about 1:25 for hepatitis B-associated chronic liver disease and about 1:10 for *Mycobacterium tuberculosis*. Although the ratio for *H. pylori*-associated gastric adenocarcinoma is lower (1:200), the morbidity and mortality associated with this disease are substantial (Monath *et al.*, 1998).

*Exposure assessment*: routes of exposure and transmission, occurrence in source water, environmental fate:

- Because *H. pylori* is such a fastidious organism, culturing in environmental and clinical samples can be challenging. PCR has been used more successfully to detect *H. pylori* DNA in environmental samples, however, little is known about the occurrence of the organism in water.
- Under conditions of stress, *H. pylori* can transform to a durable, coccoid morphology. While this form could be significant in transmission, controversy surrounds its function and viability (Cellini *et al.*, 1994; Kusters *et al.*, 1997; Ren *et al.*, 1999).
- Few data on environmental occurrence are available; however, of 42 surface water and 20 well water samples in the USA, 40% and 65% respectively, tested positive using fluorescent antibody and PCR methods (Hegarty and Baker, 1999).
- Limited studies are available indicating that *Helicobacter* spp. could be present in water distribution system biofilm (Stark *et al.*, 1999; Park *et al.*, 2001).
- While epidemiological evidence suggests transmission by multiple pathways, *H. pylori*'s exact route of transmission is unknown. *H. pylori* infection has been associated with consumption of contaminated drinking water (Klein *et al.*, 1991; Baker and Hegarty, 2001). Evidence is limited, but the waterborne route of exposure is probably important in some populations.
- Occurrence data indicate that infection clusters in families as well as group-living situations (i.e. orphanages, mental institutions). Secondary spread, particularly among children, is likely.

*Risk mitigation*: drinking-water treatment, medical treatment:

- Multiple medical therapies using a variety of antibiotics and bismuth preparations are generally 70–95% effective at eradicating *H. pylori* infection. Re-infection in adults seems to be a rare occurrence, though the re-infection rate for children is unknown. Infection eradication's ability to improve gastric carcinoma precursors is controversial; however, antibiotic treatment can successfully cause tumour regression in gastric lymphoma cases.
- The ability of conventional drinking water treatment to remove *H. pylori* from drinking water is unknown. The results of studies on the efficacy of chlorine as a disinfectant of *H. pylori* are equivocal (Johnson *et al.*, 1997; Hegarty *et al.*, 1999).
- Until the exact mode of transmission is understood, it will be difficult to launch public health intervention programmes. Meanwhile, improved sanitation and living conditions have resulted in decreased incidence.
- Though it is unlikely that drinking water is a significant source of infection in countries with adequate treatment, more research into the role of water in the spread of *H. pylori* is necessary.

# References

Akada, J.K., Shirai, M., Takeuchi, H. *et al.* (1999). The urease operon in *Helicobacter pylori* is regulated by decay of mRNA. Yamaguchi University School of Medicine, Department of Microbiology. *Unpublished communication*, T. Nakazawa, Oct 1999.

Aleljung, P., Nilsson, H.O., Wang, X. *et al.* (1996). Gastrointestinal colonisation of BALB/cA mice by *Helicobacter pylori* monitored by heparin immunomagnetic separation. *FEMS Immun Med Microbiol*, **13**: 303–309.

Alm, R.A., Ling, L.S., Moir, D.T. *et al.* (1999). Genomic-sequence comparison of two unrelated isolates of the human gastric pathogen *Helicobacter pylori*. *Nature*, **14**: 176–180.

Anderson, A.P., Elliott, D.A., Lawson, M. *et al.* (1997). Growth and morphological transformations of *Helicobacter pylori* in broth media. *J Clin Microbiol*, **35**: 2918–2922.

Antonioli, D.A. (1994). Precursors of gastric carcinoma: A critical review with a brief description of early (curable) gastric cancer. *Hum Pathol*, **25**: 994–1005.

Aruin, L.I. (1997). *Helicobacter pylori* infection is carcinogenic for humans. *Arkh Patol*, **59**(3): 74–78.

Arvanitidou, M., Stathopoulos, G.A. and Katsouyannopoulos, V.C. (1994). Isolation of *Campylobacter* and *Yersinia* spp. from drinking waters. *J Travel Med*, **1**: 156–159.

Atherton, J.C., Cao, P., Peek, R.M. Jr *et al.* (1995). Mosaicism in vacuolating cytotoxin alleles of *Helicobacter pylori*: association of specific *vacA* types with cytotoxin production and peptic ulceration. *J Biol Chem*, **270**: 17771–17777.

Baker, K.H. and Hegarty, J.P. (2001). Presence of *Helicobacter pylori* in drinking water is associated with clinical infection. *Scan J Infect Dis*, **33**(10): 744–746.

Baker, K.H., Hegarty, J.P., Redmond, B. *et al.* (2002). Effect of oxidizing disinfectants (chlorine, monochloramine, and ozone) on *Helicobacter pylori*. *Appl Environ Microbiol*, **68**: 981–984.

Bakker, E.P. (ed.) (1993). Cell $K^-$ and $K^+$ transport systems in procaryotes. In *Alkali Cation Transport Systems in Prokayotes*. Boca Raton: CRC Press, pp. 205–224.

Barer, M.R., Gribbon, L.T., Harwood, C.R. *et al.* (1993). The viable but non-culturable hypothesis and medical bacteriology. *R Med Microbiol*, **4**: 183–191.

Barton, S.G.R.G., Young, K.A., Hardie, J.M. *et al.* (1997). CagA status does not influence survival of *H. pylori* after exposure to hydrogen peroxide. European *Helicobacter pylori* Study Group, Sept 12–14, Lisbon, Portugal. 10th International Workshop Abstracts, p. A11.

Basso, D., Stefani, A., Gallo, N. *et al.* (1999). Polymorphonuclear oxidative burst after *Helicobacter pylori* water extract stimulation is not influenced by the cytotoxic genotype but indicates infection and gastritis grade. *Clin Chem Lab Med*, **37**: 223–229.

Bauerfeind, P., Garner, R., Dunn, B.E. *et al.* (1997). Synthesis and activity of *Helicobacter pylori* urease and catalase at low pH. *Gut*, **40**: 25–30.

Beier, D., Spohn, G., Rappuoli, R. *et al.* (1997). Identification and characterization of an operon of *Helicobacter pylori* that is involved in motility and stress adaptation. *J Bacteriol*, **179**(15): 4676–4683.

Biville, F., Laurent-Winter, C. and Danchin, A. (1996). *In vivo* positive effects of exogenous pyrophosphate on *Escherichia coli* cell growth and stationary phase survival. *Res Microbiol*, **147**(8): 597–608.

Blaser, M.J. (1987). Gastric *Campylobacter*-like organisms, gastritis, and peptic ulcer disease. *Gastroenterology*, **93**(2): 371–383.

Blaser, M.J. (1997). *Helicobacter pylori* eradication and its implications for the future. *Aliment Pharmacol Ther*, **1**: 103–107.

Bode, G., Mauch, F., Ditschuneit, H. *et al.* (1993). Identification of structures containing polyphosphate *in Helicobacter pylori*. *J Gen Microbiol*, **139**: 3029–3033.

Bode, G., Rothenbacher, D., Brenner, H. *et al.* (1997). Pets are no risk for *Helicobacter pylori* infection in young children. Results of a population based study in southern Germany. European *Helicobacter pylori* Study Group, Sept 12–14, Lisbon, Portugal. 10th International Workshop Abstracts, p. A37.

Borody, T.J., Cole, P., Noonan, S. *et al.* (1989). Recurrence of duodenal ulcer and *Campylobacter pylori* infection after eradication. *Med J Aust*, **151**(8): 431–435.

Bras, A.M., Chatterjee, S., Wren, B.W. *et al.* (1999). A novel *Campylobacter jejuni* two component regulatory system important for temperature-dependent growth and colonization. *J Bacteriol*, **181**(10): 3298–3302.

Brown, L.M. (2000). *Helicobacter pylori*: epidemiology and routes of transmission. *Epidemiol Rev*, **22**(2): 283–297.

Bucklin, K.E., McFeters, G.A. and Amirtharaja, A. (1991). Penetration of coliforms through municipal drinking water filters. *Water Res*, **25**: 1013–1017.

Bunn, J.E.G., Thomas, J.E., Harding, M. *et al.* (1997). Supplemental water in early infancy: A risk factor for *H. pylori*? European *Helicobacter pylori* Study Group, Sept 12–14, Lisbon, Portugal. 10th International Workshop Abstracts, p. A37.

Cantet, F., Magras, C., Marais, A. *et al.* (1999). *Helicobacter* species colonizing pig stomach: Molecular characterization and determination of prevalence. *Appl Environ Microbiol*, **65**(10): 4672–4676.

Cappelier, J.M., Minet, J., Magras, C. *et al.* (1999). Recovery in embryonated eggs of viable but nonculturable *Campylobacter jejuni* cells and maintenance of ability to adhere to HeLa cells after resuscitation. *Appl Environ Microbiol*, **65**(11): 5154–5157.

Carballo, F., Caballero, P., Parra, T. *et al.* (1997). Untreated drinking water is a source of *H. pylori* infection. European *Helicobacter pylori* Study Group, Sept 12–14, Lisbon, Portugal. 10th International Workshop, p. A39.

Cave, D.R. (1997). How is *Helicobacter pylori* transmitted? *Gastroenterology*, **113**: S14.

Cellini, L., Allocati, N., Angelucci, D. *et al.* (1994). Coccoid *Helicobacter pylori* not culturable in vitro reverts in mice. *Microbiol Immunol*, **38**(11): 843–850.

Cellini, L., Dainelli, B., Angelucci, D. *et al.* (1999). Evidence for an oral-faecal transmission of *Helicobacter pylori* infection in an experimental murine model. *APMIS*, **107**(5): 477–484.

Cellini, L., Robuffo, I., Maraldi, N.M. *et al.* (2001). Searching the point of no return in *Helicobacter pylori* life: necrosis and/or programmed death? *J Appl Microbiol*, **90**(5): 727–732.

Chen, Y., Wang, J.D. and Xu, Z.M. (1997). Using polymerase chain reaction to detect *Helicobacter heilmannii* in gastric biopsy materials. *Chung Hua Liu Hsing Ping Hsueh Tsa Chih*, **18**(4): 241–243.

Chevalier, C., Thiberge, J.M., Ferrero, R.L. *et al.* (1999). Essential role of *Helicobacter pylori* gammaglutamyltranspeptidase for the colonization of the gastric mucosa of mice. *Mol Microbiol*, **31**(5): 1359–1372.

Chin, E.Y., Whary, M.T., Fox, J.G. *et al.* (2001). Urease is required for intestinal colonization of the mouse by the pathogen *Helicobacter hepaticus*. (Personal communication from David B. Schauer, MIT.)

Chowers, M.Y., Keller, N., Tal, R. *et al.* (1999). Human gastrin: A *Helicobacter pylori*-specific growth factor. *Gastroenterology*, **17**(5): 1113–1118.

Costa, K., Bacher, G., Allmaier, G. *et al.* (1999). The morphological transition of *Helicobacter pylori* cells from spiral to coccoid is preceded by a substantial modification of the cell wall. *J Bacteriol*, **181**(12): 3710–3715.

Costerton, J.W., Stewart, P.S. and Greenberg, E.P. (1999). Bacterial biofilms: A common cause of persistent infections. *Science*, **284**: 1318–1322.

Crooke E., Akiyama M., Rao N.N. *et al.* (1994). Genetically altered levels of inorganic polyphosphate in *Escherichia coli*. *J Biol Chem*, **269**(9): 6290–6295.

Cullen, D.J., Collins, B.J. and Christiansen, K.J. (1993). When is *Helicobacter pylori* infection acquired? *Gut*, **34**: 1681–1682.

Dieterich, C., Bouzourene, H., Blum, A.L. *et al.* (1999). Urease-based mucosal immunization against *Helicobacter heilmannii* infection induces corpus atrophy in mice. *Infect Immun*, **67**(11): 6206–6209.

DiPetrillo, M.D., Tibbetts, T., Kleanthous, H. *et al.* (1999). Safety and immunogenicity of phoP/phoQ-deleted *Salmonella typhi* expressing *Helicobacter pylori* urease in adult volunteers. *Vaccine*, **18**: 449–459.

Doig, P., de Jonge, B.L., Alm, R.A. *et al.* (1999). *Helicobacter pylori* physiology predicted from genomic comparison of two strains. *Microbiol Mol Biol Rev*, 63(3): 675–707.

Dominici, P., Bellentani, S., Di Biase, A.R. *et al.* (1999). Familial clustering of *Helicobacter pylori* infection: Population based study. *Br Med J*, 319: 537–541.

Dooley, C.P., Cohen, H. and Fitzgibbons, P.L. (1989). Prevalence of *Helicobacter pylori* infection and histologic gastritis in asymptomatic persons. *New Engl J Med*, 321: 1562–1566.

Dore, M.P., Bilotta, M., Vaira, D. *et al.* (1999a). High prevalence of *Helicobacter pylori* infection in shepherds. *Dig Dis Sci*, 44(6): 1161–1164.

Dore, M.P., Sepulveda, A.R., Osato, M.S. *et al.* (1999b). *Helicobacter pylori* in sheep milk. *Lancet*, 354: 132.

Dowsett, S.A., Archila, L., Segreto, V.A. *et al.* (1999). *Helicobacter pylori* infection in indigenous families of Central America: Serostatus and oral and fingernail carriage. *J Clin Microbiol*, 37(8): 2456–2460.

Dunn, B.E. (1993). Pathogenic mechanisms of *Helicobacter pylori*. *Gastroenterol Clin North Am*, 22(1): 43–57.

Eaton, K.A., Catrenich, C.E., Makin, K.M. *et al.* (1995). Virulence of coccoid and bacillary forms of *Helicobacter pylori* in gnotobiotic piglets. *J Infect Dis*, 171(2): 459–462.

Enroth, H. and Engstrand, L. (1995). Immunomagnetic separation and PCR for detection of *Helicobacter pylori* in water and stool specimens. *J Clin Microbiol*, 33(8): 2162–2165.

Enroth, H., Wreiber, K., Rigo, R. *et al.* (1999). *In vitro* aging of *Helicobacter pylori*: Changes in morphology, intracellular composition and surface properties. *Helicobacter*, 4(1): 7–16.

Fan, X.G., Chua, A., Li, T.G. *et al.* (1998). Survival of *Helicobacter pylori* in milk and tap water. *J Gastroenterol Hepatol*, 13(11): 1096–1098.

Farinati, F., Cardin, R., Degan, P. *et al.* (1988). Oxidative DNA damage accumulation in gastric carcinogenesis. *Gut*, 42(3): 351–356.

Farr, S.B. and Kogoma, T. (1991). Oxidative stress responses in *Escherichia coli* and *Salmonella typhimurium*. *Microbiol Rev*, 55(4): 561–585.

Fawcett, P.T., Gibney, K.M. and Vinette, K.M.B. (1999). *Helicobacter pylori* can be induced to assume the morphology of *Helicobacter heilmannii*. *J Clin Microbiol*, 37(4): 1045–1048.

Foliguet, B., Vicari, F., Guedenet, J.C. *et al.* (1989). Scanning electron microscopic study of *Campylobacter pylori* and associated gastroduodenal lesions. *Gastroenterol Clin Biol*, 13: 65B–70B.

Forrest, K., Stinson, M. and Wright, S.M. (1998). The presence of *Helicobacter pylori* in sewage. American Society for Microbiology, Abstracts to the 98th General Meeting, Atlanta GA, p. 445.

Freedberg, A.S. and Barron, L.E. (1940). The presence of spirochetes in human gastric mucosa. *Am J Dig Dis*, 7: 443–445.

Friedman, B. (1998). European *Helicobacter* Pylori Study Group. XIth International Workshop on Gastroduodenal Pathology and *Helicobacter pylori*. *Gut*, 43: 2.

Fujioka, T. (1995). Virulence and pathogenesis of *Helicobacter pylori* infection. *Rinsho Byori*, 43(6): 557–561.

Fukuyama, M., Arimatu, M., Sakamato, K. *et al.* (1999). PCR detection of *Helicobacter pylori* in water samples collected from rivers in Japan. In Abstracts of 10th International CHRO Workshop, Baltimore, MD, Sept 14, p. 117.

Garner, R.M., Fulkerson, J. Jr and Mobley, H.L. (1998). *Helicobacter pylori* glutamine synthetase lacks features associated with transcriptional and posttranstational regulation. *Infect Immun*, 66(5): 1839–1847.

Ge, Z. and Taylor, D.E. (1999). Contributions of genome sequencing to understanding the biology of *Helicobacter pylori* (Review). *Ann Rev Micro*, 53: 353–387.

Geldreich, E.E. (1990). Microbiological quality of source waters for water supply. In *Drinking Water Microbiology*, McFeters, G.A. (ed.). New York: Springer, pp. 3–31.

Gibson, J.R., Ferrus, M.A., Woodward, D. *et al.* (1999). Genetic diversity in *Helicobacter pullorum* from human and poultry sources identified by an amplified fragment length polymorphism technique and pulsed-field gel electrophoresis. *J Appl Microbiol*, **87**(4): 602–610.

Goodman, K.J., Correa, P., Tengana Aux, H.J. *et al.* (1996). *Helicobacter pylori* infection in the Colombian Andes: a population-based study of transmission pathways. *Am J Epidemiol*, **144**(3): 290–299.

Goodwin, C.S., Armstrong, J.A., Chlivers, T. *et al.* (1989). Transfer of *Campylobacter pylori* and *C. mustelae* to *Helicobacter* gen. nov. as *H. pylori* comb. nov. and *H. mustelae* comb nov., respectively. *Int J Syst Bacteriol*, **39**: 397–405.

Graham, D.Y., Malaty, H.M., Evans, D.G. *et al.* (1991). Epidemiology of *Helicobacter pylori* in asymptomatic population in the U.S. *Gastroenterology*, **100**: 1495–1501.

Gribbon, L.T. and Barer, M.R. (1995). Oxidative metabolism in nonculturable *Helicobacter pylori* and *Vibrio vulnificus* cells studied by substrate-enhanced tetrazolium reduction and digital image processing. *Appl Environ Microbiol*, **61**(9): 3379–3384.

Grubel, P., Huang, L., Masubuchi, N. *et al.* (1998). Detection of *Helicobacter pylori* DNA in houseflies (*Musca domestica*) on three continents. *Lancet*, **352**: 788–789.

Gurel, S., Besisk, F., Demir, K. *et al.* (1999). After the eradication of *Helicobacter pylori* infection, relapse is a serious problem in Turkey. *J Clin Gastroenterol*, **28**(3): 241–244.

Hammermeister, I., Janus, G., Schamorowski, F. *et al.* (1992). Elevated risk of *Helicobacter pylori* infection in submarine crews. *J Clin Microbiol Infect Dis*, **11**: 9–14.

Handwerker, J., Fox, J.G. and Schauer, D.B. (1995). Detection of *Helicobacter pylori* in drinking water using polymerase chain reaction amplification. American Society for Microbiology, In Abstracts to the 95th General Meeting, Washington DC, p. 435.

Hazen, S.L., Hsu, F.F., Mueller, D.M. *et al.* (1996). Human neutrophils employ chlorine gas as an oxidant during phagocytosis. *J Clin Invest*, **98**(6): 1283–1289.

Hegarty, J.P. and Baker, K.H. (1999). Occurrence of *Helicobacter pylori* in surface water in the United States. *J Appl Microbiol*, **87**(5): 697–701.

Hegarty, J.P., Herson, D.S., Redmond, B.W. *et al.* (1999). Comparative efficacy of oxidative disinfectants on *Helicobacter pylori*. American Society for Microbiology, In Abstracts to the 99th General Meeting, Chicago, Illinois, June 1. Session 129. Paper Q-132.

Henderson, J.P., Byun, J. and Heinecke, J.W. (1999). Molecular chlorine generated by the myeloperoxidase-hydrogen peroxide-chloride system of phagocytes produces 5-chlorocytosine in bacterial RNA. *J Biol Chem*, **274**(47): 33440–33448.

Holck, S., Ingeholm, P., Blom, J. *et al.* (1997). The histopathology of human gastric mucosa inhabited by *Helicobacter heilmannii*-like (Gastrospirillum hominis) organisms, including the first culturable case. *APMIS*, **105**(10): 746–756.

Hulten, K., Han, S.W., Enroth, H. *et al.* (1996). *Helicobacter pylori* in the drinking water in Peru. *Gastroenterology*, **110**(4): 1031–1035.

Hulten, K., Enroth, H., Nystrom, T. *et al.* (1998). Presence of *Helicobacter* species DNA in Swedish water. *J Appl Microbiol*, **85**(2): 282–286.

Hunter, P.R. (1997). *Waterborne Diseases*. New York: John Wiley & Sons, pp. 202–205.

Jiang, X. and Doyle, M.P. (1998). Effect of environmental and substrate factors on survival and growth of *Helicobacter pylori*. *J Food Prot*, **61**(8): 929–933.

Johnson, C.H., Rice, E.W. and Reasoner, D.J. (1997). Inactivation of *Helicobacter pylori* by chlorination. *Appl Environ Microbiol*, **63**(12): 4969–4970.

Kamat, A.H., Mehta, P.R., Natu, A.A. *et al.* (1998). Dental plaque: An unlikely reservoir of *Helicobacter pylori*. *Indian J Gastroenterol*, **17**(4): 138–140.

Kelly, D.J. (1998). The physiology and metabolism of the human gastric pathogen *Helicobacter pylori*. *Adv Microb Physiol*, **40**: 137–189.

Kelly, S.M., Pitcher, M.C., Farmery, S.M. *et al.* (1994). Isolation of *Helicobacter pylori* from feces of patients with dyspepsia in the United Kingdom. *Gastroenterology*, **107**(6): 1671–1674.

Klein, P.D., Graham, D.Y., Gaillour, A. *et al.* (1991). Water source as risk factor for *Helicobacter pylori* infection in Peruvian children. Gastrointestinal Physiology Working Group. *Lancet*, **337**(8756): 1503–1506.

Kolter, R., Siegele, D.A. and Tortno, A. (1993). The stationary phase of the bacterial life cycle. *Annu Rev Microbiol*, **47**: 855–874.

Kornberg, A. (1999). Inorganic polyphosphate: A molecule of many functions. *Prog Mol Subcell Biol*, **23**: 1–18.

Kuipers, E.J. (1999). Exploring the link between *Helicobacter pylori* and gastric cancer. *Aliment Pharmacol Ther*, **13**(Suppl.): 3–11.

Kung, N.N., Sung, J.J., Yuen, N.W. *et al*. (1997). Anti-*Helicobacter pylori* treatment in bleeding ulcers: randomized controlled trial comparing 2-day versus 7-day bismuth quadruple therapy. *Am J Gastroenterol*, **92**: 438–441.

Kuo, C.H., Poon, S.K., Su, Y.C. *et al*. (1999). Heterogeneous *Helicobacter pylori* isolates from *H. pylori*-infected couples in Taiwan. *J Infect Dis*, **180**(6): 2064–2068.

Kurokawa, M., Nukina, M., Nakanishi, H. *et al*. (1999). Resuscitation from the viable but nonculturable state of *Helicobacter pylori*. *Kansenshogaku Zasshi*, **73**(1): 15–19.

Kusters, J.G., Gerrits, M.M., Van Strijp, J.A. *et al*. (1997). Coccoid forms of *Helicobacter pylori* are the morphologic manifestation of cell death. *Infect Immun*, **65**(9): 3672–3679.

Kusters, J.G. and Kuipers, E.J. (1998). Non-*pylori Helicobacter* infections in humans. *Eur J Gastroenterol Hepatol*, **10**(3): 239–241.

Leung, W.K., Sung, J.J., Ling, T.K. *et al*. (1999). Use of chopsticks for eating and *Helicobacter* pylori infection. *Dig Dis Sci*, **44**(6): 1173–1176.

Loffeld, R.J. and Arends, J.W. (1993). The role of *Helicobacter pylori* in non-ulcer dyspepsia and gastritis. *Neth J Med*, **42**(1–2): 73–79.

Luman, W., Alkout, A.M., Blackwell, C.C. *et al*. (1996). *Helicobacter pylori* in the mouth: Negative isolation from dental plaque and saliva. *Eur J Gastroenterol Hepatol*, **8**(1): 11–14.

Mackay, W.G., Gribbon, L.T., Barer, M.R. *et al*. (1999). Are drinking water biofilms a source of *Helicobacter pylori*? *J Appl Microbiol*, **85**(Suppl.): 52S–59S.

Mackerness, C.W., Colbourne, J.S., Dennis, P.J.L. *et al*. (1993). Formation and control of coliform biofilms in drinking water distribution systems. In *Society for Applied Bacteriology Technical Series*, vol. 30, Denyer, S.P. (ed.). London: Blackwell Scientific, pp. 217–227.

Malaty, H.M. and Graham, D.Y. (1994). Effect of childhood socio-economic status on the current prevalence of *Helicobacter pylori* infection. *Gut*, **35**: 742–745.

Mapstone, N.P., Lewis, F.A., Tompkins, D.S. *et al*. (1993). PCR identification of *Helicobacter pylori* from gastritis patients. *Lancet*, **341**: 447.

Marais, A., Mendz, G.L., Hazell, S.L. *et al*. (1999). Metabolism and genetics of *Helicobacter pylori*: the genome era. *Microbiol Mol Biol Rev*, **63**(3): 642–674.

Marshall, B.J. and Warren, J.R. (1984). Unidentified curved bacilli in the stomach of patients with gastritis and peptic ulceration. *Lancet*, i: 1311–1315.

McFeters, G.A. (1990). Enumeration, occurrence and significance of injured bacteria in drinking water. In *Drinking Water Microbiology: Progress and Recent Developments*, McFeters, G.A. (ed.). New York: Springer, pp. 478–492.

McGee, D.J., May, C.A., Garner, R.M. *et al*. (1999). Isolation of *Helicobacter pylori* genes that modulate urease activity. *J Bacteriol*, **81**(8): 2477–2484.

McGowan, C.C., Necheva, A.S. and Cover, T.L. (1997). Acid-induced expression of oxidative stress protein homologs in *H. pylori*. European *Helicobacter pylori* Study Group, 10th International Workshop Abstracts, Sept 12–14.

McKeown, I., Orr, P., Macdonald, S. *et al*. (1999). *Helicobacter pylori* in the Canadian arctic: Seroprevalence and detection in community water samples. *Am J Gastroenterol*, **94**(7): 1823–1829.

McNulty, C.A. and Dent, J.C. (1987). Rapid identification of *Campylobacter pylori* (*C pyloridis*) by preformed enzymes. *J Clin Microbiol*, **25**(9): 1683–1686.

Mendall, M.A., Goggin, P.M. and Molineaux, N. (1992). Childhood living conditions and *Helicobacter pylori* seropositivity in adult life. *Lancet*, **339**: 896–897.

Michetti, P., Kreiss, C., Kotloff, K.L. *et al*. (1999). Oral immunization with urease and *Escherichia coli* heat-labile enterotoxin is safe and immunogenic in *Helicobacter pylori*-infected adults. *Gastroenterology*, **116**(4): 804–812.

Mizoguchi, H., Fujioka, T., Kishi, K. *et al*. (1998). Diversity in protein synthesis and viability of *Helicobacter pylori* coccoid forms in response to various stimuli. *Infect Immun*, **66**(11): 5555–5560.

Mizoguchi, H., Fujioka, T. and Nasu, M. (1999). Evidence for viability of coccoid forms of *Helicobacter pylori. J Gastroenterol*, **34**(Suppl. 11): 32–36.

Mizote, T., Yoshiyama, H. and Nakazawa, T. (1997). Urease-independent chemotactic responses of *Helicobacter pylori* to urea, urease inhibitors, and sodium bicarbonate. *Infect Immun*, **65**(4): 1519–1521.

MMWR. (1999). Outbreak of *Escherichia coli* O157:H7 and *Campylobacter* among attendees of the Washington County Fair – New York. *MMWR*, **48**(36): 803–805.

Mobley, H.L.T., Island, M.D. and Hausinger, R.P. (1995). Molecular biology of ureases. *Microbiol Rev*, **59**: 451–480.

Monath, T.P., Lee, C.K., Ermak, T.H. *et al.* (1998). The search for vaccines against *Helicobacter pylori. Infect Med*, **15**(8): 534–546.

Moore, A.C., Herwaldt B.L., Craun, G.F. *et al.* (1994). Waterborne disease in the United States, 1991 and 1992. *JAWWA*, **86**: 87–99.

Nakamura, H., Yoshiyama, H., Takeuchi, H. *et al.* (1998). Urease plays an important role in the chemotactic motility of *Helicobacter pylori* in a viscous environment. *Infect Immun*, **66**(10): 4832–4837.

Namavar, F., Sparrius, M., Veerman, E.C. *et al.* (1998). Neutrophil-activating protein mediates adhesion of *Helicobacter pylori* to sulfated carbohydrates on high-molecular-weight salivary mucin. *Infect Immun*, **66**(2): 444–447.

Nedrud, J.G. and Czinn, S.J. (1999). Host, heredity and *Helicobacter. Gut*, **45**: 323–324.

Nowak, J.A., Forouzandeh, B. and Nowak, J.A. (1997). Estimates of *Helicobacter pylori* densities in the gastric mucus layer by PCR, histologic examination, and CLOtest. *Am J Clin Pathol*, **108**(3): 284–288.

Odenbreit, S., Wieland, B. and Haas, R. (1996). Cloning and genetic characterization of *Helicobacter pylori* catalase and construction of a catalase-deficient mutant strain. *J Bacteriol*, **178**(23): 6960–6967.

Oliver, J.D. (1993). Formation of viable but non-culturable cells. In *Starvation in Bacteria*, Kjelleberg, S. (ed.). New York: Plenum, pp. 239–272.

Oliver, J.D. (1995). The viable but non-culturable state in the human pathogen *Vibrio vulnificus. FEMS Microbiol Lett*, **133**: 203–208.

On, S.L. and Holmes, B. (1992). Assessment of enzyme detection tests useful in identification of Campylobacteria. *J Clin Microbiol*, **30**(3): 746–749.

Osata, M.S., Le, H.H., Ayoub, K. *et al.* (1997). Houseflies are an unlikely reservoir for *Helicobacter pylori*. European *Helicobacter pylori* Study Group, 10th International Workshop Abstracts, Sept 12–14.

Oshowo, A., Gillam, D., Botha, A. *et al.* (1998a). *Helicobacter pylori*: The mouth, stomach, and gut axis. *Ann Periodontol*, **3**(1): 276–280.

Oshowo, A., Tunio, M., Gillam, D. *et al.* (1998b). Oral colonization is unlikely to play an important role in *Helicobacter pylori* infection. *Br J Surg*, **85**(6): 850–852.

Park, S.R., Mackay, W.G. and Reid, D.C. (2001). *Helicobacter* sp. recovered from drinking water biofilm sampled from a water distribution system. *Water Res*, **35**(6): 1624–1626.

Parsonnet, J. (1998). *Helicobacter pylori. Infect Dis Clin North Am*, **12**: 185–197.

Parsonnet, J., Shmuely, H. and Haggerty, T. (1999). Fecal and oral shedding of *Helicobacter pylori* from healthy infected adults. *JAMA*, **282**(23): 2240–2245.

Pawlowski, K., Zhang, B., Rychlewski, L. *et al.* (1999). The *Helicobacter pylori* genome: From sequence analysis to structural and functional predictions. *Proteins*, **36**(1): 20–30.

Pitson, S.M., Mendz, G.L., Srinivasan, S. *et al.* (1999). The tricarboxylic acid cycle of *Helicobacter pylori. Eur J Biochem*, **260**(1): 258–267.

Popovic-Uroic, T., Patton, C.M., Nicholson, M.A. *et al.* (1990). Evaluation of the indoxyl acetate hydrolysis test for rapid differentiation of *Campylobacter, Helicobacter,* and *Wolinella* species. *J Clin Microbiol*, **28**(10): 2335–2339.

Pretolani, S., Stroffolini, T., Rapicetta, M. *et al.* (1997). Seroprevalence of hepatitis A virus and *Helicobacter pylori* infections in the general population of a developed European country (the San Marino study): Evidence for similar pattern of spread. *Eur J Gastroenterol Hepatol*, **9**(11): 1081–1084.

Ramirez-Ramos, A., Gilman, R.H., Leon-Barua, R. *et al.* (1997). Rapid recurrence of *Helicobacter pylori* infection in Peruvian patients after successful eradication.

Gastrointestinal Physiology Working Group of the Universidad Peruana Cayetano Heredia and The Johns Hopkins University. *Clin Infect Dis*, **25**(5): 1027–1031.

Rao, N.N. and Kornberg, A. (1999). Inorganic polyphosphate regulates responses of *Escherichia coli* to nutritional stringencies, environmental stresses and survival in the stationary phase. *Prog Mol Subcell Biol*, **23**: 183–195.

Ren, Z., Pang, G., Musicka, M. *et al.* (1999). Coccoid forms of *Helicobacter pylori* can be viable. *Microbios*, **97**(388): 153–163.

Rice, E.W., Rodgers, M.R., Wesley, I.V. *et al.* (1999). Isolation of *Arcobacter butzleri* from ground water. *Lett Appl Microbiol*, **28**(1): 31–35.

Rocha, E.R. and Smith, C.J. (1998). Characterization of a peroxide-resistant mutant of the anaerobic bacterium *Bacteroides fragilis*. *J Bacteriol*, **180**(22): 5906–5912.

Roszak, D.B. and Colwell, R.R. (1987). Survival strategies of bacteria in the natural environment. *Microbiol Rev*, **51**: 365–379.

Russo, A., Eboli, M., Pizzetti, P. *et al.* (1999). Determinants of *Helicobacter pylori* seroprevalence among Italian blood donors. *Eur J Gastroenterol Hepatol*, **11**(8): 867–873.

Santra, A., Chowdhury, A., Chaudhuri, S. *et al.* (2000). Oxidative stress in gastric mucosa in *Helicobacter pylori* infection. *Indian J Gastroenterol*, **19**(1): 21–23.

Sasaki, K., Tajiri, Y., Sata, M. *et al.* (1999). *Helicobacter pylori* in the natural environment. *Scand J Infect Dis*, **31**(3): 275–279.

Sato, F., Saito, N., Shouji, E. *et al.* (1999). The maintenance of viability and spiral morphology of *Helicobacter pylori* in mineral water. *J Med Microbiol*, **48**(10): 971.

Schauer, D.B., Handwerker, J., Correa, P. *et al.* (1995). Detection of *H. pylori* in drinking water using PCR amplification. *European Helicobacter pylori Study Group* 8th International Conference.

Scoarughi, G.L., Cimmino, C. and Donini, P. (1999). *Helicobacter pylori*: a eubacterium lacking the stringent response. *J Bacteriol*, **181**(2): 552–555.

Shahamat, M., Mai, U., Paszko-Kolva, C. *et al.* (1993). Use of autoradiography to assess viability of *Helicobacter pylori* in water. *Appl Environ Microbiol*, **59**(4): 1231–1235.

Shaw, S.E. and Regli, S. (1999). U.S. regulations on residual disinfection. *JAWWA*, **91**(1): 75–80.

Sibille, I., Sime-Ngando, T., Mathieu, L. *et al.* (1998). Protozoan bacterivory and *Escherichia coli* survival in drinking water distribution systems. *Appl Environ Microbiol*, **64**(1): 197–202.

Stanley, K., Cunningham, R. and Jones, K. (1998). Isolation of *Campylobacter jejuni* from groundwater. *J Appl Microbiol*, **85**(1): 187–191.

Stark, R.M., Gerwig, G.J., Pitman, R.S. *et al.* (1999). Biofilm formation by *Helicobacter pylori*. *Lett Appl Microbiol*, **28**(2): 121–126.

Steer, H.W. and Colin-Jones, D.G. (1975). Mucosal changes in gastric ulceration and their response to carbenoxolone sodium. *Gut*, **16**: 590–597.

Suzuki, H., Mori, M., Suzuki, M. *et al.* (1997). Extensive DNA damage induced by monochloramine in gastric cells. *Cancer Lett*, **115**(2): 243–248.

Suzuki, J., Muraoka, H., Kobayasi, I. *et al.* (1999). Rare incidence of interspousal transmission of *Helicobacter pylori* in asymptomatic individuals in Japan. *J Clin Microbiol*, **37**(12): 4174–4176.

Tholozan, J.L., Cappelier, J.M., Tissier, J.P. *et al.* (1999). Physiological characterization of viable-but-nonculturable *Campylobacter jejuni* cells. *Appl Environ Microbiol*, **65**: 1110–1116.

Thomas, J.E., Gibson, G.R., Darboe, M.K. *et al.* (1992). Isolation of *Helicobacter pylori* from human feces. *Lancet*, **340**(8829): 1194–1195.

Tomb, J.-F., White, O., Kerlavage, A.R. *et al.* (1997). The complete genome sequence of the gastric pathogen *Helicobacter pylori*. *Nature*, **388**(6642): 539–547.

Tytgat, G.N.J. and Rauws, E.A.J. (1990). *Campylobacter pylori* and its role in peptic ulcer disease. *Gastroenterol Clin North Am*, **19**: 183–196.

Unge, P., Gad, A., Gnarpe, H. *et al.* (1989). Does omeprazole improve antimicrobial therapy directed towards gastric *Campylobacter pylori* in patients with antral gastritis? A pilot study. *Scand J Gastroenterol Suppl*, **167**: 49–54.

Vacheron, M.J., Guinand, M., Francon, A. *et al.* (1979). Characterisation of a new endopeptidase from sporulating *Bacillus sphaericus* which is specific for the gamma-D-glutamyl-L-lysine and gamma-D-glutamyl-(L)meso-diaminopimelate linkages of peptidoglycan substrates. *Eur J Biochem*, 100: 189–196.

Vakil, N.B. (1997). Managing patients with peptic ulcer disease: Improving the value of care. *Drug Benefit Trends*, 9(4): 30–32.

Vaira, D., Holton, J., Menegatti, M. *et al.* (1998). Routes of transmission of *Helicobacter pylori* infection. *Ital J Gastroenterol Hepatol*, 30: S279–S285.

van Vliet, A.H., Baillon, M.L., Penn, C.W. *et al.* (1999). *Campylobacter jejuni* contains two Fur homologs: Characterization of iron-responsive regulation of peroxide stress defense genes by the PerR repressor. *J Bacteriol*, 181(20): 6371–6376.

Velazquez, M. and Feirtag, J.M. (1999). *Helicobacter pylori*: characteristics, pathogenicity, detection methods and mode of transmission implicating foods and water. *Int J Food Microbiol*, 53(2–3): 95–104.

Vincent, P. (1995). Transmission and acquisition of *Helicobacter pylori*: Evidences and hypothesis. *Biomed Pharmacother*, 49: 11–18.

von Freiesleben, U. and Rasmussen, K.V. (1992). The level of supercoiling affects the regulation of DNA replication in *Escherichia coli*. *Res Microbiol*, 143(7): 655–663.

Vyas, S.P. and Sihorkar, V. (1999). Exploring novel vaccines against *Helicobacter pylori*: protective and therapeutic immunization. *J Clin Pharm Ther*, 24(4): 259–272.

Wang, X., Sturegard, E., Rupar, R. *et al.* (1997). Infection of BALB/c A mice by spiral and coccoid forms of *Helicobacter pylori*. *J Med Microbiol*, 46(8): 657–663.

Warren, J.R. and Marshall, B.J. (1983). Unidentified curved bacilli on gastric epithelium in active chronic gastritis. *Lancet*, 1: 1273–1275.

Wesley, I.V., Schroeder-Tucker, L., Baetz, A.L. *et al.* (1995). *Arcobacter*-specific and *Arcobacter butlzeri*-specific 16S rRNA-based DNA probes. *J Clin Microbiol*, 33: 1691–1698.

West, A.P., Millar, M.R. and Tompkins, D.S. (1992). Effect of physical environment on survival of *Helicobacter pylori*. *J Clin Pathol*, 45(3): 228–231.

Whitaker, C.J., Dubiel, A.J. and Galpin, O.P. (1993). Social and geographical risk factors in *Helicobacter pylori* infection. *Epidemiol Infect*, 111: 63–70.

Wolf, S.G., Frenkiel, D., Arad, T. *et al.* (1999). DNA protection by stress-induced biocrystallization. *Nature*, 400: 83–85.

Wotherspoon, A.C. (1998). *Helicobacter pylori* infection and gastric lymphoma. *Br Med Bull*, 54(1): 79–85.

WQDC, Water Quality Disinfection Committee (1992). Survey of water utility disinfction practices. *JAWWA*, 82: 121–128.

Xia, H.H. and Talley, N.J. (1997). Natural acquisition and spontaneous elimination of *Helicobacter pylori* infection: Clinical implications. *Am J Gastroenterol*, 92(10): 1780–1787.

Xia, H.X., Keane, C.T. and O'Morain, C.A. (1994). Pre-formed urease activity of *Helicobacter pylori* as determined by a viable cell count technique: Clinical implications. *J Med Microbiol*, 40(6): 435–439.

Yamaguchi, H., Osaki T., Takahashi M. *et al.* (1999). Colony formation by *Helicobacter pylori* after long-term incubation under anaerobic conditions. *FEMS Microbiol Lett*, 175(1): 107–111.

Yoshimatsu, T., Shirai, M., Nagata, K. *et al.* (2000). Transmission of *Helicobacter pylori* from challanged to nonchallanged nude mice kept in a single cage. *Dig Dis Sci*, 45(9): 1747–1753.

Yoshiyama, H., Mizote, T., Nakamura, H. *et al.* (1998). Chemotaxis of *Helicobacter pylori*: A urease-independent response. *J Gastroenterol*, 33: 1–5.

Zago, A., Chugani, S. and Chakrabarty, A.M. (1999). Cloning and characterization of polyphosphate kinase and exopolyphosphatase genes from *Pseudomonas aeruginosa* 8830. *Appl Environ Microbiol*, 65(5): 2065–2071.

Zheng, P.Y., Hua, J., Ng, H.C. *et al.* (1999). Unchanged characteristics of *Helicobacter pylori* during its morphological conversion. *Microbios*, 98: 51–64.

# 9

# Other heterotrophic plate count bacteria
## (*Flavobacterium*, *Klebsiella*, *Pseudomonas*, *Serratia*, *Staphylococcus*)

## Basic microbiology

### Flavobacterium

The genus *Flavobacterium* are aerobic, Gram-negative, asporogenous bacilli that are non-motile, oxidase-positive, non-fermentative- and non-glucose oxidizing. They produce pigmented colonies (when grown at incubation temperature below 35°C) that may be yellow, orange, and red to brown. *Flavobacterium meningosepticum* is clinically the most significant of all *Flavobacterium*. It lives in moist environments, particularly the soil, and is frequently isolated from nebulizers. *Flavobacterium meningosepticum* is a common nosocomial infection in children and has been isolated from a number of meningitis outbreaks.

## Klebsiella

*Klebsiella* are lactose-fermenting, Gram-negative bacilli. *Klebsiella* are straight rods 1–2 μm long and 0.5–0.8 μm wide. They are non-motile and many strains are fimbriated. They grow best at temperatures between 12 and 43°C and are killed by moist heat at 55°C for 30 minutes. Under certain conditions they form a gelatinous encapsulation. They are facultatively anaerobic with poor growth in anaerobic conditions. This is particularly evident in the strain *Klebsiella pneumoniae*. Five species of *Klebsiella* including *K. pneumoniae*, *K. oxytoca*, *K. rhinoscleromatis*, *K. planticola* and *K. ozaenae* are known to be clinically significant.

Most *Klebsiella* strains are of environmental origin without significance to human health.

## Pseudomonas

Pseudomonads are aerobic, Gram-negative bacilli and non-spore forming. They are oxidase- and catalase-positive and motile by use of polar flagella. Of the 200 species that were originally present in the genus *Pseudomonas*, the most important medical strain is that of *Pseudomonas aeruginosa*. *Pseudomonas aeruginosa* is motile by means of one or two polar flagella and is frequently isolated from soil and water. *P. aeruginosa* produces two main types of soluble diffusible pigments, pyocyanin (blue phenazine) and, in low iron media, the fluorescent pigment pyoverdin (yellow-green). The latter is produced abundantly in media with a low-iron content functioning as a siderophore. Two additional pigments are also produced by *Pseudomonas aeruginosa*, pyorubrin (red) and melanin (brown).

*Pseudomonas aeruginosa* is renowned for its resistance to antibiotics. Its natural resistance to many antibiotics is principally due to the permeability barrier afforded by its outer membrane (lipopolysaccharides). As its natural habitat is the soil, living in association with the bacilli, actinomycetes and moulds, it has developed resistance to a variety of their naturally-occurring antibiotics. *Pseudomonas* also has antibiotic resistance plasmids and it also is able to transfer these genes to other bacteria. Aside from *Pseudomonas aeruginosa*, several other species of *Pseudomonas* have been associated with clinical specimens, including *Pseudomonas putida*, *Pseudomonas fluorescens* and *Pseudomonas stutzeri*.

To date many of the species that originally resided in the genus *Pseudomonas* now belong to new genera, namely *Burkholderia*, *Comamonas*, *Stenotrophomonas* and *Brevundimonas*. Of these new genera *Burkholderia pseudomallei* and *Burkholderia cepacia* have emerged as very important pathogens, particularly in immunocompromised patients.

## Serratia

The *Serratia* genus is composed of small, Gram-negative, motile, coccobacilli that characteristically give positive reactions for citrate, Voges-Proskauer and

ONPG. They are aerobic and facultatively anaerobic, catalase-positive but oxidase-negative. *Serratia* do not normally develop a capsule but capsular material has been shown to be formed when *Serratia* are grown in well-aerated medium with low levels of nitrogen and phosphate. *Serratia* ferment sugars including mannitol and trehalose often with the production of gas.

*Serratia marcescens* is the type species and produces a non-diffusible red pigment that is more pronounced on certain media at temperatures between 25 and 30°C. In the environment and clinical laboratories non-pigmented strains of *Serratia* are very common. As well as *Serratia marcescens* other *Serratia* have been isolated from the clinical environment and include *S. liquefaciens* and *S. odorifera*.

## Staphylococcus

Staphylococci are Gram-positive cocci. Under the microscope the cocci occur as single cells, in pairs, packet clusters, or as short chains of several individual cells. Species in this genus are usually non-motile, catalase-positive and ferment glucose. By far the most significant and pathogenic strain in the genus is *Staphylococcus aureus*. *Staphylococcus aureus* strains are coagulase-positive and often form a yellow to orange pigmentation on some media after 2–3 days of incubation. Some strains are encapsulated or form a slime layer aiding in their resistance to antimicrobial agents. It is well documented that while most species are facultative anaerobes, some strains of *Staphylococcus aureus* grow more favourably in aerobic environments.

In drinking water, *Staphylococcus aureus* are the potential pathogens of the genus. They are able to grow in the presence of 10% sodium chloride and can withstand high temperature changes. Of the coagulase-negative species of the genus *Staphylococcus*, *S. epidermidis* and *S. saprophyticus* have been associated with human infections.

## Origin of the organism

### Flavobacterium

The exact taxonomy of some of these organisms is still not resolved, principally because of atypical reactions in most biochemical identification media. Estimates of the number of *Flavobacterium* species presently known to date vary from 40 to 70. The genus comprises seven well-defined taxa that fall into three natural groups. The first group comprises three saccharolytic, indole-positive species; the second a single non-saccharolytic indole-negative species; and the third group three saccharolytic, indole-negative species. *Flavobacterium meningosepticum* is the species of major clinical significance.

## Klebsiella

The genus *Klebsiella* includes at least seven currently recognized species and 72 serotypes. Originally called Friedlander's bacillus (later called *K. pneumoniae*) the problem with *Klebsiella* was that it had to be distinguished from *Aerobacter aerogenes*, later called *K. aerogenes*. Historically many workers called *Klebsiella aerogenes K. pneumoniae*. A number of studies have concluded, from DNA-DNA hybridization studies, that the original *Klebsiella* species *aerogenes*, *ozaenae*, *pneumoniae* and *rhinoscleromatis* were in fact only biotypes of *K. pneumoniae*. Therefore it was acknowledged and in the interest of uniformity, that all subspecies of *Klebsiella* should be grouped into one species, namely *K. pneumoniae*.

## Pseudomonas

Until relatively recently *Pseudomonas aeruginosa* and *Pseudomonas fluorescens* were virtually the only species of *Pseudomonas* studied. *Pseudomonas aeruginosa* is of most significance and belongs to the bacterial family *Pseudomonadaceae*.

## Serratia

*Serratia marcescens* was the first *Serratia* to be identified and later, in 1948, *Serratia rubidaea* was described. This was found to have very similar characteristics to that of *Serratia marcescens*. Since that time other species of *Serratia* have been identified and often associated with clinical conditions, namely *S. ficaria* and *S. plymuthica*.

## Staphylococcus

Pasteur observed small spherical bacteria in the pus of furuncles and osteomyelitis and thought that these bacteria might be pathogenic.

The criteria used in the classification of *Staphylococcus* were based on both cell morphology and type of cell aggregation. Colony colour was the criterion for species classification. Staphylococci were suggested as being a group of saprophytic, tetrad-forming micrococci.

There are 32 species currently acknowledged in the genus *Staphylococcus*. *Staphylococcus* spp. are classified first of all on the basis of DNA-DNA hybridization as determined by the relative binding of DNAs in reassociation reactions conducted at non-restrictive (optimal), restrictive (stringent) conditions, or a combination of both (Kloos, 1980; Schleifer, 1986).

# Metabolism and physiology

## Flavobacterium

With *Flavobacterium* all carbohydrates are attacked slowly by oxidation or not at all. The type species is *F. aquatile*.

## Klebsiella

*Klebsiella* typically use citrate and give a negative methyl red and positive Voges-Proskauer reaction, although some strains have atypical biochemical reactions, such as fermenting glucose at 5°C or lactose fermentation at 44.5°C (faecal *Klebsiella*). An estimated 60–85% of all *Klebsiella* isolated from faeces and clinical specimens are positive in the faecal coliform test and identify as *K. pneumoniae*. As a consequence, classification of these bacteria into a clear-cut scheme has been difficult.

## Pseudomonas

Although the bacterium is respiratory and never fermentative, it will grow in the absence of $O_2$ if $NO_3$ is available as a respiratory electron acceptor. *P. aeruginosa* possesses the metabolic versatility for which pseudomonads are so renowned. Organic growth factors are not required, and it can use more than 30 organic compounds for growth. *Pseudomonas aeruginosa* is often observed growing in 'distilled water' and shampoos, evidence of its minimal nutritional requirements. While its optimum temperature for growth is 37°C, it is able to grow at temperatures as high as 42°C. It is able to tolerate a wide variety of physical conditions. *Pseudomonas aeruginosa* obtains energy from carbohydrates by an oxidative rather than a fermentative metabolism but, generally, it is only able to utilize glucose.

## Serratia

*Serratia* attack sugars fermentatively often with gas production. They are gluconate-positive and some strains produce orthithine decarboxylase.

## Staphylococcus

Staphylococci are capable of using a variety of carbohydrates as carbon and energy sources. These carbohydrates can be taken up, unmodified, and accumulated inside the cell or it can be modified during uptake. In most cases the carbohydrate is phosphorylated during the transport process mediated by the phosphoenolpyruvate (PEP)-carbohydrate phosphotransferase system (PTS). Glucose, mannose, glucosamine, fructose, lactose, galactose, mannitol, N-acetyl-glucosamine and beta-glucosides are taken up by the PTS (Reizer *et al.*, 1988).

The Embden-Meyerhof-Parnas (EMP, glycolytic) pathway and the oxidative hexose monophosphate pathway (HMP) are the two central routes used by staphylococci for glucose metabolism (Blumenthal, 1972). *S. aureus* and *S. epidermidis* metabolize glucose mainly by glycolysis. The major end product of anaerobic glucose metabolism in *S. aureus* is lactate (73–94%); smaller quantities of acetate (4–7%) and traces of pyruvate are also formed. Under aerobic conditions, only 5–10% of the glucose carbon appears as lactate and most of it appears as acetate and $CO_2$.

In *S. epidermidis*, lactate is the major end product of anaerobic glucose metabolism. The other end products are acetate, formate and $CO_2$ and these are formed in only trace amounts. *S. saccharolyticus* ferments glucose mainly to ethanol, acetic acid and $CO_2$ together with small amounts of lactic and formic acid (Kilpper-Bälz and Schleifer, 1981). In members of the *S. saprophyticus* species group and *S. intermedius*, *S. capitis*, *S. haemolyticus* and *S. warneri*, lactose and galactose are metabolized via the Leloir pathway, whereby galactose-1-phosphate is epimerized to glucose-1-phosphate (Schleifer, 1986). *S. aureus*, *S. epidermidis*, *S. hominis*, *S. chromogenes*, *S. sciuri* and *S. lentus* metabolize lactose and galactose via the tagatose-6-phosphate pathway, whereby galactose-6-phosphate is isomerized to tagatose-6-phosphate, which is further converted to tagatose-1,6-diphosphate and then cleaved to triosephosphates (Schleifer, 1986).

Most staphylococcal species are capable of synthesizing a large proportion of the different amino acids needed for growth. The amino acid requirements of approximately half the recognized species of staphylococci have been determined *in vitro* with the use of chemically defined media (Hussain *et al.*, 1991). Members of the *S. epidermidis* species group have numerous (c. 5–13) amino acid requirements. Most of the strains of species in this group also require isoleucine-valine and proline and, with the exception of *S. epidermidis*, most require histidine. *S. aureus* requires arginine and has either an absolute or partial requirement for proline. Most strains of this species also require isoleucine-valine and have either an absolute or partial requirement for cysteine and leucine. Different ecovars of *S. aureus* demonstrate some differences in their amino acid requirements (Tschäpe, 1973).

Members of the *S. saprophyticus* species require fewer amino acids than the other species and some do not even require an amino acid or an organic nitrogen source. Some strains of *S. saprophyticus* have an absolute or partial requirement for proline and isoleucine-valine.

## Clinical features

Most species in the genera found in heterotrophic plate count (HPC) bacteria have only very rarely caused disease in humans. *Pseudomonas aeruginosa*, however, has been frequently associated with diseases such as urinary tract infections, respiratory disease, and ear and eye infections. These have also been associated with bacteraemia, osteomyelitis and meningitis (Pollack, 2000).

*Acinetobacter* spp., an opportunist pathogen in hospital patients, is known to cause bacteraemia and respiratory infections (Allen and Hartman, 2000).

*Chryseobacterium (Flavobacterium) meningosepticum, Stenotrophomonas maltophilia* and *Pseudomonas paucimobilis* have also been associated with diseases in the hospital setting.

## Flavobacterium

Clinically, the most important species of genus *Flavobacterium* are *F. meningosepticum, F. breve* and *F. odoratum. F. meningosepticum* is the species most frequently involved as an opportunistic pathogen in nosocomial infections, including meningitis (particularly in infants), pneumonia, endocarditis and septicaemia.

## Klebsiella

*Klebsiella* are a common cause of urinary tract infections, sometimes giving rise to bronchopneumonia, sometimes with chronic destructive lesions and multiple abscesses. They are occasionally associated with bacteraemia often with a high mortality rate and are a major cause of nosocomial infections.

Klebsiella strains that are responsible for most infections are generally *K. pneumoniae* being associated with clinical sepsis, particularly in surgical wounds and urinary tract infections. Colonization in the respiratory tract is a very common concern, particularly in those who are receiving antibiotics. *K. pneumoniae* subspecies, *ozaenae* and *rhinoscleromatis*, cause upper respiratory tract disease (*rhinoscleromatis*). This seems to be particularly prevalent in Eastern Europe. Infections caused by *Klebsiella* suggest it to be the primary aetiological agent but more often it is found in mixed infection or as an opportunistic invader.

## Pseudomonas

*Pseudomonas aeruginosa* became apparent after the 1950s as a very important organism associated with a wide range of infections. *Pseudomonas aeruginosa* is the major cause of hospital acquired (nosocomial) infections. It infects mainly immunocompromised individuals, burn victims, and individuals on respirators or with indwelling catheters. *Pseudomonas aeruginosa* also colonizes the lungs of cystic fibrosis patients, increasing the mortality rate of individuals with the disease. Infections can lead to sepsis, pneumonia, pharyngitis, and many other problems. Rarely will *Pseudomonas* be a cause of infection in healthy individuals.

## Serratia

*S. marcescens*, *S. odorifera*, *S. rubidaea*, and the subgroup *S. liquefaciens* are recognized as potential opportunistic pathogens that may spread in epidemic proportions causing nosocomial infections in hospital patients. *S. marcescens* is a frequent cause of infections ranging from cystitis to life-threatening bloodstream and central nervous system infections. *Serratia marcescens* may cause occasional infections, particularly in children. *S. ficaria*, normally associated with figs, has been isolated from a respiratory specimen and from a leg ulcer and *S. plymuthica* has been isolated from a burn site.

## Staphylococcus

*S. aureus*, *S. epidermidis* and *S. saprophyticus* are the opportunistic pathogens associated with infections of the skin (cellulitis, pustules, boils, carbuncles and impetigo), bacteraemia, peritonitis associated with dialysis, genitourinary infections, and postoperative wound infections. *S. aureus* is also a cause of meningitis, osteomyelitis, and violent diarrhoea and vomiting from ingestion of the enterotoxin caused by the organism growing in food. Infections caused by this species are often acute and pyogenic and, if untreated, may spread to surrounding tissue or to metastatic sites involving other organs.

S. aureus is a common aetiological agent of postoperative wound infections, bacteraemia, pneumonia, osteomyelitis, acute endocarditis, mastitis, toxic shock syndrome and abscesses of the muscle, urogenital tract, central nervous system and various intra-abdominal organs. Food poisoning is frequently attributed to staphylococcal enterotoxin.

S. epidermidis and other members of the *S. epidermidis* species group are commonly associated with hospital-acquired bacteraemia, especially in patients in intensive therapy units. In most cases, the focus of infection is an intravascular catheter.

The *Staphylococcus* species of relevance to water is *S. aureus*. Concentrations in drinking water can be a health concern for individuals in contact with water for extended periods, such as from dishwashing, whirlpool therapy, and dental hygiene. Densities of 200–400 cocci/ml have been shown to set up a carrier state in the nose of 50% of newborn infants, and an *S. aureus* density of a few hundred cells per ml in a water contact may induce infection in traumatized skin.

## Pathogenicity and virulence

### Klebsiella

By far the most significant *Klebsiella* is the capsulated strain *Klebsiella pnemoniae*. Virulence, however, does not appear to depend on capsule formation.

In a study of 94 hospitals, the infection r... [text obscured]
for pathogenic *K. pneumoniae* was the caus... [text obscured]
deaths. Infections of the urinary system, lower resp... [text obscured]
wounds were the most frequent cause of *Klebsiella*-associ... [text obscured]

It is documented that between 30 and 40% of all warm-b... [text obscured]
have *Klebsiella* in their intestinal tract. *K. planticola* and *K. terrig...* [text obscured]
their origins in the environment, being found on fruit and vegetables, d... [text obscured]
products, seed embryos, internal and external tree tissues, hay and cotton. [text obscured]

*Klebsiella* strains possess a high affinity iron uptake system, one employing aerobactin the other, enterochelin. *Klebsiella* also show resistance to complement-mediated serum killing and phagocytosis if both the K and O antigens are present. The virulence of *K. pneumoniae* in animal models varies considerably and does not appear to depend on capsule formation.

## Pseudomonas

*Pseudomonas aeruginosa* is able to infect both internal and external sites. It has been associated with many superficial and mild infections such as otitis externa. Most strains of *Pseudomonas aeruginosa* produce two exotoxins, exotoxin A and exoenzyme S, as well as a variety of cytotoxins including proteases, phospholipases, rhamnolipids and also a pycocyanin. All these virulence factors depend on the site and the nature of infection. For example, the proteases are particularly important in corneal infections; exotoxin and proteases are important in burn infections; phospholipases, proteases and aliginate are important in chronic pulmonary colonization. Production of fluorescein is important as it allows *Pseudomonas aeruginosa* to compete with human iron-binding proteins.

## Serratia

Pigmented forms of *S. marcescens* as well as non-pigmented forms are found occasionally in the human respiratory tract and faeces. Most infections occur in hospitals causing urinary tract infections, respiratory tract infections, meningitis, wound infections, septicaemia and endocarditis. Some strains are endemic, particularly in hospitals, as strains are often able to multiply at room temperature.

## Staphylococcus

*Staphylococus aureus* possess a large collection of virulence mechanisms which are important to overcome the body's defences and also invade, survive and adhere to tissues. On average, over 60% of all *Staphylococcus aureus* strains produce at least five enterotoxins (A–E). These toxins are often produced alone or in combination and are found to be very resistant to heat. They

.ucing symptoms of
Staphylococcal toxic
*Staphylococcus aureus*
.ns in that they are potent
response with the release of
factor. Certain of the cell sur-
cellular proteins produced by S.
.n the pathogenicity of this organ-
.ssociated with intraepidermal blis-
.iildren's nurseries and nursing homes.
oxins is a condition known as scalded
.tering lesions occur resulting in a look of

Clinica.        *Klebsiella* are naturally resistant to ampicllin, amoxycillin and most othei        .illins. They are generally found to be sensitive to cephalosporins, gentamicin and other aminoglycosides. With infections of the urinary tract caused by *Klebsiella*, trimethoprim, nitrofurantoin, co-amoxiclav or oral cephalosporins are administered. In cases of *Klebisella pneumoniae* treatment with an aminoglycoside or a cephalosporin, such as cefotaxime, is required.

## Pseudomonas

Only a few antibiotics are effective against *Pseudomonas*. These have included fluoroquinolone, gentamicin and imipenem, but even these antibiotics are not effective against all strains. *Pseudomonas* infections associated with cystic fibrosis patients become very difficult to treat because of the high prevalence of resistant strains. Until about 1960 all pseudomonas infections were treated with polymyxins. Now treatment with gentamicin and tobramycin combined with an aminoglycoside and a beta-lactam is commonly adopted. In general ciprofloxacin shows good activity against *Pseudomonas aeruginosa*.

## Serratia

*Serratia* are commonly resistant to cephalosporins with variable resistance to ampicillin and gentamicin documented. Generally, an aminoglycoside such as gentamicin is usually the most reliable first-line defence for treating infections caused by *Serratia*. In recalcitrant cases fluoroquinolones or carbapenems are often used.

## Staphylococcus

*Staphylococcus aureus* are inherently sensitive to many antibiotics. However, in the hospital setting 90% of *Staphylococcus aureus* are resistant to benzylpenicillin. This is brought about by the production of the enzyme pencillinase which inactivates penicillin. It is methicillin-resistant *Staphylococcus aureus* (MRSA) that has become a major problem, particularly in the hospital setting. Vancomycin is being used to treat infections caused by MRSA but vancomycin-resistant *Staphylococcus aureus* (VRSA) are now being documented, with the first one isolated in the USA in June 2002. The choice of antibiotic for treating *Staphylococcus aureus* infections should principally be based on the results of sensitivity tests. Treatment initially should, in cases of severe infection, begin with flucloxacillin, unless MRSA is highly endemic, in which case, a glycopeptide should be used. In general terms, if the staphylococcus causing the infection is sensitive to penicillin, benzylpenicillin should be used but with patients who are hypersensitive to this then erythromycin, clindamycin or vancomycin should be used.

## Survival in the environment and water

### Flavobacterium

In the aquatic environment, flavobacteria are ubiquitous, being found in soil, water, sewage, vegetation and dairy products. Soil and water appear to be significant reservoirs for *Flavobacterium* with evidence that some are chlorine-resistant. Stagnation of building plumbing systems can provide opportunities for *Flavobacteri*um to adhere to water pipes, particularly in hospital water systems. *Flavobacterium* has been introduced into drinking water through microbial growth on devices connected to the water supply. On the availability of current evidence it is feasible to suggest that the route of exposure of *Flavobacterium* may be by ingestion, body contact with water supply, or by person-to-person contact in hospitals.

Current evidence suggests that conditions contributing to *Flavobacterium* regrowth in drinking water include absence of free chlorine residual, water temperatures above 15°C, accumulations of bacterial nutrients in pipe sediments, and static water conditions.

### Klebsiella

Surface water and unprotected groundwater receive *Klebsiella* from both environmental and faecal sources. *Klebsiella* has been shown to survive for 20 days in laboratory 'pure water'. Environmental strains are introduced to source water from urban and rural runoff and by discharges from industrial waters. Faecal *Klebsiella* enter the water cycle from municipal sewage and meat processing works and source discharge from farm animal waste runoff.

*Klebsiella* organisms are included in the national and international guidelines for total coliform occurrences. Generally, these standards limit total coliforms to less than one organism or their absence in 100 ml of treated drinking water. World Health Organization guidelines have established a limit of less than 10 total coliforms per 100 ml of untreated groundwater, provided no faecal coliforms are detected in the sample.

*Klebsiella* is often transmitted via body contact with water supply during bathing, ingestion or by person-to-person contact through poor hand-washing habits in hospitals but, more specifically, in senior citizen care institutions. Inhalation of moisture from vaporizers using drinking water contaminated with *Klebsiella* should also be considered a risk to some individuals. No community waterborne outbreaks have been reported to be caused by *Klebsiella* in public water supplies. Most of the *Klebsiella* waterborne occurrences do not involve faecal strains. In those infrequent situations in which the laboratory analyses reveal faecal *Klebsiella* in the distribution system, the colonization sites must be destroyed to avoid more frequent releases of this opportunistic pathogen at higher densities into the water supply. Infective dose (ID50) values for environmental and clinical isolates of *Klebsiella* have been reported to be between $3.5 \times 10$ and $7.9 \times 10^5$ cells/ml. Therefore, ingestion of 100 ml of drinking water (approximately one glass of water) containing $3.5 \times 10$ *Klebsiella* per ml could present a risk to susceptible individuals.

Klebsiellae can be controlled effectively by adequate disinfection in a clean pipe environment – these organisms can be protected by particulate material, porous pipe sediments, biological debris, macroinvertebrates and disinfection demand products. Furthermore, *Klebsiella* can encapsulate, which provides some protection from disinfectants.

## Pseudomonas

Surveys undertaken within potable water supplies have shown *Pseudomonas* to constitute around 2–3% of the total heterotrophic plate count and is responsible for some 10% of nosocomial infections. A large number of *Pseudomonas* species are often isolated from potable water. These have included *Pseudomonas aeruginosa, fluorescens, alcaligenes, mendocina, putida, cepacia, allei, maltophila, testosteroni, vescularis, flava, pseudoflava, palleroni, rhodos, echinoides, radiora* and *mesophilica*. Recreational and occupational infections are associated with pseudomonas, which include Jacuzzi or whirlpool rashes.

Molecular subtyping of isolates of *P. aeruginosa* from patients and water have proven that water is the source of many *Pseudomonas* infections. *Pseudomonas aeruginosa* is usually found in water that has been contaminated with faecal material such as surface waters. It is excreted in the faeces of many healthy adults. The occasional occurrence of *Pseudomonas aeruginosa* in drinking water often indicates deterioration of the quality of water.

It is a typical biofilm bacterium being isolated from many materials that are in contact with water.

## Serratia

*Serratia* are considered to be ubiquitous in the environment. They can be found in surface and groundwater, soil, decaying vegetation, insects, decaying meat and spoiled milk. *Serratia* strains are spread by person-to-person contact and by contaminated water from sumps in hospital equipment. *S. marcescens* infections have also been transmitted via medical solutions and peritoneal-dialysis effluents.

*Serratia* appear to occur seasonally in high-quality waters (such as private wells, distribution systems, finished reservoir supplies, and bottled water). Colonization may occur in a variety of attachment devices, including drinking water fountains, ice machines, point-of-use treated water, laboratory high quality water systems, humidifying units and haemodialysis equipment. *Serratia* have also been found in the heterotrophic bacterial population on media substrate in granular activated carbon (GAC) filters. Densities in water are variable, most often being less than 100 organisms/ml unless predominate colonization occurs in a biofilm. The persistence of *Serratia* in tap water is about 100 days, and much longer in contaminated well water. In distilled water, *Serratia* may survive for 48 days at room temperature.

## Staphylococcus

Staphylococci have been detected in the pharynx, conjunctiva, mouth, blood, mammary glands, faeces, bodily discharges, excretions and intestinal, genitourinary and respiratory tracts of their hosts. Small numbers of staphylococci have been found in air, dust, soil and water, and on inanimate surfaces or fomites, molluscs, insects and plants in areas frequented by mammals and birds. Is it any wonder the major reservoirs of staphylococci include warm-blooded animals (in the skin, nose, ear, and mucous membranes), sewage and stormwater runoff?

The major concern with *S. aureus* in water transmission is contact with cuts and scratches on the skin, infecting the ears or the eyes during bathing, or in water used to prepare uncooked foods. Ingestion is the pathway for gastrointestinal infections from contaminated foods, or for individuals on intensive antibiotic therapy. The density of coagulase-positive *Staphylococcus* (i.e. *S. aureus*) ranges from $10^2$ to $10^4$/g in the normal human faecal flora, but occurrence is quite variable (10–93%). Coagulase-negative strains of *Staphylococcus* conversely may be found in 31–59% of faeces from healthy people. It is not surprising then that *Staphylococcus* are the most numerous bacteria shed by swimmers in natural bathing waters and chlorinated swimming pools. In one study, approximately two-thirds of the staphylococci in bathing waters were *S. aureus*. Densities of *S. aureus* in both studies ranged from a few to several hundred cells/100 ml. *S. aureus* was found to be one of the two most concentrated opportunistic pathogens in urban stormwater, ranging from 10 to 1000 organisms/ml. In private water supplies *S. aureus* is present in quite high concentrations often as high as 400 organisms/ml. Private water supplies constitute a very important concern in water-related diseases.

*S. aureus* in the hospital environment and water supplies are generally classed as opportunistic pathogens, however, they may play a subordinate role in disease. No waterborne outbreaks have been caused by *Staphylococccus aureus*. However, the risk of staphylococci infection acquired from poor quality small water systems and private water has been documented. In drinking water, *Staphylococcus* may persist at 20°C for 20–30 days, provided trace amounts of organic nutrients are available. Growth in water is slow at temperatures below 20°C, and merely at subsistence rate below 10°C.

## Methods of detection

### Flavobacterium

*Flavobacterium* can be recovered easily from water with R2A spread plates incubated between 22 and 28°C for at least 7 days. Following growth all pigmented colonies should be tested using conventional biochemical tests.

### Klebsiella

Biochemical characteristics are used in *Klebsiella* speciation when purified on M-Endo agar from water. Other differential media can be used for the isolation of *Klebsiella*, such as M-Kleb agar, often in connection with the membrane filtration. On this agar *Klebsiella* form dark blue to dark grey colonies.

### Pseudomonas

*Pseudomonas aeruginosa* grow on most common culture media. For culture from soil or water a selective medium containing acetamide as the sole carbon source and nitrogen source should be used. Formal identification can be done using biochemical tests. *P. aeruginosa* isolates may produce a number of different colony types depending on the environmental selective pressure. Isolates of *Pseudomonas* identified from soil or water classically produce a small, rough colony. As a comparison, isolates from clinical samples yield one or another of two smooth colony types. One type has a fried-egg appearance, which is large, smooth, with flat edges, and an elevated appearance. Another type, frequently obtained from respiratory and urinary tract secretions, has a mucoid appearance, which is attributed to the production of alginate slime.

It is documented that the smooth and mucoid colonies play a role in colonization and virulence.

## Serratia

*Serratia* in water samples can be isolated on R2A agar when incubated at temperatures between 20 and 30°C using either the membrane filtration (MF) procedure or a spread plate technique. Colonies of *S. marcescens* are generally homogeneous for the first day or two and then often become convex, pigmented with a relatively opaque centre and an effuse, colourless, almost transparent periphery with an irregular crenated edge. Pigment is only formed in the presence of oxygen and at an appropriate temperature. The red pigment (prodigiosin) is soluble in alcohol (absolute) but insoluble in water.

## Staphylococcus

Pigmented *Staphylococcus* colonies can be seen on heterotrophic plate count (HPC) cultures performed on samples taken from the water distribution system. No specific media are available for the detection of *Staphylococcus* from water. Some success has been reported for the use of M-staphylococcus broth in a modified multiple-tube procedure or Baird-Parker agar in the membrane filter (MF) procedure. Presumptive results are verified by placing pure cultures from the MF into a commercial multi-test system for biochemical reactions. Any turbid tubes in the multiple-tube test are confirmed in Lipoviettin-salt mannitol agar streak plates.

Staphylococci produce distinctive colonies on a variety of commercial, selective and non-selective agar media. The commonly used selective media in the medical setting include mannitol-salt agar, lipase-salt-mannitol agar, phenylethyl alcohol agar, Columbia colistin-nalidixic acid (CNA) agar, and Baird-Parker agar base supplemented with egg yolk tellurite enrichment. These media inhibit the growth of Gram-negative bacteria, but allow the growth of staphylococci and certain other Gram-positive bacteria. Baird-Parker agar when supplemented with egg yolk tellurite enrichment is widely recommended for the detection and enumeration of *S. aureus* species that share some properties with *S. aureus*, it is now recognized that such a distinction. Schleifer-Krämer agar is used by some laboratories for the selective isolation and enumeration of staphylococci from foods and other heavily contaminated sources. Incubation of cultures on selective media should be for at least 48–72 hours at 35–37°C for colony development. Many of the staphylococcal species can produce abundant anaerobic growth in a semi-solid Brewer's thioglycollate medium within 24–72 hours at 35–37°C.

Staphylococci from a variety of clinical specimens are usually isolated in primary culture on blood agar (e.g. tryptic soy agar supplemented with 5% sheep blood), following an incubation period of 18–24 hours at 35–37°C. In this short time, most staphylococcal colonies will be 1–3 mm in diameter, circular, smooth and raised, with a butyrous consistency.

All staphylococcal species grow well on tryptic soy agar. Some strains, however, should be cultured on tryptic soy agar or tryptic soy agar supplemented with blood. Brain-heart infusion agar and nutrient agar support good growth of staphylococci, although not many species comparisons have been made of colonies growing on these media.

## Epidemiology and waterborne outbreaks

Epidemiological studies showing the association between HPC and human health effects are scarce (Ferley *et al.*, 1986; Zmirou *et al.*, 1987). A study by Payment has found an association between illness and total counts of bacteria using a randomized controlled trial (Payment *et al.*, 1991a). From this study it was found that there was an association between gastrointestinal illnesses and heterotrophic plate counts at 35°C. An association between HPC bacteria and human illness was also found in a second trial (Payment *et al.*, 1997). This study demonstrated no association between counts of HPC bacteria and gastrointestinal illness in humans.

Overall, from the available literature to date there is no evidence that there is an association between HPC counts in drinking water and disease in humans. There have, however, been a number of outbreaks of disease linked to water supplies. These have included an outbreak of multiresistant *Chryseobacterium (Flavobacterium) meningosepticum*, which affected eight neonates on a neonatal intensive care unit (Hoque *et al.*, 2001). *Pseudomonas aeruginosa* genotypes have been detectable in tap water causing infections in hospitals (Bert *et al.*, 1998; Ferroni *et al.*, 1998; Trautmann *et al.*, 2001). *Stenotrophomonas maltophilia* was cultured from endotracheal aspirate samples from five preterm infants in a neonatal intensive care unit of whom four were colonized and one died from septicaemia (Verweij *et al.*, 1998). Six patients in an intensive care unit (ICU) were colonized or infected with *Pseudomonas paucimobilis* (Crane *et al.*, 1981) which were subsequently recovered from the ICU hot water line.

Although many HPC bacteria have been associated infrequently with disease no epidemiological study has demonstrated an association with drinking water in the community. Severely ill hospitalized patients, such as those infected with AIDS, can acquire *Mycobacterium avium* complex from hospital water systems.

## Risk assessment

*Health effects*: occurrence of illness, degree of morbidity and mortality, probability of illness based on infection:

- No studies have shown a relationship between gastrointestinal illness and HPC bacteria in drinking water.

- Certain *Flavobacterium* species are associated with nosocomial infections, including meningitis, pneumonia, endocarditis and septicaemia.
- *Klebsiella* are a common cause of urinary tract infections. They are occasionally associated with bacteraemia often with a high mortality rate and are a major cause of nosocomial infections.
- *Pseudomonas aeruginosa* is an important organism associated with a wide range of infections. *Pseudomonas aeruginosa* is the major cause of hospital-acquired infections in the very ill. Infections can lead to sepsis, pneumonia, pharyngitis, and other problems. Rarely will *Pseudomonas* cause infection in healthy individuals.
- *Serratia* spp. are recognized as potential opportunistic pathogens that may spread in epidemic proportions causing nosocomial infections in hospital patients. *S. marcescens* is a frequent cause of infections ranging from cystitis to life-threatening bloodstream and central nervous system infections.
- *Staphylococcus aureus*, *S. epidermidis* and *S. saprophyticus* are opportunistic pathogens associated with infections of the skin, bacteraemia, peritonitis associated with dialysis, genitourinary infections and postoperative wound infections. *S. aureus* is also a cause of meningitis, osteomyelitis and violent diarrhoea and vomiting from ingestion of the enterotoxin caused by the organism growing in food. The *Staphylococcus* species of relevance to water is *S. aureus*. Concentrations in drinking water can be a health concern for individuals in contact with water for extended periods, such as from dishwashing, whirlpool therapy, and dental hygiene.
- With the possible exception of *P. aeruginosa* urinary tract infections, cases of infection due to HPC bacteria outside of hospital are very rare. Even within hospitals many HPC bacteria are only very occasionally associated with disease.

*Exposure assessment*: routes of exposure and transmission, occurrence in source water, environmental fate:

- HPC bacteria are generally found plentifully in the environment and frequently colonize the bodies of healthy people, who excrete organisms in their faeces.
- Exposure through water is typically from contact with cuts and scratches on the skin rather than ingestion, though transmission may come from ingestion, contact, or inhalation.
- Some HPC can colonize water pipes and systems like drinking fountains, humidifying units and whirlpool spas.
- Many HPC species colonize biofilms, where they can live for long periods in contact with water.
- Soil and water appear to be significant reservoirs for *Flavobacterium* with evidence some are chlorine-resistant. Current evidence suggests that conditions contributing to *Flavobacterium* regrowth in drinking water include absence of free chlorine residual, water temperatures above 15°C, accumulations of bacterial nutrients in pipe sediments and static water conditions.

- The persistence of *Serratia* in tap water is about 100 days and much longer in contaminated well water. In distilled water, *Serratia* may survive for 48 days at room temperature.
- In drinking water, *Staphylococcus* may persist at 20°C for 20–30 days, provided trace amounts of organic nutrients are available. Growth in water is slow at temperatures below 20°C, and merely at subsistence rate below 10°C.

*Risk mitigation*: drinking-water treatment, medical treatment:

- Traditional water treatment with coagulation, sedimentation, and disinfection should handle HPC bacteria in drinking water sources; however, they can regrow in the absence of residual disinfection or in treated water that contains enough nutrients for growth. Bacteria in biofilm are more difficult to treat, but maintaining adequate residual throughout the distribution system, reducing nutrients available for regrowth, and regular flushing of distribution lines should control HPC.
- HPC bacteria are often resistant to antibiotics:
  - Clinical isolates of *Klebsiella* are generally sensitive to cephalosporins, gentamicin and other aminoglycosides.
  - Only a few antibiotics are effective against *Pseudomonas*. These have included fluoroquinolone, gentamicin and imipenem.
  - Generally an aminoglycoside, such as gentamicin, is usually reliable in treating infections caused by *Serratia*. In recalcitrant cases fluoroquinolones or carbapenems are used.
  - The choice of antibiotic for treating *Staphylococcus aureus* infections should principally be based on the results of sensitivity tests. Benzylpenicillin should be used if the isolate is sensitive. Antibiotic-resistant strains of *Staphylococcus* have become a major health concern.

# References

Allen, D.M. and Hartman, B.J. (2000). *Acinetobacter* species. In *Principles and Practice of Infectious Diseases*, 5th edn, Mandell, G.L., Bennett, J.E. and Dolin, R. (eds). Philadelphia: Churchill Livingstone, pp. 2339–2342.

Bert, F., Maubec, E., Bruneau, B. *et al.* (1998). Multi-resistant *Pseudomonas aeruginosa* outbreak associated with contaminated tap water in a neurosurgery intensive care unit. *J Hosp Infect*, **39**: 53–62.

Blumenthal, H.J. (1972). Glucose catabolism in staphylococci. In *The Staphylococci*, Cohen, J.O. (ed.). New York: John Wiley, pp. 111–135.

Crane, L.R., Tagle, L.C. and Palutke, W.A. (1981). Outbreak of *Pseudomonas paucimobilis* in an intensive care facility. *J Am Med Assoc*, **246**: 985–987.

Ferley, J.P., Zmirou, D., Collin, J.F. *et al.* (1986). Etude longitudinale des risques liés à la consommation d'eaux non conformes aux normes bactériologiques. *Rev Epidemiol Sante Publ*, **34**: 89–99.

Ferroni, A., Nguyen, L., Pron, B. *et al.* (1998). Outbreak of nosocomial urinary tract infections due to *Pseudomonas aeruginosa* in a paediatric surgical unit associated with tap-water contamination. *J Hosp Infect*, **39**: 301–307.

Hoque, S.N., Graham, J., Kaufmann, M.E. *et al.* (2001). *Chryseobacterium (Flavobacterium) meningosepticum* outbreak associated with colonization of water taps in a neonatal intensive care unit. *J Hosp Infect*, **47**: 188–192.

Hussain, M., Hastings, J.G.M. and White, P.J. (1991). Isolation and composition of the extracellular slime made by coagulase-negative staphylococci in a chemically defined medium. *J Infect Dis*, **163**: 534–541.

Kilpper-Bälz, R. and Schleifer, K.H. (1981). Transfer of *Peptococcus saccharolyticus* Foubert and Douglas to the genus Staphylococcus: *Staphylococcus saccharolyticus* (Foubert and Douglas) comb. nov. *Zentralbl Bakteriol Mikrobiol Hyg Abt 1 Orig C*, **2**: 324–331.

Kloos, W.E. (1980). Natural populations of the genus Staphylococcus. *Annu Rev Microbiol*, **34**: 559–592.

Payment, P., Franco, E., Richardson, L. *et al.* (1991a). Gastrointestinal health effects associated with the consumption of drinking water produced by point-of-use domestic reverse-osmosis filtration units. *Appl Environ Microbiol*, **57**: 945–948.

Payment, P., Siemiatycki, J., Richardson, L. *et al.* (1997). A prospective epidemiological study of gastrointestinal health effects due to the consumption of drinking water. *Int J Environ Hlth Res*, 7: 5–31.

Pollack, M. (2000). *Pseudomonas aeruginosa*. In *Principles and Practice of Infectious Diseases*, 5th edn, Mandell, G.L., Bennett, J.E. and Dolin, R. (eds). Philadelphia: Churchill Livingstone, pp. 2310–2335.

Reizer, J., Saier, M.H. Jr *et al.* (1988). The phosphoenolpyruvate:sugar phosphotransferase system in Gram-positive bacteria: properties, mechanisms, and regulation. *Crit Rev Microbiol*, **15**: 297–338.

Schleifer, K.H. (1986). Taxonomy of coagulase-negative staphylococci. In *Coagulase-negative Staphylococci*, Mårdh, P.-A. and Schleifer, K.H. (eds). Stockholm: Almqvist & Wiksell International, pp. 11–26.

Trautmann, M., Michalsky, T., Wiedeck, H. *et al.* (2001). Tap water colonization with *Pseudomonas aeruginosa* in a surgical intensive care unit (ICU) and relation to Pseudomonas infections of ICU patients. *Infect Control Hosp Epidemiol*, **22**: 49–52.

Tschäpe, H. (1973). Genetic studies on nutrient markers and their taxonomic importance. In *Staphylococci and Staphylococcal Infections*, Jeljaszewicz, J. (ed.). Warsaw: Polish Medical Publishers, pp. 57–62.

Verweij, P.E., Meis, J.F., Christmann, V. *et al.* (1998). Nosocomial outbreak of colonization and infection with *Stenotrophomonas maltophilia* in preterm infants associated with contaminated tap water. *Epidemiol Infect*, **120**: 251–256.

Zmirou, D., Ferley, J.P., Collin, J.F. *et al.* (1987). A follow-up study of gastro-intestinal diseases related to bacteriologically substandard drinking water. *Am J Public Hlth*, **77**: 582–584.

# 10

# *Legionella*

## Basic microbiology

Legionellae stain very poorly as Gram-negative bacteria. They are catalase-positive, and are not able to reduce nitrate. They also do not utilize carbohydrates by either oxidation or fermentation. They are short, rod-shaped bacteria, often coccobacillary and non-spore-forming. The rods of free-living *Legionella* are irregular with non-parallel sides approximately 0.3–0.9 µm wide and 1–3 µm long (Rodgers *et al.*, 1978). *In vitro Legionella* will grow to 2–6 µm, but can form even longer filamentous forms in old cultures. All *Legionella* are motile except *L. oakridgensis*, *L. londinensis* and *L. nautarum*. This motility is facilitated by one or more polar or subpolar unsheathed flagella (Chandler *et al.*, 1980).

*Legionella* are strict aerobes. For their isolation they require an enriched agar supplemented with L-cysteine and ferric salts. Optimally they grow at 35°C.

In water and the environment *Legionella* require the presence of other bacteria or protozoa in order to grow (Wadowsky and Yee, 1985). Protozoa known to support the growth of legionellae include species of *Vahlkampfia*, *Hartmanella*, *Acanthamoeba*, *Naegleria*, *Echinamoeba* and *Tetrahymena* (Fields, 1993). The growth of *Legionella* in the environment in the absence of protozoa has not been demonstrated which suggests that protozoa are the natural reservoir for *Legionella* (Fields, 1993). It is possible that the biofilm may be an area where *Legionella* may multiply. This, however, needs further investigation. The growth of *Legionella pneumophila* also occurs in association with cyanobacteria (Tison *et al.*, 1980).

## Origin and taxonomy of the organism

It was in 1943 that the first strains of *Legionella* were isolated from guinea-pigs. In 1954, a bacterium was isolated from free-living amoebae which was not, however, classified as a species of *Legionella* until 1996 (Hookey *et al.*, 1996). The genus *Legionella* and the strain *pneumophila* were discovered following an outbreak, in 1976, during a convention of the Philadelphia Branch of the American Legion. In this outbreak 184 attendees fell ill with an apparent pneumonia. Of these 29 died. Over the next few years, the organism was identified as the cause of a number of other large outbreaks of similar illnesses. Subsequent serological studies from previous outbreaks of a similar nature allowed retrospective diagnosis, confirming that the isolate causing the pneumonia was *Legionella pneumophila*.

*Legionella* is the sole genus of the family Legionellaceae. It is composed of many species and serogroups and following 16S rRNA analysis it now belongs to the gamma-2 subgroup of the class Proteobacteria. To date, the group *Legionella* now contains 48 species consisting of 70 serogroups.

## Metabolism and physiology

Legionellae have an absolute requirement for iron and utilize amino acids for energy rather than carbohydrates. The essential amino acids utilized by *Legionella* include arginine, cysteine, methionine, serine, threonine and valine. A number of strains of *Legionella* also require isoleucine, leucine, phenylalanine and tyrosine for growth. Most species of *Legionella* are beta-lactamase positive and are able to liquefy gelatine, with the only exception being *L. micdadei*. Most species of *Legionella* do not hydrolyse hippurate, the exception being *L. pneumophila* (except serogroups 4 and 15). *Legionella* are superoxide dismutase-positive, weakly peroxidase-positive and possess the cytochromes a–d.

## Clinical features

*Legionella* has been detected in humans, guinea-pigs, rats, mice, marmosets and monkeys. Infection with *Legionella* presents as two distinct entities,

Legionnaires' disease and Pontiac fever. Legionnaires' disease, which has a high fatality rate, produces pneumonia and also affects the nervous, gastrointestinal and urinary systems (Mayock *et al.*, 1983). Pontiac fever, on the other hand, is a mild non-pneumonic, self-limited flu-like illness. One to 15% of *Legionella* can be community-acquired with up to 50% hospital-acquired (Butler and Breiman, 1998). The incubation period for Legionnaires' disease is approximately 2–14 days. The infection has, however, been documented as lasting several months.

Symptoms of *Legionella* infection include pneumonia, myalgia, malaise and headache. Other symptoms include fever and chills, a cough, chest pain and diarrhoea. The symptoms of Pontiac fever include chills, fever, myalgia and headache. The incubation period for this disease is 5–66 hours and can last up to 7 days.

Smoking and alcoholism are commonly acknowledged to be predisposing factors of acquiring *Legionella*, with immunocompromised individuals at a very high risk of developing disease. It is also well documented that infection is more common in males than females and individuals who are over 40 years of age (World Health Organization, 1990). Other risk factors have been documented including people receiving steroid treatments, patients with chronic lung disease and diabetes.

*L. pneumophila* serogroup 1 is the most commonly isolated serotype of *L. pneumophila* from infected individuals, followed by *L. pneumophila* serogroup 6 (Tang and Krishnan, 1993). *L. micdadei* is the second most frequent cause of Legionnaires' disease in the USA (Goldberg *et al.*, 1989) with *L. longbeachae* an important cause of Legionnaires' disease in Australia (Steele *et al.*, 1990).

While 15 serotypes of *L. pneumophila* are known to exist, over 70% of all culture confirmed are caused by *L. pneumophila* serogroup 1.

## Pathogenicity and virulence

Legionellosis occurs following inhalation of aerosolized droplets of *Legionella*. Once inhaled, the organism then attaches to alveolar macrophages (Horwitz, 1993). Legionellae are then phagocytosed by these macrophages. The process involves complement fragment C3 and the monocyte complement receptors CR1 and CR3. Virulent strains of *Legionella* are able to multiply inside macrophages as they are able to inhibit the fusion of phagosomes with lysosomes. With the non-virulent strains of *Legionella* it seems that multiplication in macrophages is not possible (Horwitz, 1993).

Two products of *Legionella* are thought to be the main virulence factors for *Legionella* (Fields, 1996). These include a macrophage invasion protein (MIP), which is 24 kDa, and an integral protein of the cytoplasmic membrane (113 kDa), which is the product of the *dotA* gene (Roy and Isberg, 1997). Other proteins have been suggested as virulence factors, but their role is unknown to date (Barker *et al.*, 1993; Kwaik and Engleberg, 1994).

## Treatment

*Legionella* are not detectable in body fluids or tissues but antibody levels in the blood reach high levels during an infection (Fallon *et al.*, 1993).

Antibiotics used in the treatment of Legionnaires' disease include erythromycin. Other antibiotics that have proved clinically effective include rifampicin, doxycycline, co-trimoxazole and members of the 5-fluoroquinolones. In the case of Pontiac fever most patients recover without the need for any therapy.

Newer antibiotics are now available to treat Legionnaires' disease. These include clarithromycin, ciprofloxacin, azithromycin, tetracycline, dexycycline and now azithromycin and levofloxacin.

## Survival in the environment

*Legionella* have been isolated in small numbers from a wide range of fresh water habitats including groundwater (Lye *et al.*, 1997), rivers, lakes and natural thermal pools (Verissimo *et al.*, 1991), specifically man-made water supplies. The man-made supplies include hot water systems in hospitals, hotels and, in particular, cooling towers. In fact in drinking water the long-term survival of *Legionella* has been reported.

The optimal growth temperature of *Legionella pneumophila* is 35°C, however, it is able to multiply at temperatures between 25 and 42°C. The occurrence of *Legionella* organisms in water supplies does not constitute a major cause for concern as its presence does not necessarily lead to a *Legionella* outbreak. Of concern, however, is the aerosolizing of the organism that is a risk factor for nosocomial infections, particularly in immunocompromised patients.

As mentioned previously, *Legionella* multiplies intracellularly in amoebae and ciliates of the genera *Hartmanella*, *Acanthamoeba*, *Naegleria*, *Echinamoeba*, *Tetrahymena* and *Cyclidium* (Breiman *et al.*, 1990). Under adverse conditions amoebae form a cyst entrapping *Legionella* and the cyst is blown away in the air. The amoebic cyst provides a protective environment for viable *Legionella*.

A number of other species of *Legionella* including *L. bozemanii*, *L. longbeachae*, *L. jordanis*, *L. wadsworthii*, *L. birminghamensis*, *L. cincinnatiensis*, *L. oakridgeiensis* and *L. tucsonensis* have been isolated from the environment but not generally implicated in disease.

Good management is required for the control of *Legionella* in water. Legionellae can be killed when temperatures are elevated above 50°C (Schulze-Ršbbecke *et al.*, 1987).

## Methods of detection

In environmental samples Legionellae can be concentrated by centrifugation (e.g. 12 000 $g$ for 10 minutes at about 20°C) or by membrane filtration. To reduce the presence of contaminating unwanted bacterial species heat or acid treatment could be applied. Following decontamination the sample is transferred onto buffered charcoal–yeast extract (BCYE) agar (Edelstein, 1981) and then incubated for 5 days at 37°C. On BCYE, *Legionella* are identified by their colonial morphology and the requirement of L-cysteine. After 5 days' growth on BCYE *Legionella* colonies have a characteristic 'cut glass' appearance: grey-white, glistening, convex and 3–4 mm in diameter. Isolates that grow on BCYE react with appropriate antisera allowing for confirmation of legionellae. Species within Legionellaceae can be easily distinguished using the appropriate biochemical, serology and nucleic acid analysis. Environmental and clinical isolates of *Legionella* can be subtyped by molecular techniques such as ribotyping, macrorestriction analysis by pulsed-field gel electrophoresis, or PCR-based methods (Van Belkum *et al.*, 1996).

When exposed to long-wave UV light at 366 nm, the colonies of some *Legionella* spp. autofluoresce brilliant blue-white, some species appear red or yellow-green, whereas *L. pneumophila* is not autofluorescent.

## Epidemiology

Legionnaires' disease accounts for 1–4% of all cases of pneumonia. This, however, has been documented as being as high as 30% (Macfarlane *et al.*, 1982; Fang *et al.*, 1990). Pontiac fever incidences are unknown.

Aerosols provide a means of transporting *Legionella* over immense distances, suggesting a low infective dose for *Legionella*. The infectivity of *Legionella* is enhanced if amoebae are inhaled or aspirated (Brieland *et al.*, 1996; Berk *et al.*, 1998). Aspiration following ingestion of contaminated water, ice and food has also been implicated as the route of infection in some cases (Marrie *et al.*, 1991; Venezia *et al.*, 1994; Graman *et al.*, 1997).

Outbreaks of Legionnaires' disease are not very common. However, a number of outbreaks, often sporadic, have been traced to air-conditioning systems, decorative fountains, room humidifiers, cooling towers and ultrasonic nebulizers, hot whirlpool (Castellani Pastoris *et al.*, 1999) and spa baths (Jernigan *et al.*, 1996), hot water from taps and showers, and medical devices containing water (Butler and Breiman, 1998). Most cases of *Legionella* in hospitals are associated with contaminated potable water and hot-water systems (Joseph *et al.*, 1994).

## Risk assessment

*Health effects*: occurrence of illness, degree of morbidity and mortality, probability of illness based on infection:

- *Legionella* causes legionellosis, which has two forms: Pontiac fever and Legionnaires' disease.
- Pontiac fever is a relatively mild, self-limiting, influenza-like illness. Pontiac fever presents in otherwise healthy individuals as pleuritic pain in the absence of pneumonic or multisystem manifestations.
- Pontiac fever has a short incubation period, a high attack rate, but a very low mortality ratio.
- Legionnaires' disease is a severe respiratory illness. The infection can last for weeks to months. The symptoms of infection include pneumonia with anorexia, malaise, myalgia and headache, rapid fever and chills, a cough, chest pain, abdominal pain and diarrhoea. Acute renal failure, disseminated intravascular coagulation, shock, respiratory insufficiency, coma and circulatory collapse are the major factors precipitating death.
- Legionnaires' disease has a low attack rate (1–6%), but a mortality rate of about 10%.
- Legionnaires' disease is more common in those with underlying illnesses, smokers, the elderly, or the immunocompromised (for whom the prognosis is poor).
- Legionnaires' disease accounts for 1–4% of all cases of pneumonia, although rates as high as 30% have been reported. The incidence of Pontiac fever is however unknown.

*Exposure assessment*: routes of exposure and transmission, occurrence in source water, environmental fate:

- *Legionella* are ubiquitous in the environment and have been isolated from a wide range of fresh water habitats including ground water, rivers, lakes and natural thermal pools. Higher numbers of *Legionella* have been documented in man-made water supplies. These have included air-conditioning condensers, cooling tower effluent, humidifiers, nebulizers, potable and hot water supplies, domestic and hospital showerheads, whirlpool spas, decorative fountains and vegetable misting machines.
- The association of *Legionella* with fresh water amoebae in the aquatic environment is well documented. The organism multiplies intracellularly in amoebae and ciliates, especially when water temperatures are elevated. This

relationship protects the bacteria against dry conditions, extremes of temperature, and treatment with biocides.

- Delivery of the agent to the respiratory tract in the form of water droplets 5–15 μm in diameter serves as the primary vehicle for disease transmission; aerosols containing the organism constitute a major risk factor for nosocomial infections and for those who are immunocompromised.
- Person-to-person transmission of *Legionella* is not thought to occur.
- The infectious dose is unknown.

*Risk mitigation*: drinking-water treatment, medical treatment:

- *Legionella* is susceptible to both heat and chlorine. Intermittent temperature increases in the water supply of up to 60°C, as well as the use of chlorination procedures to give a continuous 1–2 ppm residual are effective. In addition to disinfection, remedial cleaning and flushing of water systems as well as the removal of sediment from hot water tanks, coupled with water treatment and constant water monitoring are necessary.
- It is imperative that when designing the plumbing systems of new buildings and those undergoing modification, dead space volumes must be reduced, sediment build-up and stagnation prevented as these conditions favour the growth of legionellae, particularly in biofilms.
- The antibiotic most often given for legionellosis is erythromycin, however ciprofloxacin or rifampicin may be added. With Pontiac fever most patients recover without specific therapy.
- In immunocompromised individuals the death rate for Legionnaires' disease is relatively high, despite appropriate therapy.

# References

Barker, J., Lambert, P.A. and Brown, M.R.W. (1993). Influence of the intra-amoebic and other growth conditions on the surface properties of *Legionella pneumophila*. *Infect Immun*, **61**: 3503–3510.

Berk, S.G. *et al.* (1998). Production of respirable vesicles containing live *Legionella pneumophila* cells by two *Acanthamoeba* spp. *Appl Environ Microbiol*, **64**: 279–286.

Breiman, R.F., Fields, B.S. *et al.* (1990). Association of shower use with Legionnaires' disease. Possible role of amoeba. *JAMA*, **263**: 2924–2926.

Brieland, J. *et al.* (1996). Coinoculation with *Hartmannella vermiformis* enhances replicative *Legionella pneumophila* lung infection in a murine model of Legionnaires' disease. *Infect Immun*, **64**: 2449–2456.

Butler, J.C. and Breiman, R.F. (1998). Legionellosis. In *Bacterial infections of humans*, Evans, A.S. and Brachman, P.S. (eds). New York: Kluwer Academic/Plenum, pp. 355–375.

Castellani Pastoris, M. *et al.* (1999). Legionnaires' disease on a cruise ship linked to the water supply system: clinical and public health implications. *Clin Infect Dis*, **28**: 33–38.

Chandler, F.W., Roth, K. *et al.* (1980). Flagella on Legionnaires' disease bacteria: ultrastructural observations. *Ann Intern Med*, **93**: 711–714.

Edelstein, P.H. (1981). Improved semiselective medium for isolation of *Legionella pneumophila* from contaminated clinical and environmental specimens. *J Clin Microbiol*, **14**: 298–303.

Fallon, R.J. *et al.* (1993). Pontiac fever in children. In *Legionella: Current Status and Emerging Perspectives*, Barbaree, J.M., Breiman, R.F. and Dufour, A.P. (eds). Washington, DC: American Society for Microbiology, pp. 50–51.

Fang, G.D., Fine, M. *et al.* (1990). New and emerging etiologies for community acquired pneumonia with implications for therapy. A prospective multicenter study of 359 cases. *Medicine (Baltimore)*, **69**: 307–316.

Fields, B.S. (1993). *Legionella* and protozoa: interaction of a pathogen and its natural host. In *Legionella: Current Status and Emerging Perspectives*, Barbaree, J.M., Breiman, R.F. and Dufour, A.P. (eds). Washington, DC: American Society for Microbiology, pp. 129–136.

Fields, B.S. (1996). The molecular ecology of legionellae. *Trends Microbiol*, **4**: 286–290.

Goldberg, D.J. *et al.* (1989). Lochgoilhead fever: outbreak of non-pneumonic legionellosis due to *Legionella micdadei*. *Lancet*, i: 316–318.

Graman, P.S., Quinlan, G.A. and Rank, J.A. (1997). Nosocomial legionellosis traced to a contaminated ice machine. *Infect Control Hosp Epidemiol*, **18**: 637–640.

Hookey, J.V. *et al.* (1996). Phylogeny of Legionellaceae on small-subunit ribosomal DNA sequences and proposal of *Legionella lytica* comb. nov. for *Legionella*-like amoebal pathogens. *Int J Systematic Bacteriol*, **46**: 526–531.

Horwitz, M.A. (1993). Toward an understanding of host and bacterial molecules mediating *Legionella pneumophila* pathogenesis. In *Legionella: Current Status and Emerging Perspectives*, Barbaree, J.M., Breiman, R.F. and Dufour, A.P. (eds). Washington, DC: American Society for Microbiology, pp. 55–62.

Jernigan, D.B. *et al.* (1996). Outbreak of Legionnaires' disease among cruise ship passengers exposed to a contaminated whirlpool spa. *Lancet*, **347**: 494–499.

Joseph, C.A. *et al.* (1994). Nosocomial Legionnaires' disease in England and Wales, 1980–1992. *Epidemiol Infect*, **112**: 329–345.

Kwaik, Y.A. and Engleberg, N.C. (1994). Cloning and molecular characterization of *Legionella pneumophila* upon infection of macrophages. *Infect Immun*, **61**: 1320–1329.

Lye, D. *et al.* (1997). Survey of ground, surface, and potable waters for the presence of *Legionella* species by EnviroAmpÒ PCR *Legionella* kit, culture, and immunofluorescent staining. *Water Res*, **31**: 287–293.

Macfarlane, J.T., Finch, R.G. *et al.* (1982). Hospital study of adult community acquired pneumonia. *Lancet*, **2**: 255–258.

Marrie, T.J. *et al.* (1991). Control of endemic nosocomial Legionnaires' disease by using sterile potable water for high risk patients. *Epidemiol Infect*, **107**: 591–605.

Mayock, R., Skale, B. and Kohler, R.B. (1983). *Legionella pneumophila* pericarditis proved by culture of pericardial fluid. *Am J Med*, **75**: 534–536.

Pruckler, J.M. *et al.* (1995). Comparison of *Legionella pneumophila* isolates by arbitrarily primed PCR and pulsed-field gel electrophoresis: analysis from seven epidemic investigations. *J Clin Microbiol*, **33**: 2872–2875.

Rodgers, F.G., Macrae, A.D. and Lewis, M.J. (1978). Electron microscopy of the organism of Legionnaires' disease. *Nature*, **272**: 825–826.

Roy, C.R. and Isberg, P.R. (1997). Topology of *Legionella pneumophila* dotA: an inner membrane protein required for replication in macrophages. *Infect Immun*, **65**: 571–578.

Schulze-Ršbbecke, R., Ršdder, M. and Exner, M. (1987). Multiplication and killing temperatures of naturally occurring Legionellae. *Zentralbl Bakteriol Mikrobiol Hyg*, **184**: 495–500.

Steele, T.W., Lanser, J. and Sangster, N. (1990). Isolation of *Legionella longbeachae* serogroup 1 from potting mixes. *Appl Environ Microbiol*, **56**: 49–53.

Tang, P. and Krishnan, C. (1993). Legionellosis in Ontario, Canada: Laboratory aspects. In *Legionella: Current Status and Emerging Perspectives*, Barbaree, J.M., Breiman, R.F. and Dufour, A.P. (eds). Washington, DC: American Society for Microbiology, pp. 16–17.

Tison, D.L. *et al.* (1980). Growth of *Legionella pneumophila* in association with blue-green algae (*Cyanobacteria*). *Appl Environ Microbiol*, **39**: 456–459.

Van Belkum, A. *et al.* (1996). Serotyping, ribotyping, PCR-mediated ribosomal 16S–23S spacer analysis and arbitrarily primed PCR for epidemiological studies on *Legionella pneumophila*. *Res Microbiol*, **147**: 405–413.

Venezia, R.A. *et al.* (1994). Nosocomial legionellosis associated with aspiration of naso-gastric feedings diluted in tap water. *Infect Control Hosp Epidemiol*, **15**: 529–533.

Verissimo, A. *et al.* (1991). Distribution of *Legionella* spp. in hydrothermal areas in continental Portugal and the island of Sao Miguel, Azores. *Appl Environ Microbiol*, **57**: 2921–2927.

Wadowsky, R.M. and Yee, R.B. (1985). Effect of non-Legionellaceae bacteria on the multiplication of *Legionella pneumophila* in potable water. *Appl Environ Microbiol*, **49**: 1206–1210.

World Health Organization. (1990). Epidemiology, prevention and control of legionellosis: memorandum from a WHO meeting. *Bull World Hlth Org*, **68**: 155–164.

# 11

# The *Mycobacterium avium* complex

## Basic microbiology

Mycobacteria are Gram-positive, slender, non-motile, non-spore forming, rod-shaped bacilli, which may appear bent or curved. Cells are pleomorphic, ranging from a coccoid form to long slender rods, 0.2–0.6 μm wide and 1.0–10 μm long. Most mycobacteria are aerobic organisms, but some species may be microaerobic (Falkinham, 1996). All mycobacteria are catalase-positive. The high lipid content of the bacterial cell wall provides the bacteria with resistance to acid and alkali. Certain mycobacteria, particularly *M. avium paratuberculosis* (MAP), can shed their cell walls, forming a spheroplast (Thompson, 1994). It is the cell walls of bacteria that pick up stains, therefore the spheroplast form of the bacteria cannot be detected using the acid-fast stain test. Because the mycobacterial cell wall is hydrophobic *Mycobacterium* tend to grow in clumps and mycobacterial colonies float on the surface of liquid media. As a consequence detergents, such as Tween 80, are often used in culture media to help break down the clumping of organisms.

Many species of mycobacteria exist as saprophytes in soil and water and are referred to as environmental mycobacteria; they do not generally cause human infections. Some species of mycobacteria, however, do cause disease in animals and humans and are classified according to their host species, i.e. human, bovine, avian, murine and piscine, with species of mycobacteria categorized by their potential pathogenicity to humans. Mycobacterial species are very closely related to *Nocardia*, *Rhodococcus* and *Corynebacterium* spp. However, the first requirement for an unknown microorganism to be classified with the mycobacteria, is that it be 'acid fast'. *Nocardia* and *Rhodococcus* are referred to as 'partially acid fast'.

The most important disease causing *Mycobacterium* is *Mycobacterium tuberculosis* which causes tuberculosis in humans. Tuberculosis is also caused by *Mycobacterium bovis*, which causes tuberculosis in cattle and humans and *Mycobacterium africanum*, a rare cause of human tuberculosis in central Africa. *Mycobacterium tuberculosis*, *bovis* and *africanum* all display some phenotypic differences. They are genetically very closely related and generally classed as the '*Mycobacterium tuberculosis* complex'. However, epidemiological and treatment differences exist for diseases caused by *Mycobacterium tuberculosis* and *Mycobacterium bovis* – these two are still usually regarded as separate species.

Other important pathogenic mycobacteria include the '*Mycobacterium avium intracellulare*' complex and the 'atypical' mycobacteria. These groups of *Mycobacterium* are frequent opportunistic pathogens in the immunosuppressed, however, they are documented as causing disease in individuals with normal immune systems.

The genus *Mycobacterium* can also be divided into two groups on the basis of growth rate: rapid growers and slow growers. Rapid growers form visible colonies on solid media within 7 days, while slow growers require longer than 7 days to produce visible colonies. *Mycobacterium avium* complex (MAC) species are defined as slow growers. In the Runyon Classification of non-tuberculous Mycobacteria, MAC organisms are classed in group III, non-photochromogenic slow growers. There are 28 known serotypes belonging to this MAC complex (Wolinsky, 1992). They are discussed in more detail below.

## Origins of the organism

There are 71 validly named species of *Mycobacterium* and an additional three subspecies. The principal pathogens in the genus are *M. bovis*, *M. leprae* and *M. tuberculosis* but, in all, 32 species are known to be pathogenic to humans or animals. Species of *Mycobacterium* other than, *M. bovis*, *M. leprae* and *M. tuberculosis* are often referred to as 'atypical mycobacteria' (see Table 11.1).

The most commonly encountered pathogens among the atypical mycobacteria are species of the *Mycobacterium avium* complex. The *M. avium* complex (MAC) is considered to contain *M. avium*, *M. avium* subspecies *paratuberculosis*, *M. avium* subspecies *silvaticum* and *M. intracellulare*. However, poorly identified strains which show some similarity to *M. avium* are also frequently, and incorrectly, allocated to the complex. There are over 20 recognized serotypes within the *M. avium* complex.

Mycobacteria were first discovered in 1874 when Armauer Hansen found acid-fast bacilli in individuals suffering from leprosy (Hansen, 1874). *Mycobacterium tuberculosis* was discovered in culturable form in 1882 when Robert Koch, using coagulated bovine serum, isolated it (Koch, 1882). In 1896 the generic name *Mycobacterium* was recognized (Goodfellow and Wayne, 1982).

Johne first discovered *Mycobacterium avium paratuberculosis* in 1895, an organism now thought to cause the disease pseudotuberculosis enteritis (Johne and Frothingham, 1895). *Mycobacterium avium paratuberculosis* has been isolated from many species of animals. It was not until 1910 that Twort successfully isolated the first *M. avium paratuberculosis* (Twort, 1911). At this time the isolate of *Mycobacterium* was named *Mycobacterium enteriditis chronicae pseudotuberculosae bovis Johne* (Twort and Ingram, 1912). However, in 1932 the bacterium was renamed *M. paratuberculosis*. To date *Mycobacterium avium* subspecies *paratuberculosis* (MAP) is a member of the *Mycobacterium avium* complex (MAC) (Thorel *et al.*, 1990). *M. paratuberculosis* seems to be linked to Crohn's disease in humans. However, the evidence for this is very limited at present. Crohn's disease was first recognized in 1913 by Kennedy Dalziel, a surgeon from Glasgow who believed the disease must have a mycobacterial cause. However, he was unable to culture viable organisms from tissue samples (Dalziel, 1913). In 1985, the IS900 insertion sequence, unique to *M. paratuberculosis* was discovered. IS900 was the first DNA insertion sequence to be found in mycobacteria, it comprises a 1451–1453 base pair repetitive element (Green *et al.*, 1989). The finding of this species-specific marker and the amplification of the polymerase chain reaction (PCR) made specific identification of MAP possible.

**Table 11.1**  Principal types of disease in humans and causative agents

| Disease | Agent |
| --- | --- |
| Lymphadenopathy | *M. avium* complex |
| | *M. scrofulaceum* |
| Skin lesions | |
|    Post-trauma abscess | *M. chelonae* |
| | *M. fortuitum* |
| | *M. terrae* |
|    Swimming pool granuloma | *M. marinum* |
|    Buruli ulcer | *M. ulcerans* |
|    Pulmonary disease | *M. avium* complex |
| | *M. kansasii* |
| | *M. xenopi* |
| | *M. malmoense* |
| Disseminated disease | |
|    AIDS-related | *M. avium* complex |
| | *M. genevense* |
|    Non-AIDS related | *M. avium* complex |
| | *M. chelonae* |

In 1959, a classification scheme for *Mycobacterium* was developed (Runyon). This scheme enabled mycobacteria to be classified according to rate of growth (slow or rapid) and production of pigment (yellow or orange). Runyon's scheme included:

- Group I *Photochromogens* which produced pigment on exposure to light
- Group II *Scotochromogens* which produced pigment in the dark
- Group III *Non-chromogens* which produced no pigment
- Group IV *Rapid growers*.

In this scheme all strains of *Mycobacterium* in groups I, II and III grew slowly; rapid growers (group IV) may be photochromogens, scotochromogens or non-chromogens. Runyon's classification scheme separates *Mycobacteria tuberculosis* from most of the 'atypicals'. The preliminary division of the atypicals in Runyon's classification is still used today for the classification of mycobacteria species, when combined with other biochemical and growth characteristics. Identification of species of mycobacteria today is undertaken by specialist reference laboratories. These are usually based on cultural characteristics (rate and temperature of growth and pigmentation), various biochemical reactions, resistance to antimicrobials and lipid content of the cell wall.

The *Mycobacterium* which belong to group I photochromogens are colourless cultures in the dark but form bright yellow or orange pigmentation when they are exposed to light. Three important species make up this group, namely *M. kansasii*, *M. simiae* and *M. marinium*. *M. kansasii* grows well at 37°C and produces a yellow pigment, it rapidly hydrolyses tween in 3 days and has strong pyrazinamidase activity. It is an organism that is commonly isolated from chronic pulmonary disease. *M. simiae* also grows well at 37°C and hydrolyses tween slowly, taking 10 days or more to do this. Like *M. kansasii* it has a high thermostable catalase and is isolated from cases of pulmonary disease. *M. marinum*, on the other hand, is the fish tubercule bacillus that causes warty skin infection and is often referred to as swimming pool granuloma or fish tank granuloma. Its optimal growth temperature is 30–32°C with poor growth at 37°C. It is urease-positive, has no heat stable catalase, but does produce pyrazinamidase.

Mycobacteria belonging to group II are the scotochromogens and include the three species *Mycobacterium scrofulaceum*, *M. gordonae* and *M. szulgai*. *M. scrofulaceum* is associated with scrofula or cervical lymphadenitis and also causes pulmonary disease. It is a slow grower (4–6 weeks) and produces a light yellow to deep orange pigment which is independent of light exposure. It fails to hydrolyse tween, produces catalase at 68°C but does not reduce nitrate. *M. szulgai* as a comparison to *M. scrofulaceum* is an uncommon cause of pulmonary disease and bursitis. It has a temperature dependent pigment production and is a rapid grower (2 weeks at 37°C). It hydrolyses tween slowly and is intolerant to 5% sodium chloride. *M. gordonae* is also a rare cause of lung disease and is frequently found in water. However, it is often a common contaminant of clinical material. It is able to hydrolyse tween and produces a stable catalase. It is urease-negative and does not reduce nitrates. It is resistant to isoniazid and streptomycin but is susceptible to rifampicin and ethanbutol.

Group III mycobacteria are referred to as the non-chromogens and include *M. avium* and *M. intracellulare* (MAI or MAC) and *M. malmoense, M. xenopi, M. ulcerans* and *M. terrae*. The *Mycobacterium avium* complex constitutes the most prevalent and important opportunistic group of human pathogens. *M. avium* is the avian tubercule bacillus and *M. intracellulare* is the battery bacillus. Most of the MAC clinical isolates are smooth and easily emulsifiable. There are 28 serotypes delineated by agglutination by specific antisera. MAC is responsible for lymphadenitis, pulmonary lesions and disseminated disease, notably in patients with AIDS. MAC does not produce a heat stable catalase and is generally inert biochemically. *M. xenopi* was first isolated from the xenopus toad. It is a thermophile that grows at 45°C and has been isolated from pulmonary lesions. *M. malmoense* causes pulmonary disease and lymphadenitis and grows very slowly (10 weeks) and is therefore likely to be missed if cultures are not maintained for up to 12 weeks. *M. ulcerans* is the cause of Buruli ulcers. It is a very slow growing bacterium and produces a toxin which causes tissue necrosis. On inoculation into the foot pads of mice it causes disease, including swelling, ulceration and often autoamputation.

*M. haemophilum* is characterized by a growth requirement for haem or other iron sources. It is a rare cause of granulomatous or ulcerative skin lesion in xenograft recipients and other immunocompromised patients. It has an optimum growth at 28–32°C which is stimulated by 10% $CO_2$. The only biochemical test that is positive is pyrazinimidase production.

The final group of *Mycobacterium* are found in Group IV and referred to as the rapid growers. *M. chelonae* and *M. fortuitum* are the two well-recognized human pathogens found in this group. They are principally responsible for post-injection abscesses, wound infections and corneal ulcers. They occasionally cause pulmonary or disseminated disease. Both species are non-chromogenic.

The *Mycobacterium* species as an entire group have a large number of non-culturable medical forms. These are referred to as the very poorly growing species that are mostly isolated from the blood of AIDS patients. The classification of this group is based on unique base sequences in their 16S ribosomal RNA. These 'non-culturable' *Mycobacterium* include: *M. genevense, M. confluentis, M. intermedium* and *M. interjectum*.

## Clinical features

*Mycobacterium*, but more specifically the environmental mycobacteria, have been known to be opportunistic pathogens of many disseminated diseases with patients dying of AIDS, skin lesions such as post-injection abscesses, swimming pool granulomas and Buruli ulcers, tuberculosis-like lesions and also lympadenitis following infection. Thirty-two of the 71 species of

*Mycobacterium* are known to be pathogenic to humans or animals. The most common non-tuberculous, or atypical mycobacterial pathogens are species found within the *Mycobacterium avium* complex (MAC). MAC contains *M. avium*, *M. intracellulare*, *M. avium* subspecies *paratuberculosis* (MAP), and *M. avium* subspecies *silvaticum*.

MAC organisms cause disease in pigs and poultry. The organisms are continually excreted in the faeces of birds and can live in the soil for long periods of time (Inderlied *et al.*, 1993).

In normal healthy individuals MAC rarely cause disease. However, individuals who have suppressed immune systems, i.e. the elderly and the very young and AIDS suffers, are susceptible to both pulmonary and non-pulmonary infections. Individuals who are HIV positive disseminate MAC to other parts of the body including joints, skin, blood, lungs, liver and brain. MAC entry into the host is via the respiratory and gastrointestinal tracts (Chin *et al.*, 1994).

*M. paratuberculosis* is suspected to cause the chronic, inflammatory, gastrointestinal disease of humans known as Crohn's disease (Hermon-Taylor, 1998). It has been found to be present in milk, meat and faeces of infected cattle. In cattle the disease caused by MAP is called Johne's disease. Johne's disease has clinical, systemic and pathological characteristics similar to infection in Crohn's disease in humans (Chiodini, 1989).

Crohn's disease is often referred to as an autoimmune disease where the body's immune system attacks and inflames its own tissues. The place where this occurs is the gastrointestinal tract (Chiodini, 1996). As the GI tract becomes inflamed this leads to a narrowing of the digestive system leading to severe pain and uncontrollable bowel movements. A number of research papers have been published that have patients going into remission when treated with appropriate antibiotics directed at *M. paratuberculosis*. The first documented case of human disease caused by *M. paratuberculosis* was published in 1998 following a 7-year-old boy who developed cervical lymphadenitis. Following the removal of the boy's lymph nodes the biopsy was found to contain *M. paratuberculosis*. Follow up of this patient 5 years after the operation found that the boy was now suffering from a chronic inflammation of his intestine. Following treatment with antibiotics active against *M. paratuberculosis* the boy's intestinal disease went into remission (Barnes *et al.*, 1998).

MAP is known to be an intracellular pathogen, which is able to colonize and multiply in macrophages. This may well be a way that MAP is transmitted in milk (cow's milk contains white blood cells) following ingestion of contaminated milk from infected cattle. (Hermon-Taylor, 1993). Research has shown that the normal pasteurization process of 72°C for 15 seconds does not ensure the complete eradication of MAP (Grant *et al.*, 1998). Results from a study conducted in 1991 to 1993 found MAP organisms in commercially pasteurized milk samples throughout central and southern England (Millar, 1996). It is probable that as *M. paratuberculosis* is known to form aggregates or possible biofilms this may well aid in its survival during the pasteurization process.

## Treatment

Atypical mycobacteria are often found to be resistant to many anti-TB drugs *in vitro*. However, using appropriate patient management and applying a combination of drugs, non-typical mycobacteria infections can be treated. Regimens with five or six drugs are used for pulmonary disease due to MAC, *M. kansasii* and *M. xenopi*. Triple therapy using isoniazid, rifampicin and ethambutol has also been used over an 18-month period with positive results. There are multiple drug regimens for AIDS-associated MAC disease. These include clarithromycon or azithromycin with two other drugs selected from rifabutin, ethambutol, clofazimine, fluoroquinolones and amikacin. Local skin lesions, which are due to mycobacterium, are usually excised where possible. With *M. chelonae* keratitis relapses are very frequent and therefore surgical intervention is often necessary. The treatment of non-tuberculosis diseases are usually unsuccessful in a large minority of cases.

## Survival in the environment

MAC have been isolated world-wide from all natural water systems, including marine water and soil (Grange *et al.*, 1990). Man-made habitats, such as water distribution systems, are colonized by considerably high numbers of mycobacterial species. Aquatic mycobacteria will colonize biofilms at solid–water and air–water interfaces, which seem to be an important replication site in oligotrophic habitats (Schulze-Röbbecke, 1993).

Mycobacteria therefore seem to survive well in the environment. They are also documented as being able to survive in temperatures below 0°C. Iivanainen *et al.* (1995) collected samples from shallow brooks to investigate the effect of prolonged storage on mycobacteria and other heterotrophic bacteria. Water concentrates were stored in nutrient broth at −75°C for 15 months. Viable counts were taken from the fresh samples and again following storage. Their results showed a threefold increase in the numbers of mycobacteria present following storage (Iivanainen, 1995). This suggests a very sophisticated survival mechanism and one which may aid in the transmission of *Mycobacterium* in the environment. Food, in this case defrosted water from fish, has been shown to be a good medium for the transmission of mycobacteria (Mediel *et al.*, 2000). From a study conducted in California a variety of foods collected from supermarkets and market stalls were shown to contain *Mycobacterium* spp. From this study six species of non-tuberculous mycobacteria were isolated from 25 out of the 121 foods tested. *M. avium* was found to be the most frequently isolated species of *Mycobacterium* (Argueta *et al.*, 2000).

*Mycobacterium avium paratuberculosis* can also survive and possibly even replicate in natural waters and soil. However, to date this has not been determined. It is probable that MAP is able to replicate in farm animals. These infected animals then disseminate MAP in their faeces that are then reingested by other animals (Hermon-Taylor *et al.*, 1994).

Water seems to be the main reservoir for environmental mycobacteria (Wallace, 1987; Iivanainen *et al.*, 1999). However, water-treatment processes are known to help reduce the numbers of MAC organisms present in water supplies. Replication in biofilms in drinking water pipes may suggest a public health concern. Mycobacteria are able to survive the normal chlorination process used in drinking water because of the thick, lipid-rich cell wall. It is well documented that chlorine levels used in water purification are unlikely to be effective against mycobacteria, including the MAC organisms (Pelletier, 1988). Evidence for this lies in a study of nosocomial outbreaks, occurring when water supplies from two hospitals, in Boston and New Hampshire, were the source of a MAC strain, showing the same genetic pattern as MAC strains infecting groups of AIDS patients in the hospitals (Von Rehn *et al.*, 1994). Also the current water-treatment processes are unlikely to eradicate MAP organisms from the water supply (Hermon-Taylor, 1998). Domestic water systems have been associated with human non-tuberculous mycobacterial infections even at high temperatures (Schulze-Röbbecke and Buchholtz, 1992; Miyamoto *et al.*, 2000).

A number of studies have shown that mycobacteria are present in tap water. A study by Peters *et al.*, in 1995, showed the presence of *Mycobacterium* in tap waters in Berlin. In this study mycobacteria were found in 42.4% of samples; 28% were *Mycobacterium gordonae* and 1.7% MAC (Peters *et al.*, 1995).

Covert *et al.* (1999) sampled water for the presence of mycobacteria in 21 states in the USA. The water sources included potable water, water in cisterns, bottled water, drinking-water-treatment samples and ice-machine samples. The results of the study found that non-tuberculous mycobacteria were detected in 54% of the ice samples and 35% of the public drinking water samples analysed. The species isolated in all waters were: *M. gordonae*; *M. peregrinum*; *M. mucogenicum*; *M. intracellulare*; *M. scrofulaceum*; *M. gastri/kansasii*; *M. fortuitum*; *M. avium* and *M. chelonae*. *M. mucogenicum* was the isolate most commonly found (Covert *et al.*, 1999).

Another study in Los Angeles examined potable water to estimate if *M. avium* complex strains isolated from the water were related to strains isolated from AIDS patients in the Los Angeles area. This study concluded that there was a possible spread of *M. avium* to immunocompromised patients, particularly those suffering from AIDS (Aronson, 1999).

When Schulze-Röbbecke *et al.* (1992) analysed 50 biofilm samples obtained from water-treatment plants, domestic water supplies and aquaria, 90% of the samples contained mycobacteria.

A study by Falkinham and coworkers (2001) over an 18-month period found *Mycobacterium* in water and biofilm samples collected from eight water-distribution systems. Mycobacteria were isolated in only 15% of samples tested. Water-treatment processes substantially reduced the numbers of mycobacteria from the raw water. The highest numbers of mycobacteria were present in the distribution system samples. *M. intracellulare* was mostly recovered from biofilm samples, whereas *M. avium* was predominant in the water samples (Falkinham *et al.*, 2001).

Cirillo (1999) investigated the possibility that *M. avium* could interact with and replicate within *Acanthamoeba castellanii*. The results from this study showed that *M. avium* was able to replicate within the amoebae (Cirillo *et al.*, 1997). The ability of *M. avium* to enter and replicate inside cells may aid its pathogenicity. From a study by Cirillo *et al.* (1997) it was found that *M. avium* grown within amoebae had increased invasion of cells and intracellular replication when compared to *M. avium* grown in broth (Cirillo *et al.*, 1997). In addition, intracellular *M. avium* within *A. castellanii* have increased resistance to disinfectants and also antibiotics (Miltner and Bermudez, 2000).

## Detection methods

To isolate and enumerate *Mycobacterium* from the environment an environmental sample needs first to be decontaminated for the growth of *Mycobacterium*. This is because environmental samples contain a lot of fast-growing heterotrophic bacteria that will overgrow, preventing the isolation of slow growing *Mycobacterium*. The most commonly used decontaminating agent used in the isolation of *Mycobacterium* was sodium hydroxide. It was first used as a decontaminant in the culture of *M. paratuberculosis* from clinical specimens (Petroff, 1915). However, sodium hydroxide does have detrimental effects on the growth of mycobacteria. Other decontaminating agents used in the isolation of *Mycobacterium* have included sulphuric acid (Kamala *et al.*, 1994), sodium triphosphate, oxalic acid (Portaels *et al.*, 1988) and cetylpyridinium chloride (CPC) (Neumann *et al.*, 1997).

In order to grow *M. avium* and *M. intracellulare* a solid medium supplemented with egg or albumin is required. As *M. paratuberculosis* is the slowest growing of the culturable mycobacteria Mycobactin J is required as a growth factor (Portaels *et al.*, 1988). For the *M. avium* subspecies *silvaticum* any media used for its isolation must be supplemented with pyruvate and an acidic pH. MAC will grow over a temperature range of 25–45°C, however, 30°C appears to be the optimum temperature for maximum growth (Neumann *et al.*, 1997). The incubation period for the culture of environmental mycobacteria can take 6 months or longer (Iivanainen *et al.*, 1993).

One of the most popular media for the growth of *Mycobacterium* is Löwenstein-Jensen medium. It is composed of whole egg, citrate, salts, potato starch, asparagine, glycerol and malachite green. Glycerol is used as a carbon source for the growing bacteria and malachite green helps to inhibit other microbial growth and also provides a contrasting colour to enable the small, pale mycobacterial colonies to be more visible (Jenkins *et al.*, 1982). Löwenstein-Jensen medium has a pH of 7.0. Many non-tuberculous mycobacteria however have a pH optimum of pH 5.4–6.5 (*M. avium* is pH 5–5.5) (Portaels and Pattyn, 1982; Falkinham, 1996). Ogawa egg yolk medium has a lower pH (pH 6), and this medium seems to be more suitable than Löwenstein-Jensen for the growth of environmental species (Portaels *et al.*, 1988). Middlebrook 7H10 agar has also been used for the isolation of mycobacteria (Iivanainen *et al.*, 1999).

The best decontamination methods and culture media will determine the highest numbers and greatest diversity of species of mycobacteria isolated from environmental samples. A study by Iivanainen (1995) involving 15 combinations of decontamination methods established that the best method for greatest recovery of *Mycobacterium* was achieved by using egg media supplemented with glycerol at pH 6.5, following decontamination with NaOH, malachite green and cycloheximide (Iivanainen, 1995).

In Germany, Neumann *et al.* (1997), using 12 isolation methods, undertook a study to optimize methods for the isolation of mycobacteria from water samples. They analysed 109 treated water samples and 26 surface water samples. For the treated water samples cetylpyridinium chloride (CPC) was used at concentrations of 0.005% or 0.05%, with an exposure time of 30 minutes. For the surface waters, an additional method was also used. For this analysis Tryptic soy broth (TSB) was added to the sample and incubated at 37°C for 5 hours. Malachite green oxalate, cycloheximide and NaOH were then added, with an exposure time of 30 minutes, followed by the addition of hydrochloric acid. Löwenstein-Jensen media containing glycerol, Ogawa egg yolk medium and Ogawa medium containing ofloxacin and ethambutol were used for primary culture. From this study 586 mycobacterial strains were isolated. Strains were identified using mycolic acid thin-layer chromatography and PCR restriction analysis.

The best methods for isolation of mycobacterium in treated water involved decontamination in 0.005% CPC, followed by culture on Löwenstein-Jensen medium with an incubation temperature of 30°C. For surface water the best three overall methods were: decontamination in 0.05% CPC, followed by culture on Löwenstein-Jensen media and incubation at 30°C; decontamination with the TSB method, culture on Löwenstein-Jensen medium and incubation at 30°C or decontamination with the TSB method, culture on Ogawa egg yolk medium and incubation at 30°C (Neumann *et al.*, 1997).

Automated culture systems are available for the culture and isolation of mycobacterial species. The first to be developed was BACTEC 460. This system uses Middlebrook 7H9 broth supplemented with antibiotics and C-palmitic acid (Iivanainen *et al.*, 1999). The ESP Culture System II uses Middlebrook 7H9 broth supplemented with antibiotics and enrichments. This system detects mycobacterial respiration as a decrease in gas pressure. MGIT, BBL and BACTEC 9000 use an oxygen-sensitive fluorescent sensor to detect a decrease in oxygen levels indicating bacterial respiration. The BacT/Alert System measures an increase in carbon dioxide levels using a reflectometer. The BacT/Alert commenced in the USA in June 1990.

The rate of mycobacterial isolation and recovery from 5208 samples was compared between the MB/BacT culture system and Löwenstein-Jensen medium (Palacios *et al.*, 1999). Accuprobe DNA probes were used to identify any mycobacteria isolates. From this study mycobacteria were isolated from only 301 samples. This study established that the detection rate of MB/BacT was higher than with Löwenstein-Jensen medium, in fact MB/BacT was found to be more efficient and faster at isolating and detecting mycobacteria than the Löwenstein-Jensen medium (Palacios *et al.*, 1999).

The BACTEC 460 TB system has become the 'gold standard' by which other culture and detection systems are evaluated (Woods *et al.*, 1997). A number of comparisons between the BACTEC 460 TB and the MB/BacT system have taken place. A study by Roggenkemp *et al.* (1999) found the MB/BacT system had a lower sensitivity than the BACTEC 460. A Swiss study evaluated the MB/BacT system in comparison to the BACTEC 460 system and solid media. From this study the BACTEC system had a shorter time to detection with less contamination of cultures. However, the MB/BacT system is fully automated and is a closed system, reducing the risk of cross-contamination (Rohner *et al.*, 1997). Other comparisons have taken place which have found that MB/BacT system took a longer time to detect all mycobacterial isolates and had a higher contamination rate than other systems of detection (Benjamin *et al.*, 1998). Brunello *et al.* (1999) compared the sensitivity and detection time of the MB/BacT system, BACTEC 460 TB system and conventional solid media, Löwenstein-Jensen slopes. From this study there was no significant difference in the recovery rates of the three systems examined. Another study in Toronto evaluated the performance of the MB/BacT system and the BACTEC 460 TB system in comparison with the solid egg based medium Löwenstein-Gruft, in detecting mycobacterial growth. In this study the MB/BacT system recovered more mycobacteria than other systems of recovery (Laverdiere *et al.*, 2000).

A study in 2000 by Harris and coworkers involved analysing 681 clinical samples and eight control samples to compare the recovery of mycobacteria using three isolation systems, the MB/BacT rapid culture system, the BACTEC 460 radiometric system and conventional egg media. It was found that none of the systems recovered all of the positive isolates.

Once mycobacteria have been recovered by an appropriate method they then have to be identified. Mycobacteria can be identified using standard taxonomy methods. Identification between species can be very difficult and complex. It is generally achieved using gene probes or DNA analysis. Mycobacteria are usually detected using staining and fluorescence microscopy. The method involves staining *Mycobacterium* using the Ziehl-Neelsen stain, the Kinyoun method which uses light microscopy to view bacterial cells stained red by carbol fuchsin and the use of auramine O.

Mycobacteria can be identified using high-pressure liquid chromatography or gas-liquid chromatography (Butler *et al.*, 1992; Glickman *et al.*, 1994). In these methods mycolic acids, specific to mycobacterial cells, can be extracted, methylated and then analysed. Molecular techniques have been applied for the detection of mycobacteria. These have used target or non-targeted areas of

DNA, such as pulse-field gel electrophoresis and restriction fragment length polymorphism. rRNA gene sequences can be used to identify members of the MAC complex (Saito *et al.*, 1990) including *M. paratuberculosis* (Thoresen and Saxegaard, 1991). However, to distinguish between MAC and MAP a DNA probe based on IS900 can be used. Commercially available probes, AccuProbe and Gen-Probe, are used for the identification of mycobacterial species.

Monoclonal antibodies against a phosphoglycolipid found in the cell wall of MAC bacteria have been used to produce a latex agglutination test (Olano *et al.*, 1998). MAC organisms can be identified using DNA sequence analysis of 16S rRNA (Frothingham and Wilson, 1994). There are three genetic differences between MAC and MAP. The IS900 DNA insertion element, specific for *M. paratuberculosis*, is present in multiple copies, 14 to 18 per bacterium, allowing PCR methods to be used to identify positively MAP (Green *et al.*, 1989). *M. paratuberculosis* has a single copy of a genetic element, which has been termed 'GS' (Tizard *et al.*, 1998). A gene called *hsp*X is located within *M. paratuberculosis* in a third genomic region (Ellingson *et al.*, 1998).

The differentiation of *Mycobacterium* species is best achieved by using a two-step assay of gene amplification and restriction fragment length polymorphism (Plikaytis *et al.*, 1992; Roth *et al.*, 2000).

## Epidemiology

There have been a number of waterborne outbreaks due to atypical *Mycobacteria* but many of these have been referred to as pseudo-outbreaks (Sniadack *et al.*, 1993; Bennett *et al.*, 1994; Wallace *et al.*, 1998; Lalande *et al.*, 2001). From these studies the source of contamination by *Mycobacterium* was in the laboratory.

The first true outbreak of *Mycobacterium* associated with water was established in a sternal wound infection. It was found that the wound infection was due to *M. fortuitum* (Kuritsky *et al.*, 1983). From this study it was found that the same strain was isolated from both clinical samples and a number of water samples taken from the hospital environment. A similar route of infection was established during a study in India. From this study Chadha and colleagues (1998) reported an outbreak of post-surgical wound infections due to *M. abscessus*. The organism responsible for this infection was linked to contaminated tap water.

MAC infections are associated predominately with HIV disease. From a study by Von Reyn *et al.* (1994) four types of *Mycobacterium* were identified. Two types of *Mycobacterium* were simultaneously isolated from patients and their respective hospital but not from patients' homes. These findings provided circumstantial evidence that infection may be related to hospital water supplies.

Many epidemiological studies have found that potable water is not a risk factor for *Mycobacterium* acquisition. However, the strongest evidence of a transmission route of *Mycobacterium* via water relates specifically to immunocompromised individuals. This seems to be particularly relevant in the hospital environment.

## Risk assessment

*Health effects*: occurrence of illness, degree of morbidity and mortality, probability of illness based on infection:

- MAC generally causes three different types of diseases in humans: (1) pulmonary disease, (2) cervical lymphadenitis and (3) disseminated MAC, both AIDS and non-AIDS related. Data suggest that the incidence of the first two types of MAC disease continue to be increasing, and disseminated MAC disease is one of the most common opportunistic infections to strike AIDS patients.
- MAC is generally an opportunistic infection; however, the incidence of pulmonary MAC in people (especially elderly women) without risk factors and lymphadenitis in children has been increasing. Data from the late 1970s and early 1980s showed an annual rate of 1.3 cases/100 000 (O'Brien *et al.*, 1987). In AIDS patients, disseminated MAC occurs in 20–40% of the population (Horsburgh, 1991), though that number has decreased with new AIDS drug therapies.
- Pulmonary MAC can cause significant pulmonary and systemic symptoms; disseminated MAC commonly causes fever, anaemia, night sweats and weight loss. Both can decrease expected lifespan in patients, but the mortality rate is unknown.

*Exposure assessment*: routes of exposure and transmission, occurrence in source water, environmental fate:

- Routes of exposure are ingestion through the gastrointestinal tract and the inhalation of aerosols. Some believe the latter is responsible for more transmission than the former.
- The infectious dose of MAC organisms in humans is unknown.
- Mycobacteria are common in the environment and have been isolated from all natural water systems world-wide. They are very hardy in the environment and have the ability to survive temperatures below 0°C and thrive in temperatures of 52–57°C.
- Many studies have implicated water as the transmission route for MAC infection. However, MAC is ubiquitous throughout the environment, and infection undoubtedly occurs through other routes of exposure, such as food. No evidence supports any person-to-person transmission.
- MAC's colonization in biofilm of water distribution systems is a major problem and has been associated with human infection. The organism is

thermophilic and MAC has been frequently isolated from showerheads and recirculating hot water systems in institutions. They are more frequently found colonizing hot water systems rather than cold.

*Risk mitigation*: drinking-water treatment, medical treatment:

- Chlorine is an ineffective disinfectant for mycobacteria, especially at residual levels. Physical treatment, such as sedimentation, coagulation and sand filtration is effective at removing most, but not all, of the organisms.
- Pulmonary MAC and disseminated MAC are difficult to treat and are resistant to antimicrobials; symptomatic infection can be highly problematic and result in lengthy and complicated chemotherapy. Relapse is relatively common. The incidence of disseminated MAC infections has decreased dramatically due to antiretroviral treatment protocols for AIDS patients.
- Untreated, pulmonary MAC can cause numerous respiratory and systemic symptoms. Available treatment is difficult and sometimes ineffective and can last for months or years. Permanent lung damage depends on the extent of infection when the patient presents for treatment. AIDS patients who develop disseminated MAC are usually in the end-stage of their disease.

# References

Aronson, T., Holtzman, A., Glover, N. *et al.* (1999). Comparison of large restriction fragments of *Mycobacterium avium* isolates recovered from AIDS and non-AIDS patients with those of isolates from potable water. *J Clin Microbiol*, **37**(4): 1008–1012.

Argueta, C., Yoder, S., Holtzman, A.E. *et al.* (2000). Isolation and identification of non-tuberculous mycobacteria from foods as possible exposure sources. *J Food Prot*, **63**(7): 930–933.

Barnes, N., Clarke, C., Finlayson, C. *et al.* (1998). *Mycobacterium paratuberculosis* cervical lymphadenitis followed five years later by terminal ileitis similar to Crohn's disease. *Br Med J*, **316**: 449–453.

Benjamin, W.H. Jr, Waites, K.B. and Beverley, A. (1998). Comparison of the MB/BacT system with a revised antibiotic supplement kit to the BACTEC 460 system for detection of mycobacteria in clinical specimens. *J Clin Microbiol*, **36**(11): 3234–3238.

Bennett, S.N., Peterson, D.E., Johnson, D.R. *et al.* (1994). Bronchoscopy-associated *Mycobacterium xenopi* pseudoinfections. *Am J Resp Crit Care Med*, **150**: 245–250.

Brunello, F., Favari, F. and Fontana, R. (1999). Comparison of the MB/BacT and BACTEC 460 TB systems for recovery of mycobacteria from various clinical specimens. *J Clin Microbiol*, **37**(4): 1206–1209.

Butler, W.R., Thibert, L. and Kilburn, J.O. (1992). Identification of *Mycobacterium avium* complex strains and some similar species by high-performance liquid chromatography. *J Clin Microbiol*, **30**: 2698–2704.

Chadha, R., Grover, M., Sharma, A. *et al.* (1998). An outbreak of post-surgical wound infections due to *Mycobacterium abscessus*. *Pediatr Surg Internatl*, **13**: 406–410.

Chin, D.P., Hopewell, P.C., Yajko, D.M. *et al.* (1994). *Mycobacterium avium* complex in the respiratory or gastrointestinal tract and the risk of *M. avium* bacteremia in patients with the human immunodeficiency virus. *J Infect Dis*, **169**: 289–295.

Chiodini, R.J. (1989). Crohn's disease and the mycobacterioses: a review and comparison of two disease entities. *Clin Microbiol Rev*, **2**: 90–117.

Chiodini, R.J. (1996). *M. paratuberculosis* in Foods and the Public Health Implications. In *Proceedings of the Fifth International Colloquium on Paratuberculosis*, Chiodini, R.K.,

Hines, M.E. and Collins, M.T. (eds). Madison, WI: International Association for Paratuberculosis, pp. 353–365.

Cirillo, J.D. (1999). Exploring a novel perspective on pathogenic relationships. *Trend Microbiol*, 7: 96–98.

Cirillo, J.D., Falkow, S., Tompkins, L.S. *et al.* (1997). Interaction of *Mycobacterium avium* with environmental amoebae enhances virulence. *Infect Immun*, 65(9): 3759–3767.

Covert, T.C., Rodgers, M.R., Reyes, A.L. *et al.* (1999). Occurrence of nontuberculous mycobacteria in environmental samples. *Appl Environ Microbiol*, 65(6): 2492–2496.

Dalziel, T.K. (1913). Chronic interstitial enteritis. *Br Med J*, 3(2): 1068–1070.

Ellingson, J.L., Bolin, C.A. and Stabel, J.R. (1998). Identification of a gene unique to *Mycobacterium avium* subspecies *paratuberculosis* and application to diagnosis of paratuberculosis. *Mol Cell Probes*, 12: 133–142.

Falkinham, J.O. III. (1996). Epidemiology of infection by nontuberculosis mycobacteria. *Clin Microbiol Rev*, 9: 177–215.

Falkinham, J.O. III, Norton, C.D. and Le Chevallier, M.W. (2001). Factors influencing numbers of *Mycobacterium avium*, *Mycobacterium intracellulare* and other myco-bacteria in drinking water distribution systems. *Appl Environ Microbiol*, 67(3): 1225–1231.

Frothingham, R. and Wilson, K.H. (1994). Molecular phylogeny of the *Mycobacterium avium* complex demonstrates clinically meaningful divisions. *J Infect Dis*, 169: 305–312.

Glickman, S.E., Kilburn, J.O., Butler, W.R. *et al.* (1994). Rapid identification of mycolic acid patterns of mycobacteria by liquid chromatography using pattern recognition soft-ware and a *Mycobacterium* library. *J Clin Microbiol*, 32: 740–745.

Goodfellow, M. and Wayne, L.G. (1982). Taxonomy and nomenclature. In *The Biology of the Mycobacteria*, vol. 1, Ratledge, C. and Stanford, J. (eds). London: Academic Press, pp. 471–521.

Grange, J.M., Yates, M.D. and Boughton, E. (1990). Review: the avian tubercle bacillus and its relatives. *J Appl Bacteriol*, 68: 411–431.

Grant, I.R., Ball, H.J. and Rowe, M.T. (1998). Effect of high-temperature, short-time (HTST) pasteurisation on milk containing low numbers *of Mycobacterium paratuber-culosis*. *Lett Appl Microbiol*, 26: 166–170.

Green, E.P., Tizard, M.L.V., Moss, M.T. *et al.* (1989). Sequence and characteristics of IS900, an insertion element identified in a human Crohn's disease isolate of *Mycobacterium paratuberculosis*. *Nuc Acid Res*, 17: 9063–9073.

Hansen, G.A. (1874). Undersogelser angaende spendalskhedens arsager. *Norsk Magazin for Laegevidenskaben*, 4: 1–88.

Harris, G., Rayner, A., Blair, J. *et al.* (2000). Comparison of three isolation systems for the culture of mycobacteria from respiratory and non-respiratory samples. *J Clin Pathol*, 53: 615–618.

Hermon-Taylor, J. (1993). Causation of Crohn's disease: the impact of clusters. *Gastroenterology*, 104: 643–646.

Hermon-Taylor, J. (1998). The causation of Crohn's disease and treatment with antimicro-bial drugs. *Ital J Gastroenterol Hepatol*, 30(6): 607–610.

Hermon-Taylor, J., Tizard, J., Sanderson, J. *et al.* (1994). Mycobacteria and the aetiology of Crohn's disease. *Inflammatory Bowel Dis*, http://iol.ie/alank/CROHNS/paratub.htm.

Horsburgh, C.R. Jr. (1991). *Mycobacterium avium* complex infection in the acquired immunodeficiency syndrome. *New Engl J Med*, 324: 1332–1338.

Iivanainen, E. (1995). Isolation of mycobacteria from acidic forest soil samples: compari-son of culture methods. *J Appl Bacteriol*, 78: 663–668.

Iivanainen, E., Martikainen, P.J., Väänänen, P.K. *et al.* (1993). Environmental factors affecting the occurrence of mycobacteria in brook waters. *Appl Environ Microbiol*, 32: 398–404.

Iivanainen, E., Martikainen, P.J. and Katila, M.L. (1995). Effect of freezing of water sam-ples on viable counts of environmental mycobacteria. *Lett Appl Microbiol*, 21: 257–260.

Iivanainen, E., Martikainen, P.J., Väänänen, P.K. *et al.* (1999). Environmental factors affecting the occurrence of mycobacteria in brook sediments. *J Appl Microbiol*, 86(4): 673–681.

Inderlied, C.B., Kemper, C.A. and Bermudez, L.E.M. (1993). The *Mycobacterium avium* complex. *Clin Microbiol Rev*, 6(3): 266–310.

Jenkins, P.A., Pattyn, S.R. and Portaels, F. (1982). Diagnostic bacteriology. In *The Biology of the Mycobacteria*, vol. 1, Ratledge, C. and Stanford, J. (eds). London: Academic Press, pp. 441–470.

Johne, H.A. and Frothingham, L. (1895). Em eigenthuemlicher Fall von tuberkulose beim Rind. *Dtsch Ztschr Tiermed Path*, 21: 438–454.

Kamala, T., Paramasivan, C.N., Herbert, D. *et al.* (1994). Evaluation of procedures for isolation of nontuberculous mycobacteria from soil and water. *Appl Environ Microbiol*, 60(3): 1021–1024.

Koch, R. (1882). Die aetilogie der tuberculose. *Berl Klin Wochensch*, 19: 221–230.

Kuritsky, J.N., Bullen, M.G., Broome, C.V. *et al.* (1983). Sternal wound infections and endocarditis due to organisms of the *Mycobacterium fortuitum* complex. *Ann Intern Med*, 98: 938–939.

Lalande, V., Barbut, F., Varnerot, A. *et al.* (2001). Pseudo-outbreak of *Mycobacterium gordonae* associated with water from refrigerated fountains. *J Hosp Infect*, 48: 76–79.

Laverdiere, M., Poirier, L., Weiss, K. *et al.* (2000). Comparative evaluation of the MB/BacT and BACTEC 460 TB systems for the detection of mycobacteria from clinical specimens: clinical relevance of higher recovery rates from broth-based detection systems. *Diag Microbiol Infect Dis*, 36: 1–5.

Mediel, M.J., Rodriguez, V., Codina, G. *et al.* (2000). Isolation of Mycobacteria from frozen fish destined for human consumption. *Appl Environ Microbiol*, 43: 3637–3638.

Millar, D., Ford, J., Sanderson, J.D. *et al.* (1996). IS900 PCR to detect *Mycobacterium paratuberculosis* in retail supplies of whole pasteurised cows' milk in England and Wales. *Appl Environ Microbiol*, 62: 3446–3452.

Miltner, E.C. and Bermudez, L.E. (2000). *Mycobacterium avium* grown in *Acanthamoeba castellanii* is protected from the effects of antimicrobials. *Antimicrob Agents Chemother*, 44(7): 1990–1994.

Miyamoto, M., Yamaguchi, Y. and Sasatsu, M. (2000). Disinfectant effects of hot water, ultraviolet light, silver ions and chlorine on strains of *Legionella* and nontuberculous mycobacteria. *Microbios*, 101(398): 7–13.

Neumann, M., Schulze-Röbbecke, R., Hagenau, C. *et al.* (1997). Comparison of methods for isolation of mycobacteria from water. *Appl Environ Microbiol*, 63: 547–552.

O'Brien, R.J., Geiter, L.J. and Snider, D.E. Jr. (1987). The epidemiology of nontuberculous mycobacterial diseases in the United States. Results from a national survey. *Am Rev Respir Dis*, 135: 1007–1014.

Olano, J.P., Holmes, H. and Woods, G.L. (1998). Evaluation of the MycoAKT Latex Agglutination test for rapid diagnosis of *Mycobacterium avium* complex infection. *Diag Microbiol Infect Dis*, 30: 71–74.

Palacios, J.J., Ferro, J., Ruiz Palma, N. *et al.* (1999). Fully automated liquid culture system compared with Löwenstein-Jensen solid medium for rapid recovery of mycobacteria from clinical samples. *Eur J Clin Microbiol Infect Dis*, 18: 265–273.

Pelletier, P.A., Du Moulin, G.C. and Stottmeier, K.D. (1988). Mycobacteria in public water supplies: comparative resistance to chlorine. *Microbiol Sci*, 5(5): 147–148.

Peters, M., Muller, C., Rusch-Gerdes, S. *et al.* (1995). Isolation of atypical mycobacteria from tap water in hospitals and homes: is this a possible source of disseminated MAC infection in AIDS patients? *J Infect*, 31(1): 39–44.

Petroff, S.A. (1915). A new and rapid method for the isolation and cultivation of tubercle bacilli directly from the sputum and feces. *J Exp Med*, 21: 38–42.

Plikaytis, B.B., Plikaytis, B.D., Yakrus, M.A. *et al.* (1992). Differentiation of slowly growing *Mycobacterium* species, including *Mycobacterium tuberculosis*, by gene amplification and restriction fragment length polymorphism analysis. *J Clin Microbiol*, 30(7): 1815–1822.

Portaels, F. and Pattyn, S.R. (1982). Growth of mycobacteria in relation to the pH of the medium. *Ann Microbiol (Paris)*, 133B: 213–221.

Portaels, F., De Muynck, A. and Sylla, M.P. (1988). Selective isolation of mycobacteria from soil: a statistical analysis approach. *J Gen Microbiol*, 134: 849–855.

Roggenkemp, A., Hornef, M.W., Masch, A. *et al.* (1999). Comparison of MB/BacT and BACTEC 460 TB systems for recovery of mycobacteria in a routine diagnostic laboratory. *J Clin Microbiol*, **37**(11): 3711–3712.

Rohner, P., Ninet, B., Metral, C. *et al.* (1997). Evaluation of the MB/BacT system and comparison to the Bactec 460 system and solid media for isolation of mycobacteria from clinical specimens. *J Clin Microbiol*, **35**(12): 3127–3131.

Roth, A., Reischl, U., Streubel, A. *et al.* (2000). Novel diagnostic algorithm for identification of mycobacteria using genus-specific amplification of the 16S-23S rRNA gene spacer and restriction endonucleases. *J Clin Microbiol*, **38**(3): 1094–1104.

Saito, H., Tomioka, H., Sato, K. *et al.* (1990). Identification of various serovar strains of *Mycobacterium avium* complex by using DNA probes specific for *Mycobacterium avium* and *Mycobacterium intracellulare*. *J Clin Microbiol*, **28**: 1694–1697.

Schulze-Röbbecke, R. (1993). Mycobacteria in the environment. *Immun Infekt*, **21**(5): 126–131.

Schulze-Röbbecke, R. and Buchholtz, K. (1992). Heat susceptibility of aquatic mycobacteria. *Appl Environ Microbiol*, **58**(6): 1869–1873.

Schulze-Röbbecke, R, Janning, B. and Fischeder, R. (1992). Occurrence of mycobacteria in biofilm samples. *Tuber Lung Dis*, **73**(3): 141–144.

Sniadack, D.H., Ostroff, S.M., Karlix, M.A. *et al.* (1993). A nosocomial pseudo-outbreak of *Mycobacterium xenopi* due to a contaminated potable water supply: lessons in prevention. *Infect Control Hosp Epidemiol*, **14**: 636–641.

Thompson, D.E. (1994). The role of Mycobacteria in Crohn's disease. *J Med Microbiol*, **41**: 74–94.

Thorel, M.F., Krichevsky, M. and Levy-Frebault, V.V. (1990). Numerical taxonomy of mycobactin-dependent mycobacteria, emended description of *Mycobacterium avium*, and description of *Mycobacterium avium* subsp. *avium* subsp. Nov., *Mycobacterium avium* subsp. *paratuberculosis* subsp. Nov., and *Mycobacterium avium* subsp. *silvaticum* subsp. Nov. *Int J Syst Bacteriol*, **40**: 254–260.

Thoresen, O.F. and Saxegaard, F. (1991). Gen-Probe rapid diagnostic system for the *Mycobacterium avium* complex does not distinguish between *Mycobacterium avium* and *Mycobacterium paratuberculosis*. *J Clin Microbiol*, **29**: 625–626.

Tizard, M.L.V., Bull, T., Millar, D. *et al.* (1998). A low G+C content genetic island in *Mycobacterium avium* subsp. *paratuberculosis* and M. *avium* subsp. *silvaticum* with homologous genes in *Mycobacterium tuberculosis*. *Microbiology*, **144**: 3413–3423.

Twort, F.W. (1911). A method for isolating and growing the lepra bacillus of man and the bacillus of Johne's disease in cattle. *Vet J*, **3**(67): 118–120.

Twort, F.W. and Ingram, G.L.Y. (1912). A method for isolating and cultivating the *Mycobacterium enteritidis chronicae pseudotuberculosae bovis*, Johne and some experiments on the preparation of a diagnostic vaccine for pseudotuberculous enteritis of bovines. *Vet J*, **68**: 353–365.

Von Rehn, C.F., Maslow, J.N., Barber, T.W. *et al.* (1994). Persistent colonisation of potable water as a source of *Mycobacterium avium* infection in AIDS. *Lancet*, **343**: 1137–1141.

Wallace, R.J. (1987). Nontuberculous mycobacteria and water: a love affair with increasing clinical importance. *Infect Dis Clin North Am*, **1**(3): 677–686.

Wallace, R.J. Jr, Brown, B.A. and Griffith, D.E. (1998). Nosocomial outbreaks/pseudo-outbreaks caused by nontuberculous mycobacteria. *Ann Rev Microbiol*, **52**: 453–490.

Wolinsky, E. (1992). Mycobacterial diseases other than tuberculosis. *Clin Infect Dis*, **15**: 1–10.

Woods, G.L., Fish, G., Plaunt, M. *et al.* (1997). Clinical evaluation of Difco ESP culture system II for growth and detection of mycobacteria. *J Clin Microbiol*, **35**: 121–124.

# 12

# *Salmonella*

## Basic microbiology

*Salmonella*, as a group, are facultatively anaerobic, non-spore-forming, Gram-negative bacilli. On average they are 2–5 μm long and 0.8–1.5 μm wide. Motility, aided by peritrichous flagella, is a fundamental criterion for the identification of *Salmonella*. However, a large number of non-motile strains have been isolated from the clinical environments. Somatic and flagellar antigens closely relate salmonella serotypes to each other, with many strains showing diphasic variation. A large number of *Salmonella* also express type 1 fimbriae.

    *Salmonella* constitute a very pathogenic collection of species that cause infections to a large diverse collection of animals. *Salmonella* predominantly give rise to enteritis and to typhoid-like diseases.

## Origin

The group *Salmonella* is now considered to comprise two species, namely *S. enterica* and *Salmonella bongori*. There are six subspecies of *S. enterica*, the most important of which is *S. enterica* subsp. *enterica* (subspecies I) that includes the typhoid and paratyphoid bacilli. Members of the other five sub-species (II–VI), found in the natural environment, are primarily parasites of cold-blooded animals.

*Salmonella* possesses two sets of antigens; a flagellar or H antigen and a heat stable polysaccharide known as the somatic or O antigen. Some *Salmonella* produce a surface polysaccharide, one example is *Salmonella typhi* where the Vi antigen is very important in its identification. The antigenic structure of any salmonella is expressed as an antigenic formula that is made up of three parts. These are the O antigens, the phase 1 H antigens and the phase 2 H antigens.

Early identification of *Salmonella* species involved description of the disease it caused or the host with which the serotype was associated, this of course led to major problems. These problems were overcome by the introduction of a new system where each new type of *Salmonella* was named after the place in which it was first isolated. The first published table of Salmonella serotypes contained some 20 entires. This was increased in 1995 to 2399 (Popoff *et al.*, 1995).

With the introduction of more modern taxonomic techniques, Le Minor *et al.* (1982a, b) have suggested that all serotypes of *Salmonella* probably belonged to one DNA-hybridization group within which seven subgroups were identi-fied. The seven DNA-hybridization subgroups were designated as subspecies: *S. enterica* subspp. *enterica, salamae, arizonae, diarizonae, houtenae, bongori* and *indica*. However, the DNA-DNA hybridization studies of Le Minor and colleagues (Le Minor *et al.*, 1982a, 1986) suggested that DNA subgroup V (*S. enterica* subsp. *bongori*) had evolved significantly from the other six sub-species. Results from multilocus enzyme electrophoresis (MLEE) studies sup-ported that observation and the elevation of *S. enterica* subsp. *bongori* to the level of species as *S. bongori* was proposed (Reeves *et al.*, 1989).

## Metabolism and physiology

Salmonellae are facultative anaerobes. They are catalase-positive, oxidase-negative and ferment glucose and mannitol to produce acid or acid and gas. This also seems true for sorbitol. While *S. arizonae* is able to ferment lactose, this is the exception rather than the rule. As a group, *Salmonella* are able to ferment sucrose but rarely adonitol and overall do not form indole. They also do not hydrolyse urea or deaminate phenylalanine, but usually form $H_2S$ on triple sugar iron agar and use citrate as sole carbon source. They form lysine and ornithine decarboxylases, exceptions to this include *S. paratyphi* A and *S. typhi*. Salmonellae yield negative Voges-Proskauer and positive methyl red tests and do not produce cytochrome oxide. Salmonellae are not able to deaminate tryptophan or phenylalanine and are usually urease and indole negative.

However, for a more detailed identification of *Salmonella*, isolates are generally serotyped, especially for epidemiological investigations. As mentioned previously, typing of *Salmonella* is based on the recognition of bacterial surface antigens – the thermostable polysaccharide cell wall or somatic ('O') antigens and the thermolabile flagella proteins or 'H' antigens. It is also possible to subtype *Salmonella* serotypes on the basis of phage typing. These subdivisions of *Salmonella* are made possible by plasmid profiling, ribotyping or pulsed field gel electrophoresis (PFGE) of DNA fragments.

## Clinical features

There are three clinically distinguishable forms of salmonellosis that occur in humans. These include gastroenteritis, enteric fever and septicaemia. Gastroenteritis is an infection of the colon which usually occurs 18–48 hours after ingestion of *Salmonella*. Gastroenteritis is characterized by diarrhoea, fever and abdominal pain. The infection is usually self-limiting, lasting 2–5 days.

Enteric fever is most often caused by *S. typhi* (typhoid fever) and the paratyphoid bacilli, *S. paratyphi* A, B, and C. Enteric fever from *S. typhi* is more prolonged and has a higher mortality rate than paratyphoid fever. Symptoms include sustained fever, diarrhoea, abdominal pain and may involve fatal liver, spleen, respiratory and neurological damage. Typhoid fever symptoms persist for 2–3 weeks. Enteric fevers from other than *S. typhi* have a shorter incubation period, 1–10 days, compared to 7–14 days for typhoid fever, and the symptoms are less severe.

Chills, high remittent fever, anorexia and bacteraemia characterize *Salmonella* septicaemia. Organisms may localize in any organ in the body and produce focal lesions resulting in meningitis, endocarditis, pneumonia or osteomyelitis. The incubation period for human salmonellosis is usually between 12 and 36 hours, but where people have consumed a large number of cells, incubation periods of 6 hours or less have been recorded. However, it has been documented in some *Salmonella* outbreaks that 12 days have elapsed between consumption of contaminated material and clinical illness.

## Pathogenicity and virulence

Most serotypes of *Salmonella* are pathogenic in mammals and birds and generally belong to subsp. 1. When infected with *Salmonella* the disease may present in three ways. First, a few host-adapted serotypes cause systemic disease in their hosts. When the manifestations of systemic disease are mainly septicaemic, as is usually the case with typhoid and paratyphoid bacilli in humans, the clinical picture is one of enteric fever with an incubation period of

10–20 days, but with outside limits of 3 and 56 days depending on the infecting dose (Mandal, 1979). Diarrhoea, starting 3–4 days after onset of fever and lasting, on average, 6 days, may occur in 50% of cases of typhoid fever and is more common in younger, than in older children or adults in whom intestinal symptoms may be absent or insignificant (Roy *et al.*, 1985). Second, certain other serotypes – Blegdam, Bredeney, Choleraesuis, Dublin, Enteritidis, Panama and Virchow in humans and Gallinarum in adult fowl – are also invasive but tend to cause pyaemic infections and to localize in the viscera, meninges, bones, joints and serous cavities. Third, most other salmonellae, the ubiquitous serotypes found in a number of animal species, tend to cause an acute, but mild, enteritis with a short incubation period of 12–48 h, occasionally as long as 4 days.

Local abscesses and even pyaemia may develop occasionally as a late complication of a diarrhoeal disease. Serotypes that usually cause enteritis in healthy adults may cause septicaemia or pyaemic infections in young or elderly patients. The incidence of laboratory-confirmed salmonellosis in AIDS patients is 20–100 times that in the general population (Angulo and Swerdlow, 1995). Recurrent non-typhoidal salmonella septicaemia (RSS) has been included as an AIDS-defining illness since 1987. Typhimurium, generally associated with enteritis in humans, may give rise to more severe disease in other hosts. Again, the typhoid bacillus, often considered to cause mild and atypical infections in children, will give rise to severe infections with high mortality in children whose previous health and nutrition have been poor (Scragg *et al.*, 1969).

Convalescent patients may continue to excrete salmonellae, usually in the faeces or urine but also from other sites after pyaemic infections, even after treatment. Asymptomatic persons may also harbour salmonellae unknowingly. Both remain potentially infectious for weeks or months and some become life-long excreters.

The factors responsible for the virulence of salmonellae are still being researched and are far from being fully understood – progress remains slow. Because salmonellae are able to invade, survive and replicate within eukaryotic cells this forms an important virulence mechanism for successful infection. Aside from this, salmonellae possess several virulence factors that contribute to the disease process; these virulence-promoting factors are generally conserved among species, as are the mechanisms by which bacteria interact with the host cell. Such factors facilitate entry into (invasion of) non-phagocytic cells, survival in the intracellular environment and replication within host cells.

After oral ingestion, provided a sufficient number of *Salmonella* cells have been injected, disease is initiated by the adhesion of *Salmonella* onto the epithelial lining of the intestines. The next stage in the disease process is penetration of the intestinal epithelium. It is well documented that invasive salmonellae interact with the well-defined apical microvilli of epithelial cells and disrupt the brush border. Once these cells have been penetrated, the bacteria are enclosed in membrane-bound vacuoles within the host cytoplasm. Virulent salmonellae survive and multiply within these vacuoles, eventually lysing their host cells and disseminating to other parts of the body. Salmonellae also depolarize epithelial barriers by affecting the integrity of tight junctions between cells, which has the result of promoting significant cytotoxic damage to epithelial cells.

Induction of protein synthesis *de novo* is necessary for invasion and is regulated by the microenvironment (low oxygen), growth phase or the epithelial cell surface. Only viable, metabolically active salmonellae can adhere to host cell surfaces. Adherence is followed almost immediately by internalization into host cells. The bacterial genes necessary for salmonellae to enter eukaryotic cells have yet to be fully characterized (Finlay *et al.*, 1992; Guiney *et al.*, 1995).

Internalization of salmonellae through the apical surface of epithelial cells is associated with disruption of the eukaryotic brush border. After a lag period of about 4 hours, invading organisms begin to multiply within the vacuoles of host epithelial cells. For Typhimurium, it has been demonstrated that this intracellular replication is essential for pathogenicity. Salmonellae remain within membrane-bound inclusions and their survival appears to involve blockage of phagosome–lysosome fusion events.

## Treatment

The treatment for *Salmonella* can be divided into three areas namely: enteric fever, gastroenteritis and salmonella bacteraemia. In enteric fever many patients are treated with oral chloramphenicol. With very ill patients intravenous treatment may be necessary. With the use of chloramphenicol comes the problem of bone marrow toxicity and the emergence of plasmid-mediated resistance. Because of this, the drug of choice is ciprofloxacillin for adult typhoid. With gastroenteritis, the management requires fluid and electrolyte replacement and control of nausea, vomiting and pain. The role of antibiotics is limited but is necessary where a patient has an increased risk of bacteraemia and salmonella invasion. This is particularly important in children under 3 months of age, ulcerative colitis and patients who are immunocompromised. Patients with salmonella bacteraemia also require antimicrobial treatment with chloramphenicol, co-trimoxazole, and high doses of ampicillin or ciprofloxacin.

## Survival in the environment

Reservoirs of *Salmonella* are domestic and wild animals, including poultry, swine, cattle, birds, dogs, rodents, tortoises, turtles and cats. Humans also serve as a reservoir, e.g. convalescent carriers and those with asymptomatic infections. The occurrence of chronic carriers is rare in humans, but is common in birds and animals. Infection from *Salmonella* occurs through ingestion of food, milk or water contaminated with faeces from infected hosts or by ingestion of the infected meat products. *S. typhi* and *S. paratyphi* are not widely distributed in nature with colonization only occurring in humans. Human infection with these organisms indicates exposure to human faeces. Contrary to this, non-typhoidal *Salmonella* spp. are widely distributed in nature and closely associated with animals.

The most common source of transmission of *Salmonella* seems to be the consumption of contaminated poultry and meat products. Contamination of meat products occurs particularly when exposed to faecal matter during slaughter. Once the meat is contaminated, improper storage or undercooking allows *Salmonella* to proliferate. When *Salmonella* are excreted in faeces, contamination of food and water permits transmission of the infection to humans. Person-to person, faecal-oral transmission does occur and has been a problem in health care facilities traced to inadequate hand washing.

Studies undertaken to investigate the survival of enteric pathogens in aquatic environments have shown that *Salmonella* may enter a 'viable but non-culturable' physiological state. The vast majority of detection and enumeration techniques require culturing *Salmonella* using selective media. As a result if one just relies on culturable techniques for the detection of viable organisms a gross underestimation of the true extent of *Salmonella* viability in the environment may be realized. Salmonellae are frequently isolated in polluted waters and can persist in high-nutrient waters. Wastewater treatment reduces but does not completely eliminate *Salmonella*. Reported *Salmonella* levels in non-chlorinated wastewater effluents range from 1 to 1100/cells per ml.

With the exception of *S. typhi*, water is not often implicated as the vehicle in outbreaks of human infection. Nevertheless, faecal contamination of groundwater or surface waters, and insufficient treatment or inadequate disinfection of drinking water have been identified as causes of waterborne outbreaks of salmonellosis. However, food is a much more important source of exposure to *Salmonella* infection. It seems though that contaminated water is important indirectly when it is used in food processing or preparation.

Natural water courses are important in the transmission of infection between herds of food animals and therefore play a role in zoonotic transmission.

## Survival in water

Salmonellae are associated with faecal pollution, and may be found in any water due to contamination. They have been isolated from sewage-polluted surface waters (fresh, estuarine and marine) and groundwater. A significant factor in some outbreaks has been contamination of well water (subsequently consumed untreated) by sewage, runoff from agricultural land, leaking domestic drains, and seepage from septic tanks. Most *Salmonella* serotypes are capable of prolonged survival in water and may be able to grow in heavily polluted water in the warmer months. Standard disinfection used in drinking water treatment procedures, in light of the documented evidence, are active against salmonellae, although there is some evidence that they are more resistant to inactivation than coliforms. However, absence of coliforms and *E. coli* from treated drinking water should not be relied upon as adequate assurance that salmonellae will be absent. Regrowth, in the form of a biofilm, of salmonellae in potable water distribution systems is thought to be possible. If *Salmonella* are evident in drinking water biofilms, it is probable they may survive for long periods due to protection from disinfection within the biofilm matrix.

In drinking water there are no permissible levels for *Salmonella*. This is primarily due to the problem that so many waterborne pathogens are difficult to detect or enumerate in water and the methods are often impractical to use for routine monitoring. In general, because of these problems, both the US Environmental Protection Agency (US EPA) and World Health Organization have adopted total coliform, faecal coliform and *Escherichia coli* microbiological standards for drinking water. It is hoped that these organisms will indicate the efficacy of a water-treatment plant, distribution system integrity, and recent faecal contamination. However, the viable but non-culturable concept needs addressing more extensively.

## Methods of detection

Salmonellae are facultative anaerobes and grow readily on ordinary media. Salmonellae are known to grow over a wide temperature and pH range, 7–48°C and pH 4–8, respectively (Baird-Parker, 1991). Most salmonellae are prototrophic growing readily on minimal salts medium incorporating an appropriate carbon source. Auxotrophic strains of prototrophic serotypes are quite common, e.g. around 7% of Typhimurium strains (Duguid *et al.*, 1975). These auxotrophic serotypes will grow on a minimal medium supplemented with appropriate growth factor(s). On general laboratory agar some salmonellae, particularly Paratyphi B, produce mucoid colonies. These have been shown to develop at low temperature, low humidity and high osmolarity (Anderson and Rogers, 1963). It is well documented that the presence of a mucoid surface layer on the cell surface, the M antigen of Kauffmann, inhibits O and H agglutinability. Because the M antigen is identical in all salmonella serotypes and is like the colanic acid materials of other Enterobacteriaceae, it is unimportant diagnostically.

If *Salmonella* is being isolated from mixed bacterial samples (i.e. faeces, food and environmental), selective enrichment is needed. Within the diagnostic laboratory three selective enrichment media, which are commonly used, are tetrathionate broth, selenite broth, and Rappaport-Vassiliadis (RV) medium. Colonies of *Salmonella* forming on this agar are generally circular with a smooth surface and an even edge, or flat with an uneven surface and serrated edge. While the temperature required for the optimal growth of *Salmonella* is 37°C, they are able to grow at temperatures ranging from 10 to 43°C.

After primary isolation on selective media, presumptive *Salmonella* isolates can be tested with commercial identification systems or screened with triple sugar iron agar, urea broth, and lysine iron agar. Any isolate that has a biochemical pathway for *Salmonella* should then be tested with commercial polyvalent O group, H, and Vi antisera.

With environmental samples, a large sample volume usually should be examined (1 litre or more). Concentration of the organisms can be accomplished using Moore swabs, membrane filtration, diatomaceous earth, or large-volume samplers. Although salmonellae can be recovered readily from water, sewage and related environments they may be present in low numbers and may be sublethally injured by exposure to such environments. It is normal practice for water samples to be incubated in 'pre-enrichment' media, such as buffered peptone water (BPW) or lactose broth, for 18–24 hours at 37°C, although some studies using naturally contaminated sewage samples have shown 43°C to be superior. Volumes of pre-enrichment media are added to a selective broth such as Rappaport-Vassiliadis (RV), selenite or tetrathionate broths. Following selective culture the broths are plated onto selective/diagnostic agars. As with the liquid media, there are many agars commercially available. It is common practice for two media to be used and most laboratories will choose two from xylose lysine deoxycholate (XLD), deoxycholate citrate (DCA) or brilliant green (BG) agars. Newer plating media such as Rambach and XLT4 agars have been shown to improve isolation rates of *Salmonella*.

*Salmonella*-like colonies on selective agars need to be confirmed by biochemical testing, and serology and phage-typing where appropriate. It may take up to 5 days to complete sample examination. This delay has led to the investigation of more rapid methods of *Salmonella* detection. These tend to be genome-based and make use of techniques such as the polymerase chain reaction (PCR) where specific sections of *Salmonella* DNA are targeted.

## Epidemiology of waterborne outbreaks

With the exception of host-specific salmonellae, such as *S. typhi* in man and *S. gallinarum* in poultry, salmonellosis is regarded as a zoonosis. Human illness is usually the result of the consumption of contaminated foods or milk, although contaminated drinking and recreational waters have been implicated. *Salmonella* outbreaks have also occurred when contaminated water has been used to cool cans of food after heat treatment. The number of bacteria required to cause infection in humans will vary depending on the individuals at risk, their age, general health and the presence of underlying disease. The vast majority of waterborne outbreaks of salmonellosis are classified as acute gastrointestinal illness of unknown aetiology. Waterborne outbreaks usually involve poor-quality source water, inadequate treatment, or contamination of the distribution system (e.g. cross-connections) (Angulo *et al.*, 1997). Large outbreaks of waterborne salmonellosis have not been reported and would suggest that these organisms are not a major problem as a waterborne pathogen in correctly treated water. Despite this, salmonellosis represents a major communicable world-wide disease problem. The annual occurrence of typhoid fever is estimated at 17 million cases, with 600 000 deaths. In contrast to typhoid fever, with humans being the sole source of the organism, animals and animal products are major sources of other *Salmonella*. Large numbers of *Salmonella* may be present in contaminated surface water and waste treatment plant influents and effluents. Studies have shown that 80% of activated sludge effluents and 58% of contaminated surface waters may contain *Salmonella*.

## Risk assessment

*Health effects*: occurrence of illness, degree of morbidity and mortality, probability of illness based on infection:

- Three clinically distinguishable forms of salmonellosis occur in humans. These include gastroenteritis, enteric fever and septicaemia.
- *Salmonella enteritidis* causes gastroenteritis, with fever, cramps and diarrhoea. This is the most common infection in developed countries.
- Enteric fever is most often caused by *S. typhi* (typhoid fever) and the paratyphoid bacilli, *S. paratyphi* A, B and C. Enteric fever from *S. typhi* is more prolonged and has a higher mortality rate than paratyphoid fever. Symptoms include sustained fever, abdominal pain and may involve fatal liver, spleen, respiratory and neurological damage.
- Death rates for typhoid fever range from 12 to 30%.
- *Salmonella* septicaemia is characterized by chills, high remittent fever, anorexia and bacteraemia.
- Two per cent of cases result in chronic arthritis.
- According to the US Centers for Disease Control and Prevention, 1.4 million cases of salmonellosis occur yearly in the USA.
- The number of bacteria required to cause infection in humans will vary depending on the individuals at risk, their age, general health and the presence of underlying disease.

*Exposure assessment*: routes of exposure and transmission, occurrence in source water, environmental fate:

- *Salmonella* spp. have been isolated and cultured from the soil, water, wastes, plants and the normal flora of animals. At present there are some 2000 *Salmonella* serotypes that are human pathogens.
- Studies have shown that 80% of activated sludge effluents and 58% of contaminated surface waters may contain *Salmonella*.
- Most *Salmonella* serotypes are capable of prolonged survival in water and may be able to grow in heavily polluted water in the warmer months.
- Regrowth of *Salmonella* in distribution systems is thought to be possible, especially if levels of assimilable organic carbon are high.

- Bacterial survival in aquatic ecosystems is affected by numerous factors, including the presence of protozoa, antibiosis, organic matter, algal toxins, dissolved nutrients, ultraviolet light, heavy metals and temperature.
- Many kinds of domestic and wild animals serve as reservoirs of *Salmonella*, including poultry, swine, cattle and birds. Humans with asymptomatic infections can also serve as reservoirs.
- Infection from *Salmonella* occurs through ingestion of food, milk or water contaminated with faeces from infected hosts or by ingestion of the infected meat and egg products or handling animals (e.g. pet turtles and lizards). Consumption of contaminated poultry and meat products is the most common source of transmission. A person-to-person transmission occurs, but not as frequently.
- Faecal contamination of groundwater or surface waters, and insufficient treatment of drinking water have been identified as causes of waterborne outbreaks of salmonellosis. However, food is a much more important source of exposure.

*Risk mitigation*: drinking-water treatment, medical treatment:

- Disinfection is generally effective on *Salmonella*. Chlorine residuals of at least 0.2 mg/l are sufficient to handle *Salmonella* in the distribution system.
- In enteric fever, many patients are treated with oral chloramphenicol. However, the drug of choice is ciprofloxacin for adult typhoid.
- With gastroenteritis, the management requires fluid and electrolyte replacement and control of nausea, vomiting and pain. Antibiotics are usually not necessary.
- Patients with *Salmonella* bacteraemia also require antimicrobial treatment with chloramphenicol, co-trimoxazole and high doses of ampicillin or ciprofloxacin.
- An increasing number of *Salmonella enteritidis* isolates have shown antibiotic resistance.

## References

Anderson, E.S. and Rogers, A.H. (1963). Slime polysaccharides of the Enterobacteriaceae. *Nature (London)*, **198**: 714–715.

Angulo, F.J. and Swerdlow, D.L. (1995). Bacterial enteric infections in persons infected with human immunodeficiency virus. *Clin Infect Dis*, **21**(Suppl. 1): S84–S93.

Angulo, F.J. *et al.* (1997). A community waterborne outbreak of salmonellosis and the effectiveness of a boil water order. *Am J Public Hlth*, **87**(4): 580–584.

Baird-Parker, A.C. (1991). *Foodborne Salmonellosis, Lancet Review of Foodborne Illness*. London: Edward Arnold, pp. 53–61.

Duguid, J.P., Anderson, E.S. *et al.* (1975). A new biotyping scheme for *Salmonella typhimurium* and its phylogenetic significance. *J Med Microbiol*, **8**: 149–166.

Finlay, B.B., Leung, K.Y. *et al.* (1992). Salmonella interactions with the epithelial cell. *ASM News*, **58**: 486–489.

Guiney, D.G., Fang, F.C. *et al.* (1995). Biology and clinical significance of virulence plasmids in Salmonella serovars. *Clin Infect Dis*, **21**(Suppl. 2): S146–S151.

Le Minor, L., Veron, M. and Popoff, M. (1982a). Taxonomie des Salmonella. *Ann Microbiol (Paris)*, **133B**: 223–243.

Le Minor, L., Veron, M. and Popoff, M. (1982b). Proposition pour une nomenclature des Salmonella. *Ann Microbiol (Paris)*, **133B**: 245–254.

Le Minor, L., Popoff, M.Y. *et al.* (1986). Individualisation d'une septième sous-espèce de Salmonella. *Ann Microbiol (Paris)*, **137B**: 211–217.

Mandal, B.K. (1979). Typhoid and paratyphoid fever. *Clin Gastroenterol*, 8: 715–735.

Popoff, M.Y., Bockemühl, J. and Hickman-Brenner, F.W. (1995). Supplement 1994 (no. 38) to the Kauffmann-White scheme. *Res Microbiol*, **146**: 799–803.

Reeves, M.W., Evins, G.M. *et al.* (1989). Clonal nature of *Salmonella typhi* and its genetic relatedness to other salmonellae as shown by multilocus enzyme electrophoresis, and proposal of *Salmonella bongori* comb. Nov. *J Clin Microbiol*, **27**: 313–320.

Roy, S.K., Speelman, P. *et al.* (1985). Diarrhea associated with typhoid fever. *J Infect Dis*, **151**: 1138–1143.

Scragg, J., Rubidge, C. and Wallace, H.L. (1969). Typhoid fever in African and Indian children in Durban. *Arch Dis Child*, **44**: 18–28.

# 13

# *Shigella*

## Basic microbiology

Shigellae are Gram-negative, non-motile rods. As a group they do not produce gas from carbohydrates. They reside presently in the family Enterobacteriaceae and show very similar characteristics to *Escherichia coli*. Unlike other members in the Enterobacteriaceae group, Shigellae are non-lactose fermenting on MacConkey agar or desoxycholate citrate agar after a period of incubation of 24 hours.

Shigellae, by definition, are non-motile and do not decarboxylate lysine or hydrolyse arginine. Except for *S. sonnei*, certain serovars within the other species, and certain strains within these serovars, they do not produce gas from glucose or decarboxylate ornithine, nor do they use sodium acetate or produce indole from tryptophan.

## Origin and taxonomy

Kiyoshi Shiga first documented shigella in 1898 (Shiga, 1898). However, there is evidence that Ogata (1892) in Japan and Chantemesse and Widal (1888) in

France may have actually isolated Shigella much earlier. The genus name *Shigella* was published in 1919 by (Castellani and Chalmers, 1919). Numerous publications on *Shigella* occurred after this date with a significant paper on its grouping in 1954 (Enterobacteriaceae Subcommittee, 1954; Rowe and Gross, 1981).

*Shigella*, based on both biochemical and serological evidence, can be classified into four major serological groups, namely: Group A, *Shigella dysenteriae*, which includes at least ten serotypes; Group B, *Shigella flexneri*, includes six serotypes; Group C, *Shigella boydii*, which includes 15 serotypes and Group D *Shigella sonnei*, which includes only one serotype. The type species is *S. dysenteriae* 1.

## Metabolism and physiology

*Shigella* are facultatively anaerobic organisms that grow poorly under anaerobic conditions. The optimal temperature for the growth of *Shigella* is 37°C, in media containing 1% peptone as a carbon and nitrogen source. Shigellae are killed at a temperature of 55°C within 1 hour (Rowe, 1990).

While shigellae are able to tolerate extremely acid conditions (pH 2.5) for short periods they prefer to be grown at neutral or slightly alkaline pH (pH 7.0–7.4).

## Clinical features

Apart from chimpanzees and monkeys, bacillary dysentery is specifically a human disease characterized by a type of diarrhoea in which the stools contain blood and mucus. In extreme cases it becomes associated with heavy inflammation of the colonic mucosa. Shigellosis, principally a self-limiting disease in healthy adults, has been known to cause fatalities, particularly in young children (Salyers and Whitt, 1994).

Shigellae are transmitted by the direct faecal-oral route with infected individuals typically excreting $10^5$–$10^9$ shigellae per gram of wet faeces, with symptomless carriers excreting $10^2$–$10^6$ per gram (Thomas, 1955; Dale and Mata, 1968). Because of this food has the potential to be contaminated through the soiled fingers of patients or carriers. The transfer of shigellae by flies breeding on faeces has been established as a very important transmission route during some outbreaks.

In a study of endemic shigellosis in Bangladesh, 2.1% of children 5 years of age and under, were found to be asymptomatic carriers (Hossain *et al.*, 1994). Similar results were reported from Mexico where 55% of infants, 2 years of age and under, infected with *Shigella* were asymptomatic (Guerrero *et al.*, 1994).

The severity of disease due to *Shigella* depends on the virulence of the infecting strain, aside from the host's immune system. The disease caused by *S. sonnei* tends to be mild and of short duration, whereas that caused by *S. flexneri* tends to be more severe. *S. boydii* and *S. dysenteriae* produce disease of varying severity but *S. dysenteriae* has often caused epidemics of severe infections.

The infective dose for Shigella is generally quite low when compared to other pathogens such as *E. coli* and *Vibrio cholerae*. The median infective dose ($ID_{50}$) for *Shigella* is around $10^4$ (Dupont *et al.*, 1972) in healthy adults compared to *Vibrio cholerae*, which is around $10^7$ or higher. Shaughnessy *et al.* (1946) found that a dose of $10^8$ organisms of *Shigella flexneri* were required to induce disease in volunteers who had previously ingested 2 g of sodium bicarbonate.

The incubation period following ingestion of *Shigella* ranges from 36 to 72 hours, but can be as short as 12 hours, with frank dysentery appearing within 2 days.

Symptoms of shigellosis usually develop suddenly, often as an abdominal colic, followed by watery diarrhoea often accompanied by fever and malaise. Some individuals do go onto develop abdominal cramps, tenesmus and the frequent passage of small volumes of stool (bloody mucus). Symptoms of dysentery can last for about 4 days. In severe cases up to 10 days has been documented. Patients who are recovering from infection often continue to excrete *Shigella* for a month following infection. There is some documented evidence that some patients can still excrete *Shigella* for over a year following recovery (Du Pont *et al.*, 1970). This carriage may be more important under conditions of poor hygiene.

*Shigella* is known to produce an exotoxin, initially described as a neurotoxin, but it also has a fluid transuding effect on the intestinal mucosa. To date the role of the toxin in the pathogenesis of dysentery is uncertain. *Shigella dysenteriae* type 1 is responsible for many cases of haemolytic uraemic syndrome (HUS), first described in 1955 (Gasser *et al.*, 1955), accompanying outbreaks of dysentery and in some parts of the world is one of the commonest forms of acute renal failure in children. Although it is an invasive disease, *Shigella* usually do not reach tissue beyond the lamina propria and, therefore, they very rarely cause bacteraemia or systemic infections except under very special circumstances (Struelens *et al.*, 1985).

## Pathogenicity and virulence

The two main virulence factors in *Shigella* are their invasive techniques and toxigenic properties, however, the exact role of the toxin produced by *Shigella* has yet to be defined.

As with most enteric pathogens, the first step in the invasion of host cells by shigellae, as demonstrated in HeLa cells, is attachment. In the case of *Shigella* this process does not seem to be mediated through interaction with a specific receptor (Clerc and Sansonetti, 1987). There is, however, mounting evidence which suggests that the invasion plasmid antigen (Ipa) D may be involved as an adhesive. The proteins, IpaB and IpaC, encoded by genes located on the virulence plasmid, have been found on the bacterial surface suggesting some importance in the pathogenicity of *Shigella* (Menard *et al.*, 1994). It has been suggested that these two proteins are essential for phagocytosis. It is thought that the proteins rupture phagocytic vesicles allowing for intercellular spread. This intercellular spread (ICS), mediated by two proteins, IcsA (also called VirG; Bernardini *et al.*, 1989) and IcsB, leads to cell death and inflammatory response in the host cells. The process of cell death, induced by these proteins is unknown to date, but seems to be independent of the production of Shiga toxin (Salyers and Whitt, 1994).

S. *dysenteriae* is known to produce a Shiga toxin, encoded on the chromosome, which belongs to the family of A1/B5 toxins. This Shiga toxin, whose main action is on blood vessels, is only released during cell lysis and not actively secreted from the cell (Rowe, 1990).

## Treatment

Most cases of shigella dysentery, due to *Shigella sonnei*, are mild and do not require any antibiotic treatment. Hydration of the patient using oral salt rehydration is generally the best treatment regimen. If antibiotics need to be administered ampicillin, tetracycline, co-trimoxazole, or ciprofloxacin are an appropriate choice for treatment. Antibiotics are usually administered because they shorten the duration of illness and decrease the relapse rate (Hruska, 1991).

The first reported case of multiple-drug-resistant *Shigella* occurred in 1956 in Japan. To date, *Shigella* is acquiring resistance against many different drugs, constituting a major concern. For example, between 1979 and 1982 the percentage of resistant S. *sonnei* strains isolated in a hospital in Madrid increased from 39.6 to 97.9% for ampicillin, from 34.4 to 96.9% for co-trimoxazole (SXT), from 6.3 to 18.0% for tetracycline and from 1.6 to 15.1% for chloramphenicol (Lopez-Brea *et al.*, 1983). The first SXT-resistant strain of S. *dysenteriae* type 1 in Bangladesh was isolated in 1982 (Shahid *et al.*, 1985) with SXT resistance also evident in Finnish travellers (Heikkilä *et al.*, 1990).

Antibiotic therapy for the treatment of shigellosis should always be based on proper antibiotic susceptibility testing. If this is not possible, blind therapy should include a quinolone as the majority of strains are susceptible.

## Survival in the environment

Despite the public health significance of *Shigella* their presence and also persistence in the environment is not well documented when compared to other species in the family Enteriobacteriaceae. Humans are the only important reservoir of *Shigella*. Excretion of *Shigella* in stools is highest during the acute phase of dysenteric illness. During this phase the environment is contaminated and the organisms can survive for weeks in cool and humid locations (Rowe and Gross, 1981). They survive for 5–46 days when dried on linen and kept in the dark, and for 9–12 days in soil at room temperature (Roelcke, 1938). Although shigellae tolerate a low pH (<3) for short periods, they soon perish, but remain alive for days if specimens are kept alkaline and are prevented from drying (Rowe, 1990).

Of all Shigella, *S. sonnei* appears to be more resistant to harsh conditions when compared to *S. dysenteriae* and *S. flexneri*. It has been shown that *Shigella sonnei* can survive for over 3 hours on fingers and up to 17 days on wooden toilet seats (Huchinson, 1956). In a study by McGarry and Stainforth (1978), *Shigella dysenteriae* was shown to survive for up to 17 days in biogas plant effluent (11–28°C) but less than 30 hours in the biogas plant itself (14–24°C).

*Shigella* may be spread in aerosol droplets, possibly by flushing toilets and spray irrigation systems. A study by Newson (1972) showed that by flushing a suspension ($10^{10}$ cells) of *Shigella sonnei* an aerosol of about 39 bacteria/mm$^3$ air was produced and that the *Shigella* could be recovered from the splashes and could survive for up to 4 days.

The effectiveness of the sewage treatment plant in inactivating *Shigella* is limited. However, research has shown that *Shigella* removal is very similar to that of *E. coli*.

## Survival in water

*Shigella* can be found in surface waters and also within contaminated drinking water. This is highly significant as a mode of transmission in developing countries. As a rule drinking water will not contain *Shigella* unless it is untreated or if there are problems with the water-treatment process (Green *et al.*, 1968).

There have been studies on the survival of *Shigella* in water (Feachem *et al.*, 1980). From this and a number of other studies it is found that survival of *Shigella* depends upon concentrations of other bacteria, nutrients, oxygen and temperature. Within clean water, survival times are less than 14 days at temperatures >20°C but in waters of less than 10°C they have been shown to survive for weeks. The half-life of *Shigella* was found to be 24 hours. Talayeva (1960) found that *Shigella flexneri* survived for up to 21 days in clean river water at a temperature of 19–24°C, up to 47 days in autoclaved river water, up to 9 days in well water, up to 44 days in autoclaved tap water and up to 6 days in polluted well water. Shrewsbury and Barson (1957) found that *Shigella dysenteriae* could survive for between 2.5 and 29 months in sterile but faecally contaminated water at 21°C. Hendricks (1972) reported that *Shigella flexneri* was also able to multiply in sterilized river water.

Infections with *Shigella* spp. are often acquired by drinking water contaminated with human faeces or by eating food washed with contaminated water. Food-borne outbreaks of shigellosis occur, especially in the tropics and less frequently in the developed countries (Coultrip *et al.*, 1977). In addition to contamination from faeces by food-handlers who have poor hygiene, flies may sometimes act as vectors in tropical countries (Khalil *et al.*, 1994).

There is evidence for *Shigella* infection acquired by swimming in sewage-contaminated recreational waters (Rosenberg *et al.*, 1976; Makintubee *et al.*, 1987; Sorvillo *et al.*, 1988). In developed countries infections are usually associated with recent travel (Parsonnet *et al.*, 1989; Lüscher and Altwegg, 1994) to countries with insufficient sanitary facilities.

Outbreaks of *Shigella* have occurred in day-care centres associated with direct person-to-person transmission by the faecal-oral route, which is facilitated by the low infectious dose (10–200 organisms) necessary to induce infection. Shigellae are the most often cause of laboratory-acquired infections (Aleksic *et al.*, 1981; Grist and Emslie, 1989).

Natural disasters and wars are frequently associated with shigellosis outbreaks and mass encampments become breeding places, causing a high incidence of illness and fatalities (Centers for Disease Control, 1994b; Sharp *et al.*, 1995).

## Methods of detection

In pure cultures *Shigella* form circular, glistening, translucent or slightly opaque colonies on nutrient agar. They are able to grow on blood agar but with no haemolysis. The colony size varies with small colonies in *S. dysenteriae*, but most *Shigella* serovars produce colonies of 1–2 mm after 18–24 hours of incubation at 37°C. *S. sonnei* may dissociate in smooth and larger, flatter colonies showing an irregular edge, often referred to as phase 1 colonies. 'Phase 2' colonies are associated with the loss of the 120 000–140 000 kDa virulence plasmid and a change of the antigenic specificity. Phase 2 strains still grow homogeneously in broth culture but they may partially sediment after boiling at 100°C or autoagglutinate in 3.5% NaCl solution (Rowe, 1990; Bockemühl, 1992). On Leifson's deoxycholate citrate agar, Salmonella-Shigella agar, or xylose-lysine-deoxycholate and on MacConkey agar, shigellae grow with colourless, translucent, smooth colonies of 1–2 mm after 18–24 hours of incubation at 37°C. After prolonged incubation (>48 hours) growth of *S. sonnei* becomes pinkish due to delayed fermentation of lactose and sucrose.

For the isolation of shigellae, a freshly passed stool specimen during the acute stage of illness is the material of choice. If present, mucus or blood-stained portions should be selected for culture and, if desired, for microscopic examination for the presence of faecal leucocytes. If the specimens cannot be cultured within 2–4 hours, they should be preserved in a transport medium.

Shigellae are easily overgrown by the concomitant aerobic intestinal flora. Enrichment of stool specimens in Gram-negative broth (Hajna) or selenite broth for about 6 hours at 37°C can be tried.

After growth on an appropriate agar, *Shigella* spp. are identified by biochemical reactions, combined with agglutination in group- or serovar-specific antisera for shigellae.

## Epidemiology and waterborne outbreaks

A number of outbreaks, primarily food related, have been due to *Shigella*. These have included poi (Lewis *et al.*, 1972), tuna (Bowen, 1980) and contaminated salads. In fact, during 1961–1975 there were 10 648 cases (72 outbreaks) of shigellosis reported in the USA, principally due to contaminated salads (Black *et al.*, 1978). The contamination of food is possibly the most important route of transmission of *Shigella* (Barrel and Rowland, 1979).

*S. dysenteriae* 1 has caused major epidemics in Central America (1969–70), Bangladesh (1972) and East Africa (since 1991), whereas *S. boydii* is mainly prevalent in South Asia and the Middle East. In Europe and North America *S. sonnei* is by far the predominant, followed by *S. flexneri* (Aleksic *et al.*, 1987; Lee *et al.*, 1991; Centers for Disese Control, 1994a) and infections due to *S. boydii* and *S. dysenteriae* are almost exclusively imported by travellers or foreign-born citizens returning from visits to their home countries (Aleksic *et al.*, 1987).

In the UK and the rest of Europe, *Shigella dysenteriae*, common during the First World War, is rare. Between 1920 and 1930 both *Shigella flexneri* and *Shigella sonnei* were endemic and of approximately equal incidence, but by 1940 *Shigella sonnei* had become dominant, increasing in incidence annually to a peak of over 49 000 (99% of all shigellae notified) in 1956. The incidence of Sonne dysentery then declined steadily in the UK to an annual average of about 3000 notified cases between 1970 and 1990. However, the numbers rose sharply in 1991 when there were several widespread community outbreaks and continued to rise to a peak of 17 000 cases in 1992. The incidence has since fallen but there were still more than 4550 cases in 1995. Infections due to other shigellae, usually imported, have remained constant during the last decade at about 800–900 a year.

Similar changes have taken place in the USA. Up to 1968 *Shigella flexneri* and *Shigella sonnei* were equally common but *Shigella sonnei* now accounts for 65% of cases and *Shigella flexneri* for about 30%.

In tropical areas shigellosis is endemic. It has been estimated that 5 million cases of shigellosis require hospital treatment. Of these 5 million cases about 600 000 die every year.

There is general agreement in the literature that the maintenance of endemic shigellosis has little or no relationship to water quality, but that it is strongly related to water availability and associated hygienic behaviour. However, there will always be specific exceptions to this; for instance, Sultanov and Solodovnikov (1977) considered that the maintenance of dysentery was due to the widespread use of polluted surface water for domestic purposes.

Some epidemics of bacillary dysentery are waterborne. An outbreak of 2000 cases of shigellosis due to *Shigella sonnei* occurred in 1966 in Scotland when the chlorination plant on the town's water supply broke down (Green *et al.*, 1968). During 1961–75, 38 waterborne outbreaks of shigellosis were reported in the USA (Black *et al.*, 1978). Most of these outbreaks involved semi-public or individual water systems and were usually the result of inadequate or interrupted chlorination of water contaminated by faeces. Such water-borne epidemics are usually dramatic but they can be terminated very quickly when the water supply is adequately treated.

## Risk assessment

*Health effects*: occurrence of illness, degree of morbidity and mortality, probability of illness based on infection:

- *Shigella* can be classified into four major serological groups. Group A, *Shigella dysenteriae*, Group B, *Shigella flexneri*, Group C, *Shigella boydii*, and Group D *Shigella sonnei*, which includes only one serotype. *Shigella sonnei* accounts for most cases of dysentery in the developed world.
- *Shigella* infection is characterized by watery or bloody diarrhoea, abdominal pain, fever, and malaise. Symptoms of dysentery can last for about 4 days but, in severe cases, up to 10 days has been documented. Shigellosis is principally a self-limiting disease in otherwise healthy adults; it has been known to cause fatalities, particularly in malnourished infants.
- *Shigella dysenteriae* type 1 is responsible for many cases of haemolytic uraemic syndrome.
- The severity of disease depends on the virulence of the infecting strain, aside from the host's immune system. The disease caused by *S. sonnei* tends to be mild and of short duration whereas that caused by *S. flexneri* tends to be more severe.
- Children and infants generally suffer a higher rate of infection and more severe disease course.

*Exposure assessment*: routes of exposure and transmission, occurrence in source water, environmental fate:

- Shigellosis is specifically a human disease that is transmitted by the direct or indirect faecal-oral route.
- Infections with *Shigella* spp. are often acquired by drinking water contaminated with human faeces or by eating food washed with contaminated

water. Transfer of shigellae by flies has been very important during some outbreaks.

- Helped by the low infectious dose (10–200 organisms) necessary to induce infection, person-to-person contact in day-care centres, institutions, and other places where hygiene may not be the best have resulted in outbreaks.
- *Shigella* can be found in surface waters and also within contaminated drinking water. This is significant as a mode of transmission in developing countries.
- Within clean water, survival times are less than 14 days at temperatures >20°C but, in waters of less than 10°C, they have been shown to survive for weeks. They do not survive at acid pH.
- *S. sonnei* appears to be more resistant to detrimental environmental conditions than *S. dysenteriae* and *S. flexneri*. *Shigella dysentaeriae* has been shown to survive for at least 6 days in septic tank effluent.

*Risk mitigation*: drinking-water treatment, medical treatment:

- Drinking water treatment that includes disinfection is sufficient to remove *Shigella*. Waterborne outbreaks have generally resulted from inadequate treatment.
- Most cases of Shigella dysentery (i.e. *S. sonnei*) are mild and do not require any antibiotic treatment. Treatment by means of hydration using an oral salt rehydration is all that is required. Ampicillin, co-trimoxazole, tetracycline or ciprofloxacin are appropriate antibiotic choices.

# References

Aleksic, S., Bockemühl, J. and Aleksic, V. (1987). Serologisch-epidemiologische Untersuchungen an 908 Shigella-Stämmen von Patienten in der Bundesrepublik Deutschland, 1978–1985. *Bundesgesundheitsblatt*, **30**: 207–210.

Barrell, R.A.E. and Rowland, M.G.M. (1979). Infant foods as a potential source of diarrhoeal illness in rural West Africa. *Trans Roy Soc Trop Med Hyg*, **73**: 85–90.

Bernardini, M.L., Mounier, J. *et al.* (1989). Identification of icsA, a plasmid locus of *Shigella flexneri* that governs intra- and intercellular spread through interaction with F-actin. *Proc Natl Acad Sci USA*, **86**: 3867–3871.

Black, R.E., Craun, G.F. and Blake, P.A. (1978). Epidemiology of common-source outbreaks of shigellosis in the United States, 1961–1975. *Am J Epidemiol*, **108**: 47–52.

Bockemühl, J. (1992). Enterobacteriaceae. In *Mikrobiologische Diagnostik*, Burkhardt, F. (ed.). Stuttgart: Thieme Verlag, pp. 119–153.

Bowen, G.S. (1980). An outbreak of shigellosis among staff members of a large urban hospital. In *Epidemic Intelligence Service 29th Annual Conference*. Atlanta, Georgia: Centers for Disease Control, p. 46.

Castellani, A. and Chalmers, A.J. (1919). *Manual of Tropical Medicine*, 3rd edn. New York: Williams Wood & Co.

Centers for Disease Control. (1994a). Summary of notifiable diseases, United States 1994. *MMWR*, **43**: 54.

Centers for Disease Control. (1994b). Health status of displaced persons following civil war – Burundi, December 1993–January 1994. *MMWR*, **43**: 701–703.

Chantemesse, A. and Widal, F. (1888). Sur les microbes de la dysentérie épidémique. *Bull Acad Méd Sér 3*, **19**: 522–529.

Clerc, P. and Sansonetti, P.J. (1987). Entry of *Shigella flexneri* into HeLa cells: evidence for directed phagocytosis involving actin polymerization and myosin accumulation. *Infect Immun*, **55**: 2681–2688.

Coultrip, R.L., Beaumont, W. and Siletchnik, M.D. (1977). Outbreak of shigellosis – Fort Bliss, Texas. *MMWR*, **26**: 107–108.

Dale, D.C. and Mata, L.J. (1968). Studies of diarrheal disease in Central America. XI. Intestinal bacterial flora in malnourished children with shigellosis. *Am J Trop Med Hyg*, **17**: 397–403.

Du Pont, H.L., Gangarosa, E.J., Reller, L.B. *et al.* (1970). Shigellosis in custodial institutions. *Am J Epidemiol*, **92**: 172–179.

DuPont, H.L., Hornick, R.B. *et al.* (1972). Immunity in shigellosis. II. Protection induced by live oral vaccine or primary infection. *J Infect Dis*, **125**: 12–16.

Enterobacteriaceae Subcommittee Reports. (1954). *Int Bull Bacteriol Nomencl Taxon*, **4**: 1–94.

Feachem, R.T.G.A., Bradley, D.J., Garelick, H. *et al.* (1980). Health aspects of excreta and sullage management: A state of the art review. Appropriate technology for water supply and sanitation, vol. 2. Washington, DC: The World Bank, Transportation, Water and Telecommunications Department.

Gasser, C., Gautier, E. *et al.* (1955). Hämolytisch-urämische Syndrome: bilaterale Nierenrindennekrosen bei akuten erworbenen hämolytischen Anämien. *Schweiz Med Wochenschr*, **85**: 905–909.

Green, D.M., Scott, S.S., Mowat, D.A.E. *et al.* (1968). Waterborne outbreak of viral gastroenteritis and Sonne dysentery. *J Hyg*, **66**: 383–392.

Grist, N.R. and Emslie, J.A.N. (1989). Infections in British clinical laboratories, 1986–7. *J Clin Pathol*, **42**: 677–681.

Guerrero, L., Calva, J.J. *et al.* (1994). Asymptomatic Shigella infections in a cohort of Mexican children younger than two years of age. *Pediatr Infect Dis J*, **13**: 597–602.

Heikkilä, E., Siitonen, A. *et al.* (1990). Increase of trimethoprim resistance among Shigella species. *J Infect Dis*, **161**: 1242–1248.

Hendricks, C.W. (1971). Enetric bacterial metabolism of stream sediment eluates. *Can J Microbiol*, **17**: 551–556.

Hossain, M.A., Hasan, K.Z. and Albert, M.J. (1994). Shigella carriers among non-diarrhoeal children in an endemic area of shigellosis in Bangladesh. *Trop Geogr Med*, **46**: 40–42.

Hruska, J.F. (1991). Gastrointestinal and intraabdominal infections. In *A Practical Approach to Infectious Diseases*, 3rd edn, Reese, R.E. and Betts, R.F. (eds). Boston: Little, Brown, pp. 305–356.

Huchinson, R.I. (1956). Some observations on the method of spread of Sonne dysentery. *Month Bull Ministr Hlth Public Hlth Lab Serv*, **15**: 110–118.

Khalil, K., Lindblom, G.B. *et al.* (1994). Flies and water as reservoirs for bacterial enteropathogens in urban and rural areas around Lahore, Pakistan. *Epidemiol Infect*, **113**: 435–444.

Lee, L.A., Shapiro, C.N. *et al.* (1991). Hyperendemic shigellosis in the United States: a review of surveillance data for 1967–1988. *J Infect Dis*, **164**: 894–900.

Lewis, J.N., Loewenstein, M.S., Guthrie, L.C. *et al.* (1972). *Shigella sonnei* outbreak on the island of Maui. *Am J Epidemiol*, **96**: 50–58.

Lopez-Brea, M., Collado, L. *et al.* (1983). Increasing antimicrobial resistance of *Shigella sonnei*. *J Antimicrob Chemother*, **11**: 598.

Lüscher, D. and Altwegg, M. (1994). Detection of shigellae, enteroinvasive and enterotoxigenic *Escherichia coli* using the polymerase chain reaction (PCR) in patients returning from tropical countries. *Mol Cell Probes*, **8**: 285–290.

Makintubee, S., Mallonee, J. and Istre, G.R. (1987). Shigellosis outbreak associated with swimming. *JAMA*, **236**: 1849–1852.

McGarry, M.G. and Stainforth, J. (eds) (1978). Compost, fertilizer and Biogas. Production from human and farm wastes in the People's Republic of China. Ottawa: International Development Research Centre.

Menard, R., Sansonetti, P. and Parsot, C. (1994). The secretion of the *Shigella flexneri* Ipa invasins is activated by epithelial cells and controlled by IpaB and IpaD. *EMBO J*, **13**: 5293–5302.

Newson, S.W.B. (1972). Microbiology of hospital toilets. *Lancet*, **2**: 700–703.

Ogata, M. (1892). Zur ätiologie der Dysenterie. *Zentralbl Bakteriol Parasitenkd Infektionskr Hyg*, **11**: 264–272.

Parsonnet, J., Greene, K.D. *et al.* (1989). *Shigella dysenteriae* type 1 infections in US travellers to Mexico. *Lancet*, **2**: 543–545.

Roelcke, K. (1938). Über die Resistenz verschiedener Ruhrkeime. *Z Hyg Infektionskr*, **120**: 307–314.

Rosenberg, M.L., Hazlet, K.K., Schaefer, J. *et al.* (1976). Shigellosis from swimming. *JAMA*, **236**: 1849–1852.

Rowe, B. (1990). Shigella. In *Topley & Wilson's Principles of Bacteriology, Virology and Immunity*, 8th edn, vol. 2, Parker, M.T. and Collier, L.H. (eds). London: Edward Arnold, pp. 455–468.

Rowe, B. and Gross, R.J. (1981). The genus Shigella. In *The Prokaryotes*, vol. II, Starr, M.P., Stolp, H. *et al.* (eds). Berlin, Heidelberg and New York: Springer-Verlag, pp. 1248–1259.

Salyers, A.A. and Whitt, D.D. (1994). *Bacterial Pathogenesis: a Molecular Approach*. Washington, DC: ASM Press.

Shahid, N.S., Rahaman, N.M. *et al.* (1985). Changing pattern of resistant Shiga bacillus (*Shigella dysenteriae* type 1) and *Shigella flexneri* in Bangladesh. *J Infect Dis*, **152**: 1114–1119.

Sharp, T.W., Thornton, S.A. *et al.* (1995). Diarrheal disease among military personnel during Operation Restore Hope, Somalia, 1992–1993. *Am J Trop Med Hyg*, **52**: 188–193.

Shaughnessy, D.J., Olson, R.C., Bass, K. *et al.* (1946). Experimental human bacillary dysentery. *JAMA*, **132**: 362–368.

Shiga, K. (1898). Über den Dysenteriebacillus (Bacillus dysentericus). *Zentralbl Bakteriol Parasitenkd Infektionskr Hyg*, **24**: 817–828, 870–874, 913–918.

Shrewsbury, J.F.D. and Barson, G.J. (1957). On the absolute viability of certain pathogenic bacteria in a synthetic well water. *J Pathol Bacteriol*, **74**: 215–220.

Sorvillo, F.J., Waterman, S.H. *et al.* (1988). Shigellosis associated with recreational water contact. *Am J Trop Med Hyg*, **38**: 613–617.

Struelens, M.J., Patte, D. *et al.* (1985). Shigella septicemia: prevalence, presentation, risk factors, and outcome. *J Infect Dis*, **152**: 784–790.

Sultanov, G.V. and Solodovnikov, Y.P. (1977). Significance of water factor in epidemiology of dysentery. *Zh Mikrobiol Epidemiol Immunobiol*, no. 6 (June), 99–101.

Talayeva, J.G. (1960). Survival of dysentery bacteria in water according to the results of a reaction with haptenes. *Microbiol Immunol*, **4**: 314–320.

Thomas, S. (1955). The numbers of pathogenic bacilli in faeces in intestinal diseases. *J Hyg*, **53**: 217–224.

# 14

# *Vibrio cholerae*

## Basic microbiology

Vibrio are Gram-negative, short rods (0.5 by 1.3–3 μm) which are often curved or comma shaped. They are non-sporulating, non-capsulated, facultative anaerobes, catalase-positive and motile by means of a single polar flagellum. In liquid media all vibrios show vigorous darting motility. Most species are oxidase-positive and reduce nitrates to nitrites.

By far the most significant of all vibrios is *V. cholerae*, which is 0.5–0.8 μm by 1.5–2.5 μm in size and the cause of cholera. To date, *V. cholerae* have been classified into 206 'O' serogroups (Yamai *et al.*, 1997). Initially *V. cholerae* was divided into O1 strains and non-O1 strains, however, there is now evidence of an O139 strain which will be discussed in more detail below. The O1 strains form two biotypes, namely classical and El Tor. These are further subdivided into three serotypes, namely Inaba, Ogawa and Hikojima (which is rare). It is probable that these three serotypes, over many years, have undergone genetic switching and possibly constitute variants of the same strain. In fact, this serotype conversion has been shown to take place both *in vivo* and *in vitro*.

The first six cholera pandemics have been caused by the classical biotype, whereas the seventh pandemic was caused by the El Tor biotype. The first reports of a new epidemic of *Vibrio cholerae* occurred in 1993. At first, the organism responsible for the outbreak was referred to as non-O1 *V. cholerae*. It was later established that the organism belonged to a new serogroup, O139 Bengal (Shimada *et al.*, 1993).

There are other potentially pathogenic *Vibrio*, aside from *V. cholerae*, such as *V. parahaemolyticus* which is a major cause of food-borne illness in South-East Asia, particularly Japan. Other species of *Vibrio* occasionally isolated from humans include *V. alginolyticus* and *V. vulnificus*. *V. alginolyticus* is a halophile and termed biotype 2 of *V. parahaemolyticus*. *Vibrio vulnificus* is a highly invasive vibrio species affecting immunocompromised persons who have consumed seafood. Other vibrios of human significance include *V. damsela*, associated with wound infections, *V. hollisae*, associated with diarrhoea, *V. mimicus*, associated with gastroenteritis and *V. fluvialis* which is a cause diarrhoea and fever.

## Origin and taxonomy

Pacini first described the infection caused by *V. cholerae* in 1854 and it was described later by Koch in 1884 (Koch, 1884). *Vibrio cholerae* O1, biotype El Tor was first detected in 1934 in Indonesia, however, prior to this it was isolated but not truly identified in Sinai in 1905. It remained pandemic in Asia until the 1960s and was detected in 1970 in Russia and South Korea. The first case of *Vibrio cholerae* in the Americas occurred in Peru in 1991 spreading to other countries in South America within weeks. Until 1992 the toxigenic O1 serogroup had been associated with cholera epidemics and pandemics. The non-O1 serogroup was mainly associated with extraintestinal infections and limited outbreaks of gastroenteritis.

To date there are presently 36 species in the genus *Vibrio* with 12 species of these potentially pathogenic to humans.

## Metabolism and physiology

Vibrios have a requirement for salt, the concentration of which ranges for the different species (Baumann *et al.*, 1984). This difference provides a means of separating pathogenic vibrios into the non-halophilic species, consisting of *V. cholerae* and *V. mimicus*, which grow on nutrient agar, and the halophilic species that require a salt supplement in the growth media.

Vibrios are able to grow over a wide temperature range (20 to >40°C). They grow better in alkaline conditions, although most species of *Vibrio* grow between a pH range of 6.5 and 9.0.

Vibrios catabolize D-glucose anaerobically via the Embden-Meyerhof pathway, producing formic, lactic, acetic, succinic acids, ethanol and pyruvate

(Baumann *et al.*, 1984). D-glucose and many other sugars are transported into *Vibrio* internally via the phosphoenolpyruvate:carbohydrate phosphotransferase system. In the case of D-glucose it is transported and subsequently phosphorylated to glucose-6-phosphate (Sarker *et al.*, 1994).

## Clinical features

*V. cholerae* causes infections ranging from asymptomatic to very serious, profuse watery diarrhoea known as 'rice water stools' (watery, colourless stools with a fishy odour and flecks of mucus). Cholera can also be characterized by a sudden onset of effortless vomiting, which leads to rapid and severe dehydration and possibly death within 1–5 days. Under these conditions, rehydration therapy is needed. If death does occur this arises because of fluid and electrolyte imbalance. The incubation period is short and the clinical signs of cholera develop within 0.5–5 days after infection.

The reservoirs of *V. cholerae* are asymptomatic human carriers and diseased people who shed the microorganisms in their faeces. In fact convalescent and asymptomatic individuals may excrete $10^2$–$10^5$ *V. cholerae*/g faeces whereas an active case excretes $10^6$–$10^9$/ml of 'rice water stool'.

Infective doses of *V. chloreae* are high in healthy individuals. Reports of $10^8$ classical *V. cholerae* in water produced diarrhoea in 50% of adult volunteers (Hornick *et al.*, 1971) and $10^{11}$ organisms produced 'rice water stools'. With the addition of 2 g of sodium bicarbonate the $ID_{50}$ was found to be $10^4$ for diarrhoea and $10^8$ for cholera-like diarrhoea. Although more than $10^8$ *V. cholerae* cells are required to induce infection and diarrhoea the administration of sodium bicarbonate ($NaHCO_3$) reduces the infectious dose to less than $10^4$ organisms (Cash *et al.*, 1974; Levine *et al.*, 1988).

## Pathogenicity and virulence

Infection due to *V. cholerae* begins with the ingestion of contaminated water or food. *Vibrio* then colonizes the epithelium of the small intestine using the toxin-coregulated pili (Taylor *et al.*, 1987). Other colonization factors such as the different haemagglutinins, accessory colonization factor, and core-encoded pilus are all thought to play a role in the adhesion process. Once adhered the enterotoxin is produced. In the small intestine *V. cholerae* produces an enterotoxin known as the cholera toxin (CT). CT consists of one A subunit (holotoxin, MW 27.2 kDa, two polypeptide chains linked by a disulphide bond) and five B subunits. It is the B subunit that is involved in attaching the toxin to a ganglioside receptor on the villi cell wall and also crypts in the intestine. The B subunit enters the host cell membrane forming a hydrophilic trans-membrane channel. This allows subunit A, which is toxic, to enter the cytoplasm. Once in the cytoplasm the toxin causes the transfer of adenosine diphosphoribose (ADP ribose) from nicotinamide adenine dinucleotide (NAD) to a regulatory protein that is responsible for the generation of intracellular cyclic adenosine monophosphate (cAMP). Over-activation of cAMP occurs due to activation of adenylate cyclase that then causes inhibition of the uptake of $Na^+$ and $Cl^-$ ions and water. Overall, there is a net outflow of water across the mucosal cells and ultimately there is extensive loss of electrolytes and water.

Cholera enterotoxin (CT) was first suggested by Robert Koch in 1884, however, its actual existence was not confirmed until 1959 (De, 1959). Another factor, which is thought to contribute to the disease process, is haemolysin/cytolysin (Honda and Finkelstein, 1979). Haemolysin has been shown to cause accumulation of bloody fluid in ligated rabbit ileal loops. Other toxins produced by *V. cholerae* include the shiga-like toxin, a heat-stable enterotoxin (Takeda *et al.*, 1991), sodium channel inhibitor (Tamplin *et al.*, 1987), thermostable direct haemolysin-like toxin (Nishibuchi *et al.*, 1992), and a nonmembrane-damaging cytotoxin (Saha and Nair, 1997).

## *V. cholerae* O1

As mentioned, *V. cholerae* O1 produces a potent and sometimes lethal cytotonic enterotoxin, known as cholera toxin (CT). The CT seems to be both structurally and functionally related to the heat-labile enterotoxin of *E. coli* (Spangler, 1992).

The structural genes encoding both toxin subunits (ctxA and ctxB) have been identified. The ctxB operon is located on a portion of the bacterial chromosome termed the core region (Ottemann and Mekalanos, 1994). In classical *V. cholerae* strains two copies of the ctx element are widely separated on the chromosome; for El Tor strains multiple copies are tandemly arranged (Mekalanos, 1983). The B subunit serves to bind the toxin to the cell receptor and the A subunit provides the toxigenic activity intracellularly after proteolytic cleavage into two peptides, A1 and A2 (Kaper *et al.*, 1995). The A1 peptide is the active portion of the molecule acting as an ADP-ribosyltransferase from NAD to a G protein, named Gs. Activation of Gs results in increased intracellular levels of cAMP, which ultimately leads to protein kinase activation, protein dephosphorylation, altered ion transport and diarrhoeal disease (Kaper *et al.*, 1995). *In vitro* CT acts as a cytotonic (non-lethal) enterotoxin causing rounding of Y1 adrenal cells or CHO cell elongation. Removal of CT from culture supernatant or preincubation of CT with antitoxin to CT causes Y1 and CHO cells to retain their original cell morphology.

Expression of CT is regulated by a 32 kDa integral membrane protein called ToxR (Ottemann and Mekalanos, 1994). Certain amino acids, osmolarity and temperature help to regulate ToxR (gene toxR) expression (Parsot and Mekalanos, 1990; Ottemann and Mekalanos, 1994). In addition to CT, ToxR also regulates several other factors. One of these is the toxin co-regulated pilus, TcpA (gene tcpA), a 20.5 kDa protein that makes up the major subunit of the *V. cholerae* pilus (Taylor *et al.*, 1987). A β-haemolysin is expressed by most *V. cholerae* O1 El Tor strains. The gene (hlyA) encodes for a mature 84 kDa protein with haemolytic and cytolytic activity (Rader and Murphy, 1988).

## *V. cholerae* O139

*V. cholerae* O139, is genetically similar to *V. cholerae* O1 El Tor (Albert, 1994). There are, however, a number of differences between *V. cholerae* O1 and O139. In O139 there is evidence of a polysaccharide capsule and LPS (Waldor *et al.*, 1994). The structure of the O139 capsule is composed of one residue each of N-acetylglucosamine, N-acetylquinovosamine, galacturonic acid and galactose and two molecules of 3,6-dideoxyxylohexose. LPS of O139 contains colitose, glucose, l-glycero-d-manno-heptose, fructose, glucosamine and quinovosamine in its polysaccharide (Hisatsune *et al.*, 1993). In O1 LPS perosamine is present but in O139 strains it is absent.

## *V. cholerae* non-O1

Non-O1 *V. cholerae* cause mild diarrhoea which is often bloody and in extreme cases it can be severe. Non-O1 *V. cholerae* have also been reported in wound infections, meningitis and bacteraemia. However, unlike O1 and O139 *V. cholerae* non-O1 strains lack the capability to cause epidemic and pandemic cholera. However, in some adult volunteers some strains have been shown to produce mild to moderate gastroenteritis. The non-O1 strains are documented as surviving better than *V. cholerae* O1 in a wide range of foods. It has also been documented that some rare non-O1 strains (<4%) possess CT. A common phenotypic feature to almost all non-O1 isolates is the production of a β-haemolysin.

## Treatment

Oral administration of fluid and electrolytes is necessary for an individual who has cholera. The formula of a rehydration solution is sodium chloride, 3.5 g, potassium chloride, 1.5 g, sodium citrate, 2.9 g and glucose, 20 g all dissolved in 1 litre of clean drinking water. In extreme cases, to reduce the excretion of *V. cholerae*, patients may be given tetracycline in order to reduce the chances of cross-contamination and environmental contamination.

## Survival in the environment

*V. cholerae* has been shown to survive for contradictory time periods in environmental waters (Feachem *et al.*, 1981). The toxigenic O1 strains of *Vibrio cholerae* have been shown to survive in aquatic environments for years, possibly residing in a biofilm. It is probable that these environmental biofilms function as a reservoir for *V. cholerae*; research in this area is warranted. While the O1 serogroup of *Vibrio cholerae* has frequently been isolated from aquatic environments most do not produce the cholera toxin (CT).

When present in the environment, *V. cholerae* are impossible to grow using conventional culture techniques. Work by Colwell *et al.* (1994) found that *V. cholerae* O1 enters a state of dormancy in response to nutrient deficiency, high levels of saline, and low temperatures. *Vibrio cholerae* have also been found in association with a wide range of aquatic life, including cyanobacteria (Islam *et al.*, 1989) and diatoms (*Skeletonema costatum*) (Martin and Bianchi, 1980) to name but a few.

### Survival in water

Water is important in the transmission of cholera, with properly treated public water supplies not generally considered to be a risk factor. In fact *V. cholerae* can survive longer in the environment than other faecal organisms suggesting a public health concern when issues of risk of disease are based on the coliform index. *Vibrio cholerae* has been isolated from surface water and drinking water and has been shown to be viable from one hour to 13 days in these environments (Pesian, 1965).

It is probable that *V. cholerae* exists in the marine environment in several forms. These include a free-living state, particularly during elevated water temperatures and nutrient concentrations; an epibiotic phase, association of vibrios with specific substrates, e.g. chitin of shellfish, and the 'viable but non-culturable' state (Colwell and Huq, 1994). It is thought that *Vibrio* form microvibrio that have altered morphologies resulting in a decrease in size and metabolic requirements/activities. These are formed under adverse conditions (Hood *et al.*, 1984).

## Methods of detection

All environmental samples suspected of containing *V. cholerae* should be transported to the laboratory at 4–10°C, inside a sterilized container and processed within 6 hours (Donovan and van Netten, 1995). For the isolation and detection of *V. cholerae*, specifically from water and the environment, a qualitative enrichment medium of alkaline peptone water (APW) is often used.

However, a number of nutrient-rich modifications of APW, such as blood–APW and egg–APW have been documented (Donovan and van Netten, 1995). Following enrichment samples are plated onto thiosulphate-citrate-bile-salts-sucrose agar (TCBS), a selective differential medium, or taurocholate-tellurite-gelatine agar and incubated at 37°C for 18–24 hours. In the case of TCBS it is known to suppress the growth of Gram-positive bacteria, pseudomonads, coliforms and aeromonads (Kobayashi *et al.*, 1963). Other media that have also been designed for the selective isolation or differentiation of *Vibrio* species have included thiosulphate-chloride-iodide agar (Beazley and Palmer, 1992) and polymyxin-mannose-tellurite agar (Shimada *et al.*, 1990).

Following the incubation of TCBS agar plates at the appropriate temperatures all suspected colonies of *V. cholerae* strains are identified by means of biochemical tests. These tests are those that are used for the identification of members of the Enterobacteriaceae and Vibrionaceae families.

For the effective identification of *Vibrio* colonies, following growth on agar, the genus can be separated into two major groups based upon their ability to utilize sucrose. Those colonies that are able to ferment sucrose form yellow colonies, indicative of the possible presence of *V. cholerae*, *V. alginolyticus*, or *V. fluvialis* and green (sucrose-negative) colonies are observed when *V. parahaemolyticus*, *V. vulnificus* or *V. mimicus* are present.

Once *V. cholerae* has been identified it is then important to establish the correct strain. For the identification of *V. cholerae* O1 or O139 an agglutination test is essential.

Blood agar appears to be a useful medium in the isolation, recognition and identification of members of the family Vibrionaceae. Many *Vibrio* species produce zones of β-haemolysis on blood agar plates after overnight incubation (35–37°C). For *V. cholerae*, haemolysis of sheep red blood cells has been traditionally used as one of several tests to distinguish the two biotypes of *V. cholerae* O1. The classic biotype is non-haemolytic while the El Tor biotype is haemolytic; non-O1 *V. cholerae* strains are usually haemolytic.

Molecular diagnostic tests have now been developed for both clinical and environmental monitoring of *V. cholerae* O1 and O139. Polymerase chain reaction (PCR), using primer pairs corresponding to the genes of the *rfb* complex, which encode the O antigen, have been designed for the detection of O1 (Hoshino *et al.*, 1998) and O139 (Albert *et al.*, 1997). These primers have been developed to detect *V. cholerae* from stool specimens. There are also specific probes for the detection of the A and B subunit genes of CT (Wright *et al.*, 1992). Molecular epidemiological techniques such as restriction fragment length polymorphism of the enterotoxin gene have been used to study outbreaks of *V. cholerae* strains (Yam *et al.*, 1991).

## Epidemiology of waterborne outbreaks

It seems that contaminated food is one of the most prevalent modes of transmission for *V. cholerae*. If we consider the USA, most cases of cholera have been associated with the consumption of partially or undercooked seafood, such as shellfish and oysters. However, water as both a direct and an indirect means of transportation of *V. cholerae* is of great significance, particularly in the developing world. Contaminated drinking water is a vector for *V. cholerae* in areas of the world that do not practise drinking water disinfection, or where treated water is susceptible to post-treatment contamination.

The source of *Vibrio* is usually the faeces of carriers or patients with cholera. However, there are reported cases of cholera being acquired from the natural water environments.

Cholera is considered an infection of over-crowding where poor standards of hygiene are prevalent. Therefore the prevention of waterborne outbreaks of *V. cholerae* could be brought about by good sanitation practices, including protection of water resource quality, sewage treatment and effective treatment of water supplies.

## Risk assessment

*Health effects*: occurrence of illness, degree of morbidity and mortality, probability of illness based on infection:

- The most significant *Vibrio* pathogen is *Vibrio cholerae*, though there are several other pathogenic species including *Vibrio parahaemolyticus* and *Vibrio vulnificus*.
- *V. cholerae* causes infections ranging from asymptomatic to deadly. Illness is marked by profuse watery diarrhoea. Sometimes there is a sudden onset of vomiting, which leads to rapid and severe dehydration and possibly death from electrolyte imbalance within 1–5 days.
- Some cholera species are associated with wound infections, bacteraemia and meningitis.
- Illnesses occur rarely in developed countries.

*Exposure assessment*: routes of exposure and transmission, occurrence in source water, environmental fate:

- The reservoirs of *V. cholerae* are asymptomatic human carriers and sick people who shed the microorganisms in their faeces. Cholera is transmitted through contaminated food or water. Person-to-person transmission is unlikely because of the high infective dose.
- Infective doses of *V. cholerae* are high in healthy people. Reports of $10^8$ organisms of classical *V. cholerae* in water produced diarrhoea in 50% of adult volunteers.
- The ability of vibrios to survive in the environment varies. It is possible that *V. cholerae* can survive for years in an aquatic environment, possibly in biofilms. More information is needed.

- *Vibrio cholerae* has been found in surface and drinking water in areas where the disease is endemic and has been shown to be viable from 1 hour to 13 days in these environments. They survive better in saline water. They can also survive for long periods in low-acid food.

*Risk mitigation*: drinking-water treatment, medical treatment:

- The vibrios are susceptible to chlorination, so countries with adequately disinfected water supplies are not susceptible to outbreaks. The prevention of waterborne outbreaks of *V. cholerae* depends on adequate sanitation practices, including sewage treatment, protection of water resource quality, and effective processing of water supplies.
- Oral administration of fluid and electrolytes is necessary for a person with cholera. Adjunct antibiotic treatment can sometimes be helpful.

# References

Albert, M.J. (1994). *Vibrio cholerae* O139. *J Clin Microbiol*, **32**: 2345–2349.

Albert, M.J. *et al.* (1997). Rapid detection of *Vibrio cholerae* O139 Bengal from stool specimens by PCR. *J Clin Microbiol*, **35**: 1663–1665.

Baumann, P., Furniss, A.L. and Lee, J.V. (1984). *Bergey's Manual of Systematic Bacteriology*, vol. 1, Krieg, N.R. and Holt, J.G. (eds). Baltimore: Williams & Wilkins, pp. 518–538.

Beazley, W.A. and Palmer, G.G. (1992). TCI – a new bile free medium for the isolation of Vibrio species. *Austr J Med Sci*, **5**: 25–27.

Cash, R.A. *et al.* (1974). Response of man to infection with *Vibrio cholerae*. I. Clinical, serologic, and bacteriologic responses to a known inoculum. *J Infect Dis*, **129**: 45–52.

Colwell, R.R. and Huq, A. (1994). Vibrio cholerae *and Cholera: Molecular to Global Perspectives*. Washington, DC: ASM Press, pp. 117–135.

Colwell, R.R. *et al.* (1994). Ecology of pathogenic vibrios in Chesapeake Bay. In *Vibrios in the Environment*, Colwell, R.R. (ed.). New York: Wiley, pp. 367–387.

De, S.N. (1959). Enterotoxicity of bacteria-free culture filtrate of *Vibrio cholerae*. *Nature*, **183**: 1533–1534.

Donovan, T.J. and van Netten, P. (1995). Culture media for the isolation and enumeration of pathogenic *Vibrio* species in foods and environmental samples. *Int J Food Microbiol*, **26**: 77–91.

Feachem, R., Miller, C. and Drasar, B. (1981). Environmental aspects of cholera epidemiology. II. Occurrence and survival of *Vibrio cholerae* in the environment. *Trop Dis Bull*, **78**: 865–880.

Finkelstein, R.A. and Lospalluto, J.J. (1969). Pathogenesis of experimental cholera: preparation and isolation of choleragen and choleragenoid. *J Exp Med*, **130**: 185–202.

Hisatsune, K., Kondo, S. *et al.* (1993). O-antigenic lipopolysaccharide of Vibrio cholerae O139 Bengal, a new epidemic strain for recent cholera in the Indian subcontinent. *Biochem Biophys Res Commun*, **196**: 1309–1315.

Honda, T. and Finkelstein, R.A. (1979). Purification and characterization of a haemolysin produced by *V. cholerae* biotype El Tor: another toxic substance produced by cholera vibrios. *Infect Immun*, **26**: 1020–1027.

Hood, M.A., Ness, G.E. *et al.* (1984). *Vibrios in the Environment*. New York: John Wiley, pp. 399–409.

Hornick, R.B., Music, S.I., Wenzel, R. *et al.* (1971). The broad street pump revisited: response of volunteers to ingested cholera vibrio. *Bull NY Acad Med*, **47**: 1181–1191.

Hoshino, K. *et al.* (1998). Development and evaluation of a multiplex PCR assay for rapid detection of toxigenic *Vibrio cholerae* O1 and O139. *FEMS Immunol Med Microbiol*, **20**: 201–207.

Islam, M.S., Drasar, B.S. and Bradley, D.J. (1989). Attachment of toxigenic *Vibrio cholerae* O1 to various freshwater plants and survival with a filamentous green alga, *Rhizoclonium fontanum*. *J Trop Med Hyg*, **92**: 396–401.

Kaper, J.B., Morris, J.G. Jr and Levine, M.M. (1995). Cholera. *Clin Microbiol Rev*, **8**: 48–86.

Koch, R. (1884). An address on chlorea and its bacillus. *Br Med J*, August 30: 403–407, 453–459.

Kobayashi, T.S. *et al.* (1963). A new selective medium for vibrio group on a modified Nakanishi's medium (TCBS agar medium). *Jap J Bacteriol*, **18**: 387–392.

Levine, M.M. *et al.* (1988). Volunteers studies of deletion mutants of *Vibrio cholerae* O1 prepared by recombinant techniques. *Infect Immun*, **56**: 161–167.

Martin, Y.P. and Bianchi, M.A. (1980). Structure, diversity and catabolic potentialities of aerobic heterotropic bacterial population associated with continuous cultures of natural marine phytoplankton. *Microb Ecol*, **5**: 265.

Mekalanos, J.J. (1983). Duplication and amplification of toxin genes in *Vibrio cholerae*. *Cell*, **35**: 253–263.

Mekalanos, J.J. (1985). Cholera toxin: genetic analysis, regulation, and role in pathogenesis. *Curr Top Microbiol Immunol*, **118**: 97–118.

Nishibuchi, M. *et al.* (1992). Enterotoxigenicity of *Vibrio parahaemolyticus* with and without genes encoding thermostable direct hemolysin. *Infect Immun*, **60**: 3539–3545.

Ottemann, K.M. and Mekalanos, J.J. (1994). Vibrio cholerae *and Cholera: Molecular to Global Perspectives*. Washington, DC: ASM Press, pp. 177–187.

Pacini, F. (1854). Osservazionemicroscopiche e deduzioni pathologiche sul Cholera Asiatico. *Gaz Med Ital Toscana Firenza*, **6**: 405–412.

Parsot, C. and Mekalanos, J.J. (1990). Expression of ToxR, the transcriptional activator of the virulence factors in *Vibrio cholerae*, is modulated by the heat shock response. *Proc Natl Acad Sci USA*, 9898–9902.

Pesian, T.P. (1965). Studies on the viability of El Tor vibrios in contaminated foodstuffs, fomites and in water. In Proceedings of the cholera research symposium. *PHS PUB*, **1328**: 317–332.

Rader, A.E. and Murphy, J.R. (1988). Nucleotide sequences and comparison of the hemolysin determinants of *Vibrio cholerae* El Tor RV79(Hly+) and RV79(Hly−) and Classical 569B(Hly−). *Infect Immun*, **56**: 1414–1419.

Saha, P.K. and Nair, G.B. (1997). Production of monoclonal antibodies to the non-membrane damaging cytotoxin (NMDCY) purified from *Vibrio cholerae* O26 and distribution of NMDCY among strains of *Vibrio cholerae* and other enteric bacteria determined by monoclonal-polyclonal sandwich enzyme-linked immunosorbent assay. *Infect Immun*, **65**: 801–805.

Sarker, R.I., Ogawa, W. *et al.* (1994). Characterization of a glucose transport system in *Vibrio parahaemolyticus*. *J Bacteriol*, **176**: 7378–7382.

Shimada, T.E. *et al.* (1993). Outbreaks of *Vibrio cholerae* non-O1 in India and Bangladesh. *Lancet*, **341**: 1347.

Shimada, T., Sakazaki, R. *et al.* (1990). A new selective, differential agar medium for isolation of *Vibrio cholerae* O1: PMT (polymyxin-mannose-tellurite) agar. *Jpn J Med Sci Biol*, **43**: 37–41.

Spangler, B.D. (1992). Structure and function of cholera toxin and the related *Escherichia coli* heat-labile enterotoxin. *Microbiol Rev*, **56**: 622–647.

Takeda, T. *et al.* (1991). Detection of heat-stable enterotoxin in a cholera toxin gene-positive strain of *V. cholerae* O1. *FEMS Microbiol Lett*, **80**: 23–28.

Tamplin, M.L. *et al.* (1987). Sodium channel inhibitors produced by enteropathogenic *Vibrio cholerae* and *Aeromonas hydrophila*. *Lancet*, **i**: 975.

Taylor, R.K., Miller, V.L. *et al.* (1987). Use of phoA gene fusions to identify a pilus colonization factor coordinately regulated with cholera toxin. *Proc Natl Acad Sci USA*, **84**: 2833–2837.

Waldor, M.K., Colwell, R.R. and Mekalanos, J.J. (1994). The *Vibrio cholerae* O139 serogroup antigen includes an O-antigen capsule and lipopolysaccharide virulence determinants. *Proc Natl Acad Sci USA*, **91**: 11388–11392.

Wright, A.C. *et al.* (1992). Development and testing of a nonradioactive DNA oligonucleotide probe that is specific for *Vibrio cholerae* cholera toxin. *J Clin Microbiol*, **30**: 2302–2306.

Yam, W.C. *et al.* (1991). Restriction fragment length polymorphism analysis of *Vibrio cholerae* strains associated with a cholera outbreak in Hong Kong. *J Clin Microbiol*, **28**: 1058–1059.

Yamai, S. *et al.* (1997). Distribution of serogroups of *Vibrio cholerae* non-O1 non-O139 with specific reference to their ability to produce cholera toxin and addition of novel serogroups. *J Jap Assoc Infect Dis*, **71**: 1037–1045.

# 15

# *Yersinia*

## Basic microbiology

*Yersinia* comprises three important species of human significance. Each one is essentially an animal parasite known to infect man. *Yersinia enterocolitica* is found in wild and also domestic animals and known to cause gastroenteritis. *Yersinia pseudotuberculosis* is a parasite of rodents and is known to infect man occasionally. *Yersinia pestis* is the agent responsible for the plague.

*Yersinia* are oxidase-negative, catalase-positive, straight, Gram-negative rods (or coccobacilli) often measuring 0.8–3.0 μm by 0.8 μm. They are facultative anaerobes, predominantly mesophilic though all exhibit growth at low temperatures (e.g. 0–4°C). *Yersinia* are facultative anaerobes that ferment glucose in addition to other sugars without the production of gas. *Yersinia* are motile at 22–30°C, but not at 37°C.

As *Yersinia* possess the ability to grow under extreme ranges in temperature they are well-adapted to survival in the environment. The species of *Yersinia* of most significance to water transmission is *Yersinia enterocolitica* which will be dealt with in this chapter primarily as it is excreted in faeces. Information on *Y. pseudotuberculosis* is also documented as a comparison.

## Origins

*Yersinia*, derived from the French bacteriologist Alexander Yersin, was first isolated in Hong Kong in 1894 (Gyles and Thoen, 1993). *Yersinia* was not isolated in the USA until 1923 and not recognized as a human pathogen until its first human case in 1963. To date *Yersinia* is responsible for about 1% of acute cases of gastroenteritis in Europe and parts of America.

Initially members of the genus *Yersinia* were included in the genus *Pasteurella*. *Yersinia* was removed from this group in the late 1960s despite requests to change this in 1954 (Bercovier and Mollaret, 1984).

To date the genus *Yersinia* is now classified as genus XI of the family Enterobacteriaceae. Included in the genus *Yersinia* are three significant human pathogens, *Y. pestis*, *Y. pseudotuberculosis*, after inoculation of guinea-pigs with material isolated from a skin lesion of a child who died of meningitis, and *Y. enterocolitica*. *Y. enterocolitica* was first described in 1939 under the name *Bacterium enterocoliticum* (Cover and Aber, 1989), later *Pasteurella pseudotuberculosis rodentium* and then *Pasteurella X* and finally *Y. enterocolitica* in 1964 (Bottone, 1977).

*Yersinia* is classified into 11 species, of these *Y. enterocolitica* and *Y. pseudotuberculosis* are human pathogens, whereas the other species, with the exception of *Y. ruckeri* which is a fish pathogen, are environmental, non-pathogenic organisms (Romalde, 1993).

*Yersinia enterocolitica* possess 50 serotypes of which O3, O8 and O9 are the most significant to humans. *Y. enterocolitica* was initially divided into five biogroups based on a number of biochemical reactions namely, indole production, hydrolysis of aesculin and salicin, lactose oxidation, acid from xylose, trehalose, sucrose, sorbose and sorbitol, o-nitrophenyl-βb-d-galactopyranoside (ONPG), ornithine decarboxylase, Voges-Proskauer reaction and nitrate reduction (Wauters, 1981). Following modification of the classification of *Yersinia* there are now six biotypes. These include biogroup 1 which was divided into two groups, namely Group 1A, which include the environmental strains that lack virulence plasmids and pyrazinamidase, aesculin and β-d-glucosidase positive. Biogroup 1B are found to be aesculin, pyrazinamidase and β-d-glucosidase negative and belong to one of the following serogroups: O:4, O:8, O:13, O:18, O:20 or O:21.

## Metabolism and physiology

*Yersinia* are facultative anaerobes. Acid, but not gas, is produced from *d*-glucose and polyhydroxylalcohols. Fructose, galactose, maltose, mannitol, mannose, N-acetylglucosamine and trehalose are fermented. *Y. enterocolitica* produces acetoin when incubated at 28°C but not at 37°C and produces catalase, but not oxidase.

# Clinical features

The most common clinical presentation in *Y. enterocolitica* infections, but rare in *Y. pseudotuberculosis* infections, is acute enteritis. *Y. enterocolitica* is also known to cause severe gastroenteritis. Other disease presentations less frequently associated with infection by these agents include systemic disorders, focal abscesses of liver, spleen, kidney and lung, erythema nodosum, polyarthritis, Reiter's syndrome, myocarditis, pneumonia, meningitis, conjunctivitis, septicaemia, pustules, osteomyelitis; and local manifestations such as cellulitis and wound infections and panophthalmitis (Bottone, 1992) with endocarditis (Urbano-Marquez *et al.*, 1983), pericarditis (Lecomte *et al.*, 1989) and osteitis (Fisch *et al.*, 1989). In children infection symptoms often include fever, diarrhoea, abdominal pain and vomiting.

*Y. pseudotuberculosis* infections in humans present as severe typhoid-like illness often with purpura, fever and enlargement of the spleen and liver. It is frequently a cause of mesenteric lymphadenitis which often simulates subacute appendicitis and is most common in children and young adults. Terminal ileitis is infrequently seen in *Y. pseudotuberculosis* infection, but is very common and often severe in *Y. enterocolitica* infections. Cases of fatal septicaemia caused by *Y. pseudotuberculosis* have been reported.

*Y. enterocolitica* have been reported as a special hazard when blood donors have a bacteraemia. This is due to the fact that the organisms can grow during cold temperature storage in blood and cause severe infection in transfusion recipients (Arduino *et al.*, 1989).

# Pathogenesis and virulence

Pathogenicity of *Yersinia* species is aided by a 70–75 kb plasmid (pYV) and two additional plasmids. These plasmids are known to control four major virulence factors in this genus. These factors include the excreted antiphagocytic proteins (Yops), proteins involved in processing and excretion of the Yops (Ysc), regulatory proteins (Lcr) and adhesin/invasin proteins. Enterotoxins are also thought to play a role in virulence, and are documented as causing disease in humans.

In virulent *Y. enterocolitica*, the presence of a 72 kb low calcium response plasmid is very important. Other plasmids in *Y. enterocolitica* encode for eight or more *Yersinia* outer-membrane proteins. Of significance are the Yops, protein 1 ('P1'), which is associated with resistance to killing by serum, hydrophobicity, autoagglutination in fluid media, and production of a fibrillar adhesin which enhances attachment of the bacteria to epithelial cells of the host. Many of the other Yops function to inhibit phagocytosis by mammalian hosts (Forsberg *et al.*, 1994). Yops produced by *Y. enterocolitica* include YadA, Yops B, C, D, E and H. As for *Y. pseudotuberculosis*, YopE and YopH are essential for full virulence (Sodeinde *et al.*, 1988; Brubaker, 1991). Chromosomally generated virulence factors include an invasin and a second factor that allows specific invasion of various cell types (Miller and Falkow, 1988). Researchers have established that M cells in the Peyer's patches are the primary target cells of invading yersiniae, principally *Y. enterocolitica*. *Y. enterocolitica* and *Y. pseudotuberculosis* colonization of these surfaces and invasion through mucosal surfaces is based principally on toxin Yst. This toxin is possibly responsible for the production of diarrhoea but to date it has not been determined. An exotoxin, YPM, has also been identified from *Y. pseudotuberculosis*. This toxin acts as a superantigen by activation of specific human T cells in the presence of immune cells bearing MHC class II molecules. Another invasion factor in *Yersinia* aside from invasin is Ail. Ail (product of the Ail gene) is of significance during adherence, invasion and serum resistance in *Y. enterocolitica* (Wachtel and Miller, 1995). Invasin and YadA bind to integrins in intestinal tissue (Pepe and Miller, 1993; Skurnik *et al.*, 1994). YadA has also been shown to bind to fibronectin (Tertti *et al.*, 1992).

Once *Yersinia* has colonized the mucosal surfaces it has to avoid phagocytosis. The ability to avoid phagocytosis is under the control of basically 11 Yops (Straley *et al.*, 1993). Yops function by interfering with signal transduction of the phagocytes or directly attacking the host cells. This will inhibit phagocytosis by macrophages (Straley *et al.*, 1993).

*Yersinia* have been found to produce both a siderophore, which is active at low temperature, and an iron storage system which is induced at 37°C (Gyles and Thoen, 1993).

*Yersinia* species produce a fibrillar protein, specifically in pathogenic serotypes of *Y. enterocolitica*. This protein is called Myf and is composed of MyfA (a 21 kDa protein), MyfB (a chaperone) and MyfC (an outer-membrane protein) (Cornelis, 1994; Iriarte and Cornelis, 1995). It is probable that these proteins, together with Yst, aid in the adhesion of *Yersinia* to a host cell.

Another virulence mechanism of *Yersinia* is possibly the production of stress proteins that help in its survival within phagocytes. It is probable that the stress proteins neutralize toxic products produced by host cells.

## Treatment

*Y. enterocolitica* is found to be sensitive to many antibiotics. These include, and are by no means exhaustive, aminoglycosides, chloramphenicol and tetracycline. Treatment is only given with severe infections and tetracycline is usually the antibiotic of choice. Penicillin is not used to treat *Y. enterocolitica* infection as strains are resistant. Treatment of *Y. pseudotuberculosis* septicaemia requires the use of tetracycline or ampicillin.

# Environment

Both *Y. enterocolitica* and *Y. pseudotuberculosis* can survive for long periods in the environment due to their requirements for minimal nutrition and their ability to remain metabolically active at extremes of temperature. Most of the environmental isolates of *Y. enterocolitica* are not pathogenic and are often called atypical *Y. enterocolitica*-like organisms (*Y. intermedia*, *Y. frederiksenii* and *Y. kristensenii*) (Lassen, 1972; Kapperud, 1977; Langeland, 1983; Aleksic and Bockem, 1988; Hellberg and Lofgren, 1988). Little is truly known about the occurrence and survival of *Y. enterocolitica* in the environment and research on its ecology is needed. Work by Dominowska and Malottke (1971) studied the survival of *Y. enterocolitica* in water and found that the average time for the survival in unfiltered surface water was 38 days in spring and 7 days in summer. Conversely in filtered water it was found that the bacteria survived for 197 days in spring and 184 days in the summer. In the laboratory, *Yersinia* was found to survive, following an initial incoculum of $10^3$ cfu/ml, for about 7 days in tap water and for 28 days in lake water. Back in 1972, *Y. enterocolitica* was isolated from 10 of 50 drinking water supplies in Norway (Lassen, 1972). They have also been isolated from wells (Schiemann, 1978), lakes and streams (Kapperud, 1977). *Y. enterocolitica* has also been isolated from soil in California and Germany (Botzler, 1979, 1987). Most of the German strains were not serotypable using serotyping reagents available at the time; the one *Y. enterocolitica* that could be serotyped was O:6,33, not commonly involved in human infection. The California soil isolates were of serotypes O:4,32; O:5, O:17 and O:20, also not associated in human infections.

Wild animals such as foxes, hares and shrews form the natural reservoir of *Yersinia enterocolitica*. Domestic animals such as sheep, cattle, pigs and dogs have also been shown to harbour the bacterium. As with many water organisms oysters and mussels have been shown to harbour *Yersinia enterocolitica*.

*Y. pseudotuberculosis* is found in a large array of mammalian and avian hosts. Many of these infections are subclinical or chronic; the infected hosts shed the organisms into the environment over long periods of time. A single report has incriminated water as the source of infection of more than 200 humans with *Y. pseudotuberculosis* serotypes 1b and 4b in Japan (Fukushima *et al.*, 1989).

# Water

*Y. enterocolitica* are quite common contaminants of water supplies. The presence of *Y. enterocolitica* in drinking water, within the USA, was first reported by Botzler and colleagues (Botzler *et al.*, 1976). Lassen (1972) and Wauters (1972) were the first to document cases in Europe. *Y. enterocolitica* has also been isolated in wells (Highsmith *et al.*, 1977). In this study the strains that were serotypable (5 of 14 isolates) were of types not commonly involved in human disease. Wetzler *et al.* (1978) reported numerous isolates from a variety of water sources and wastewaters in Washington State.

## Methods of detection

*Yersinia* are able to grow at temperatures ranging from 4 to 43°C and over a pH range of 4–10, although the optimum pH is 7.2–7.4. *Y. pseudotuberculosis* and *Y. enterocolitica* will grow in up to 5% salt and all species can grow on nutrient agar. Haemolysis is not observed when *Yersinia* is grown on blood agar and most *Yersinia* will grow on MacConkey medium, although growth of *Y. pseudotuberculosis* is variable. *Y. pseudotuberculosis* will grow as small grey to black colonies on tellurite medium.

Infection by *Yersinia* is diagnosed by isolating the organism from lymph nodes, blood or faeces on blood or MacConkey's agar. Cold enrichment of cultures may enhance the isolation rates for *Y. pseudotuberculosis* in heavily contaminated specimens. This may be accomplished in tetrathionate broth or selenite F broth or simply in physiological saline or a nutrient broth, all of which are kept in the refrigerator at 4°C for up to 6 weeks with periodic subculturing for recovery of the agents. Most isolates of the enteropathogenic yersiniae grow well on MacConkey agar and eosin-methylene blue agar; their growth is often inhibited or delayed on Salmonella-Shigella agar and deoxycholate agar.

Among the more commonly used selective media for rapid identification of the enterocolitica group of yersiniae are CIN agar (Schiemann, 1979) and VYE agar (virulent *Y. enterocolitica* agar) (Fukushima, 1987). Specimens not grossly contaminated on an enriched medium, such as blood agar or brain-heart infusion agar may facilitate recovery of *Yersinia*. Recovery (especially of virulent organisms) may also be enhanced if incubation after the first 24 hours is conducted at room temperature (24–28°C).

Identity of *Yersinia* is determined by biochemical test and the serotype is confirmed by slide agglutination with rabbit antisera. *Y. enterocolitica* and *Y. pseudotuberculosis* are motile, with peritrichous or paripolar flagella, when grown at 22°C, but not at 37°C. It has been demonstrated biochemically and by visualization under the electron microscope that at room temperature strains of *Y. enterocolitica* could go through a transition to a spheroplast type L-form which then could revert back to an irregular-shaped structure with an intact cell wall (Pease, 1979).

All species of *Yersinia* have a requirement of iron for growth. Virulent strains of *Y. enterocolitica* and *Y. pseudotuberculosis* also have a requirement for a low concentration of calcium in order to express some of the antigens involved in the virulence of these species. The concentration of calcium in the medium is critical for the expression of virulence factors of Y. *pseudotuberculosis* and Y. *enterocolitica*.

## Epidemiology

The primary method of transmission for Y. *enterocolitica* is via the faecal-oral route but it may also occasionally occur via contaminated water sources (Lassen, 1972; Saari and Quan, 1976; Saari and Jansen, 1977; Wetzler *et al.*, 1978) and contaminated meats (Aulisio *et al.*, 1983) and poultry (DeBoer *et al.*, 1982). Uncommon infections have resulted from the transfusion of contaminated units of blood from bacteraemic, though apparently healthy, donors (Jacobs *et al.*, 1989). Several outbreaks of Y. *enterocolitica* infections have been traced to the consumption of pasteurized milk (Aulisio *et al.*, 1983; Shayegani *et al.*, 1983; Tacket *et al.*, 1984) and chocolate milk (Black *et al.*, 1978) as well as of raw milk and cheese and contaminated foods (Aulisio *et al.*, 1983) and meats (Doyle *et al.*, 1981; Shayegani *et al.*, 1983). Several outbreaks of gastrointestinal illness have been traced to the ingestion of contaminated foods such as raw and pasteurized milk (Moustafa *et al.*, 1983; Tacket *et al.*, 1984) and tofu (Aulisio *et al.*, 1983).

Person-to-person transmission has not been well documented; the major mode of transmission does seem to be the faecal-oral route. Domestic pets are being increasingly associated with the occurrence of human disease (Gutman *et al.*, 1973; Fantasia *et al.*, 1985; Fukushima *et al.*, 1988).

Y. *enterocolitica* strains associated with human disease are mainly of serotypes 3 and 9. This is presently the case in Europe. In Japan and the USA other serotypes are common.

## Risk assessment

*Health effects*: occurrence of illness, degree of morbidity and mortality, probability of illness based on infection:

- *Yersinia enterocolitica* causes gastroenteritis and is the most significant *Yersinia* species related to water transmission.
- *Yersinia* is responsible for about 1% of acute cases of gastroenteritis in Europe and parts of America.
- Y. *enterocolitica* causes mainly acute enteritis, but systemic infections, such as bacteremia, joint pain, and rashes have occasionally resulted. It can also cause inflammation of the mesenteric lymph glands, which is often confused with appendicitis. The most common symptoms include fever, diarrhoea and abdominal pain.

- Young children are most likely to become infected and ill, though pseudoappendicitis is more often seen in older children and adults.

*Exposure assessment*: routes of exposure and transmission, occurrence in source water, environmental fate:

- *Y. enterocolitica* is spread through the faecal-oral route – most often by contaminated food – and sometimes by water. It is rarely transmitted person-to-person. Contact with animals may be a route of transmission as well.
- Many wild and domestic animals are reservoirs for *Y. enterocolitica*.
- *Y. enterocolitica* can survive a long time in the environment, because it is resistant to the extremes of temperature.
- The organisms have been found in surface and groundwater sources, but little overall is known about their occurrence in the environment. Those found in drinking water sources have mostly been strains non-pathogenic to humans.

*Risk mitigation*: drinking-water treatment, medical treatment:

- *Yersinia* seems to be as susceptible to chlorine as *E. coli*, so adequately disinfected water supplies should control *Y. enterocolitica*.
- Uncomplicated cases of diarrhoea due to *Y. enterocolitica* are usually self-limiting. However, in more severe infections, treatment may be required, and *Y. enterocolitica* is sensitive to many antibiotics. Tetracycline is usually the antibiotic of choice.

# References

Aleksic, S. and Bockem, H.L.J. (1988). Serological and biochemical characteristics of 416 Yersinia strains from well water and drinking-water plants in the Federal Republic of Germany: lack of evidence that these strains are of public health importance. *Zentralbl Bakteriol Mikrobiol Hyg Reiche B*, **185**: 527–533.

Arduino, M.J., Bland, L.A. *et al.* (1989). Growth and endotoxin production of *Yersinia enterocolitica* and *Enterobacter agglomerans* in packed erythrocytes. *J Clin Microbiol*, **27**: 1483–1485.

Aulisio, C.C.G., Stanfield, J.T. *et al.* (1983). Yersinioses associated with tofu consumption: serological, biochemical, and pathogenicity studies of *Yersinia enterocolitica* isolates. *J Food Prot*, **46**: 226–230.

Bercovier, H. and Mollaret, H.H. (1984). *Bergey's Manual of Systematic Bacteriology*, vol. 1, Krieg, N.R. and Holt, J.G. (eds). Baltimore: Williams & Wilkins, p. 498.

Black, R.E., Jackson, R.J. *et al.* (1978). Epidemic *Yersinia enterocolitica* infection due to contaminated chocolate milk. *New Engl J Med*, **298**: 76–79.

Bottone, E.J. (1977). *Yersinia enterocolitica*: a panoramic view of a charismatic microorganism. *Crit Rev Mirobiol*, **5**: 211–41.

Bottone, E.J. (1992). The genus Yersinia (excluding *Yersinia pestis*). In *The Prokaryotes*, 2nd edn, Balows, A., Trüpper, H.G. *et al.* (eds). New York: Springer-Verlag, pp. 2862–2887.

Botzler, R.G. (1979). Yersiniae in the soil of an infected wapiti range. *J Wildl Dis*, **15**: 529–532.

Botzler, R.G. (1987). Isolation of *Yersinia enterocolitica* and *Y. frederiksenii* from forest soil, Federal Republic of Germany. *J Wildl Dis*, **23**: 311–313.

Botzler, R.G., Wetzler, T. and Cowan, A.B. (1976). *Yersinia enterocolitica* and Yersinia-like organisms isolated from frogs and snails. *Bull Wildl Dis Assoc*, **4**: 110–115.

Botzler, R.G., Wetzler, T. *et al.* (1976). Yersiniae in pond water and snails. *J Wildl Dis*, **12**: 492–426.

Brubaker, R.R. (1991). Factors promoting acute and chronic diseases caused by yersiniae. *Clin Microbiol Rev*, **4**: 309–324.

Cornelis, G.R. (1994). Yersinia pathogenicity factors. *Curr Topics Microbiol Immunol*, **192**: 243–263.

Cover, T.L. and Aber, R.C. (1989). *Yersinia enterocolitica*. *New Engl J Med*, **321**: 16–24.

DeBoer, E., Hartog, B.J. and Oosterom J. (1982). Occurrence of Yersinia in poultry products. *J Food Prot*, **45**: 322–325.

Dominowska, C. and Matlotte, R. (1971). Survival of Yersinia in water samples originating from various sources. *Bull Inst Marine Med Gdansk*, **22**: 173–182.

Doyle, M.P., Hugdahl, M.B. and Taylor, S.L. (1981). Isolation of virulent *Yersinia enterocolitica* from porcine tongues. *Appl Environ Microbiol*, **42**: 661–666.

Fantasia, M., Grazia Mingrone, M. *et al.* (1985). Isolation of *Yersinia enterocolitica* biotype 4 serotype O3 from canine sources in Italy. *J Clin Microbiol*, **22**: 314–315.

Fisch, A., Prazuck, T. *et al.* (1989). Hematogenous osteitis due to *Yersinia enterocolitica*. *J Infect Dis*, **160**: 554.

Forsberg, A., Rosqvist, R. and Wolf-Watz, H. (1994). Regulation and polarized transfer of the Yersinia outer proteins (Yops) involved in antiphagocytosis. *Trends Microbiol*, **2**: 14–19.

Fukushima, H. (1987). New selective agar medium for isolation of virulent *Yersinia enterocolitica*. *J Clin Microbiol*, **25**: 1068–1073.

Fukushima, H., Gomyoda, M. *et al.* (1988). *Yersinia pseudotuberculosis* infection contracted through water contaminated by a wild animal. *J Clin Microbiol*, **26**: 584–585.

Fukushima, H., Gomyoda, M. *et al.* (1989). Cat-contaminated environmental substances lead to *Yersinia pseudotuberculosis* infection in children. *J Clin Microbiol*, **27**: 2706–2709.

Gutman, L.T., Wilfert, C.M. and Quan, T.J. (1973). Susceptibility of *Yersinia enterocolitica* to trimethoprim-sulfamethoxazole. *J Infect Dis*, **128S**: 538.

Gyles, C.L. and Thoen, C.O. (1993). *Pathogenesis of bacterial infections in animals*, 2nd edn. Ames, Iowa: Iowa State University Press, pp. 226–235.

Hellberg, B. and Lofgren, S. (1988). Gastrointestinal disturbances due to an established bacterial contamination of a water distribution system. *Vatten*, **44**: 277–281.

Highsmith, A.K., Feeley, J.C. *et al.* (1977). Isolation of *Yersinia enterocolitica* from well water and growth in distilled water. *Appl Environ Microbiol*, **34**: 745–750.

Iriarte, M. and Cornelis, G.R. (1995). MyfF, an element of the network regulating the synthesis of fibrillae in *Yersinia enterocolitica*. *J Bacteriol*, **177**: 738–744.

Jacobs, J., Jamaer, D. *et al.* (1989). *Yersinia enterocolitica* in donor blood: a case report and review. *J Clin Microbiol*, **27**: 1119–1121.

Kapperud, G. (1977). *Yersinia enterocolitica* and Yersinia-like microbes isolated from mammals and water in Norway and Denmark. *Acta Pathol Microbiol Scand: Section B: Microbiol Immunol*, **85**: 129–135.

Langeland, G. (1983). *Yersinia enterocolitica* and *Yersinia enterocolitica*-like bacteria in drinking-water and sewage sludge. *Acta Pathol Microbiol Scand B*, **91**: 179–185.

Lassen, J. (1972). *Yersinia enterocolitica* in drinking water. *Scand J Infect Dis*, **4**: 125–127.

Lecomte, F., Eustache, M. *et al.* (1989). Purulent pericarditis due to *Yersinia enterocolitica*. *J Infect Dis*, **159**: 363.

Miller, V.L. and Falkow, S. (1988). Evidence for two genetic loci in *Yersinia enterocolitica* that can promote invasion of epithelial cells. *Infect Immun*, **56**: 1242–1248.

Moustafa, M.K., Ahmed, A.A.-H. and Marth, E.H. (1983). Occurrence of *Yersinia enterocolitica* in raw and pasteurized milk. *J Food Prot*, **46**: 276–278.

Pease, P. (1979). Observations on L-forms of *Yersinia enterocolitica*. *J Med Microbiol*, **12**: 337.

Pepe, J.C. and Miller, V.L. (1993). *Yersinia enterocolitica* invasin: a primary role in the initiation of infection. *Proc Natl Acad Sci USA*, **90**: 6473–6477.

Saari, T.N. and Jansen, G.P. (1977). Waterborne *Yersinia enterocolitica* in Wisconsin. *Abstr Bacteriol Proc*, New Orleans.

Saari, T.N. and Quan, T.J. (1976). Waterborne *Yersinia enterocolitica* in Colorado. *Abstr Bacteriol Proc*, Atlantic City.

Schiemann, D.A. (1978). Isolation of *Yersinia enterocolitica* from surface and well water in Ontario. *Can J Microbiol*, **24**: 1048–1052.

Schiemann, D.A. (1979). Synthesis of selective agar medium for isolation of *Yersinia enterocolitica*. *Can J Microbiol*, **25**: 1298–1304.

Shayegani, M., Morse, D. *et al.* (1983). Microbiology of a major foodborne outbreak of gastroenteritis caused by *Yersinia enterocolitica* serogroup O:8. *J Clin Microbiol*, **17**: 35–40.

Skurnik, M., Tahir, Y. *et al.* (1994). YadA mediates specific binding of enteropathogenic *Yersinia enterocolitica* to human intestinal submucosa. *Infect Immun*, **62**: 1252–1261.

Sodeinde, O.A., Sample, A.K. *et al.* (1988). Plasminogen activator/coagulase gene of *Yersinia pestis* is responsible for degradation of plasmid encoded outer membrane proteins. *Infect Immun*, **56**: 2743–2748.

Straley, S.C., Skrzypek, E. and Plano, G.V.E. (1993). Yops of *Yersinia* spp. pathogenic for humans. *Infect Immun*, **61**: 3105–3110.

Tacket, C.O., Narain, J.P. *et al.* (1984). A multistate outbreak of infections caused by *Yersinia enterocolitica* transmitted by pasteurized milk. *JAMA*, **251**: 483–486.

Tertti, R., Skurnik, M. *et al.* (1992). Adhesion protein YadA of Yersinia species mediates binding of bacteria to fibronectin. *Infect Immun*, **60**: 3021–3024.

Urbano-Marquez, A., Estruch, R. *et al.* (1983). Infectious endocarditis due to *Yersinia enterocolitica*. *J Infect Dis*, **148**: 940.

Wachtel, M.R. and Miller, V.L. (1995). *In vitro* and *in vivo* characterization of an ail mutant of *Yersinia enterocolitica*. *Infect Immun*, **63**: 2541–2548.

Wauters, G. (1972). Souches de *Yersinia enterocolitica* isolées de l'eau. *Rev Ferm Ind Ailment*, **7**: 18.7.

Wauters, G. (1981). Antigens of *Yersinia enterocolitica*. In *Yersinia enterocolitica*. Boca Raton: CRC Press, pp. 41–53.

Wetzler, T.F., Rea, J.T. *et al.* (1978). *Yersinia enterocolitica*. in waters and waste waters. Presented at 106th Annual Meeting, American Public Health Association, Los Angeles, CA, October 18, 1978.

# Part 3

# Protozoa

# 16

# *Acanthamoeba* spp.

## Basic microbiology

Several species of the free-living amoebae *Acanthamoeba* (Phylum *Sarco-mastigophora*, Order *Amoebida*, Family *Acanthamoebidae*), which are commonly found in soil and water, have been implicated in human infections and disease, including *Acanthamoeba culbertsoni*, *Acanthamoeba polyphaga*, *Acanthamoeba castellanii*, *Acanthamoeba astronyxis*, *Acanthamoeba hatchetti*, *Acanthamoeba griffini*, *Acanthamoeba lugdenensis* and *Acanthamoeba rhysodes* (Bottone, 1993). These have been divided into three morphological groups, I to III (Pussard and Pons, 1977). However, the taxonomy of the genus is uncertain since isoenzymatic and genetic assays have shown similarities among strains traditionally assigned to different species on the basis of morphology (Kilvington and Beeching, 1996). Cyst morphology has been shown to vary even within clonal populations of the species (Kilvington *et al.*, 1991). Additionally, the pathogenicity of some species is poorly established. Those mainly associated with human disease are *A. culbertsoni*, which causes the rare, but fatal granulomatous amoebic encephalitis (GAE) and *A. polyphaga* and *A. castellanii* which are more usually associated with infections (keratitis) of the eye (Visvesvara and Stehr-Green, 1990). GAE is not regarded as water-related while keratitis is.

The life cycle of *Acanthamoeba* spp. is commonly referred to as 'simple', comprising a feeding and dividing trophozoite stage and a resistant cyst stage (Visvesvara, 1991). These amoebae favour aquatic habitats such as water, mud and soil (Anon, 1989) where the active, aerobic trophozoites feed on bacteria, fungi and other protozoa. The slender trophozoites (25–40 μm) possess spine-like processes called acanthopodia, which facilitate a slow gliding movement. They divide by binary fission using a mitotic process, and do not possess a flagellate stage. In adverse environments, including anaerobic conditions, the trophozoites form dormant, resistant cysts. Cysts are wrinkled, double walled and of a variety of shapes, measuring 10–30 μm in diameter. The cyst comprises an outer ectocyst and an inner endocyst, linked by a pore with a mucoid plug.

The transmission and epidemiology of disease caused by *Acanthamoeba*, particularly keratitis, is largely driven by host behaviour, mainly from improper use and disinfection of contact lenses.

## Origin of the organism

*Acanthamoeba* spp. were originally ascribed to the genus *Hartmannella*, but subsequent studies of both the cyst and trophozoite forms showed distinct differences between the two genera. Additionally, *Hartmannella* spp. are not pathogenic to humans, whereas the pathogenicity of *Acanthamoeba* for humans was suggested following experimental animal infections in 1958 (Culbertson *et al.*, 1958).

GAE was first recognized in humans in 1971 (Kenney, 1971), since when over 100 cases have been reported world-wide, mainly in patients immunosuppressed by chemotherapy, alcohol abuse, chronic disease and GAE has been recorded as the primary cause of death in AIDS patients (Martinez, 1991). GAE is thus regarded as a rare, fatal, opportunistic infection of immunocompromised hosts. The main risk to humans from *Acanthomoeba* in terms of numbers of cases is from keratitis. The first case of acanthamoeba keratitis was diagnosed in a Texan farmer in 1973 (Jones *et al.*, 1975). Suffering from ocular trauma caused by straw fragments, he used tap water to rinse the affected eye. Two further cases were reported in the UK (Nagington *et al.*, 1974) and the condition was regarded as a rare opportunistic infection resulting from injury to the eye (Ma *et al.*, 1981) until the mid-1980s when increased reports prompted examination and review of previous cases in the USA and UK. Of 208 cases reported in the USA between 1973 and 1988, three-quarters occurred from 1985 (Visvesvara and Stehr-Green, 1990). Details of 189 cases were available and showed that 160 (85%) were in contact lens wearers, particularly those using soft contact lenses. A review of 72 consecutive cases between 1984 and 1992 in the UK showed that 64 patients were contact lens wearers, 28 of which wore disposable lenses (Bacon *et al.*, 1993). This identified the need for better education of contact lens wearers regarding cleaning and disinfection practices.

## Clinical features

*Acanthamoeba* spp. cause two distinct diseases in humans: GAE affects the central nervous system and keratitis affects the cornea of the eye. GAE has a sudden onset and can last from 8 days to several months, invariably resulting in death. Symptoms of GAE include fever, headache, seizures, meningitis and visual abnormalities (Martinez, 1991).

A. *polyphaga* and A. *castellanii* have been associated with chronic granulomatous lesions of the skin, with or without secondary invasion of the CNS, and infection of the eye (conjunctivitis) and cornea (keratoconjunctivitis). *Acanthamoeba* keratitis is characterized by intense pain and ring-shaped infiltrates in the corneal stroma, which can progress to hypopyon, scleritis, glaucoma and cataract formation (Bacon *et al.*, 1993). Corneal perforation may occur. One eye only is usually affected, although bilateral disease has been reported (Auran *et al.*, 1987). As recognition of the disease is heightened, early diagnosis is prompted when infiltration of the superficial epithelium only may be involved (Larkin *et al.*, 1992). Symptoms may be non-specific and falsely diagnosed as herpes simplex virus infection (Auran *et al.*, 1987).

## Pathogenicity and virulence

*Acanthamoeba* are widespread in the environment and hence antibodies are common in human sera (Bottone, 1993). Parasites have been found in the respiratory tract of healthy people, and subclincial, self-limiting infection may be common in immune-competent hosts (Martinez, 1993). Immunity probably involves both cell-mediated and humoral systems, and skin lesions caused by *Acanthamoeba* may progress to invasion of the brain and meninges of immunocompromised individuals and occasionally immune intact people, causing GAE (Bottone, 1993; Martinez, 1993). The lower respiratory tract may be a route of invasion through inhalation and adherence to mucosal surfaces. The trophozoites probably reach the meninges by haematogenous spread from the site of primary infection. Although cases of GAE have been reported in immunocompetent hosts, immunosuppressed individuals are at greater risk, including those undergoing chemotherapy, AIDS patients, drug abusers and alcoholics.

Infection with A. *polyphaga* and A. *castellanii* can occur in healthy people and rarely cause GAE but, more commonly, leads to conjunctivitis and keratoconjunctivitis, the latter resulting in blindness. These infections occur primarily in wearers of soft contact lenses. Although the mechanism of disease within the cornea is unclear, cytopathic enzymes including a variety of proteases (Mitro *et al.*, 1994), such as collagenase (He *et al.*, 1990), and proteolytic activity are probably important.

## Treatment

There is no treatment for GAE and it is invariably fatal. If untreated, *Acanthamoeba* keratitis can lead to permanent blindness. However, management is difficult since the cysts are resistant to most antimicrobials at concentrations tolerated by the cornea and a prolonged course of treatment is often required to achieve elimination. Treatment includes ketoconazole, slotrimazole and propamidine isethionate (Martinez, 1993). Chlohexidine gluconate and polyhexamethylene biguanide (PHMB) are effective, particularly PHMB administered following prompt diagnosis.

## Survival in the environment

*Acanthamoeba* are ubiquitous in the natural and man-made environment, including tap water, swimming, hydrotherapy and spa pools (De Jonckheere, 1987; Kilvington *et al.*, 1990) and have even been isolated from eyewash stations (Paszilo-Kolua *et al.*, 1991). *Acanthamoeba* spp. have a growth temperature range of 12–45°C, and all pathogenic species have an optimum of 30°C, but most also grow at 37°C, although this is uncertain for members of morphological group II. Therefore, during laboratory diagnosis, growth at both 32°C and 37°C is recommended. The cysts are highly resistant to a number of environmental pressures, including drying, freezing to −20°C, moist heat at 56°C and 50 ppm free chlorine (Kilvington, 1989; Kilvington and Price, 1990), and have been found in nearly all soil and aquatic environments (Page, 1988).

### Survival in water

Trophozoites are killed by saline concentrations >1%, although the cysts can survive and have been detected in seawater. Commercial contact lens disinfectant solutions based on chlorhexidine, hydrogen peroxide or moist heat will kill *Acanthamoeba* cysts providing the correct time and temperature (where relevant) exposure conditions are met (Ludwig *et al.*, 1986; Davies *et al.*, 1988). Cysts can be removed from lenses by commercial lens-cleaning agents (Kilvington and Larkin, 1990). A survey of contact lens cases belonging to healthy wearers showed that 43% contained >$10^6$ ml/l viable bacteria and seven contained *Acanthamoeba* (Larkin *et al.*, 1990). It is therefore likely that infection can be acquired through primary contamination of the lens storage case, which becomes contaminated from rinsing lenses in tap water or non-sterile saline solutions or through wearing lens while bathing or swimming in lakes or ponds (Visvesvara, 1993). Non-contact lens-associated keratitis clearly also occurs in the minority of cases and is associated with ocular trauma or environmental contamination. Despite cyst survival in 50 ppm free chlorine, and detection in swimming pools, no infection has been reported to have been directly acquired in a chlorinated swimming pool according to USA data.

# Methods of detection

Clinical diagnosis of GAE is by microscopic examination of histological sections. Examination at high power can help avoid misidentification as macrophages or *Entamoeba histolytica*. Diagnosis of eye infections with *Acanthamoeba* spp. is by similar examination, direct smears or culture of eye and skin lesion scrapings, swabs or aspirates on non-nutrient agar seeded with suitable *Enterobacter* spp. (including *Escherichia coli* and *Klebsiella aerogenes*) (Martinez, 1993). Trophozoites can be observed following Gram or Giemsa stains (Bottone, 1993) and show sluggish motility. Cysts can be observed by a variety of stains including calcofluor, which stains the chitin and cellulose in the cyst wall (Bottone, 1993). This stain has also been used to demonstrate *Acanthamoeba* on contact lens surfaces (Johns *et al.*, 1991).

*Acanthamoeba* has been isolated from a variety of environments, including soil, natural and man-made aquatic environments (including chlorinated pools), and potable water supplies (Kilvington and White, 1994). The organism is readily cultured on non-nutrient agar plates seeded with *Escherichia coli* from environmental samples using prior concentration by sedimentation or filtration (Kilvington *et al.*, 1990). Prior concentration of the water sample by centrifugation or by membrane filtration and elution of captured cells may be required before culture on an *E. coli* lawn at 32°C and 37°C. Once growth on seeded bacterial lawns is established, strains can be adapted to axenic (bacteria-free) culture media. For detection, trophozoite plaques are subcultured from the *E. coli* lawn to Page's saline in a microtitre plate (Anon, 2002), incubated at 32°C for 1–3 hours and examined microscopically.

Cyst morphology within a species varies with cultural conditions and so other characteristics are used for species identification, including isoenzyme profiles (Costas and Griffiths, 1980) and genetic analyses (Bolger *et al.*, 1982; McLaughlin *et al.*, 1988). Since there is difficulty in the use of morphological criteria for identification of isolates, molecular methods have been developed. Many of these are based on the polymerase chain reaction for amplification prior to identification (Vodkin *et al.*, 1992) and differentiation of pathogenic species (Howe *et al.*, 1997). Since *Acanthamoeba* can be cultured to produce a relatively pure suspension from both clinical and environmental samples, eukaryotic-specific primers can also be used with confidence. However, having to produce a culture prior to PCR can add a number of days to the analysis process. In their studies, Schroeder *et al.* (2001) developed a genus-specific

PCR to amplify DNA from all known *Acanthamoeba* 18s rDNA genotypes, that did not amplify DNA from closely related organisms, making it particularly applicable to environmental samples. Amplicons can then be used for further analysis for genetic relatedness.

## Epidemiology

The study of the epidemiology of cases of *Acanthamoeba* keratitis in the USA in the late 1980s (Visvesvara and Stehr-Green, 1990) identified that using home-made saline solution to rinse soft contact lenses was a major risk factor for disease. This practice was enabled by the then current sale of saline tablets and distilled water precisely to create home-made solutions. Prior contamination of the distilled water or the saline solution due to atmospheric exposure put users at risk of infection, and the practice has been halted by the banning of the sale of saline tablets and distilled water for this purpose. This has led to a decrease in reported cases of disease in the USA (Visvesvara, 1993). In the UK cases are also predominantly among contact lens wearers and have risen since the mid-1980s, even at specialist units which have a history of expert diagnosis. In a case control study to identify risk factors for *Acanthoamoeba* keratitis in contact lens wearers, Radford and colleagues (1995) showed that failure to disinfect daily-wear soft contact lenses, and the use of chlorine release lens disinfection systems, were major risk factors.

## Risk assessment

There appears to be no risk to healthy individuals from the ingestion of *Acanthamoeba*. However, it is recommended that immunocompromised persons boil drinking water before consumption. The health risk from *Acanthamoeba* spp. for people with intact immune systems lies not with consumption of water but through exposure via lesions, particularly contact lens wearers who account for the vast majority of cases in the USA and UK (Visvesvara and Stehr-Green, 1990; Bacon *et al.*, 1993). Recognized risk factors are poor hygiene practices such as washing and or storing lenses in non-sterile solutions or tap water, and inadequate disinfection. In non-contact lens wearers, infection usually follows trauma of the eye with environmental contamination: *A. polyphaga* and *A. castellanii* are most frequently identified in environmental samples (Page, 1988).

*Health effects*: occurrence of illness, degree of morbidity and mortality, probability of illness based on infection:

- The types of *Acanthamoeba* mainly associated with human disease are *A. culbertsoni*, which causes the rare but fatal granulomatous amoebic encephalitis (GAE) and *A. polyphaga* and *A. castellanii*, which are more usually associated with eye infections (keratitis). GAE is not regarded as water-related while keratitis is.

- *Acanthamoeba* keratitis is characterized by intense pain and ring-shaped infiltrates in the corneal stroma, which can progress to hypopyon, scleritis, glaucoma and cataract formation. Corneal perforation may occur. One eye only is usually affected, although bilateral disease has been reported.
- Based on sera studies, people are commonly exposed to *Acanthamoeba*, but rarely affected. Parasites have been found in the respiratory tract of healthy people, and subclinical, self-limiting infection may be common in immunocompetent hosts.
- There appears to be no risk to healthy individuals from the ingestion of *Acanthamoeba*.

*Exposure assessment*: routes of exposure and transmission, occurrence in source water, environmental fate:

- *Acanthamoeba* can enter the skin through a cut, wound, or through the nostrils.
- The transmission of keratitis caused by *Acanthamoeba*, is largely driven by risky behaviour in contact lens wearers. Recognized risk factors for keratitis are poor hygiene practices such as washing and/or storing contact lenses in non-sterile solutions or tap water and inadequate disinfection of lenses. Infection can also be acquired through primary contamination of the lens storage case or through wearing lenses while swimming in lakes or ponds.
- Non-contact lens-associated keratitis occurs rarely and is associated with ocular trauma or environmental contamination.
- *Acanthamoeba* are ubiquitous in the natural and man-made environment, including tap water and swimming and spa pools. The amoebae favour aquatic habitats such as water, mud and soil.
- The cysts are highly resistant to environmental forces, including drying, freezing, heat, and 50 ppm free chlorine.

*Risk mitigation*: drinking-water treatment, medical treatment:

- Prevention of *Acanthamoeba* keratitis in contact lenses is key; commercial lens-cleaning agents can remove cysts from lenses.
- If untreated, *Acanthamoeba* keratitis can lead to permanent blindness. Management is difficult since the cysts are resistant to most antimicrobials at concentrations tolerated by the cornea. A prolonged course of treatment is often required. Propamidine isethionate, chlohexidine gluconate, polyhexamethylene biguanide are effective, particularly when administered following prompt diagnosis.

# Future implications

Prevention of GAE is undetermined. Prevention of keratitis can be achieved through public health messages aimed at those groups identified as at risk and through rapid diagnosis of infection in these groups and following ocular trauma.

Protozoa such as *Acanthamoeba* can support the intracellular growth of other organisms pathogenic to humans, and have implications for their survival, resistance to disinfection regimens, ecology and dissemination and even their increased virulence following passage through protozoa.

# References

Anon. (1989). Isolation and identification of *Giardia* cysts, *Cryptosporidium* oocysts and freeliving pathogenic amoebae in water etc. In *Methods for the Examination of Waters and Associated Materials*. London: HMSO.

Anon. (2002). Isolation and identification of *Acanthamoeba* species, PHLS SOP W12, Issue 2. Colindale, London: PHLS HQ.

Auran, J.D., Starr, M.B. and Jakobiec, F.A. (1987). *Acanthamoeba keratitis*: a review of the literature. *Cornea*, 6: 2–26.

Bacon, A.S., Frazer, D.G., Dart, J.K.G. *et al.* (1993). A review of 72 consecutive cases of *Acanthamoeba* keratitis, 1984–1992. *Eye*, 7: 719–725.

Bolger, S.A., Zarley, C.D., Burianek, L.L. *et al.* (1982). Interstrain mitochondrial DNA polymorphism detected in *Acanthamoeba* by restriction endonuclease analysis. *Mol Biochem Parasitol*, 8: 145–163.

Bottone, E.J. (1993). Free-living amebas of the genera *Acanthamoeba* and *Naegleria*: an overview and basic microbiological correlates. *Mount Sinai J Med*, 60: 260–270.

Costas, M. and Griffiths, A.J. (1980). The suitability of starch gel electrophoresis of esterases and acid phosphotases for the study of *Acanthamoeba* taxonomy. *Arch Prostisterk*, 123: 2727–2729.

Culbertson, C.D., Smith, J.W. and Minner, J.R. (1958). *Acanthomoeba*: observations on animal pathogenicity. *Science*, 127: 1506.

Davies, D.J.G., Anthony, Y., Meakin, B.J. *et al.* (1988). Anti-acanthamoeba activity of chlorhexidine and hydrogen peroxide. *Trans Br Contact Lens Assoc*, 5: 80–82.

De Jonckheere, J.F. (1987). Epidemiology. In *Amphizic Amoebae, Human Pathology*, Rondanelli, E.G. (ed.). Padua, Italy: Piccin Nuova Libraria, pp. 127–147.

He, Y.G., Niederkorn, J.Y., McCulley, J.P. *et al.* (1990). *In vivo* and *in vitro* collagenolytic activity of *Acanthamoeba castellani*. *Ophthalmol Vis Sci*, 31: 2235–2240.

Howe, D.K., Vodkin, M.H., Novak, R.J. *et al.* (1997). Identification of two genetic markers that distinguish pathogenic and non-pathogenic strains of *Acanthamoeba*. *Parasitol Res*, 83: 345–348.

Johns, K.J., Head, W.S., Robinson, R.D. *et al.* (1991). Examination of the contact lens with light microscopy; an aid in diagnosis of *Acanthamoeba* keratitis. *Rev Infect Dis*, 13: S425.

Jones, D.B., Visvesvara, G.S. and Robinson, N.M. (1975). *Acanthamoeba polyphaga* and *Acanthamoeba uveitis* associated with a fatal meningitis. *Trans Ophthalmol Soc UK*, 95: 221–232.

Kenney, M. (1971). The Micro-Kolmer complement fixation test in routine screening for soil ameba infection. *Hlth Lab Sci*, 8: 5–10.

Kilvington, S. (1989). Moist-heat disinfection of pathogenic *Acanthamoaeba* cysts. *Lett Appl Microbiol*, 9: 187–189.

Kilvington, S. and Larkin, D.F.P. (1990). *Acanthamoeba* adherence to contact lenses and removal by cleaning agents. *Eye*, 4: 589–593.

Kilvington, S. and Price, J. (1990). Survival of *Legionella pneumophila* with *Acanthamoeba polyphaga* cysts following chlorine exposure. *J Appl Bacteriol*, 68: 519–525.

Kilvington, S. and White, D.G. (1994). *Acanthamoeba*: biology, ecology and human disease. *Rev Med Microbiol*, **5**: 12–20.

Kilvington, S., Beeching, J.R. and White, D.G. (1991). Differentiation of *Acanthamoeba* strains from infected corneas and the environment using restriction endonuclease digestion of whole cell DNA. *J Clin Microbiol*, **29**: 310–314.

Kilvington, S., Larkin, D.F.P., White, D.P. *et al.* (1990). Laboratory investigation of *Acanthamoeba* keratitis. *J Clin Microbiol*, **28**: 2722–2725.

Larkin, D.F.P., Kilvington, S. and Dart, J.K.G. (1992). Treatment of *Acanthamoeba* keratitis with polyhexamethylene biguanide. *Ophthalmology*, **99**: 185–191.

Larkin, D.F.P., Kilvington, S. and Easty, D.L. (1990). Contamination of contact lens storage cases by Acanthamoeba and bacteria. *Br J Ophthalmol*, **74**: 133–135.

Ludwig, I.H., Meiser, D.M., Ritherford, I. *et al.* (1986). Susceptibility of *Acanthamoeba* to soft contact lens disinfection systems. *Invest Ophthalmol Vis Sci*, **27**: 626–628.

Ma, P., Willaert, E., Juechter, K.B. *et al.* (1981). A case of keratitis due to *Acanthamoeba* in New York and features of 10 cases. *J Infect Dis*, **143**: 662–667.

Martinez, A.J. (1991). Infections of the central nervous system due to *Acanthamoeba*. *Rev Infect Dis*, **13**: S399–402.

Martinez, A.J. (1993). Free-living amebas: infection of the central nervous system. *Mount Sinai J Med*, **60**: 271–278.

McLaughlin, G.L., Brabdt, F.H. and Visvesvara, G.S. (1998). Restriction fragment length polymorphism of the DNA of selected *Naegleria* and *Acanthamoeba* amebae. *J Clin Microbiol*, **26**: 1655–1658.

Mitro, K., Bhagavathiammai, A., Zhou, O.-M. *et al.* (1994). Partial characterisation of the proteolytic secretions of *Acanthamoeba polyphaga*. *Exp Parasitol*, **78**: 377–385.

Nagington, J., Watson, P.G., Playfair, T.J. *et al.* (1974). Amoebic infections of the eye. *Lancet*, **ii**: 1537–1540.

Page, F.C. (1988). *A new key to freshwater and soil gymnamoebae*. Cumbria: The Fresh water Biological Association. The Ferry House, Ambleside, Cumbria.

Paszilo-Kolua, C., Yamamoto, H., Shahamat, M. *et al.* (1991). Isolation of amoebae and *Pseudomonas* and *Legionella* spp. from eyewash stations. *Appl Environ Microbiol*, **57**: 163–167.

Pussard, M. and Pons, R. (1977). Morphologie de la paroi kystique et taxonomie du genre Acanthomoeba (Protozoa, Amoebida). *Protistologica*, **13**: 557–598.

Radford, C.F., Woodward, E.F.G. and Stapleton, F. (1993). Contact lens hygiene compliance in a university population. *J Brit Contact Lens Assn*, **16**: 105–111.

Schroeder, J.M., Booton, G.C., Hay, J. *et al.* (2001). Use of subgenic 18s ribosomal DNA PCR and sequencing for genus and genotype identification of Acanthamoebae from humans with keratitis and from sewage sludge. *J Clin Microbiol*, **39**: 1903–1911.

Visvesvara, G.S. (1991). Classification of Acanthamoeba. *Rev Infect Dis*, **13**: S369–S372.

Visvesvara, G.S. (1993). Epidemiology of infections with free-living amebas and laboratory diagnosis of microsporidiosis. *Mount Sinai J Med*, **60**: 283–288.

Visvesvara, G.S. and Stehr-Green, J.K. (1990). Epidemiology of free-living amoeba infections. *J Protozool*, **37**: 25S–33S.

Vodkin, M.H., Howe, D.K., Visvesvara, G.S. *et al.* (1992). Identification of *Acanthamoeba* at the generic and specific levels using the polymerase chain reaction. *J Protozool*, **39**: 378–385.

# 17

# *Balantidium coli*

## Basic microbiology

*Balantidium coli* is a large, ciliated protozoan (Phylum *Ciliophora*, Order *Trichostomatida*, Family *Balantidiidae*) which, while having a world-wide distribution, is a rare cause of human infection (Arean and Koppisch, 1956). Not only is it the only ciliated protozoan pathogenic to man, it is also the largest protozoan parasite of humans. *B. coli* can cause severe, life-threatening colitis and transmission of infection is via the cyst stage shed in faeces. Following ingestion, the cysts excyst in the small intestine and subsequently trophozoites reside in the lumen of the large intestine where they feed on bacteria. Trophozoites (30–150 μm by 25–120 μm) replicate by lateral transverse binary fission, and it is also speculated that conjugation may occur, but while this has been observed in culture it has never been demonstrated in nature (Sargeaunt, 1971). At the anterior end of the trophozoite is a mouth (cytosome) that leads into the cytopharynx, which occupies about a third of the parasite length. At the anterior end is the anus (cytophage). The trophozoite is binucleate. Some trophozoites invade the wall of the colon and multiply. Trophozoites undergo encystation to produce infective cysts, which are up to 30–200 μm by 20–120 μm. The cysts also contain a micro- and macro-nucleus. The cysts are shed in the faeces and are responsible for transmission of infection. While remaining trophozoites may disintegrate in the gut lumen, they can be readily detected in stools during acute infection.

The prevalence of human infection is generally low, but higher in tropical and subtropical regions, particularly where the principal animal reservoir, pigs, are raised. For example, human balantidiasis is most common in the Philippines, but is also reported in Central and South America, Iran and Papua New Guinea (Barnish and Ashford, 1989), although prevalence is rarely above 1%. Other animal reservoirs exist including rodents and non-human primates. *B. coli* is not host-specific but may not be readily transmitted between some hosts since adjustment to the symbiotic flora of the new host appears to be required. However, the transmission of human-derived isolates to piglets and monkeys has been demonstrated (Yang *et al.*, 1995). Once adapted to the new host, it can flourish and become a serious pathogen, particularly in humans. Biological features of *B. coli* affect its transmission and epidemiology, particularly in the ecological relationship between hosts and the survival of cysts.

## Clinical features

Although most infections are asymptomatic, acute balantidiasis manifests as persistent diarrhoea, abdominal pain, weight loss, tenesmus, nausea, vomiting and occasionally dysentery which may resemble amoebiasis (Baskerville, 1970). Chronic infection and disease also occurs, characterized by intermittent diarrhoea and occasional blood in the stools. Symptoms of balantidiasis can be severe in debilitated people. Misdiagnosis can result in failure to treat the infection and subsequent case fatality. Rapid progression, with fever and prostration leads to death, usually due to peritonitis from colonic perforation. Rare cases of balantidial appendicitis have been reported (Dodd, 1991).

## Pathogenicity, virulence and causation

The pig is regarded as the primary host for *B. coli*, in which it is a commensal organism and rarely associated with the mucosa. However, in man, *B. coli* produces ulcers ranging from superficial to deep, and associated dysentery. Only rarely are other tissues, such as the liver, invaded. On invasion and penetration of the distal ileal and colonic mucosa and submucosa, the trophozoite causes mucosal inflammation and ulceration. The enzyme hyaluronidase, produced by the parasite, is thought to facilitate this penetrative invasion. Inflammation is caused by liberation of other products by the parasite and possibly by the recruitment of mucosal inflammatory cells, particularly neutrophils. Studies of trophozoite populations from pigs with balantidiasis, by cytophotometric studies, have shown differences in nucleic acids from populations infecting asymptomatic pigs (Skotarczak and Zielinski, 1997).

## Treatment

Balantidiasis is treatable by antibiotic therapy, including tetracycline, iodoquinol, bacitracin, ampicillin, paromomycin and metronidazole (Garcia-Laverde and DeBonilla, 1975). In fulminant disease surgery may be required.

## Survival in the environment and in water

*B. coli* has been detected in sewage sludge (Amin, 1988) and in water storage tanks in different parts of Hyderabad, India (Jonnalagagga and Bhat, 1995). The cysts can survive for several weeks in moist conditions, such as pig faeces or water, but are rapidly destroyed in dry heat and by pH <5. While the cysts are resistant to levels of chlorination used to treat drinking water, they are killed by boiling. The impact of water-treatment processes on removal of *B. coli* has not been measured, but due to its large size it is likely that standard treatment involving coagulation and filtration would be effective.

## Methods of detection

Although trophozoites rapidly disintegrate they are more frequently detected than cysts in stools during acute infection and therefore should be sought for diagnosis. Trophozoites are oval in shape and about 17 µm × 15 µm, and covered in cilia that propel the organism within the intestinal lumen. Shedding in stools is intermittent and therefore repeat stools should be collected and examined immediately or preserved prior to examination. Trophozoites can also be detected in material from the margins of ulcers seen in the rectum by sigmoidoscopy. Parasites from swine faeces were examined for autofluorescence. Cysts of *B. coli* have been shown to emit light after excitation with UV light, which may facilitate diagnosis (Daugschies *et al.*, 2001). While culture methods are available for *B. coli*, these are more suited to research than diagnostic laboratories (Clark and Diamond, 2002).

## Epidemiology

Human infection is generally regarded as a zoonosis, transmitted by faecal-oral contact with swine faeces. Waterborne epidemics have occurred in areas of poor sanitation and, although they do not suffer enteric disease, swine are an important reservoir. A large epidemic involving over 100 human cases of balantidiasis followed a severe typhoon which caused gross contamination of ground and surface water supplies by pig faeces (Walzer *et al.*, 1973).

Person-to-person transmission may occur, particularly if personal hygiene is difficult. In a survey of residents of psychiatric institutions in Italy, *B. coli* was detected in the stools of 97/234 (40.8%) (Giacometti *et al.*, 1997) and parasitic infections were associated with diarrhoea and other gastrointestinal symptoms.

## Risk assessment

The risk of transmission from swine to humans appears to be greatest not only where these reservoir animals are kept but also where sanitation is poor. A cross-sectional study of the prevalence of *B. coli* in pigs on a Danish research farm showed the prevalence of infection increased from 57% in suckling piglets to 100% in most pig groups ≥4 weeks old (Hindsbo *et al.*, 2000). However, no human cases have been published in Denmark indicating that either the strain was not infectious for man or that proper control measures are effective in preventing zoonotic spread.

Surveys of water supplies for *B. coli* have been rarely reported, but in a survey of stored drinking water in Hyderabad City, India, 61/232 samples indicated the presence of pathogenic parasites which include protozoans (cysts of *Giardia lamblia*, *Entamoeba histolytica*, *Balantidium coli*) and nematode eggs (*Enterobius vermicularis*, *Ascaris lumbricoides*, *Trichuris trichiura*), rhabditiform and filariform larvae and adult stages of *Strongyloides stercoralis* and *Enterobius vermicularis*) (Jonnalagagga and Bhat, 1995). Interestingly, hand washings from food handlers also showed the presence of pathogenic parasites, although the original water used for such washings were free from contamination.

### Overall risk assessment

*Health effects*: occurrence of illness, degree of morbidity and mortality, probability of illness based on infection:

- *Balantidium coli* occurs world-wide, but is a rare cause of human infection.
- Although most infections are asymptomatic, acute balantidiasis manifests as persistent diarrhoea, abdominal pain, weight loss, tenesmus, nausea, vomiting, and occasionally dysentery which may resemble amoebiasis.
- Chronic infection and disease also occurs, characterized by intermittent diarrhoea and occasional blood in the stools.
- Symptoms of balantidiasis can be severe in debilitated people. *B. coli* can cause severe, life-threatening colitis.
- The prevalence of human infection is generally low, but higher in tropical and subtropical regions, particularly where the principal animal reservoir, pigs, are raised. Prevalence is rarely above 1%.

*Exposure assessment*: routes of exposure and transmission, occurrence in source water, environmental fate:

- Infective cysts are excreted in faeces of humans and pigs. The route of transmission is faecal-oral. Person-to-person transmission may occur.
- Though pigs are the primary animal reservoir, other animal reservoirs include rodents and non-human primates.
- Human infection is generally regarded as a zoonosis, transmitted by faecal-oral contact with swine faeces.
- Waterborne epidemics have occurred in areas of poor sanitation.
- *B. coli* has been detected in sewage sludge and in water storage tanks.
- The cysts can survive for several weeks in moist conditions, such as water, but are rapidly destroyed in dry heat and by pH <5.

*Risk mitigation*: drinking-water treatment, medical treatment:

- Like other protozoan cysts, *B. coli* cysts are resistant to levels of chlorination used to treat drinking water, however, they are killed by boiling.
- The impact of water-treatment processes on removal of *B. coli* has not been measured, but because they are large (30–200 µm by 20–120 µm), it is likely that standard treatment involving coagulation and filtration would be effective.
- Balantidiasis is treatable by antibiotic therapy, including tetracycline, iodoquinol, bacitracin, ampicillin, paromomycin and metronidazole.

## Future implications

Given that *B. coli* has a cyst stage and can survive in water, it is possible that waterborne transmission can occur following contamination by faeces from infected reservoir hosts or from infected people. However, where the prevalence is low the likelihood is reduced and control can be achieved by good sanitation.

## References

Amin, O.M. (1988). Pathogenic micro-organisms and helminths in sewage products, Arabian Gulf, country of Bahrain. *Am J Public Hlth*, **78**: 314–315.

Arean, V.M. and Koppisch, E. (1956). Balantidiasis: a review and report of cases. *Am J Pathol*, **32**: 1089–1115.

Barnish, G. and Ashford, R.W. (1989). Occasional parasitic infections of man in Papua New Guinea and Irian Jaya (New Guinea). *Ann Trop Med Parasitol*, **83**: 121–135.

Baskerville, L. (1970). *Balantidium colitis*. Report of a case. *Am J Dig Dis*, **15**: 727–731.

Clark, C.G. and Diamond, L.S. (2002). Methods for cultivation of luminal parasitic protists of clinical importance. *Clin Microbiol Rev*, **15**: 329–341.

Daugschies, A., Bialek, R., Joachim, A. *et al.* (2001). Autofluorescence microscopy for the detection of nematode eggs and protozoa, in particular *Isospora suis*, in swine faeces. *Parasitol Res*, **87**: 409–412.

Dodd, L.G. (1991). *Balantidium coli* infestation as a cause of acute appendicitis (Letter). *J Infect Dis*, **163**: 1392.

Garcia-Laverde, A. and de Bonilla, L. (1975). Clinical trials with metronidazole in human balantidiasis. *Am J Trop Med Hyg*, **24**: 781–783.

Giacometti, A., Cirioni, O., Balducci, M. *et al.* (1997). Epidemiologic features of intestinal parasitic infections in Italian mental institutions. *Eur J Epidemiol*, **13**: 825–830.

Hindsbo, O., Nielsen, C.V., Andreassen, J. *et al.* (2000). Age-dependent occurrence of the intestinal ciliate *Balantidium coli* in pigs at a Danish research farm. *Acta Vet Scand*, **41**: 79–83.

Jonnalagagga, P.R. and Bhat, R.V. (1995). Parasitic contamination of stored water used for drinking/cooking in Hyderabad. *SE Asian J Trop Med Public Hlth*, **26**: 789–794.

Sargeaunt, P.G. (1971). The size range of *Balantidium coli. Trans Roy Soc Trop Med Hyg*, **65**: 428.

Skotarczak, B. and Zielinski, R. (1997). A comparison of nucleic acid content in *Balantidium coli* trophozoites from different isolates. *Folia Biol*, **45**: 121–124.

Walzer, P.D., Judson, F.N., Murphy, K.B. *et al.* (1973). Balantidiasis outbreak in Truk. *Am J Trop Med Hyg*, **22**: 33–41.

Yang, Y., Zeng, L., Li, M. *et al.* (1995). Diarrhoea in piglets and monkeys experimentally infected with *Balantidium coli* isolate from human faeces. *J Trop Med Hyg*, **98**: 69–72.

# 18

# *Cryptosporidium* spp.

## Basic microbiology

*Cryptosporidium* is an obligate intracellular parasite (Phylum *Apicomplexa*, Order *Eucoccidiida*, Family *Cryptosporidae*). It therefore requires host cells in which to continue and finish its life cycle, which is completed in a single host. Although currently classified as an eimeriid coccidian, there is mounting phenotypic and molecular phlyogenetic evidence for closer relationship with the gregarines and reclassification has been proposed but not yet adopted (Tenter *et al.*, 2002). Many *Cryptosporidium* species have been confirmed by genetic analyses (Table 18.1) and they infect a wide range of hosts. Although the majority of human disease is caused by *Cryptosporidium parvum* (*syn. C. parvum* genotype 2, 'cattle' or 'C') and *Cryptosporidium hominis* (*syn. C. parvum* genotype 1, 'human' or 'H') (Fayer *et al.*, 2000; Morgan-Ryan *et al.*, 2002), other species are also detected, albeit more rarely, in both immunocompetent and immunocompromised patients (Fayer *et al.*, 2000; Chalmers *et al.*, 2002a) (Table 18.1).

Human infection with *Cryptosporidium* occurs following the ingestion and excystation of oocysts in the small intestine, when four motile sporozoites are released. Five further developmental events follow: merogony, gametogony,

**Table 18.1** Currently accepted species of *Cryptosporidium*

| *Cryptosporidium* species | Original host | Mean oocyst sizes μm (range) in original host | Confirmed in humans by genetic analysis? | |
|---|---|---|---|---|
| | | | Immunocompetent | Immunocompromised |
| Intestinal species | | | | |
| C. parvum | Mice | 5.0 × 4.5 (3.8–6.0 × 3.0–5.3) | Yes | Yes |
| C. hominis | Man | 4.9 × 5.2 (4.4–5.4 × 4.4–5.9) | Yes | Yes |
| C. felis | Cat | 4.6 × 4.0 (3.2–5.1 × 3.0–4.0) | Yes | Yes |
| C. canis | Dog | 5.0 × 4.7 (3.7–5.9 × 3.7–5.9) | Yes | Yes |
| C. wrairi | Guinea-pig | 5.4 × 4.6 (4.8–5.6 × 4.0–5.0) | No | No |
| C. nasorum | Fish | 4.3 × 3.3 (3.5–4.7 × 2.5–4.0) | No | No |
| C. saurophilum | Lizard | 5.0 × 4.7 (4.4–5.6 × 4.2–5.2) | No | No |
| Gastric species | | | | |
| C. muris | Mice | 7.0 × 5.0 (6.5–8.0 × 5.0–6.5) | No | Yes |
| C. andersoni | Cattle | 7.4 × 5.5 (6.6–8.1 × 5.0–6.5) | No | No |
| C. serpentis | Snakes | 6.2 × 5.3 (5.6–6.6 × 4.8–5.6) | No | No |
| Multi-site species | | | | |
| C. meleagridis | Turkeys | 5.2 × 4.6 (4.5–6.0 × 4.2–5.3) | Yes | Yes |
| C. baileyi | Chickens | 6.2 × 4.6 (5.6–6.3 × 4.5–4.8) | No | No |

From: Tyzzer, 1910; Chalmers *et al.*, 1994; Arrowood, 1997; Fayer *et al.*, 1997, 2000, 2001; Koudela and Modrý, 1998; Lindsay *et al.*, 2000; Morgan-Ryan *et al.*, 2002.

fertilization, oocyst wall formation and sporogony (Current and Long, 1983; Current and Hayes, 1984; Current and Reese, 1986). Excystation is initiated by reducing conditions and exposure to secretions such as bile salts and pancreatic enzymes, although *Cryptosporidium* oocysts do excyst in warm aqueous solutions, which may enable infection of extraintestinal sites including the respiratory tract, conjunctiva of the eye, and reproductive system. Sporozoites infect the apical portion of epithelial cells (usually enterocytes in the small intestine) when the anterior end adheres to their surface and becomes surrounded by microvilli, uniquely occupying an intracellular but extracytoplasmic location in the host cell. Cycles of asexual and then sexual multiplication follow culminating in the production of immature oocysts. These mature and sporulate inside the host and oocysts containing four infectious sporozoites are shed in the faeces. These oocysts are thick-walled, resistant to many environmental pressures, and are the only exogenous stage of the life cycle. Thin walled oocysts are not detected in faeces but initiate a cycle of autoinfection within the host (Current and Reese, 1986), causing prolonged disease in those unable to clear the infection. Direct faecal-oral host-to-host transmission can occur because the parasites are immediately infective upon release in the faeces. Equally important is the robust nature of the oocysts which contributes to indirect transmission following contamination of the food or water environment and of fomites.

*Cryptosporidium* has a world-wide distribution with greater numbers of cases of cryptosporidiosis reported among children than adults. The majority of human infections are with either of two species: *C. hominis* is the anthroponotic genotype that is largely restricted to humans, and *C. parvum* is the zoonotic genotype that causes both human and animal disease (Fayer *et al.*, 2000; Morgan-Ryan *et al.*, 2002).

Thus the detection of *C. hominis* is indicative of a human source of infection or contamination and *C. parvum* of either an animal or a human source. Application of genotyping techniques has also led to the characterization of additional *Cryptosporidium* spp. and genotypes, and it is has become clear that other species are also found infecting both immunocompetent and immunocompromised patients (Fayer *et al.*, 2000; Chalmers *et al.*, 2002a).

A number of biological features of *Cryptosporidium* affect its transmission and epidemiology:

- the life cycle does not require dual or multiple hosts
- the oocyst stage is shed in a fully sporulated state so direct transmission can occur between hosts
- autoinfection enables persistent disease in immunocompromised hosts
- there is a large livestock and human reservoir
- the thick-walled oocysts are resistant to a wide range of pressures and can survive for long periods in the environment
- the infectious dose is low and so small numbers of contaminating organisms are significant
- hosts can shed large numbers of oocysts
- there is a lack of specific drug therapy to clear infection efficiently.

Of additional hindrance to risk assessment, environmental testing is difficult since sampling for and detection of oocysts is problematic, and viability/infectivity assessment currently inadequate from such samples.

## Origin of the organism

The first person to recognize, describe and name *Cryptosporidium* was E.E. Tyzzer, who, in 1907, published the asexual, sexual and oocyst stages of a parasite he frequently found in the gastric glands and faeces of laboratory mice (Tyzzer, 1907). He proposed the murine gastric isolate *Cryptosporidium muris*, the type strain (Tyzzer, 1910) and, in 1912, published a description of a new, smaller species found in the small intestine of laboratory mice and rabbits, which he called *C. parvum* (Tyzzer, 1912). Tyzzer's remarkable observations, including the proposal of autoinfection, established the life cycle of the parasite, which electron microscopy has served to confirm with the only additional observation of the extracellular developmental stages (merozoites and microgametes). In 1929, Tyzzer also described endogenous stages of *Cryptosporidium* in chicken caecal epithelium (Tyzzer, 1929).

In 1955 a new species, *Cryptosporidium meleagridis*, was reported causing illness and death in young turkeys (Slavin, 1955) and, in 1971, a report was published where *Cryptosporidium* was associated with bovine diarrhoea (Panciera *et al.*, 1971). While this stimulated veterinary investigations of the parasite, human cases were not identified until 1976 when two reports were published. One described an otherwise healthy 3-year-old girl with symptoms of vomiting, watery diarrhoea and abdominal pains (Nime *et al.*, 1976). Diagnosis was made by rectal biopsy, and the patient recovered after 2 weeks' illness. In contrast, the other described a severely dehydrated immunosuppressed patient with chronic watery diarrhoea (Meisel *et al.*, 1976). Diagnosis was by histological examination of jejunal biopsy. The patient recovered following withdrawal of immunosuppressive treatment and subsequent restoration of T-cell function.

It was not until the 1980s that the role of *Cryptosporidium* in human disease and its impact on human health really began to be recognized. Contributing to the emergence of *Cryptosporidium* as a human pathogen were the AIDS epidemic and consequent increase in the number of immunocompromised individuals unable to clear *Cryptosporidium* infections, and the occurrence of a number of waterborne outbreaks of disease in developed countries. There was clearly an inconsistency in the perception of the parasite as an opportunist zoonotic infection and its occurrence in primarily urban, male AIDS patients (Casemore and Jackson, 1984). Improved laboratory methods by veterinary workers for the detection of oocysts in faeces led to increased ascertainment and recognition of the parasite in clinical laboratories, and important epidemiological studies during the early 1980s showed that cryptosporidiosis also occurred in otherwise healthy subjects, particularly children (Casemore *et al.*, 1985). Widespread reporting of microbiological results to disease surveillance schemes contributed to the recognition of *Cryptosporidium* as a cause of acute, self-limiting gastroenteritis in the general population and of potentially fatal infection in the immunocompromised.

## Clinical features

A range of clinical features characterize cryptosporidiosis, varying in severity from asymptomatic carriage to severe, life-threatening illness. The predominant feature of cryptosporidiosis is watery, sometimes mucoid, diarrhoea with dehydration, weight loss, anorexia, abdominal pain, fever, nausea and vomiting. The incubation period has been reported from outbreaks as a mean of 7 days from exposure (range 1–14 days) (MacKenzie *et al.*, 1995). In experimental human infection diarrhoeal symptoms appeared at mean 9 days and median 6.5 days in subjects with no serological evidence of previous infection (DuPont *et al.*, 1995) and median 5 days (range 3–12 days) in subjects with evidence of prior infection (Chappell *et al.*, 1999).

In immunocompetent patients symptoms usually last for about 1–2 weeks. The duration of diarrhoeal illness in sporadic cases who visited primary health care in the UK and submitted a faecal specimen has been reported as mean 9 days, median 7 days (mode 7 days, range 1–90 days) (Palmer and Biffin, 1990). However, a recent study in Melbourne, Australia showed that similarly selected patients reported symptoms of mean 22 days (range 1–100 days) and in Adelaide mean 19 days (range 2–120 days) (Robertson *et al.*, 2002). In experimental infection, duration of gastrointestinal illness in subjects without evidence of prior exposure to *Cryptosporidium* was 6.5 days (DuPont *et al.*, 1995) and in subjects with prior exposure 3.1 days (Chappell *et al.*, 1999). It is likely that cases identified through passive surveillance represent the more severe end of the spectrum of disease in that they experienced symptoms which prompted them to seek primary health care and to submit a faecal specimen for testing.

Oocysts may continue to be shed in the faeces following cessation of diarrhoea for 7 days (range 1–15 days) (Jokipii and Jokipii, 1986). Asymptomatic carriage has been reported in natural and experimental infections (Checkley *et al.*, 1997; Chappell *et al.*, 1999). While the diarrhoeic phase may pose a greater risk of onward transmission due to unpredictable faecal release, risks from asymptomatic shedders have not been fully evaluated, but may be significant since hygiene precautions may be more relaxed.

Immunocompetent patients are able to resolve cryptosporidiosis spontaneously, albeit after prolonged diarrhoea compared with many other gastrointestinal illnesses. However, in patients with impaired cell-mediated immunity, particularly with reduced lymphocyte and CD4 T cell counts of <200/mm$^3$, gastrointestinal symptoms are chronic, severe, debilitating and indeed life threatening. In a study of HIV patients, fulminant infection, where patients passed 2 litres of watery diarrhoea per day, occurred with CD4 counts of <50 cells/mm$^3$ (Blanshard *et al.*, 1992). Prior to the introduction of highly active antiretroviral therapy (HAART) in 1996, cryptosporidiosis affected 10–15% of AIDS patients, causing death in 50% of cases (Clifford *et al.*, 1990). Colonization throughout the gastrointestinal tract has also been reported in patients with primary immunodeficiencies, such as common variable immunodeficiency, hypogammaglobulinaemia, severe combined immunodeficiency, X-linked hyper IgM syndrome (CD40 ligand deficiency) or gamma interferon deficiency, in secondary immunodeficiencies due to HIV/AIDS, organ transplantation and immunosuppressive drugs, haematological malignancies and anti-cancer chemotherapy (Farthing, 2000). Extraintestinal and extra-abdominal cryptosporidiosis has been reported and resulted in pancreatitis, chronic cholangiopathy, sclerosing cholangitis, cirrhosis, cholangiocarcinoma and respiratory involvement (Farthing, 2000).

Chronic health effects of cryptosporidiosis include Reiter's syndrome (Shepherd et al., 1988). This is a reactive arthritis that has been identified following bouts of diarrhoea caused by a number of aetiological agents, and has been linked to cryptosporidiosis in children (Cron and Sherry, 1995). Long-term effects of childhood cryptosporidiosis have been measured, particularly in children in developing countries. For example, in Guinea-Bissau undernourished children below 3 years of age suffered significant weight loss and impaired growth, which was not followed by subsequent catch-up growth (Molbak et al., 1997). However, in the investigation of gastrointestinal infection, it is often hard to unravel the elements of the malnutrition-infection cycle and the immune consequences of malnutrition. Studies in Peru have shown that a measurable effect occurred in the growth of children who were not severely or acutely malnourished following cryptosporidial infections, even in the absence of diarrhoea (Checkley et al., 1998). While catch-up growth was reported in older children this was age-related, and reduced in younger children. However, this did not occur in infants who were under 5 months at the acquisition of Cryptosporidium.

## Pathogenicity

Cryptosporidium diarrhoea is caused largely by intestinal transport defects resulting from infection of enterocytes in the small intestine (Clark and Sears, 1996). Blunting of the microvilli and villous tip damage causes malabsorption of sodium ions, but there is also evidence for increased secretion of chlorine ions stimulated by cell infection or infiltration of inflammatory cells into the lamina propria. The production of alpha interferon by infiltrating macrophages and release of inflammatory mediators, such as exogenous prostaglandin $E_2$ production, may stimulate chlorine ion secretion. Additionally, disruption of the epithelium due to hyperplasia of the intestinal crypt cells, cell defects and cell death can lead to increased paracellular permeability. However, the exact role of inflammatory cells, enteric nerves, cytokines or hormones in the pathogenesis of Cryptosporidium diarrhoea are as yet unknown.

Speculation of a cholera-like toxin has been made because of the profuse nature of the watery diarrhoea experienced by some patients and the detection of enterotoxin-like activity in the filtered faecal supernate from C. parvum-infected calves (Guarino et al., 1994). However, studies have failed to differentiate between parasite-derived enterotoxin and host-derived factors and, in further studies, between other possible causes of diarrhoea.

# Virulence

The infectious dose has been shown in human infectivity studies to be low, with infection occurring following challenge of subjects without evidence of prior infection with 30 oocysts (DuPont *et al.*, 1995). The median infective dose in that study was 132 oocysts. Although not statistically significant, higher doses of oocysts resulted in occurrence of one or more gastrointestinal symptoms, shorter incubation periods and longer duration of illness. In similar studies of volunteers with prior exposure, infection and diarrhoea were associated with higher challenge doses and the $ID_{50}$ was over 20-fold higher (1880) than that in seronegative subjects (Chappell *et al.*, 1999).

Human infectivity studies have thus far been undertaken using *C. parvum* isolates, and where the infectivity of different isolates has been compared, variation in $ID_{50}$, attack rates and duration of diarrhoea has been observed (Okhuysen *et al.*, 1999). The median infectious dose of three different *C. parvum* isolates ranged from nine to 1042 oocysts. The apparently limited host range of *C. hominis* indicates that this may be a human-adapted species with differing infectivity for humans from *C. parvum*.

Although there are differences in pathogenicity and infectivity between isolates and differences in antigenic profile and in host immunoreactivity have been observed, the molecular basis for this is poorly understood in *Cryptosporidium*. Factors responsible for the initiation, establishment and perpetuation of infection are poorly defined. Potential candidate molecules for virulence factors include those involved in locomotion, adhesion, fusion, invasion, and the investigation of heat shock proteins, toxins and host cell apoptosis and have been reviewed by Okhuysen and Chappell (2002). Genome surveys and the development of cell culture methods, particularly long-term maintenance, will assist in identifying and determining the relative importance of expressed and deleted virulence factors.

# Causation

The clinical consequences of *C. parvum* infection are directly linked to the immune function of the host. Autoinfective oocysts and recycling of type 1 meronts are probably features contributing to persistent disease in immunocompromised patients who are not repeatedly exposed to environmentally resistant oocyst forms. AIDS patients have been most studied, particularly in the USA and UK, and have shown a significant relationship between immunosuppression evaluated by total lymphocyte and CD4 counts, with fulminant infection occurring in patients with a CD4 count of <50 cells/mm$^3$ (Blanshard *et al.*, 1992). Median time of survival of this group of patients was 5 weeks. In developing countries, particularly sub-Saharan Africa, the relationship between

CD4 count and severe infection is less clear with more profound pathology and clinical manifestations, and differing background enteropathy suggests that other host-related factors may be important in determining the outcome of infection (Kelly *et al.*, 1997).

Avoidance of exposure to the parasite is imperative among high-risk groups. In the UK, patients whose T-cell function is compromised and those with specific T-cell deficiencies are advised by the Department of Health to boil and then cool all their drinking water (including that used for making ice), which may have helped limit the number of infections in these groups. In the USA, guidance created by the USEPA and CDC also recommends boiling, but alternatives include filtering through absolute 1 μm filters.

## Treatment

There is no established curative therapy for cryptosporidiosis. Management of diarrhoea can be supported by fluid replacement and electrolyte balance. While cryptosporidiosis in otherwise healthy patients may be self-limiting, immunocompromised patients and those in poor health or suffering malnutrition are at high risk of severe illness. Many of the drugs investigated for efficacy against *Cryptosporidium* have been the ones used in the treatment of infection with eimeriid coccidians. However, differences between these organisms and *Cryptosporidium* are being recognized and may explain the apparent insensitivity to anti-coccidial drugs.

In the absence of reliably effective curative chemotherapy, suppression of proliferation of the parasite has been reported, particularly with paromomycin (White *et al.*, 1994; Flanigan *et al.*, 1996), albendazole and nitazoxanide in HIV-negative children (Rossignol *et al.*, 1998, 2001; Amadi *et al.*, 2002) and paromomycin in combination with azythromycin (Smith *et al.*, 1998), although many clinical and other trials have been inconclusive (Clinton Wight *et al.*, 1994; Hoepelman, 1996; Theodos *et al.*, 1998; Hewitt *et al.*, 2000). This may in part be due to the varying reactions in different patient groups: paromomycin has been used to suppress the parasite in HIV patients, while nitazoxanide only showed efficacy in HIV-negative children. Additionally, the cryptosporidia isolated in drug trials are rarely characterized. Multi-drug regimens that reduce viral load and increase CD4 T lymphocytes, such as highly active antiretroviral therapy (HAART) prescribed to AIDS patients in developed countries, have led to the resolution of symptoms and reduction of severe health effects of cryptosporidiosis in this patient group. Where HAART is available AIDS-related cryptosporidiosis is a decreasing problem. However, this treatment is not available in developing countries, and rapid relapse has been reported after discontinuation of HAART (Carr *et al.*, 1998; Maggi *et al.*, 2000). Additional approaches to treatment of cryptosporidiosis, particularly for vulnerable groups of patients, have little supporting evidence of success but include administration of hyper-immune bovine colostrums, and new approaches to be explored further include glutamane, blocking prostaglandin and ligand/receptor blocking.

## Survival in the environment

*Cryptosporidium* oocysts have been detected in a variety of environmental matrices including farmyard manure, leachate, slurry, and soil (Kemp *et al.*, 1995), as well as various water matrices (see below) and foodstuffs (Rose and Slifco, 1999). Oocysts have been detected in sewage discharge from treatment works and data show that neither removal nor inactivation by primary and secondary treatment can be guaranteed (Robertson and Gjerde, 2000). If the integrity of the thick, two-layered oocyst wall is maintained, *Cryptosporidium* oocysts are robust and resistant to a variety of environmental pressures particularly under cool, moist conditions. Rose and Slifco (1999) have reviewed the survival of *Cryptosporidium* under various food preservation conditions. Extremes of temperature and reduction in water activity including freeze-drying, temperatures above 60°C and below −20°C for 30 minutes will kill *Cryptosporidium* (Anderson, 1985), as will low temperature, long time and high temperature, short time pasteurization conditions (Harp *et al.*, 1996). Blewett (1989) demonstrated, using excystation studies, that oocysts are killed by exposure to moist heat at 60°C and above for 5 minutes. Survival in manure heaps and slurry stores is adversely affected by the pH, temperature and ammonia generated (Bukhari *et al.*, 1995; Jenkins *et al.*, 1998). Although many common disinfectants used on farms, in hospitals or veterinary surgeries have little effect on *Cryptosporidium*, both hydrogen peroxide and ammonia inactivate oocysts (Blewett, 1989; Casemore *et al.*, 1989).

However, survival of oocysts, for example in food processing procedures, and the efficacy of disinfectants proposed for water treatment such as ozone or UV, is poorly evaluated due to lack of reliable, routine methods to assess infectivity and viability (Casemore and Watkins, 1999). Additionally, the oocysts and isolates used in survival or disinfection studies have often been poorly characterized in terms of source, age, storage, purification method and viability. While the gold standard might be human infectivity, animal models are available but the mouse model is not applicable to *C. hominis* isolates. Additionally, animal infectivity is not suitable for routine assessment. Alternative methods include vital dyes, excystation and cell culture. *In vitro* assays tend to overestimate viability when compared with a mouse model (Clancy *et al.*, 2000) and further evaluation of cell culture with different isolates is currently required. It is important that a valid assessment of viability and infectivity is made for the proper evaluation of disinfection interventions.

### Survival in water

The occurrence of *Cryptosporidium* oocysts in environmental waters depends on the land use within the catchment (since greater concentrations of oocysts are detected in waters receiving animal and sewage discharges), climate (since rainfall can influence the numbers of oocysts in surface waters and temperature affects their survival) and community factors such as watershed management (Rose *et al.*, 2002). In a study undertaken in New Zealand, the highest sample prevalence of *Cryptosporidium* was in areas of intensive livestock farming (Ionas *et al.*, 1998). In an on-farm study in the UK, *Cryptosporidium* was detected in surface waters throughout the year but with the highest frequency and maximum concentrations during the autumn and winter, coinciding with calving and peaks in wildlife populations, but not in that study with rainfall or slurry spreading (Bodley-Tickell *et al.*, 2002). The authors also studied a small pond at the top of a catchment which was not under the influence of livestock, in which the only source of oocysts detected could have been wildlife. Other studies, however, have detected a link between climatic factors and (oo)cyst concentrations, and noted the distribution in the number of human cases often follows a seasonal (rainfall-related) pattern (Rose *et al.*, 2002). Urban waters are also vulnerable to contamination from diverse sources, including sewage outfalls. In La Palta, Argentina, much of the endemic cryptosporidiosis is probably caused by the high level of contamination through the discharge of raw sewage into the city's main water source, the La Palta river, despite conventional, but probably overloaded, treatment (Basualdo *et al.*, 2000).

*Cryptosporidium* occurs frequently in raw waters world-wide. In the USA, surface waters were monitored under the EPA's Information Collection Rule between 1996 and 1998. Data show that 20% of the 5858 samples contained *Cryptosporidium* oocysts, providing a mean national estimate of 2 oocysts/100 l water, although the sampling was based on relatively low volumes of water (median 3 l) (Messner and Wolpert, 2000). While water monitoring data from a range of countries have previously been summarized and shown that the mean number of wastewater samples containing *Cryptosporidium* oocysts was 62%, the mean for surface waters was 46% and the mean number of drinking waters was 24% (Smith and Rose, 1998), further data from additional countries are now available. For example, in Russia, 23/87 (26%) raw river waters throughout the European Russian region were found to contain *Cryptosporidium* oocysts (Egorov *et al.*, 2002). In Japan, one study showed that all 13 raw water samples entering a treatment works contained *Cryptosporidium* oocysts (Hashimoto *et al.*, 2002), while more widespread sampling showed that 74/156 river water samples contained *Cryptosporidium* (Ono *et al.*, 2001). Ground waters have traditionally been considered protected from contamination but those under the influence of surface waters or other routes of contamination are vulnerable and oocysts have been detected in such waters and outbreaks have occurred in communities supplied from such water sources (Hancock *et al.*, 1998; Willocks *et al.*, 1998).

Numbers of oocysts detected and, indeed, frequency of detection will depend on sampling strategies and methods of oocyst recovery and detection. In the UK, continuous monitoring of treated water is legally required at those treatment plants considered as being at risk of having *Cryptosporidium* in the treated water, based on a structured risk assessment under the Water Supply (Water Quality) (Amendment) Regulations 1999 and Water Supply (Water Quality) Regulations 2000. Monitoring is based on continual sampling at a rate of 40 l/hour, using foam filters, with a treatment standard of 1 oocyst/10 l. Data show that between April 2000 and June 2002, oocysts were detected in 5305/109704 samples from 166/204 sites. However, 91% of the detections were at less than 10% of the treatment standard, which was violated in just seven samples from two sites (Peter Marsden, Drinking Water Inspectorate, personal communication). The methods used for this monitoring do not include viability assessment (in fact, methods for oocyst recovery may themselves affect viability, and fixing oocysts on microscope slides renders them non-viable). Although *Cryptosporidium* oocysts experience die-off under environmental pressures, this is relatively slow and variable. Survival of up to 176 days has been reported in drinking water or river water with inactivation of 89–99% of the oocyst population (Robertson *et al.*, 1992). Oocysts are resistant to chlorine at levels used to treat both drinking water and in swimming pools (Rose *et al.*, 1997; Carpenter *et al.*, 1999) and so there is no defence should oocysts breach physical barriers present in full water treatment (flocculation, sedimentation/flotation and filtration) or if such removal treatment is absent. Chlorine dioxide has shown some efficacy against *Cryptosporidium*, inactivating about 90% of oocysts, but the most successful treatments that may be of use in large-scale treatment are ozone and UV (Peeters *et al.*, 1989). Medium and low pressure UV light are capable of inactivating *Cryptosporidium* oocysts. Relatively low doses of 9 mJ/ml achieve >3 log reduction in viability (Bukhari *et al.*, 1999; Craik *et al.*, 2001) and although repair of UV-damaged DNA has been reported, oocysts did not recover infectivity (Shin *et al.*, 2001). By producing free radicals that affect the permeability of the oocyst wall and its DNA, ozone has proven to be highly effective against *Cryptosporidium*, particularly at lower temperatures (Peeters *et al.*, 1989; Rennecker *et al.*, 1999). Sequential inactivation has been demonstrated using ozone followed by free chlorine (Li *et al.*, 2001). Efficacy tests for the treatment of small volumes of drinking water have shown that iodine disinfection is not effective (Gerba *et al.*, 1997).

## Methods of detection

Despite the absence of specific therapy to eliminate infection, there is a need for differential diagnosis to prevent inappropriate application of drug therapy, to support disease surveillance and outbreak investigations and ensure appropriate public health measures are in put in place to prevent onward transmission. Clinical diagnosis is most commonly by identification of oocysts in stools by examination of stained smear on microscope slides (Arrowood, 1997). Acid-fast stains and auramine phenol stains are most commonly used in clinical laboratories, although increased sensitivity and specificity can be achieved by use of immunofluorescence microscopy. Enzyme immunoassays and other commercial test kits are available (Petry, 2000) and widely used in private laboratories.

The oocysts of many *Cryptosporidium* species are indistinguishable by microscopy, and the antibodies in test kits are not generally species specific (Anon, 1997a). The detection of *Cryptosporidium* oocysts by these methods should therefore only be reported as '*Cryptosporidium* spp.' and molecular methods are required for species determination. These have contributed greatly to our understanding of the heterogeneity of the organism and its epidemiology. Characterization of isolates using DNA amplification-based methods is advantageous over phenotypic methods since relatively few organisms are required (Gasser and O'Donoghue, 1999). Genotyping methods are largely PCR-based and have included analysis of repetitive DNA sequences, randomly amplified polymorphic DNA (RAPD), direct polymerase chain reaction (PCR) with DNA sequencing, and PCR restriction fragment length polymorphism (RFLP) analysis (Clark, 1999; Morgan *et al.*, 1999). The two distinct *C. parvum* genotypes, now recognized as separate species (*C. hominis* and *C. parvum*) have been consistently differentiated at a variety of gene loci, including *Cryptosporidium* oocyst wall protein (COWP), ribonuclease reductase, 18S rDNA (syn. small subunit ribosomal RNA), internal transcribed rDNA spacers (ITS1 and ITS2), acetyl-CoA synthetase, dihydrofolate reductase-thymidylate synthase (dhfr-ts), thrombospondin-related adhesive proteins (TRAP-C1 and TRAP-C2), the $\alpha$ and $\beta$ tubulin and 70 kDa heat shock protein (hsp70) (Fayer *et al.*, 2000). It is evident that some primer pairs are more species specific than others, such as those for TRAP-C2 which are specific for *C. parvum* and *C. hominis* (Elwin *et al.*, 2001), while others amplify related protozoan parasites, and that some PCR-RFLPs differentiate species/genotypes more readily than others (Sulaiman *et al.*, 1999). While PCR-RFLP is widely used for characterization, and allows many specimens to be analysed and compared, only bases at the restriction enzyme sites are examined. Sequence analysis provides the 'gold standard' since all the bases within the target sequence at the locus are examined. The importance of sequence confirmation of RFLP patterns was illustrated by Chalmers *et al.* (2002b) who identified a novel RFLP pattern, similar to *C. hominis*, in the COWP gene of isolates from sheep, but sequence data clearly differentiated the isolate. Therefore, careful primer selection and PCR product analysis are required for detection and characterization, particularly from environmental specimens where a wide range of cryptosporidia and other organisms may be present. It must also be noted that oocyst recovery and genotyping methods have yet to be standardized.

Detection of *Cryptosporidium* oocysts in non-clinical samples involves concentration of the oocysts from large volumes of the sample matrix (usually by centrifugation, filtration or flocculation), separation from the sample matrix (by density gradient centrifugation or immunomagnetic separation) and detection by immunofluorescence microscopy (Anon, 1999a,b). The development of immunomagnetic separation techniques has contributed greatly to the improvement of methods for detection in environmental matrices, including water, slurry and food, and enhanced our evaluation of foodborne risks. Using IMS/IFA and improved washing procedures, Robertson and Gjerde (2000) reported mean recovery efficiencies for *Cryptosporidium* of 44% from iceberg lettuce, 48% from green lollo lettuce, 41% from Chinese leaves, 38% from Autumn salad mix, 39% strawberries and 22–35% from bean sprouts.

The methods for the detection of *Cryptosporidium* in non-clinical samples based on immunodetection generally do not differentiate *C. parvum* from other *Cryptosporidium* species, nor do they differentiate live from dead oocysts. While the PCR-based genotyping analyses described above indicate the presence of a strain that may cause human/animal infection, they cannot determine if infectivity is maintained. The 'gold standard' is host infectivity studies, which have been undertaken in humans and animals, but this is obviously not practical for routine assessment or during disinfection studies. While animal models for *C. parvum* exist, there is no practical animal model for *C. hominis*. Cell culture methods have been used to determine viability, but application of cell culture for disinfection efficacy trials requires further evaluation for sensitivity and specificity (Gasser and O'Donoghue, 1999). Viability assessment as a surrogate for infectivity has been undertaken *in vitro* using vital dyes and excystation, but these do not consistently correlate with *in vivo* studies (Clancy *et al.*, 2000). The viability of *C. parvum* oocysts has been determined by induction of hsp70 and subsequent detection of messenger RNA by RT-PCR but has not been developed as a quantitative assay, and may result in the detection of residual mRNA (Gasser and O'Donghue, 1999). Fluorescence *in situ* hybridization (FISH), specific for *C. parvum*, has been developed for detection and viability assay (Vesey *et al.*, 1998) and can combine target-specific detection with confirmation and viability estimation in one test. However, specificity and sensitivity, as well as sampling difficulties must be taken into account if *Cryptosporidium* viability assays are to be applied to environmental samples.

## Epidemiology

The epidemiology of cryptosporidiosis differs between developed and developing countries: in some developing countries, infection is common in infants aged less than 1 year, while in developed countries incidence is most common in 1–5 year olds, with a secondary peak in young adults (Palmer and Biffin,

1990). The peak in adult disease is rarely seen in developed countries where more frequent exposure may generate greater immunity. During waterborne outbreaks, a relative increase in adult cases is often seen (Meinhardt *et al.*, 1996). As with many gastrointestinal infections, cryptosporidiosis is more frequently reported in boys than girls, but adult males and females are generally affected with equal frequency. Several studies world-wide have shown seasonal peaks in reported disease (Casemore, 1990), particularly in spring and autumn, which do not necessarily both occur in any one locality, nor recur year by year. They coincide generally with agricultural activities such as muck spreading, lambing and calving, with maximal rainfall, and following foreign travel.

Substantial outbreaks of cryptosporidiosis have been caused by a variety of sources of infection and routes of transmission, and reflect the ubiquity of the organism. These have included consumption of drinking water from both surface and ground water sources contaminated with human sewage and animal manure, contaminated natural and man-made recreational waters, animal contact during farm visits, person-to-person spread within institutions and day-care centres including children's nurseries, and food-borne outbreaks (Casemore *et al.*, 1997; Rose *et al.*, 1997). Person-to-person spread in households and institutions is important and has been estimated at 60%. Contact with another person with diarrhoea, particularly children, has been identified as a risk factor in case control studies (Puech *et al.*, 2001; Robertson *et al.*, 2002) and has been demonstrated in outbreaks (Hannah and Riordan, 1988).

In developed countries, foreign travel is frequently associated with illness, attributable to exposure of naive subjects to new risks and isolates, often as a result of poor hygiene. The prevalence of *Cryptosporidium* varies throughout the world and is higher in developing countries (6.1% diarrhoeic patients; 1–5.2% non-diarrhoeic) than developed countries (2.2% diarrhoeic patients; 0.2% non-diarrhoeic) (Guerrant, 1997). In developing countries cryptosporidiosis rarely occurs in indigenous adult populations. In neonates in these countries there appears to be an association with cryptosporidiosis and bottle-feeding, possibly due to a combination of contaminated water supplies and the absence of any protective effect of breast-feeding (Casemore *et al.*, 1997). By contrast, in developed countries, although more cases are reported in children than adults, cryptosporidiosis is also a disease of adults.

The application of molecular tools to collections of *Cryptosporidium* isolates has further elucidated the epidemiology of human cryptosporidiosis, and shown that regional and seasonal differences exist, probably reflecting differing exposures and behaviours (McLauchlin *et al.*, 2000; Anon, 2002). For example, regional differences may reflect urban/rural or human/zoonotic cycles of transmission. Seasonal differences in the UK may be linked to animal husbandry and reproduction, resulting in a spring increase in human *C. parvum* infections. An increase in *C. hominis* in the late summer is linked to reports of recent foreign travel during the summer holidays. This, however, is worthy of further investigation to identify more precisely the risks during foreign travel. Furthermore, species variations are observed when the data are analysed by countries visited (Anon, 2002). Thus far, little is known of the epidemiology of non-*C. parvum*, non-*C. hominis* infections in humans, of which *C. meleagridis* predominates.

Analysis of outbreak samples has confirmed that urban transmission is not restricted to *C. hominis* but can occur with *C. parvum*. For example, in an outbreak associated with an indoor swimming pool in England, where the likely source of contamination was human faecal material, *C. parvum* was confirmed in 34/41 cases (Anon, 2000a). The identification of the species causing human illness has also been of benefit during outbreaks in the form of the provision of advice regarding appropriate control measures. For example, during an outbreak in Belfast, Northern Ireland, epidemiologically linked to the drinking water supply, the source of contamination was initially assumed to be livestock since the source water arose in a rural area and flowed through an aqueduct under agricultural land prior to distribution. However, PCR-RFLP of the COWP gene identified *C. hominis* in the human cases indicating human sanitation failure as the source of contamination and infection (Glaberman *et al.*, 2002). Inspection of the aqueduct showed a point of ingress of domestic sewage and remedial action was taken.

Food-borne illness has been attributed to *Cryptosporidium* and food items associated with outbreaks of illness by descriptive and analytical epidemiology include fresh-pressed apple juice (Millard *et al.*, 1994; Anon, 1997b), chicken salad (Besser-Wick *et al.*, 1996) and improperly pasteurized milk (Gelletli *et al.*, 1997). Infected food handlers have been linked to two outbreaks (Quinn *et al.*, 1998; Quiroz *et al.*, 2000) and in one case isolates from cases of illness and the food handler were indistinguishable by PCR-RFLP (Quiroz *et al.*, 2000). Additionally, surveys of raw foods have demonstrated oocysts in molluscan shellfish in Ireland (Chalmers *et al.*, 1997), Hawaii (Johnson *et al.*, 1995), Chesapeake Bay, USA (Fayer *et al.*, 1998; Graczyk *et al.*, 1999) and Spain (Gomez-Bautista *et al.*, 2000), raw market vegetables in Costa Rica (Monge and Chinchilla, 1996) and Peru (Ortega *et al.*, 1997).

The sources of oocysts on foodstuffs may be direct faecal contamination, infected food handlers, wastewater used in irrigation or water used in food processing. Uncooked produce requires proper washing to remove surface contamination. The impact of *Cryptosporidium* on food-borne illness is probably underestimated, as is the impact of food-borne protozoal disease generally. A substantial number of outbreaks of gastrointestinal illness are not attributed to an aetiological agent, and the lack of sensitive methods for the detection of cryptosporidial oocysts in food matrices may have contributed to this. The introduction of immunomagnetic separation has facilitated greater recovery of oocysts from the sample matrix, but variable recoveries highlight the need for improved methods for detection of *Cryptosporidium* from foods (Robertson and Gjerde, 2000).

Recreational activities, such as those undertaken in natural recreational waters and man-made swimming pools, have been associated with outbreaks of cryptosporidiosis. The parasite has been increasingly recognized as a problem in swimming pools since the first two reported outbreaks associated with swimming pools occurred in 1988. One of these outbreaks was in Doncaster, UK where plumbing defects were identified allowing ingress of sewage into the circulating pool water (Joce *et al.*, 1991). The other outbreak was in Los Angeles at a pool where one of the filters was inoperative (Sorvillo *et al.*, 1992). Sources of *Cryptosporidium* in natural waters can be human or animal, but in swimming pools human sources predominate, either from breach of the pool water by sewage or faecal accidents in the pool. A survey of 54 pools in Germany identified *Cryptosporidium* oocysts in 16/94 (17%) samples of filter back wash water from swimming pools where cryptosporidiosis had not been reported, further demonstrating the widespread nature of the organism (Marcic *et al.*, 2000). Toddler pools in particular were among the positive pools. In New South Wales, Australia, a case-control study identified swimming at a public pool and swimming in a dam, river or lake as associated with having cryptosporidiosis (Puech *et al.*, 2001). Contact with a person with diarrhoea was identified as a risk factor in rural areas, and swimming in a public pool in urban areas. In the UK, 18 outbreaks of cryptosporidiosis associated with swimming pools were reported for the 10 years from 1989 to 1999, but seven were during 1999 alone (Anon, 2000b). The reasons for the apparent increase may be genuine or as a result of improved outbreak investigation as well as increased awareness of swimming pools as a risk factor for cryptosporidiosis. Since pool water filtration systems were not designed with awareness of *Cryptosporidium*, it is likely that they are not efficient at its removal and the main public health measure is to keep faecal material and hence *Cryptosporidium* out of the pool. This can be achieved by public education, encouraging people with diarrhoea not to swim, improved pool facilities (pre-swim showers, toilets and hand washing) and policies on dealing with faecal accidents available to all pool operators. Outbreaks of cryptosporidiosis have not been linked to seawater, but oocysts have been detected in marine waters (Johnson *et al.*, 1995; Fayer *et al.*, 1998) and marine mammals (Johnson *et al.*, 1995).

Despite the many sources and routes of transmission of *Cryptosporidium*, drinking water has attracted the greatest attention since the first waterborne outbreak of cryptosporidiosis was documented in 1985 in the USA (D' Antonio *et al.*, 1985). Traditionally, groundwater sources of drinking water are considered protected from contamination, but in the spring of 1997 an outbreak associated with a deep borehole supply occurred in North Thames, England (Willocks *et al.*, 1998) and has implications for the understanding of water quality for such supplies. Drinking waterborne outbreaks of cryptosporidiosis have been the largest in terms of numbers of human cases, and expose all members of the community who use potable water. The largest documented waterborne outbreak occurred in Milwaukee in April 1993, and was first detected because of the high level of absenteeism in schools and among staff at local hospitals (MacKenzie *et al.*, 1995). Although 739 cases were confirmed by laboratory diagnosis of *Cryptosporidium* in a stool specimen, a telephone survey of households was undertaken and estimated, from the numbers of people with watery diarrhoea, that the extent of the outbreak was 403 000 cases.

Drinking waterborne outbreaks of cryptosporidiosis have been associated with both *C. parvum* and *C. hominis*. Sporadic and outbreak cases, descriptive data and epidemiological evidence have demonstrated the possibility of human sewage and animal sources of contamination in source waters (Casemore, 1998). These can come from direct breaches into the water or via agricultural or natural run-off and effluent such as abattoir waste. In a recent outbreak in Northern Ireland, subtyping methods also illustrated the presence of the same *C. hominis* subtype in the human cases and in the suspected (human) source of contamination (Glaberman *et al.*, 2002). The application of genotyping techniques is providing further information about the role of animals in human cryptosporidiosis and has demonstrated that microscopical detection of oocysts alone from possible sources is not proof of source of infection (Chalmers *et al.*, 2002b). The public health significance of oocysts when detected in environmental samples must be investigated and one study has shown that a variety of species and genotypes were detected in surface and wastewaters, some of which are not known to be infectious for humans (Xiao *et al.*, 2001a). This has implications for both monitoring and for outbreak investigations.

Private water supplies (i.e. supplies not managed by a water company) are often in rural areas, inadequately protected from grazing animals and surface run off, and many do not receive any treatment. Risks from private water supplies may be underestimated since clusters of cases may be within families and not reported as outbreaks and, indeed, people who have been drinking the water for years may have generated some immunity. However, visitors may be at risk and with increasing diversification of commercial enterprise in the countryside are likely to be an increasing population.

Epidemiological studies to identify exposures during outbreaks can be affected by underlying immunity in the community generated by frequent exposure (Meinhardt *et al.*, 1996) and may be a particular problem in analytical studies of outbreaks associated with surface waters (Hunter and Quigley, 1998). Public health measures to control drinking-water outbreaks of cryptosporidiosis include changes in the source of the water provided and notice to boil water for consumption (Hunter, 2000). Boil water notices themselves impact on the community with increased risk of scald injuries (Mayon-White and Frankenberg, 1989), increased energy demands, problems for local industries who use mains water in food manufacturing, and for services such as hospitals. There may also be an adverse effect on tourism. Risk of disease is only reduced if the notice is put in place while pathogens are still present in the water, and if compliance occurs. Non-compliance was reported in over 50% people in the target area during one outbreak (O'Donnell *et al.*, 2000). Implicit to the imposition of a boil water notice must be the criteria for lifting it. As a result of water-related illness, and activities to control it, loss of public confidence in the water supply can be substantial.


## Risk assessment

The potential risk of waterborne cryptosporidiosis lies with the biological properties of the organism, particularly the robust oocyst, and environmental and climatic factors affecting introduction of oocysts into source waters and their survival, water treatment affording removal and disinfection and community factors in the exposed population. The exact significance of environmental oocysts for human health depends upon their viability and infectivity for humans, and protective immunity afforded by prior exposure in the population. Mathematical models for quantitative risk assessment of waterborne cryptosporidiosis have been explored in a number of different ways. Dose response models have estimated the number of oocysts required to cause illness after infection, which show similarity to volunteer infectivity studies, except at very low doses (Teunis *et al.*, 1999; Messner *et al.*, 2001) and demonstrate a near linear dose-response curve. This negates the need for further modelling doses for risk prediction since an arithmetic mean exposure is sufficient. However, it remains to be seen whether this assumption will hold true as further genetic analyses reveal more about the population structure of *Cryptosporidium*. There is evidence for recombination within species which could lead to the generation of more virulent genotypes. Transmission models, incorporating the multiple pathways involved in the transmission of *Cryptosporidium*, have been used to estimate risk from drinking water (Teunis and Havelaar, 1999; Chick *et al.*, 2001) and Chick and colleagues additionally explored the effect of secondary transmission. Routes of exposure for consideration in mathematical risk assessment models are illustrated with the UK regulatory framework by Gale (2002), since the aim of risk assessment is to identify worthwhile interventions. It is important, however, to note that risk assessment models must acknowledge variability (the intrinsic heterogeneity of a variable) and uncertainty (ignorance caused by methodological limitations). For *Cryptosporidium* there is considerable variation in the numbers of oocysts in source waters and in their removal by water treatment, and uncertainty particularly impacts on accurate measurement leading to best estimates of oocyst numbers and distribution in raw waters, removal processes and performance, disinfection performance, and detection parameters in treated waters (recovery, infectivity and viability) (Gale, 2001).

The further differentiation of subtypes within *Cryptosporidium* genotypes provides additional resolution for epidemiological investigations (Glaberman *et al.*, 2002), and information about the infectivity of 'strains' for humans. A variety of tracking tools are being investigated and evaluated for further segregation including the application of microsatellite typing (Caccio, 2000; Blasdall *et al.*, 2001). Sequence analysis of small double-stranded extra chromosomal RNAs in *C. parvum* (Xiao *et al.*, 2001b) and of a highly polymorphic gene encoding a 60 kDa glycoprotein (Strong *et al.*, 2000), analysis of single strand conformation polymorphisms and mutation scanning (Gasser *et al.*, 2003) also offer potential as tracking tools.

Despite many attempts to correlate oocyst detection and counts in water with both traditional and novel indicator organisms, no reliable surrogate for *Cryptosporidium* has been widely accepted. Recently, there has been better identification of risks leading to the detection of oocysts in water supplies, and event-based sampling is being investigated. However, approaches to monitoring and legislation differ world-wide. The USA Environment Protection Agency's (EPA) surface water treatment rule under the National Primary Drinking Water Regulations as of 1 January 2002, for systems using surface water or groundwater under the direct influence of surface water requires disinfection or filtration to meet the criterion of 99% removal/inactivation. Methods are prescribed by the EPA methods 1622 (*Cryptosporidium*) (Anon, 1999a) and 1623 (*Cryptosporidium* and *Giardia*) (Anon, 1999b) for the assessment of the occurrence of these parasites in raw surface source waters. These specify the testing of 10-litre volumes of water and the use of capsule filtration, immunomagnetic separation and immunofluorescence antibody staining with confirmation through vital dye inclusion and differential interference contrast microscopy. Alternative oocyst recovery methods such as membrane filtration, vortex flow and continuous flow centrifugation are permitted with appropriate evaluation and quality control.

In Australia, following the Sydney water crisis during which increased numbers of oocysts were detected in the water supply but no rise in the number of cases of cryptosporidiosis in the community was detected, a risk-based framework has been developed, assessing the systems in place from catchment to tap (Fairley *et al.*, 1999). This is in line with current WHO revisions of guidelines for drinking water incorporating source-to-customer risk assessment. By contrast, in England and Wales, the Water Supply (Water Quality) (Amendment) Regulations 1999 came in to force in June 1999 and were replaced, in January 2001, by the Water Supply (Water Quality) Regulations 2000. Water undertakers must conduct risk assessments with respect to *Cryptosporidium* on all water-treatment works and set out the results of the assessment. Sites with a 'significant risk' classification, based on consideration of the source water, catchment characteristics and treatment provided, are obliged, under the regulations, to treat the water to ensure that the standard is maintained. This 'treatment standard', i.e. an average of less than 1 oocyst in 10 litres of treated water supplied, measured by continuous sampling of at least 40 litres of water per hour, must be met and compliance demonstrated by continual monitoring and reporting of results. Significant risk works do not have to be monitored if all particles >1 μm are continuously removed. The USA approach monitors the removal of *Cryptosporidium* from the water while the UK approach measures what is in the treated water and it is a criminal offence to breach this treatment standard. While this is an operational treatment standard, not a health-related standard, the implications for public health of *Cryptosporidium* in water are hard to ignore.

It has been suggested, from risk assessment models that a threshold value of 10–30 oocysts per 100 l be applied to drinking water for action to prevent an outbreak of disease (Haas and Rose, 1995). However, it is very difficult at present to set a health-related standard for *Cryptosporidium* in drinking water in routine samples because the significance of oocysts in water samples has not been defined. Using current monitoring methods, there is no species differentiation and host specificity cannot be established, viability or infective dose is not measured, and therefore infectivity for humans is not established. Additionally, the status of the herd immunity of the local population, generated by prior exposure, will influence the significance of the numbers of oocysts for causing waterborne disease. It has been shown that individuals with prior exposure vary in their responses to re-infection (Okhuysen *et al.*, 1998). The relationship between oocyst counts in water supplies and cases of illness has not been established. In the past samples were collected retrospectively and it was rare to have samples taken prior to the outbreak. An exception is one outbreak where 34 oocysts were detected per 10 litres (Anon, 1999c). Conversely, high oocyst counts have been demonstrated without subsequent outbreaks of illness. While justification of the cost of continuous monitoring has been questioned (Fairley *et al.*, 1999), it is desirable that oocysts are kept out of the drinking water supply, and while it is impossible to characterize and assess the viability and infectivity of oocysts in water, treatment-related standards will also help improve water quality standards generally. The data also contribute to the historical picture for that water supply and trends in oocyst counts are probably more important than individual numbers. The epidemiology of cryptosporidiosis has been advanced by the identification and characterization of subtypes, which has also led to important developments in identifying sources of infection and action for prevention. However, the public health significance of oocysts in environmental samples depends on their viability and infectivity for humans, and the relationship between oocysts numbers and health risk needs to be evaluated in this context.

## Overall risk assessment

*Health effects*: occurrence of illness, degree of morbidity and mortality, probability of illness based on infection:

- *Cryptosporidium* has a world-wide distribution with more cases of cryptosporidiosis reported among children than adults.

- The primary feature of cryptosporidiosis is profuse watery, sometimes mucoid, diarrhoea, which can be accompanied by dehydration, weight loss, anorexia, abdominal pain, fever, nausea and vomiting. In immunocompetent patients symptoms usually last for about 1–2 weeks.
- *Cryptosporidium* causes acute, self-limiting gastroenteritis in the general population and potentially fatal infection in the immunocompromised.
- Some people who are infected have no symptoms of the disease.
- Individuals with prior exposure vary in their responses to re-infection. It is thought that low levels of exposure are protective against infection and illness.

*Exposure assessment*: routes of exposure and transmission, occurrence in source water, environmental fate:

- Routes of exposure have included drinking water from both surface and groundwater sources contaminated with human sewage and animal manure, contaminated natural and man-made recreational waters, animal contact, person-to-person spread within institutions and day-care centres, and food-borne outbreaks. Person-to-person spread in households and institutions is important and has been estimated at 60%.
- Groundwater under the influence of surface water has been known to become contaminated, and waterborne outbreaks have resulted from groundwater drinking water sources.
- *Cryptosporidium* occurs frequently in raw waters world-wide. The occurrence of *Cryptosporidium* oocysts in environmental waters depends on the source water's susceptibility to animal manure and sewage, climate (since rainfall can influence the numbers of oocysts in surface waters and temperature affects their survival) and community factors such as watershed management. There is uncertainty associated with the actual occurrence, because it is difficult to test for the oocysts.
- Because of the thick oocyst wall, *Cryptosporidium* oocysts are robust and resistant to a variety of environmental pressures particularly under cool, moist conditions.

*Risk mitigation*: drinking-water treatment, medical treatment:

- It is difficult to determine how effective drinking water treatment is because of the uncertainty associated with analytical methods to estimate oocyst numbers, infectivity and viability.
- Oocysts are resistant to chlorine at levels used to treat both drinking water and swimming pools.
- The most successful treatments that may be of use in large-scale treatment are ozone and UV.
- There is no current curative medical treatment for cryptosporidiosis. Dehydration from diarrhoea can be treated with fluid replacement and electrolyte balance. While cryptosporidiosis in otherwise healthy patients may be self-limiting, immunocompromised patients and those in poor health or suffering malnutrition are at high risk of severe illness or death.

- Among high-risk groups, such as the immunocompromised, avoiding exposure to the parasite is imperative. Immunocompromised patients are advised to boil their water or use appropriate filtration devices.

## Future implications

It is clear from the study of reported outbreaks of cryptosporidiosis that emerging risks are still being identified. For example, swimming pool-associated outbreaks have attracted much attention, particularly since information regarding best practice for control and prevention have previously been lacking and are currently under refinement. Issues of control of nosocomial infection, and advice for immunocompromised individuals are constantly under review: is boiling water the best advice? Guidance for food manufacturers using water during product processing is relatively new (see for example Anon, 2000c). The use of new water sources/supplies in water-poor areas (which are ever increasing) may pose a threat to human health.

Application of risk assessment methodologies to *Cryptosporidium* may require new definitions of disease, since infection without diarrhoea can have long-term health effects. Additionally, unknown or unaccounted factors may impact on risk assessment, such as the varying infectivity of isolates for humans, effects of cross-immunity, possible type variation in survival and the effect of natural waters on survival and disinfection of oocysts. Better markers of viability/infectivity are required for adequate assessment. Development and application of typing techniques provides the potential for better identification of point sources of contamination but better understanding of the dynamics of genetic exchange is required to complement progress in risk assessment.

## References

Amadi, B., Mwlya, M., Musuku, J. *et al.* (2002). Effect of nitazoxanide on mortality in Zambian children with cryptosporidiosis: a randomised conrolled trial. *Lancet*, **360**: 1375–1380.

Anderson, B.C. (1985). Moist heat inactivation of *Cryptosporidium* sp. *Am J Public Hlth*, **75**: 1433–1444.

Anon. (1997a). *Cryptosporidium* and water: a public health handbook. Working Group on Waterborne Cryptosporidiosis. CDC, Atlanta, Georgia.

Anon. (1997b). Outbreaks of *Escherichia coli* O157 infection and cryptosporidiosis associated with drinking unpasteurised apple-cider – Connecticut and New York, October 1996. *MMWR*, **46**: 4–8.

Anon. (1999a). Method 1622: Cryptosporidium in water by filtration/IMS/FA. United States Environmental Protection Agency.

Anon. (1999b). Method 1623: Cryptosporidium and Giardia in water by filtration/ IMS/FA. United States Environmental Protection Agency.

Anon. (1999c). Outbreak of cryptosporidiosis in the North West of England. *CDR Wkly*, **9**: 175–178.

Anon. (2000a). Surveillance of waterborne disease and water quality: January to June 2000, and summary of 1999. *CDR Wkly*, **10**: 319–322.

Anon. (2000b). Review of outbreaks of cryptosporidiosis in swimming pools and advice on proceedings of strategic workshop on viability tests and genetic typing. Final report to DEFRA: Drinking Water Inspectorate Foundation for Water Research, Marlow, Bucks, UK (http://www.fwr.org/).

Anon. (2000c). Water quality for the food industry: management and microbiological issues. Guideline no. 27. Campden and Chorleywood Research Association Group.

Anon. (2002). The development of a national collection for oocysts of *Cryptosporidium*. Final report to DEFRA: Drinking Water Inspectorate. Foundation for Water Research, Marlow, Bucks, UK (http://www.fwr.org/).

Arrowood, M.J. (1997). Diagnosis of *Cryptosporidium* and cryptosporidiosis. In Cryptosporidium *and Cryptosporidiosis*, Fayer, R. (ed.). Boca Raton: CRC Press, pp. 43–46.

Basualdo, J., Pezzani, B., de Luca, M. *et al.* (2000). Screening of the municipal water system of La Palta, Argentina, for human intestinal parasites. *Int J Hyg Environ Hlth*, **203**: 177–182.

Besser-Wick, J.W., Forfang, J., Hedberg, C.W. *et al.* (1996). Foodborne outbreak of diarrhoeal disease associated with *Cryptosporidium parvum* – Minnesota 1995. *MMWR*, **45**: 783.

Blewett, D.A. (1989). Disinfection and oocysts. In *Cryptosporidiosis: Proceedings of the First International Workshop*, Angus, K. and Blewett, D.A. (eds). Edinburgh: Animal Disease Research Association, pp. 107–115.

Blanshard, C., Jackson, A.M., Shanson, D.C. *et al.* (1992). Cryptosporidiosis in HIV-seropositive patients. *Q J Med*, **85**: 813–823.

Blasdall, S.A., Ongerth, J.E. and Ashbolt, N.J. (2001). Differentiation of *Cryptosporidium parvum* subtypes by a novel microsatellite-telomere PCR with PAGE. Proceedings of *Cryptosporidium* from Molecules to Disease, 7–12 October 2001, Esplanade Hotel Fremantle, Western Australia. Murdoch University, Perth.

Bodley-Tickell, A.T., Kitchen, S.E. and Sturdee, A.P. (2002). Occurrence of *Cryptosporidium* in agricultural surface waters during an annual farming cycle in lowland UK. *Water Res*, **36**: 1880–1886.

Bukhari, Z., Smith, H.V., Humphreys, S.W. *et al.* (1995). Comparison of excretion and viability patterns in experimentally infected animals: potential for release of viable *C. parvum* oocysts into the environment. In *Protozoan Parasites and Water*, Betts, W.B., Casemore, D.P., Fricker, C. *et al.* (eds). London: The Royal Society of Chemistry, pp. 188–191.

Bukhari, Z., Clancy, J.L., Hary, T.M. *et al.* (1999). Medium pressure UV for oocysts inactivation. *JAWWA*, **91**: 86–94.

Caccio, S., Homan, W., Camilli, R. *et al.* (2000). A microsatellite marker reveals population heterogeneity within human and animal genotypes of *Cryptosporidium parvum*. *Parasitology*, **120**: 237–244.

Carpenter, C., Fayer, R., Trout, J. *et al.* (1999). Chlorine disinfection of recreational water for *Cryptosporidium parvum*. *Emerg Infect Dis*, **5**: 579–584.

Carr, A., Marriott, D., Field, A. *et al.* (1998). Treatment of HIV-1-associated microsporidiosis and cryptosporidiosis with combination antiretroviral therapy. *Lancet*, **351**: 256–261.

Casemore, D.P. (1990). Epidemiological aspects of human cryptosporidiosis. *Epidemiol Infect*, **104**: 1–28.

Casemore, D.P. (1998). *Cryptosporidium* and the safety of our water supplies. *Commun Dis Public Hlth*, **4**: 218–219.

Casemore, D.P. and Jackson, B. (1984). Hypothesis: cryptosporidiosis in human beings is not primarily a zoonosis. *J Infect*, **9**: 153–156.

Casemore, D.P. and Watkins, J. (1999). Review of disinfection and associated studies on Cryptosporidium. London: Report to DEFRA: DWI.

Casemore, D.P., Sands, R.L. and Curry, A. (1985). *Cryptosporidium* species a 'new' human pathogen. *J Clin Pathol*, **38**: 1321–1336.

Casemore, D.P., Blewett, D.A. and Wright, S.E. (1989). Cleaning and disinfection of equipment for gastro-intestinal flexible endoscopy. *Gut*, **30**: 1156.

Casemore, D.P., Wright, S.E. and Coop, R.L. (1997). Cryptosporidiosis – human and animal epidemiology. In Cryptosporidium *and Cryptosporidiosis*, Fayer, R. (ed.). Boca Raton: CRC Press, pp. 65–92.

Chalmers, R.M., Sturdee, A.P., Casemore, D.P. *et al.* (1994). *Cryptosporidium muris* in wild house mice (*Mus musculus*): first report in the UK. *Eur J Protistol*, **30**: 151–155.

Chalmers, R.M., Sturdee, A.P., Mellors, P. *et al.* (1997). *Cryptosporidium parvum* in environmental samples in the Sligo area, Republic of Ireland: a preliminary report. *Lett Appl Microbiol*, **25**, 380–384.

Chalmers, R.M., Elwin, K., Thomas, A. and Joynson, D.H.M. (2002a). Unusual types of cryptosporidia are not restricted to immunocompromised patients. *J Infect Dis*, **185**: 270–271.

Chalmers, R.M., Elwin, K., Reilly, W.J. *et al.* (2002b). *Cryptosporidium* in farmed animals: the detection of a novel isolate in sheep. *Int J Parasitol*, **32**: 21–26.

Chappell, C.L., Okhuysen, P.C., Sterling, C.R. *et al.* (1999). Infectivity of Cryptosporidium parvum in healthy adults with pre-existing anti-*C. parvum* serum immunoglobulin G. *Am J Trop Med Hyg*, **60**: 157–164.

Checkley, W., Epstein, L.D., Gilman, R.H. *et al.* (1998). Effects of *Cryptosporidium parvum* in Peruvian children: growth faltering and subsequent catch up growth. *Am J Epidemiol*, **148**: 497–506.

Chick, S.E., Koopman, J.S., Soorapanth, S. *et al.* (2001). Infection transmission system models for microbial risk assessment. *Sci Total Environ*, **274**: 197–207.

Clancy, J.L., Bukhari, Z., McCuin, R.M. *et al.* (2000). *Cryptosporidium* viability and infectivity methods. Denver, CO: AWWA Research Foundation.

Clark, D.P. (1999). New insights into human cryptosporidiosis. *Clin Microbiol Rev*, **12**: 554–63.

Clark, D.P. and Sears, C.L. (1996). The pathogenesis of cryptosporidiosis. *Parasitol Today*, **12**: 221–225.

Clifford, C.P., Crook, D.W.M., Conlon, C.P. *et al.* (1990). Impact of waterborne outbreak of cryptosporidiosis on AIDS and renal transplant patients. *Lancet*, **i**: 1455–1456.

Clinton-Wight, A., Chappell, C.L., Sikander Hayat, C. *et al.* (1994). Paromomycin for cryptosporidiosis in AIDS: a prospective double-blind trial. *J Infect Dis*, **170**: 419–424.

Craik, S.A., Weldon, D., Finch, G.R. *et al.* (2001). Inacitivation of *Cryptosporidium parvum* oocysts using medium and low pressure ultraviolet radiation. *Water Sci Res*, **35**: 1387–1398.

Cron, R.Q. and Sherry, D.D. (1995). Reiter's syndrome associated with cryptosporidial gastroenteritis. *J Rheumatol*, **22**: 1962–1963.

Current, W.L. and Hayes, T.B. (1984). Complete development of *Cryptosporidium* in cell culture. *Science*, **224**: 603–605.

Current, W.L. and Long, P.L. (1983). Development of human and calf *Cryptosporidium* in chicken embryos. *J Infect Dis*, **148**: 1108–1113.

Current, W.L. and Reese, N.C. (1986). A comparison of endogenous development of three isolates of *Cryptosporidium* in suckling mice. *J Protozool*, **33**: 98–108.

D'Antonio, R.G., Winn, R.E., Taylor, J.P. *et al.* (1985). A waterborne outbreak of cryptosporidiosis in normal hosts. *Ann Intern Med*, **103**: 886–888.

DuPont, H.L., Chappell, C.L., Sterling, C.R. *et al.* (1995). The infectivity of *Cryptosporidium parvum* in healthy volunteers. *New Engl J Med*, **332**: 855–859.

Egorov, A., Paulauskis, J., Petrova, L. *et al.* (2002). Contamination of water supplies with *Cryptosporidium parvum* and *Giardia lamblia* and diarrhoeal illness in selected Russian Cities. *Int J Hyg Environ Hlth*, **205**: 281–289.

Elwin, K., Chalmers, R.M., Roberts, R. *et al.* (2001). The modification of a rapid method for the identification of gene-specific polymorphisms in *Cryptosporidium parvum*, and application to clinical and epidemiological investigations. *Appl Environl Microbiol*, **67**: 5581–5584.

Fairley, C.K., Sinclair, M.I. and Rizak, S. (1999). Monitoring drinking water: the receeding zero. *MJA*, **171**: 397–398.

Farthing, M.J.G. (2000). Clinical aspects of human cryptosporidiosis. In *Cryptosporidiosis and Microsporidiosis*, Petry, F. (ed.). Basel: Karger, 6, 50–74.

Fayer, R., Speer, C.A. and Dubey, J.P. (1997). The general biology of *Cryptosporidium*. In Cryptosporidium *and Cryptosporidiosis*, Fayer, R. (ed.). Boca Raton: CRC Press, pp. 1–41.

Fayer, R., Graczyk, T.K., Lewis, E.J. *et al.* (1998). Survival of infectious *Cryptosporidium parvum* oocysts in seawater and Eastern oysters (*Crassostrea virginica*) in the Chesapeake Bay. *Appl Environ Microbiol*, **64**: 1070–1074.

Fayer, R., Morgan, U. and Upton, S.J. (2000). Epidemiology of *Cryptosporidium*: transmission, detection and identification. *Int J Parasitol*, **30**: 1305–1322.

Fayer, R., Trout, J.M., Xiao, L. *et al.* (2001). *Cryptosporidium canis n. sp* from domestic dogs. *J Parasitol*, **87**: 1415–1422.

Flanigan, T.P., Ramratnam, B., Graeber, C. *et al.* (1996). Prospective trial of paromomycin for cryptosporidiosis in AIDS. *Am J Med*, **100**: 370–372.

Gale, P. (2001). Developments in microbiological risk assessment for drinking water. *J Appl Microbiol*, **91**: 191–205.

Gale, P. (2002). Using risk assessment to identify future research requirements. *JAWWA*, September: 30–38.

Gasser, R.B. and O'Donoghue, P. (1999). Isolation, propagation and characterization of *Cryptosporidium*. *Int J Parasitol*, **29**: 1379–1413.

Gasser, R.B., Abs El-Osta, Y.G. and Chalmers, R.M. (2003). An electrophoretic analysis of genetic variability within *Cryptosporidium parvum* from imported and autochthonous cases of human cryptosporidiosis in the United Kingdom. *Appl Environ Microbiol*, **69**: 2719–2730.

Gelletli, R., Stuart, J., Soltano, N. *et al.* (1997). Cryptosporidiosis associated with school milk. *Lancet*, **350**: 1005–1006.

Gerba, C.P., Johnson, D.C. and Hasan, M.N. (1997). Efficacy of iodine water purification tablets against *Cryptospordium* oocysts and *Giardia* cysts. *Wilderness Environ Med*, **8**: 96–100.

Glaberman, S., Moore, J.E., Lowery, C.J. *et al.* (2002). Three drinking-water-associated cryptosporidiosis outbreaks, Northern Ireland. *Emerging Infect Dis*, **8**: 631–633.

Gomez-Bautista, M., Ortega-Mora, L.M., Tabares, E. *et al.* (2000). Detection of infectious *Cryptosporidium parvum* oocysts in mussels (*Mytilus galloprovincialis*) and cockles (*Cerastoderma edule*). *Appl Environ Microbiol*, **66**: 1866–1870.

Graczyk, T.K., Fayer, R., Lewis, E.J. *et al.* (1999). *Cryptosporidium* oocysts in Bent mussels (*Ischadium recurvum*) in the Chesapeake Bay. *Parasitol Res*, **85**: 518–520.

Guarino, A., Canani, R.B., Pozio, E. *et al.* (1994). Enterotoxic effect of stool supernatant of Cryptosporidium-infected calves on human jejunum. *Gastroenterology*, **106**: 28–34.

Guerrant, R.L. (1997). Cryptosporidiosis: An emerging and highly infectious threat. *Emerging Infect Dis*, **3**: 51–57.

Haas, C. and Rose, J. (1995). Developing an action level for Cryptosporidium. *JAWWA*, **87**: 81–84.

Hancock, C.M., Rose, J.B. and Callahan, M. (1998). The prevalence of *Cryptosporidium* and *Giardia* in US groundwaters. *JAWWA*, **90**: 58.

Hannah, J. and Riordan, T. (1988). Case to case spread of cryptosporidiosis: evidence from a day nursery outbreak. *Public Hlth*, **102**: 539–544.

Harp, J.A., Fayer, R., Pesch, B.A. *et al.* (1996). Effect of pasteurisation on infectivity of *Cryptosporidium parvum* oocysts in water and milk. *Appl Environ Microbiol*, **62**: 2866–2868.

Hashimoto, A., Kunikane, S. and Hirata, T. (2002). Prevalence of *Cryptosporidium* oocysts and *Giardia* cysts in the drinking water supply in Japan. *Water Res*, **36**: 519–526.

Hewitt, R.G., Yiannoutsos, C.T., Higgs, E.S. *et al.* (2000). Paromomycin: no more effect than placebo for treatment of cryptosporidiosis in patients with advanced human immunodeficiency virus infection. AIDS clinical trial group. *Clin Infect Dis*, **31**: 1084–1092.

Hoepelman, A.L. (1996). Current therapeutic approaches to cryptosporidiosis in immuno-compromised patients. *J Antimicrob Chemother*, **37**: 871–880.

Hunter, P.R. (2000). Advice on the response from public and environmental health to the detection of cryptosporidial oocysts in treated drinking water. *Commun Dis Public Hlth*, **3**: 24–27.

Hunter, P.R. and Quigley, C. (1998). Investigation of an outbreak of cryptosporidiosis associated with treated surface water finds limits to the value of case control studies. *Commun Dis Public Hlth*, **1**: 234–238.

Ionas, G., Learmonth, J.J., Keys, E.A. *et al.* (1998). Distribution of *Giardia* and *Cryptosporidium* in natural water systems in New Zealand – a nationwide survey. *Water Sci Technol*, **38**: 57–60.

Jenkins, M.B., Bowman, D.D. and Ghiorse, W.C. (1998). Inactivation of *Cryptosporidium parvum* oocysts by ammonia. *Appl Environ Microbiol*, **64**: 784–788.

Joce, R.E., Bruce, J., Kiely, D. *et al.* (1991). An outbreak of cryptosporidiosis associated with a swimming pool. *Epidemiol Infect*, **107**: 497–508.

Johnson, D.C., Reynold, K.A., Gerba, C.P. *et al.* (1995). Detection of *Giardia* and *Cryptosporidium* in marine waters. *Water Sci Technol*, **5–6**: 439–442.

Jokipii, L. and Jokipii, A.M.M. (1986). Timing of symptoms and oocyst excretion in human cryptosporidiosis. *New Engl J Med*, **315**: 1643–1647.

Kelly, P., Davies, S.E., Mandanda, B. *et al.* (1997). Enteropathy in Zambians with HIV related diarrhoea: Regression modelling of potential determinants of mucosal damage. *Gut*, **41**: 811–816.

Kemp, J.S., Wright, S.E., Coop, R.L. *et al.* (1995). Protozoan, bacterial and viral pathogens, farm wastes and water quality protection. Final report to MAFF (CSA 2064).

Koudela, B. and Modrý, D. (1998). New species of *Cryptosporidium* (Apicomplexa: Cryptosporidiidae). *Fol Parasitol*, **45**: 93–100.

Li, H., Finch, G.R., Smith, D.W. *et al.* (2001). Sequential inactiviation of *Cryptosporidium parvum* using ozone and chlorine. *Water Res*, **35**: 4339–4348.

Lindsay, D.S., Upton, S.J., Owens, D.S. *et al.* (2000). *Cryptosporidium andersoni* n. sp. (Apicomplexa: Cryptosporiidae) from Cattle, *Bos taurus*. *J Eukaryotic Microbiol*, **47**: 91–95.

MacKenzie, W.R., Schell, W.L., Blair, K.A. *et al.* (1995). Massive outbreak of waterborne *Cryptosporidium* infection in Milwaukee, Wisconsin: recurrence of illness and risk of secondary transmission. *Clin Infect Dis*, **21**: 57–62.

McLauchlin, J., Amar, C., Pedraz-Diaz, S. *et al.* (2000). Molecular epidemiological analysis of *Cryptosporidium* spp. in the United Kingdom: results of genotyping *Cryptosporidium* spp. in 1705 fecal samples from humans and 105 fecal samples from livestock animals. *J Clin Microbiol*, **38**: 3984–3990.

Maggi, P., Larocca, A.M., Quarto, M. *et al.* (2000). Effect of antiretroviral therapy on cryptosporidiosis and microsporidiosis in patients infected with human immunodeficiency virus type 1. *Eur J Clin Microbiol Infect Dis*, **19**: 213–217.

Marcic, A., Potyka, J., Siegfriedt, D. *et al.* (2000). Toddlers and small children: a source for cryptosporidia in swimming-pools. HRWM conference, Paris 2000.

Mayon-White, R.T. and Frankenberg, R.A. (1989). Boil the water. *Lancet*, **ii**: 216.

Meinhardt, P.L., Casemore, D.P. and Miller, K.B. (1996). Epidemiologic aspects of human cryptosporidiosis and the role of waterborne transmission. *Epidemiol Rev*, **18**: 118–136.

Meisel, J.L., Perera, D.R., Meligro, B.S. *et al.* (1976). Overwhelming watery diarrhoea associated with *Cryptosporidium* in an immunosuppressed patient. *Gastroenterology*, **70**: 1156–1160.

Messner, M.J. and Walpert, R.L. (2000). Occurrence of Cryptosporidium in the national drinking water sources – ICR data analysis. In *Water Quality Technology Conference Proceedings*. Denver, CO: AWWA, 2000.

Messner, M.J., Chappell, C.L. and Okhuysen, P.C. (2001). Risk assessment for *Cryptosporidium*: a hierarchical Bayesian analysis of human dose response data. *Water Res*, **35**: 3934–3940.

Millard, P., Gensheimer, K., Addis, D. *et al.* (1994). An outbreak of cryptosporidiosis from fresh-pressed apple cider. *JAMA*, **272**: 1592–1596.

Molbak, K., Andersen, M., Aaby, P. *et al.* (1997). *Cryptosporidium* infection in infancy as a cause of malnutrition: A community study from Guinea-Bissau, West Africa. *Am J Clin Nutr*, **65**: 149–152.

Monge, R. and Chinchilla, M. (1996). Presence of *Cryptosporidium* oocysts in fresh vegetables. *J Food Protect*, **59**: 202–203.

Morgan, U.M., Xiao, L., Fayer, R. *et al.* (1999). Variation in *Cryptosporidium*: towards a taxonomic revision of the genus. *Int J Parasitol*, **29**: 1733–1751.

Morgan-Ryan, U.M., Fall, A., Ward, L.A. *et al.* (2002). *Cryptosporidium hominis* n. sp. (Apicomplexa: Cryptosporidiidae) from *Homo sapiens*. *Eukaryotic Microbiol*, **49**: 433–440.

Nime, F.A., Burek, J.D. and Page, D.L. (1976). Acute enterocolitis in a human being infected with the protozoan *Cryptosporidium*. *Gastroenterology*, **70**: 592–598.

O'Donnell, M., Platt, C. and Aston, R. (2000). Effect of a boil water notice on behaviour in the management of a water contamination incident. *Commun Dis Public Hlth*, **3**: 56–59.

Okhuysen, P.C. and Chappell, C.L. (2002). Cryptosporidium virulence determinants – are we there yet? *Int J Parasitol*, **32**: 517–525.

Okhuysen, P.C., Chappell, C.L., Sterling, C.R. *et al.* (1998). Susceptibility and serologic response of healthy adults to reinfection with *Cryptosporidium parvum*. *Infection and immunity*, **66**: 441–443.

Okhuysen, P.C., Chappell, C.L., Crabb, J.H. *et al.* (1999). Virulence of three distinct *Cryptosporidium parvum* isolates for healthy adults. *J Infect Dis*, **180**: 1275–1281.

Ono, K., Tsuji, H., Rai, S.K. *et al.* (2001). Contamination of river water by *Cryptosporidium parvum* oocysts in Western Japan. *Appl Environ Microbiol*, **67**: 3832–3836.

Ortega, Y.R., Roxas, C.R., Gilman, R.H. *et al.* (1997). Isolation of *Cryptosporidium parvum* and *Cyclospora cayetanensis* from vegetables collected in markets in an endemic region in Peru. *Am J Trop Med Hyg*, **57**: 683–686.

Palmer, S.R. and Biffin, A. and the Public Health Laboratory Service Study Group. (1990). Cryptosporidiosis in England and Wales: prevalence and clinical and epidemiological features. *Br Med J*, **30**: 774–777.

Panciera, R.J., Thomassen, R.W. and Garner, F.M. (1971). Cryptosporidial infection in a calf. *Vet Pathol*, **8**: 479–484.

Peeters, J.E., Mazas, E.A., Masschelein, W.J. *et al.* (1989). Effect of disinfection of drinking water with ozone or chlorine dioxide on survival of *Cryptosporidium parvum* oocysts. *J Appl Environ Microbiol*, **55**: 1519–1522.

Petry, F. (2000). Laboratory diagnosis of *Cryptosporidium parvum* infection. In *Cryptosporidiosis and Microsporidiois*, Petry, F. (ed.). Basel: Karger, **6**: 33–49.

Puech, M.C., McAnulty, J.M., Lesjak, M. *et al.* (2001). A statewide outbreak of cryptosporidiosis in New South Wales associated with swimming at public pools. *Epidemiol Infect*, **126**: 389–396.

Quinn, K., Baldwin, G., Stepak, P. *et al.* (1998). Foodborne outbreak of cryptosporidiosis–Spokane, Washington, 1997. *MMWR*, **47**: 565–567.

Quiroz, E.S., Bern, C., MacArthur, J.R. *et al.* (2000). An outbreak of cryptosporidiosis linked to a food handler. *J Infect Dis*, **181**: 695–700.

Rennecker, J.L., Marinas, B.J., Owens, J.H. *et al.* (1999). Inactivation of *Cryptosporidium parvum* oocysts with ozone. *Water Res*, **33**: 2481–2488.

Robertson, L.J. and Gjerde, B. (2000). Isolation and enumeration of *Giardia* cysts, *Cryptosporidium* oocysts, and Ascaris eggs from fruits and vegetables. *J Food Protect*, **63**: 775–778.

Robertson, L.J., Campbell, A.T. and Smith, H.V. (1992). Survival of *Cryptosporidium parvum* oocysts under various environmental pressures. *Appl Environ Microbiol*, **55**: 1519–1522.

Robertson, L.J., Paton, C.A., Campbell, A.T. *et al.* (2000). *Giardia* cysts and *Cryptosporidium* oocysts at sewage treatment works in Scotland, UK. *Water Res*, **34**: 2310–2322.

Robertson, B., Sinclair, M.I., Forbes, A.B. *et al.* (2002). Case-control studies of sporadic cryptosporidiosis in Melbourne and Adelaide, Australia. *Epidemiol Infect*, **128**: 419–431.

Rose, J.B. and Slifco, T.R. (1999). *Giardia*, *Cryptosporidium*, and *Cyclospora* and their impact on foods: a review. *J Food Protect*, **62**: 1059–1070.

Rose, J.B., Lisle, J.T. and LeChevallier, M. (1997). Waterborne cryptosporidiosis: incidence, outbreaks, and treatment strategies. In Cryptosporidium *and cryptosporidiosis*, Fayer, R. (ed.). Boca Raton: CRC Press, pp. 93–109.

Rose, J.B., Huffman, D.E. and Gennaccaro, A. (2002). Risk and control of waterborne cryptosporidiosis. *FEMS Microbiol Rev*, **26**: 113–123.

Rossignol, J.F., Hidalgo, H., Feregrino, M. *et al.* (1998). A double-blind placebo-controlled study of nitazoxanide in the treatment of cryptosporidial diarrhoea in AIDS patients in Mexico. *Trans Roy Soc Trop Med Hyg*, **92**: 663–666.

Rossignol, J.F., Ayoub, A. and Ayers, M.S. (2001). Treatment of diarrhoea caused by *Cryptosporidium parvum*: a prospective, randomised, double-blind, placebo-controlled study of nitazoxanide. *J Infect Dis*, **184**: 103–106.

Shepherd, R.C., Sinha, G.P., Reed, C.L. *et al.* (1988). Cryptosporidiosis in the West of Scotland. *Scot Med J*, **33**: 365–368.

Shin, G.A., Linden, K.G., Arrowood, M.J. *et al.* (2001). Low-pressure UV inactivation and DNA repair potential of *Cryptosporidium parvum* oocysts. *Appl Environ Microbiol*, **67**: 3029–3032.

Slavin, D. (1955). *Cryptosporidium meleagridis* (sp. nov.). *J Comp Pathol*, **65**: 262–266.

Smith, H.V. and Rose, J.B. (1998). Waterborne cryptosporidiosis: current status. *Parasitol Today*, **14**: 14–22.

Smith, N.H., Cron, S., Valdez, L.M. *et al.* (1998). Combination drug therapy for cryptosporidiois in AIDS. *J Infect Dis*, **178**: 900–903.

Sorvillo, F.J., Fujioka, K., Nahlen, B. *et al.* (1992). Swimming-associated cryptosporidiosis. *Am J Public Hlth*, **82**: 742–744 .

Strong, W.B., Gut, J. and Nelson, R.G. (2000). Cloning and sequence analysis of a highly polymorphic *Cryptosporidium parvum* gene encoding a 60-kilodalton glycoprotein and characterization of its 15- and 45-kilodalton zoite surface antigen products. *Infect Immun*, **68**: 4117–4134.

Sulaiman, I.M., Xiao, L. and Lal, A.A. (1999). Evaluation of *Cryptosporidium parvum* genotyping techniques. *Appl Environ Microbiol*, **65**: 4431–4435.

Tenter, A.M., Barta, J.R., Beveridge, I. *et al.* (2002). The conceptual basis for a new classification of the coccidian. *Int J Parasitol*, **32**: 595–616.

Teunis, P.F.M. and Havelaar, A.H. (1999). Cryptosporidium in drinking water: evaluation of the ILSI/RSI quantitative risk assessment framework. Report no. 284 550 006. The Netherlands: National Institute of Public Health and the Environment, Bilthoven, 1999.

Teunis, P.F.M., Nagelkerke, N.J.D. and Haas, C.N. (1999). Dose response models for infectious gastroenteritis. *Risk Analysis*, **19**: 1251–1260.

Theodos, C.M., Griffiths, J.K., D'Onfro, J. *et al.* (1998). Efficacy of nitazoxanide against *Cryptosporidium parvum* in cell culture and in animal models. *Antimicrob Ag Chemother*, **42**: 1959–1965.

Tyzzer, E.E. (1907). A sporozoan found in the peptic glands of the common mouse. *Proc Soc Exp Med*, **5**: 12–13.

Tyzzer, E.E. (1910). An extracellular coccidium, *Cryptosporidium muris* (gen.et sp. nov.) of the gastric glands of the common mouse. *J Med Res*, **23**: 487–509.

Tyzzer, E.E. (1912). *Cryptosporidium parvum* (*sp. nov.*), a coccidium found in the small intestine of the common mouse. *Arch protistenkd*, **26**: 394–412.

Tyzzer, E.E. (1929). Coccidiosis in gallinaceous birds. *Am J Hyg*, **10**: 269–383.

Vesey, G., Ashbolt, N., Fricker, E.J. *et al.* (1998). The use of a ribosomal RNA targeted oligonucleotide probe for fluorescent labeling of viable *Cryptosporidium parvum* oocysts. *J Appl Microbiol*, **85**: 429–440.

White, A.C., Chappell, C.L., Hayat, C.S. *et al.* (1994). Paromomycin for cryptosporidiosis in AIDS: a prospective, double blind trial. *J Infect Dis*, **170**: 419–424.

Willocks, L., Crampin, A., Milne, L. *et al.* (1998). A large outbreak of cryptosporidiosis associated with a public water supply from a deep chalk borehole. *Comm Dis Public Hlth*, **1**: 239–243.

Xiao, L., Singh, A., Limor, J. *et al.* (2001a). Molecular characterisation of *Cryptosporidium* oocysts in samples of raw surface water and wastewater. *Appl Environ Microbiol*, **67**: 1097–1101.

Xiao, L., Limor, J., Bern, C. *et al.* (2001b). Tracking *Cryptosoridium parvum* by sequence analysis of small double-stranded RNA. *Emerging Infect Dis*, **7**: 141–145.

## 19

# *Cyclospora cayetanensis*

## Basic microbiology

*Cyclospora* is a genus of obligate intracellular coccidian protozoan parasites (Phylum *Apicomplexa*, Order *Eucoccidiorida*, Family *Eimeriidae*). Although there are many species in the genus, the only one believed to infect humans is *Cyclospora cayetanensis* (Ortega *et al.*, 1994). The life cycle is initiated by the sporulation of spherical oocysts (8–10 μm) after a period of maturation in the environment (see below), when two sporocysts develop within the oocyst, each containing two sporozoites (Ortega *et al.*, 1993). When the mature oocysts are ingested, the sporozoites excyst and invade the enterocytes of the small intestine (Bendall *et al.*, 1993; Sun *et al.*, 1996; Ortega *et al.*, 1997). Reproduction follows, involving asexual and sexual stages, resulting in the production of immature oocysts (Ortega *et al.*, 1997). Endogenous stages have been identified in the jejunum and duodenum and inhabit an intracytoplasmic parasitopherous vacuole (Bendall *et al.*, 1993; Sun *et al.*, 1996). Asexual stages (trophozoites, type 1 meronts with eight to twelve merozoites and type II meronts with four merozoites) and sexual stages (gametocytes) have been detected in the same patient, demonstrating that the life cycle can be completed in a single host (Ortega *et al.*, 1997).

Since the oocysts are unsporulated when they are shed in the faeces, they are non-infectious at this stage. However, they mature and sporulate during environmental exposure and become infectious. *C. cayetanensis* infection has not been confirmed in any hosts other than humans nor has the organism been experimentally transmitted to other hosts (Eberhard *et al.*, 2000). Based on comparative sequence analysis of small subunit ribosomal DNA, the genus *Cyclospora* is most closely related to *Eimeria*, which are also host-species specific and complete their life cycle in a single host (Relman *et al.*, 1996).

Distribution of the parasite is world-wide but infection appears to be endemic throughout the tropics (Soave, 1996). In developing countries, the epidemiology and clinical presentation vary according to social factors. About 30% naturally infected children in the shanty towns of Lima, Peru experience diarrhoea of short duration, and symptomatic infection is rarely reported in adults (Ortega *et al.*, 1993). Upper class Peruvians, tourists and expatriates in Nepal report more disease than the local population and diarrhoea can last for over 1 month (Taylor *et al.*, 1988). In developed countries, cases are most frequently among people reporting foreign travel. However, these data may be skewed by the use of foreign travel as a selection criterion for laboratory testing (Cann *et al.*, 2000). Despite this, cases have indeed been reported among people who have not travelled outside the UK, USA and Germany. Although the exact routes of transmission of *C. cayetanensis* have yet to be elucidated, food- and waterborne disease have both been reported.

A number of features of the biology of *C. cayetanensis* affect its epidemiology and transmission. The oocysts are shed in the faeces unsporulated and require a period of maturation in the environment. This means that direct person-to-person transmission is unlikely. However, the oocysts are robust and can survive for long periods. Transmission is therefore likely to occur through food and water, contaminated by human faeces.

## Origin of the organism

Since the genus was created in 1881 for parasites found in myriapods (Schneider, 1881), *Cyclospora* spp. have been described in many animal hosts (Levine, 1982). Although not identified as such at the time, the first reports of *C. cayetanensis* in humans were in 1979 when Ashford reported three cases that occurred in the previous 2 years. Working in Papua New Guinea, he recorded unsporulated oocysts in the faeces of two ill people and one well person, which took some days to sporulate and thus be recognized as a coccidian protozoan (Ashford, 1979). He postulated that the organism was *Isospora*, but observed that until sporulated, the bodies could be mistaken for fungal spores. Over the next decade few reports were made, but included those of an 'unsporulated, coccidian body but a fungal spore could not be ruled out' (Soave *et al.*, 1986) and '*Cryptospordium muris*-like objects' (Naranjo *et al.*, 1989), and 'cyanobacterium-like or coccidian-like body' (Long *et al.*, 1991). The latter was to be adopted as the reporting nomenclature, and coccidian-like bodies (CLBs) 8–10 μm were identified world-wide using acid-fast staining and autofluorescing under UV light. The morphology of sporulated oocysts was described in detail by Ortega *et al.* (1993) and this description provided evidence for classification as *Cyclospora*. The morphological characteristics, patient symptoms and apparent failure of conventional antimicrobial chemotherapy linked *Cyclospora* to CLBs reported throughout the world, and was reinforced by sporulation studies of various isolates (Ortega *et al.*, 1993). The species name *cayetanensis* was proposed and adopted, referring to the location of the studies at the Universidad Peruana Cayetano Heredia, Lima, Peru (Ortega *et al.*, 1994). Increased use of acid-fast stains for the detection of *Cryptosporidium* in stool specimens during the 1980s facilitated better recognition of *C. cayetanensis* in primary diagnostic laboratories. Widespread outbreaks in North America during the 1990s, often associated with the consumption of raspberries imported from Guatemala, raised the profile of this organism and the disease it causes.

## Clinical features

The incubation period for cyclosporiasis has been determined from outbreaks, and has a mean of 7 days (range 2–11 days) (Soave, 1996). The onset of diarrhoeal illness is usually abrupt, although it can be preceded by a prodromal illness of several days of flu-like symptoms (Soave, 1996). Stools are frequent and sometimes explosive, and other symptoms include anorexia, nausea, vomiting, abdominal bloating and cramps, weight loss, fatigue, low grade fever and body aches. Fatigue is often reported to be profound. Although self-limiting in immunocompetent patients, illness is often prolonged, and the duration of diarrhoea averages 5 days to 15 weeks in untreated patients (Brown and Rotschafer, 1999). Symptoms can be relapsing-remitting, including alternate diarrhoea and constipation (Soave *et al.*, 1998). Therefore, patients may not have diarrhoea at the time of presentation to medical practitioners.

Oocyst shedding has been reported 3 months after initial detection but this could represent prolonged infection or intermittent re-infection (Eberhad *et al.*, 1999a). Asymptomatic infection occurs in areas where *C. cayetanensis* is endemic (Madico *et al.*, 1997), and can provide an unevaluated source for the transmission of disease.

Cyclosporiasis has occurred among immunocompromised patients, predominantly those with HIV/AIDS. In endemic countries, the prevalence is higher in AIDS patients than in diarrhoeic patients without AIDS (Chacin-Bonilla *et al.*, 2001), and prolonged, severe illness with a mean duration of 4 months has been reported (Pape *et al.*, 1994). There is also some evidence for biliary tract infection in AIDS patients, leading to acalculous cholangitis and cholecystitis (Sifuentes-Osornio *et al.*, 1995; de Górgolas *et al.*, 2001), including histological evidence in the gallbladder epithelium and acalculous cholecystitis in a patient with HIV who required a cholecystectomy (Zar *et al.*, 2001). Therefore HIV/AIDS patients are considered as having an elevated risk of cyclosporiasis.

Although symptoms usually resolve with parasite eradication, inflammatory changes may persist. A myelin-like material has been identified as a marker for persistent inflammation but requires further definition (Connor *et al.*, 1999). Reiter's syndrome, a triad of ocular inflammation, inflammatory oligoarthritis and sterile urethritis that usually occurs several weeks after a triggering infection (Konttinen *et al.*, 1988), has been reported following cyclosporiasis in a 31-year-old male in the USA (Connor *et al.*, 2001). *Cyclospora* infection has also been proposed as a trigger for Guillain-Barré syndrome (Richardson *et al.*, 1998). Post-infectious complications could be minimized by prompt, effective treatment with appropriate anti-*Cyclospora* drugs.

## Pathogenicity and virulence

Impaired absorption of D-xylose (Shlim *et al.*, 1991; Connor *et al.*, 1993) implies involvement of the proximal small intestine and endogenous stages have been identified in the epithelial cells of the jejunum and duodenum (Bendall *et al.*, 1993; Sun *et al.*, 1996; Ortega *et al.*, 1997). How *C. cayetanensis* causes diarrhoea is not fully understood, but invasion of epithelial cells by micro-organisms releases cytokines which in turn activate and recruit phagocytes from the blood (Powell, 1995). Phagocytes release factors including histamine and prostaglandin and platelet aggregating factors that increase intestinal secretion of chloride and water and inhibit absorption (Ciancio and Chang, 1992). Varying degrees of villous blunting, atrophy, crypt hyperplasia and inflammation of the lamina propria have been reported (Bendall *et al.*, 1993; Connor *et al.*, 1993). In a controlled study, inflammatory changes caused by T cells, proteases and oxidants secreted by mast cells were also associated with *C. cayetanensis* infection (Connor *et al.*, 1993). While intestinal inflammation can be severe, numbers of oocysts present in stools often appear to be low. The pathology of the infection has been likened to tropical sprue for which *C. cayetanensis* has been suggested as a trigger organism (Bendall *et al.*, 1993). The infectious dose is not known but, based on evidence from similar parasites such as *Cryptosporidium*, and the high attack rate during outbreaks, even those associated with foods consumed in small quantities, it is likely to be low.

## Causation

The immune response to *C. cayetanensis* infection plays a critical role in the clinical course and outcome but has not been characterized. Although immunocompromised patients, particularly those with AIDS, appear to harbour larger numbers of parasites than immunocompetent hosts, the prophylactic use of trimethoprim sulphamethoxazole against *Pneumocystis carinii* has probably contributed to the low prevalence of *C. cayetanensis* in AIDS patients in North America and Europe, while a high prevalence has been reported in Haiti where this prophylaxis is not widely available (Pape *et al.*, 1994). Antibodies have been detected in patients and shown to increase during convalescence (Long *et al.*, 1991), while others failed to detect them in convalescent sera (Clark and McIntyre, 1997). In contrast, a 10-fold increase in serum IgM has been reported in convalescent patients compared with acute phase sera from the same patients (Wurtz, 1994). That protective immunity may be achieved is indicated by studies in Peru. Children who live in the shanty towns of Lima and have poor sanitation may have more than one infection episode, which is usually mild or asymptomatic (Bern *et al.*, 2002a), but infection and illness are rarely detected in those over 11 years of age (Madicio *et al.*, 1997). This contrasts with prolonged and more severe illness reported in adults from more wealthy areas with good sanitation. Immunologically naive adults experience more severe disease than those in whom infection and immunity are established early in life in endemic areas. Thus adults from non-endemic areas are at higher risk of cyclosporiasis than adults from endemic areas.

## Treatment

Although cyclosporiasis is generally self-limiting, the duration of illness is prolonged and there is the possibility of chronic sequelae. Therefore differential diagnosis and treatment are desirable. Rehydration may be necessary and can be undertaken orally, but antidiarrhoeals are ineffective. Trimethoprim sulphamethoxazole provides rapid, effective anti-parasitic treatment (Soave and Johnson, 1995), first demonstrated in a double-blind placebo controlled trial in Nepal (Hoge *et al.*, 1995), and is effective in both immunocompetent and immunocompromised patients (Pape *et al.*, 1994). Although there is no recommended alternative treatment for sulph-allergic patients and treatment with trimethoprim alone is ineffectual (Brown and Rotschafer, 1999), ciprofloxacin, while not as effective as trimethoprim sulphamethoxazole, is acceptable (Verdier *et al.*, 2000).

## Survival in the environment

Although *Cyclospora*-like bodies and *C. cayetanensis* have been detected in environmental samples, including water, wastewater and foods, methods lack sensitivity and few prevalence studies have been undertaken (see below). The development of methods to assess the viability/infectivity of *C. cayetanensis* has been hampered by the lack of an animal model, and cell culture methods have yet to be established, although propagation has been reported in HCT-8 and Henle 407 cell monolayers (Miliotis *et al.*, 1997). Surrogate methods such as the inclusion/exclusion of vital dyes used for *Cryptosporidium* have not been developed for *Cyclospora*. The ability to sporulate, with subsequent excystation, has been used as a viability indicator (Ortega *et al.*, 1994). Excystation can be induced *in vitro* by exposure to bile salts, sodium tauro-cholate and mechanical pressure (Ortega *et al.*, 1994). The rate of sporulation is probably influenced by environmental factors. For example, oocyst suspensions in 2.5% potassium dichromate kept at 25°C and 32°C showed 20% sporulation by day 5 and complete sporulation between 7 and 13 days (Ortega *et al.*, 1993). Suspensions maintained at 37°C only showed contraction and darkening of central mass after 5 days. However, the age of the organisms on initiation of the experiment may affect the results. Although data are sparse, it appears that sporulation does not occur following exposure to −20°C for 24 hours or to 60°C for 1 hour thus rendering oocysts non-infectious (Smith *et al.*, 1997). However, an outbreak has been epidemiologically linked with the consumption of a cake containing raspberries at a wedding, the raspberry filling having been stored frozen, although the freezer temperature was just −3.3°C (Ho *et al.*, 2002). More rapid methods for assessing viability, such as electrorotation (which also determines sporulation state) have been explored (Dalton *et al.*, 2001) using a purified suspension of oocysts. However, applicability to parasites recovered from food, water or environmental sources has yet to be demonstrated.

If *Cyclospora* behaves in a similar way to related parasites, and using information from vehicles of infection implicated in outbreaks, a moist environment is more likely to encourage survival than a dry one. Given the problems in estimating the survival of *C. cayetanensis* due to the lack of an animal model or surrogate methods, and paucity of material for experimentation, *Eimeria acervulina* has been used as a surrogate organism to test decontamination treatments (Lee and Lee, 2001). Chick-feeding experiments with *E. acervulina* showed that freezing to −18°C and heat treatment at >80°C for 60 minutes rendered oocysts non-infectious. Gamma irradiation was completely effective at 1.0 kGy or higher. *Toxoplasma gondii* has also been suggested as a surrogate (Kniel *et al.*, 2002), but the oocysts are highly infectious for humans and could present a health and safety risk to laboratory personnel.

## Survival in water

That a waterborne outbreak occurred following consumption of drinking water containing acceptable levels of chlorine (Rabold *et al.*, 1994) indicates that *C. cayetanensis* is resistant to levels of chlorine used to treat potable water. CLBs, *C. cayetanensis* oocysts and DNA have been detected in water

Table 19.1  Detection of *Cyclospora cayetanensis* in water and wastewater

| Study area | Sample type | Recovery method | Detection method | Results | Reference |
|---|---|---|---|---|---|
| Chicago, USA | Municipal drinking water from a surface water supply (Lake Michigan) | Not stated | Microscopy | CLBs seen | Wurtz *et al.* (1994) |
| Utah, USA | Sewer drain pipe effluent | Not stated | Microscopy | CLBs seen | Hale *et al.* (1994) |
| Nepal | Chlorinated municipal water supply | Membrane filtration | Light microscopy | *Cyclospora* oocysts seen | Rabold *et al.* (1994) |
| Lima, Peru | Wastewater | Envirocheck filter capsules and Haniffin polypropylene cartridge filters | UV epifluorescence microscopy and PCR | *Cyclospora* oocysts seen and DNA amplicons detected | Sturbaum *et al.* (1998) |
| Guatemala | River water | Calcium carbonate flocculation | Microscopy | *Cyclospora* oocysts seen | Bern *et al.* (1999) |
| Nepal | Sewage water Drinking water | Centrifugation | Microscopy | *Cyclospora* oocysts seen in sewage water | Sherchand *et al.* (1999) |

and wastewater samples (Table 19.1) but difficulties in the detection of this parasite present a challenge to the analytical laboratory (see below).

## Methods of detection

Clinical diagnosis is based upon the detection of oocysts in faeces (Eberhard *et al.*, 1997). The oocysts are spherical and 8–10 μm in diameter, and show variable staining and a granular appearance with acid-fast stains, particularly the modified Ziehl Neelsen stain used to detect *Cryptosporidium* oocysts. Other suitable stains include Kinyoun and safranin. Not all oocysts present in a specimen take up the stain using cold acid-fast stains and it has been suggested that conventional heating can improve the uptake of the stain (Jayshree *et al.*, 1998). Microwave heating of faecal smears improves the uptake of a safranin-based stain (Visvesvara *et al.*, 1997). Oocysts can also be detected by phase- or interference contrast examination of a wet mount, and particular advantage can be made of their ability to autofluoresce: when examined using blue epi-illumination (using a 450–490 nm dichroic mirror exciter filter) they exhibit green autofluorescence, and when examined using a 365 nm dichroic mirror filter they display blue autofluorescence. In a comparative study of detection methods, using modified acid-fast stain as the gold standard, wet mount sensitivity was 75%, safranin O was 30% and auramine rhodamine was 23% (Pape *et al.*, 1994), while the use of autofluorescence improves the detection over conventional examination of wet mounts (Berlin *et al.*, 1998).

Oocysts are often present in low numbers even in samples received at the laboratory from clinically ill patients (Eberhard *et al.*, 1999a), and therefore concentration by the formalin-ethyl acetate technique may be required. However, this is often difficult due to resource constraints in primary testing laboratories. Other problems with diagnosis and surveillance include selective screening in the first place, the relatively non-descript appearance of unsporulated oocysts which may be overlooked or passed off as fungal spores, misidentification in acid-fast stains due to variable staining and incorrect ID, often mistakenly as *Cryptosporidium*, due to failure to take into account the size of the oocysts. Pseudo-outbreaks have been reported (Anon, 1997) and laboratory proficiency in identification is relatively poor (Cann *et al.*, 2000), resulting in under ascertainment. Demonstration of sporulation and excystation, with observation of two sporozoites from within each of two sporocysts confirms the diagnosis, but sporulation can take 7–13 days when stored at room temperature in 2.5% potassium dichromate.

Demonstration of parasite DNA by PCR has been applied as a research tool. To differentiate from other *Cyclospora* spp. or closely related *Eimeria* spp., DNA sequence analysis, restriction fragment length polymorphisms, or PCR with mismatched primers are required (Relman *et al.*, 1996). Suggestions that *Cyclospora* should be considered a member of the genus *Eimeria* on the basis of molecular analyses (Pieniazek and Herwaldt, 1997) remain unresolved, but using structural and ultrastructural definitions according to zoological nomenclature the current taxonomy remains. A molecular diagnostic assay for identifying *C. cayetanensis* in faeces was developed by Yoder *et al.* (1996), based on a nested PCR amplifying a segment of the 18S rDNA gene. While this assay does not differentiate between *Cyclospora* and *Eimeria*, only the former is found in humans. Molecular characterization differentiates *Cyclospora* spp. found in non-human primates from those found in humans (Eberhard *et al.*, 1999b; Lopez *et al.*, 1999) but multiple sequences of the intervening transcribed spacer region 1 (ITS1) have been detected in human isolates indicating the presence of either multiple clones in single clinical isolates or variability within single clones (Adam *et al.*, 2000; Olivier *et al.*, 2001).

Methods for the recovery of *Cyclospora* from water and similar sample matrices have often been based on those developed to detect *Cryptosporidium*. For example, membrane filtration and light microscopy were used to detect *Cyclospora* in chlorinated water supplied to homes of cases during an outbreak in Nepal in 1994 (Rabold *et al.*, 1994). The Envirocheck capsule and Haniffin polypropylene cartridge filters followed by UV epifluorescence microscopy and molecular tools were used to detect *Cyclospora* in wastewater in Peru (Sturbaum *et al.*, 1998). A nested PCR amplifying a 294 bp portion of the 18s rDNA gene was used with RFLP fragment digestion to differentiate *Cyclospora* from *Eimeria* in water samples (Yoder *et al.*, 1996). One of eight samples was PCR-positive while four samples were microscopy-positive. The calcium carbonate flocculation method (Vesey *et al.*, 1993) has also been successfully used to isolate *C. cayetanensis* from river (Bern *et al.*, 1999) and wastewater (unpublished observation).

*Cyclospora* has been detected in prospective studies of fruit and vegetables from markets in Peru, Egypt and Nepal. A variety of vegetables were collected in three sample rounds in a peri-urban slum in Lima, Peru where *C. cayetanensis* is endemic, and prepared by washing steps and centrifigation and detected by wet mount observation, acid-fast staining and autofluorescence (Ortega *et al.*, 1997). None of 35 samples collected in a preliminary study were positive, 2/110 (1.8%) vegetables collected at the end of the high incidence season were positive, and 1/62 vegetables collected at the beginning of the high incidence season were positive. The vegetables concerned were Yerba buena, huacatay and lettuce. However, experimental inoculation showed that the recovery was just 13–15% and scanning electron microscopy demonstrated the presence of oocysts on vegetables after washing. *Cyclospora* oocysts have also been detected on lettuce in Egypt (Abou el Naga, 1999) and green leafy vegetables in Nepal (Sherchand *et al.*, 1999). The use of antibody-coated paramagnetic beads has improved the recovery of *Cryptosporidium* and *Giardia* cysts from various matrices but have not been developed (due to the absence of suitable antibodies) for *Cyclospora* (Rose and Slifco, 1999). In an attempt to improve recovery efficiencies, Robertson *et al.* (2000) used lectin-coated paramagnetic beads. Oocyst recoveries from mushrooms, lettuce and raspberries were around 12% with no significant difference with or without the beads, although microscopy was greatly facilitated by their use. Recovery from bean sprouts remained very low at 4%. Studies of food items are important in elucidating the transmission of *Cyclospora* since many foods are produced using poor-quality and wastewaters during, for example, irrigation.

Molecular techniques have been applied to the detection of *Cyclospora* in food and are particularly useful where the nature of the food reduces the utility of microscopy. Jinneman *et al.* (1999) used an oligonucleotide-ligation assay that differentiates between *Cyclospora* and *Eimeria*. Sensitivity was 19 oocysts in an optimized template format and 25 oocysts on the addition of raspberry extract. An ITS region was used by Adam *et al.* (2000), which also provides genotype data of epidemiological value. Implicated foods have been tested and confirmed by oocyst detection and PCR in basil (Lopez *et al.*, 2001) and PCR only in raspberry filling in a wedding cake (Ho *et al.*, 2002).

## Epidemiology

Most of the epidemiological data have been generated by studies in Peru, Haiti, Guatemala and Nepal where cyclosporiasis is endemic. Cyclosporiasis has a marked but geographically variable seasonality. In Lima, Peru, which has a desert, coastal location with minimal rainfall, prevalence is highest during the warmer summer months of December to May (Madico *et al.*, 1997) and cases are rarely detected in the cooler June to November period (Bern *et al.*, 2002a). In Nepal, which is at altitude, cases cluster prior to and during the warm monsoon season (usually May to August) but decrease before the monsoon ends (Shlim *et al.*, 1991; Hoge *et al.*, 1993, 1995). In Guatemala prevalence of infection increases in May and peaks in June, coinciding with the beginning of the rainy season (Bern *et al.*, 1999). In Haiti, peak prevalence is during the dry, cool winter months of January to March (Eberhard *et al.*, 1999a). Oocyst viability and infectivity may be affected by ambient temperature and humidity, while transmission may be favoured by rainfall. The varying combination of these factors may account for the differing geographical picture. However, few survival studies have been undertaken because of the lack of a good model for viability or infectivity.

Surveillance in Guatemala has shown that 126/5552 (2.3%) specimens screened over a 1-year period from patients submitting faecal specimens to two health centres contained *C. cayetanensis*, while in patients without gastroenteritis the prevalence was 1.1% (Bern *et al.*, 1999). The detection rate peaked in June when 6.7% specimens were positive. Oocysts were detected more frequently among children aged 1.5–9 years and among people with gastroenteritis, but rarely detected in children <18 months of age. Incidence was significantly higher among HIV patients screened at a local clinic. Another study found oocysts in the stools of 1.5% subjects, although the presence or absence of symptoms was not stated (Pratdesaba *et al.*, 2001). In Haiti, prevalence of the organism in cohorts of mothers and children in a rural community peaked at 16% in March, but the organism was not detected in every month (Eberhard *et al.*, 1999a). Infection was more common in children than adults, and there was no statistical difference in detection rates between diarrhoeic and non-diarrhoeic specimens. In a study of diarrhoeic children in Nepal, 6/50 (12%) children aged <18 months were infected with *Cyclospora*, whereas none of 74 over this age were (Hoge *et al.*, 1995). In a cross-sectional study in a periurban shanty town in Lima, Peru the prevalence of *C. cayetanensis* infection in children under 18 years was 1% and was highest in children aged 2–4 years (Madico *et al.*, 1997). Infection was most prevalent during the summer when 3–4% of children were infected. One-third of infections were apparently asymptomatic. The prevalence was lower than that reported during previous studies commencing some 4 years earlier in the same area (Ortgea *et al.*, 1993), but sanitary conditions had improved. Due to the seasonal variation in incidence, longitudinal surveillance data and studies are required fully to understand the epidemiology of *C. cayetanensis* carriage and disease.

The observed age relationship may be due to waning maternal antibodies or increased environmental exposure coincident with weaning. Additionally, the development of acquired immunity may occur, and concurring with this is evidence that length of stay and therefore possibly increased exposure among travellers to Nepal has been associated with decreased symptoms (Hoge *et al.*, 1993). Thus there are consistent themes in the epidemiology of endemic cyclosporiasis: marked seasonality, high prevalence in children compared with adults and higher rates in those with than those without gastrointestinal symptoms (Bern *et al.*, 2002a).

The first case report suggesting food-borne transmission of *Cyclospora* was in an airline pilot flying between Port-au-Prince, Haiti and New York in 1995 (Connor and Shlim, 1995). During the 1990s many outbreaks of cyclosporiasis were reported in North America associated with the consumption of fresh produce, including raspberries, other berries, mesclun lettuce, basil and fruit salad (Herwaldt, 2000). An outbreak has also been reported in Germany implicating mixed whole leaf salad inclusive of fresh herbs sourced from wholesalers in France and Italy (Brockman *et al.*, 2001). However, two outbreaks, one in the USA and one in Nepal, have been associated with the consumption of drinking water.

The first outbreak of cyclosporiasis to be associated with drinking water was in 1990 among staff at a hospital in Chicago, USA (Huang *et al.*, 1995). At the time organisms identified in stools of cases were described as algal-like but later identified as *C. cayetanensis*. Nine out of 14 resident house staff and 1/7 other staff who met the case definition had stools positive for *Cyclospora*. All either lived in the house-staff accommodation or had attended a party there. Epidemiological investigations implicated the water supply but oocysts were not detected in water samples (Kocka *et al.*, 1991). The second outbreak was among British soldiers and their dependents in June 1994 in Pokhara, Nepal (Rabold *et al.*, 1994). Twelve out of 14 at risk became ill and *C. cayetanensis* was identified in 6/8 specimens from clinical cases. The water supply was a mixture of river water and municipal water, mixed and stored in a tank. The water was chlorinated to microbiologically acceptable levels in the tank and supplied to homes in a sealed pipe. Structures morphologically resembling *C. cayetanensis* were found in a 2l sample of water from the tank. Although the mode of contamination was not identified, it is plausible that the river water became contaminated from a human source.

Previously, studies in Nepal revealed a cluster of more than 50 laboratory confirmed cases of cyclosporiasis, mainly among expatriate visitors, between May and November during 1989 (Hoge *et al.*, 1993). No further cases were detected until May the following year when 85 cases were confirmed among visitors and 6/184 local people. Cases were detected until October, and the first cases each year coincided with the monsoon period. In 1992 a case control study among travellers and expatriates at two outpatient clinics in Kathmandu identified drinking untreated water as a risk factor for cyclosporiasis and oocyst-like bodies were also detected in the tap water from the household of one case (Hoge *et al.*, 1993). Cases of illness were more likely to have drunk reconstituted milk than controls. Although significant, only 28% cases had drunk untreated tap water and other routes of transmission may also have been involved. An outbreak centred on a golf club in New York in June 1995 was epidemiologically linked to drinking water from water coolers but incomplete information meant that a food-borne route could not be ruled out (Carter *et al.*, 1996).

Clusters of cases have also been reported, including a family cluster in Peru where four family members reported drinking from the same canal (Zerpa *et al.*,

1995). CLBs were also detected in faeces from a symptomatic duck but infection was unconfirmed. In a case control study in Guatemala of cases largely recruited from patients presenting at the health centres, risk factors identified in multivariate analysis included drinking untreated water in the 2 weeks before illness (OR = 4.2, 95% CI = 1.4–12.5) and contact with soil among children <2 years old (OR = 19.8, 95% CI = 2.2–182) (Bern *et al.*, 1999).

Individual case reports have also linked cases to waterborne sources but again cannot rule out other exposures. These have included a man in Utah, USA who cleaned up his basement when it became flooded with sewage, and where CLBs were also detected in the effluent from the sewer drain pipe (Hale *et al.*, 1994); a child who swam in Lake Michigan, and water from there to the Chicago municipal supply contained presumptive CLBs (Wurtz, 1994). The consumption of well water was implicated in a case in Massachussetts, USA (Ooi *et al.*, 1995). Indigenous cyclosporiasis in developed countries needs closer examination. One study in the USA has linked infection to gardening and working with soil (Koumans *et al.*, 1996).

Prospective sampling of wastewater in sewage lagoons in Lima, Peru which receive waste from shanty towns where *Cyclospora* is endemic, confirmed the presence of *C. cayetanensis* oocysts by microscopy and PCR (Sturbaum *et al.*, 1998). The water from these lagoons is used to irrigate pasture land, cornfields and trees, while water from similar lagoons in other parts of Lima is used to irrigate vegetable crops. It is possible that the use of water contaminated with human faeces in the production of crops is one of the routes by which *C. cayetanensis* is transmitted. In a study of river water in Guatemala, 2/30 (7%) specimens from two different rivers contained *Cyclospora* oocysts (Bern *et al.*, 1999). Interestingly, the river water was positive during May, coinciding in the seasonal rise in human cases which peaks in June, which also coincides with the spring raspberry harvest.

## Risk assessment

One of the problems of evaluating the waterborne risk presented by *Cyclospora* is that many of the fundamental data are missing. The relative insensitivity and lack of standardization of detection methods has made the determination of exact modes of transmission difficult. However, there is both epidemiological and microbiological evidence for a waterborne route. While outbreaks in developed countries have been primarily caused by contaminated produce, in some developing countries the main vehicle of infection is drinking water and the use of contaminated water in crop production is a biologically plausible route of contamination. Following the food-borne outbreaks in North America during the 1990s, Chalmers and colleagues (2000) evaluated the risk of similar exposures in the UK and identified different importation patterns. However, such work needs to be revisited in the light of outbreaks elsewhere and the identification of broadening importation patterns in terms of foodstuffs and countries of origin. Similarly, proper evaluation of the proportions of imported versus indigenous cases of illness needs to be undertaken.

However, sources of oocysts have not been fully identified: although CLBs have been detected in poultry, ducks and non-human primates (Ashford *et al.*, 1993; Zerpa *et al.*, 1995; Garcia-Lopez *et al.*, 1996; Smith *et al.*, 1996) convincing evidence for a non-human host of *C. cayetanensis* has not yet been provided and humans appear to be the only host species. Although the source of infection is therefore most likely to be humans, direct person-to-person transmission is unlikely since a period of maturation in the environment is required for oocysts to sporulate and become infective. Food and water are, however, likely vehicles, and it is possible that contaminated water plays an important role in food-borne cyclosporiasis.

Known risk factors for cyclosporiasis are consumption of contaminated water or produce (particularly if eaten raw and of a type difficult to clean) and environmental sources in association with avians in Guatemala (Bern *et al.*, 1999) and Peru (Bern *et al.*, 2002a). Association of infection with keeping guinea-pigs and rabbits is unexplained, but could represent a marker for some other identified risk factor, particularly since this occurred within migrant families recently relocated from rural to urban areas. Such families come from regions of lower endemnicity of cyclosporiasis and may not have developed immunity. So far as drinking water is concerned, removal by conventional coagulation and filtration should be equal to or greater than for *Cryptosporidium* since the oocysts are twice as big. Chlorine resistance has not been evaluated but is probably similar to that exhibited by *Cryptosporidium*.

In order to identify principal transmission routes and interventions, it is essential that the sources of contamination are identified. However, to enable this, methods for detection in food produce and water must be developed and applied along with environmental studies at sites of production.

## Overall risk assessment

*Health effects*: occurrence of illness, degree of morbidity and mortality, probability of illness based on infection:

- *Cyclospora* is distributed world-wide, but infection appears to be endemic throughout the tropics. Outbreaks have been sporadically reported in developed countries.
- The symptoms from *Cyclospora* infection include diarrhoea which is sometimes explosive. Other symptoms include anorexia, nausea, vomiting, abdominal bloating and cramps, weight loss, fatigue, low grade fever and body aches. Although cyclosporiasis is generally self-limiting, the duration of illness is prolonged and there is the possibility of chronic sequelae.

- Immunocompromised patients, predominantly those with HIV/AIDS are at high risk of cyclosporiasis.

*Exposure assessment*: routes of exposure and transmission, occurrence in source water, environmental fate:

- Infected persons excrete oocysts of *Cyclospora* in their faeces. Oocysts are not immediately infectious after they are excreted and may require from days to weeks to become infectious.
- Although the exact routes of transmission of *C. cayetanensis* are not yet clear, outbreaks linked to contaminated water, as well various types of fresh produce, have been reported in recent years.
- The insensitivity of current detection methods has made determining the exact mode of transmission difficult. Although *Cyclospora*-like bodies and *C. cayetanensis* have been detected in environmental samples, including water, wastewater, and foods, few prevalence studies have been undertaken because of the lack of sufficient detection methods.
- The *Cyclospora* oocysts are robust and can probably survive for long periods.

*Risk mitigation*: drinking-water treatment, medical treatment:

- Removal by conventional coagulation and filtration should be at least as efficient as for *Cryptosporidium*, since the oocysts are twice as big.
- Chlorine resistance has not been evaluated, but a waterborne outbreak occurred following consumption of chlorinated drinking water indicating that *C. cayetanensis* is resistant to levels of chlorine used to treat potable water.
- Trimethoprim sulphamethoxazole provides rapid, effective anti-parasitic treatment for those with cyclosporiasis.

## Future implications

The major risks are probably posed by inadequately treated water for consumption and/or cultivation of fresh produce. Large volumes of foods cross international borders from an ever-increasing range of countries of origin, but generally the laboratory and epidemiological tools are lacking for investigation.

Removal from food as a control measure is problematic since many implicated foods are difficult to wash and contain crevices in which oocysts can reside. Oocysts have been shown to remain on foods after washing (Ortega *et al.*, 1997) and although minimum time/temperature for inactivation is yet to be determined, pasteurization and freezing should both be effective but may damage the aesthetic nature of fresh produce. Future means of control may include ionizing radiation, although this faces a lack of consumer acceptance (Monk *et al.*, 1995).

# References

Abou el Naga, I.F. (1999). Studies on a newly emerging protozoal pathogen: *Cyclospora cayetanensis*. *J Egyptian Soc Parasitol*, **29**: 575–586.

Adam, R.D., Ortega, Y.R., Gilman, R.H. *et al.* (2000). Intervening transcribed spacer region 1 variability in *Cyclospora cayetanensis*. *J Clin Microbiol*, **38**: 2339–2343.

Anon. (1997). Outbreaks of pseudo-infection with *Cyclospora* and *Cryptosporidium* – Florida and New York, 1995. *MMWR*, **46**: 354–358.

Ashford, R.W. (1979). Occurrence of an undescribed coccidian in man in Papua New Guinea. *Ann Trop Med Parasitol*, **73**: 497–500.

Ashford, R.W., Warhurst, D.C. and Reid, G.D.F. (1993). Human infection with cyanobacterium-like bodies. *Lancet*, **341**: 1034.

Bendall, R.P., Lucas, A., Moody, A. *et al.* (1993). Diarrhoea associated with cyanobacterium-like bodies: A new coccidian enteritis of man. *Lancet*, **341**: 590–592.

Berlin, O.G., Peter, J.B., Gagne, C. *et al.* (1998). Autofluorescence and the detection of *Cyclospora* oocysts. *Emerging Infect Dis*, **4**: 127–128.

Bern, C., Hernandez, B., Lopez, M.B. *et al.* (1999). Epidemiologic studies of *Cyclospora cayetanensis* in Guatemala. *Emerging Infect Dis*, **5**: 766–774.

Bern, C., Ortega, Y., Checkley, W. *et al.* (2002a). Epidemiologic differences between cyclosporiasis and cryptosporidiosis in Peruvian Children. *Emerging Infect Dis*, **8**: 581–585.

Bern, C., Arrowood, M.J., Eberhard, M. *et al.* (2002b). *Cyclospora* in Guatemala: further considerations. *J Clin Microbiol*, **40**: 731–732.

Brockmann, S., Döller, C.R., Dreweck, C. *et al.* (2001). Cyclosporiasis in Germany. *Euro Surveillance Weekly*, **5**: 1.

Brown, G.H. and Rotschafer, J.C. (1999). *Cyclospora*: review of an emerging pathogen. *Pharmacotherapy*, **19**: 70–75.

Cann, K.J., Chalmers, R.M., Nichols, G. *et al.* (2000). *Cyclospora* infections in England and Wales: 1993 to 1998. *Commun Dis Public Hlth*, **3**: 46–49.

Carter, R.J., Guido, F., Jacquette, G. *et al.* (1996). Outbreak of cyclosporiasis associated with drinking water (abstract). In *Program of the 30th Interscience Conference on antimicrobial agents and Chemotherapy, New Oreans 1996*, p. 259.

Chacin-Bonilla, L., Estevez, J., Monsalve, F. *et al.* (2001). *Cyclospora cayetanensis* infections among diarrheal patients from Venezuela. *Am J Trop Med Hyg*, **65**: 351–354.

Chalmers, R.M., Rooney, R. and Nichols, G. (2000). Foodborne outbreaks of cyclosporiasis have arisen in North America. Is the United Kingdom at risk? *Commun Dis Public Hlth*, **3**: 50–55.

Ciancio, M. and Chang, E. (1992). Epithelial secretory response to inflammation. *Ann NY Acad Sci*, **664**: 210–221.

Clark, S.C. and McIntyre, M. (1997). An attempt to demonstrate a serological immune response in patients infected with *Cyclospora cayetanensis*. *Br J Biomed Sci*, **54**: 73.

Connor, B.A., Shlim, D.R., Scholes, J.V. *et al.* (1993). Pathologic changes in the small bowel in nine patients with diarrhoea associated with a coccidian-like body. *Ann Int Med*, **119**: 377–382.

Connor, B.A. and Shum, O.R. (1995). Foodborne transmission of *Cyclospora*. *Lancet*, **346**: 1634.

Connor, B.A., Reidy, J. and Soave, R. (1999). Cyclosporiasis: clinical and histopathologic correlates. *Clin Infect Dis*, **28**: 1216–1222.

Connor, B.A., Johnson, E. and Soave, R. (2001). Reiter syndrome following protracted symptoms of *Cyclospora* infection. *Emerging Infect Dis*, **7**: 453–454.

Dalton, C., Goater, A.D., Pethig, R. *et al.* (2001). Viability of *Giardia intestinalis* cysts and viability and sporulation state of *Cyclospora cayetanensis* determined by electrorotation. *Appl Environ Microbiol*, **67**: 586–590.

De Górgolas, M., Fortés, J. and Guerrero, M.L.F. (2001). *Cyclospora cayetanensis* cholecystitis in a patient with AIDS. *Ann Intern Med*, **143**: 166.

Eberhard, M.L., Pieniazek, N.J. and Arrowood, M.J. (1997). Laboratory diagnosis of Cyclospora infections. *Arch Pathol Lab Med*, **121**: 792–797.

Eberhard, M.L., Nace, E.K., Freeman, A.R. *et al.* (1999a). *Cyclospora cayetanensis* infections in Haiti: a common occurrence in the absence of watery diarrhoea. *Am J Trop Med Hyg*, **60**: 584–586.

Eberhard, M.L., Da Silva, A.J., Lilley, B.G. *et al.* (1999b). Morphologic and molecular characterisation of new Cyclospora species from Ethiopian monkeys: *C. cercopitheci* sp.n., *C. colobi* sp.n. and *C. papionis* sp.n. *Emerging Infect Dis*, **5**: 561–658.

Eberhard, M.L., Ortega, Y.R., Hanes, D. *et al.* (2000). Attempts to establish experimental *Cyclospora cayetanensis* infection in laboratory animals. *J Parasitol*, **86**: 577–582.

Garcia-Lopez, H.L., Rodriquez-Tovar, L.E. and Mdeina-De la Garza, C.E. (1996). Identification of *Cyclospora* in poultry. *Emerging Infect Dis*, **2**: 356–357.

Hale, D., Aldeen, W. and Carroll, K. (1994). Diarrhoea associated with cayenobacteria-like bodies in an immunocompetent host. An unusual epidemiological source. *JAMA*, **271**: 144–145.

Herwaldt, B.L. (2000). *Cyclospora cayetanensis*: a review, focussing on the outbreaks of cyclosporiasis in the 1990s. *Clin Infect Dis*, **31**: 1040–1057.

Ho, A.Y., Lopez, A.S., Eberhart, M.G. *et al.* (2002). Outbreak of cyclosporiasis associated with imported raspberries, Philadelphia, Pennsylvania, 2000. *Emerging Infect Dis*, **8**: 783–788.

Hoge, C.W., Shlim, D.R., Rajah, R. *et al.* (1993). Epidemiology of diarrhoeal illness associated with coccidian-like organism among travellers and foreign residents in Nepal. *Lancet*, **341**: 1175–1179.

Hoge, C.W., Shlim, D.R., Ghimire, M. *et al.* (1995). Placebo-controlled trial of co-trimoxazole for cyclospora infections among travellers and foreign residents in Nepal. *Lancet*, **345**: 691–693.

Huang, P., Weber, J. and Sosin, D.M. (1995). The first reported outbreak of diarrheal disease associated with *Cyclospora* in the United States. *Ann Int Med*, **123**: 409–414.

Jayshree, R.S., Acharya, R.S. and Sridhar, H. (1998). *Cyclospora cayetanensis*-associated diarrhoea in a patient with acute myeloid leukaemia. *J Diarrhoeal Dis Res*, **16**: 254–255.

Jinneman, K.C., Wetherington, J.H., Hill, W.E. *et al.* (1999). An oligonucleotide-ligation assay for differentiation between *Cyclospora* and *Eimeria* spp. polymerase chain reaction amplification products. *J Food Protect*, **62**: 682–685.

Kniel, K.E., Lindsay, D.S., Sumner, S.S. *et al.* (2002). Examination of attachment and survival of *Toxoplasma gondii* oocysts on raspberries and blueberries. *J Parasitol*, **88**: 790–793.

Kocka, F., Peters, C., Dacumos, E. *et al.* (1991). Outbreaks of diarrhoeal illness associated with cyanobacteria (Blue-Green Algae)-like bodies – Chicago and Nepal, 1989 and 1990. *MMWR*, **40**: 325–327.

Konttinen, Y.T., Nordstrom, D.C., Bergroth, V. *et al.* (1988). Occurrence of different ensuing triggering infections preceding reactive arthritis: a follow-up study. *Br Med J*, **296**: 1644–1645.

Koumans, E.H., Katz, D., Malecki, J.A. *et al.* (1996). Novel parasite and mode of transmission: *Cyclospora* infection – Florida 1996. Annual Epidemic Intelligence Service Conference, p. 4560.

Lee, M.B. and Lee, E.H. (2001). Coccidial contamination of raspberries: mock contamination with *Eimeria acervulina* as a model for decontamination treatment studies. *J Food Protect*, **64**: 1854–1857.

Levine, N.D. (1982). Taxonomy and life cycles of coccidian. In *The Biology of the Coccidian*, Long, P.L. (ed.). London: Edward Arnold, pp. 1–3.

Long, E.G., White, E.H., Carmichael, W.W. *et al.* (1991). Morphologic and staining characteristics of a cycanobacterium-like organism associated with diarrhoea. *J Infect Dis*, **164**: 199–202.

Lopez, A.S., Dodson, D.R., Arrowood, M.J. *et al.* (2001). Outbreak of cyclosporiasis associated with basil in Missouri in 1999. *Clin Infect Dis*, **32**: 1010–1017.

Lopez, F.A., Manglicmot, J., Schmidt, T.M. *et al.* (1999). Molecular characterisation of Cyclospora-like organisms from baboons. *J Infect Dis*, **179**: 670–676.

Madico, G., McDonald, J., Gilman, R. *et al.* (1997). Epidemiology and treatment of *Cyclospora cayetanensis* infection in Peruvian children. *Clin Infect Dis*, **24**: 977.

Miliotis, M.D., Hanes, D.E., Tall, B.D. *et al.* (1997). Food and Drug Administration Science Forum Poster Abstract 1997 (http://www.cfsan.fda.gov/~frf/forum97/97L10.htm).

Monk, J., Beuchat, L. and Doyle, M. (1995). Irradiation inactivation of food-borne micro-organisms. *J Food Protect*, 58: 197–208.

Naranjo, J., Sterling, C., Gilman, R. *et al.* (1989). Cryptosporidium-muris-like objects from faecal samples of Peruvians (abstract 324). In *Program and abstracts of the 38th Annual Meeting of the American Society of Tropical Medicine and Hygiene (Honolulu) 10–14 December 1989*, p. 243.

Olivier, C., van de Pas, S., Lepp, P.W. *et al.* (2001). Sequence variability in the first internal transcribed spacer region within and among *Cyclospora* species is consistent with polyparasitism. *Int J Parasitol*, 31: 1475–1487.

Ooi, W.W., Zimmerman, S.K. and Needham, C.A. (1995). *Cyclospora* species as a gastrointestinal pathogen in immunocompetent hosts. *J Clin Microbiol*, 33: 1267–1269.

Ortega, Y.R., Sterling, C.R., Gilman, R.H. *et al.* (1993). *Cyclospora* species: a new protozoan pathogen of humans. *New Engl J Med*, 328: 1308–1312.

Ortega, Y.R., Gilman, R.H. and Sterling, C.R. (1994). A new coccidian parasite (Apicomplexa: Eimeriidae) from humans. *J Parasitol*, 80: 625–629.

Ortega, Y.R., Roxas, C.R., Gilman, R.H. *et al.* (1997). Isolation of *Cryptosporidium parvum* and *Cyclospora cayetanensis* from vegetables collected in markets of an endemic region in Peru. *Am J Trop Med Hyg*, 57: 683–686.

Pape, J.W., Verdier, R.I. and Boncy, M. *et al.* (1994). *Cyclospora* infection in adults infected with HIV. *Ann Int Med*, 121: 654–657.

Pieniazek, N.J. and Herwaldt, B.L. (1997). Reevaluating the molecular taxonomy: is human-associated *Cyclospora* a mammalian Eimeria species? *Emerging Infect Dis*, 3: 381–383.

Powell, D. (1995). Approach to the patient with diarrhoea. In *Textbook of Gastroenterology*, 2nd edn, Yamada, T. (ed.). Philadelphia: JB Lippincott, pp. 820–824.

Pratdesaba, R.A., Gonzalez, M., Piedrasanta, E. *et al.* (2001). *Cyclospora cayetanensis* in three populations at risk in Guatemala. *J Clin Microbiol*, 39: 2951–2953.

Rabold, J.G., Hoge, C.W. and Shim, D.R. (1994). Cyclospora outbreak associated with chlorinated drinking water. *Lancet*, 344: 1360–1361.

Relman, D.A., Schmidt, T.M., Gajadhar, A. *et al.* (1996). Molecular phylogenetic analysis of Cyclospora, the human intestinal pathogen, suggests that it is closely related to Eimeria species. *J Infect Dis*, 173: 440.

Richardson, R.F., Remler, B.F., Katirji, B. *et al.* (1998). Guillain-Barré syndrome after Cyclospora infection. *Muscle Nerve*, May: 669–671.

Robertson, L.J., Gjerde, B. and Campbell, A.T. (2000). Isolation of Cyclospora oocysts from fruits and vegetables using lectin-coated paramagnetic beads. *J Food Protect*, 63: 1410–1414.

Rose, J.B. and Slifko, T.R. (1999). Giardia, Cryptosporidium and Cyclospora and their impact on foods: a review. *J Food Protect*, 62: 1057–1070.

Schneider, A. (1881). Sur les psorospemies oviformes ou coccidies, espèces nouvelles ou peu connues. *Arch Zool Exp Gén*, 9: 387–404.

Sherchand, J.B., Cross, J.H., Jimba, M. *et al.* (1990). Study of *Cyclospora cayetanensis* in health care facilities, sewage water and green leafy vegetables in Nepal. *SE Asian J Trop Med Public Hlth*, 30: 58–63.

Shlim, D.R., Cohen, M.T., Eaton, M. *et al.* (1991). An alga-like organism associated with an outbreak of prolonged diarrhoea among foreigners in Nepal. *Am J Trop Med Hyg*, 45: 383–389.

Sifuentes-Osornio, J., Porrras-Cortés, G., Bendall, R.P. *et al.* (1995). *Cyclospora cayetanensis* infection in patients with and without AIDS: biliary disease as another clinical manifestation. *Clin Infect Dis*, 21: 1092–1097.

Smith, H.V., Paton, C.A., Girdwood, R.W.A. *et al.* (1996). Cyclospora in non-human primates in Gombe, Tanzania (letter). *Vet Rec*, 138: 528.

Smith, H.V., Paton, C.A., Mtambo, M.M. *et al.* (1997). Sporulation of *Cyclospora* sp. Oocysts. *Appl Environ Microbiol*, 63: 1631–1632.

Soave, R. (1996). *Cyclospora*: an overview. *Clin Infect Dis*, 23: 429–437.

Soave, R. and Johnson, W.D. (1995). Cyclospora: conquest of an emerging pathogen. *Lancet*, **345**: 667–668.

Soave, R., Dubey, J.P., Ramos, L.J. *et al.* (1986). A new intestinal pathogen? (abstract). *Clin Res*, **34**: 533A.

Soave, R., Herwaldt, B.L. and Relman, D.A. (1998). Cyclospora. *Infect Dis Clin N Am*, **12**: 1–12.

Sturbaum, G.D., Ortega, Y.R., Gilman, R.H. *et al.* (1998). Detection of *Cyclospora cayetanensis* in wastewater. *Appl Environ Microbiol*, **64**: 2284–2286.

Sun, T., Ilardi, C.F., Asnis, D. *et al.* (1996). Light and electron microscopic identification of Cyclospora species in the small intestine: evidence of the presence of asexual life cycle in a human host. *Am J Clin Microbiol*, **105**: 216–220.

Taylor, D.N., Houston, R., Shlim, D.R. *et al.* (1988). Etiology of diarrhoea among travelers and foreign residents in Nepal. *JAMA*, **260**: 1245–1248.

Verdier, R.I., Fitzgerald, D.W., Johnson, W.D. *et al.* (2000). Trimethoprim-sulfamethoxazole compared with ciprofloxacin for treatment and prophylaxis of *Isospora belli* and *Cyclospora cayetanensis* infection in HIV-infected patients. A randomised, controlled trial. *Ann Int Med*, **132**: 885–888.

Vesey, G., Slade, J.S., Byrne, M. *et al.* (1993). A new method for the concentration of Cryptosporidium oocysts from water. *J Appl Bacteriol*, **75**: 82–86.

Visvesvara, G.S., Moura, H., Kovacs-Nace, E. *et al.* (1997). Uniform staining of *Cyclospora* oocysts in fecal smears by a modified safranin technique with microwave heating. *J Clin Microbiol*, **35**: 730–733.

Wurtz, R. (1994). *Cyclospora*: a newly identified intestinal pathogen of humans. *Clin Infect Dis*, **18**: 620–623.

Yoder, K.E., Sethabutr, O. and Relman, D.A. (1996). PCR-based detection of the intestinal pathogen *Cyclospora*. In *PCR Protocols for Emerging Infectious Diseases, a Supplement to Diagnostic Molecular Biology: Principles and Applications*, Persing, D.H. (ed.). Washington, DC: ASM Press, pp. 169–176.

Zar, F.A., El-Bayoumi, E. and Yungbluth, M.M. (2001). Histologic proof of acalculous cholecystitis due to *Cyclospora cayetanensis*. *Clin Infect Dis*, **33**: 140–141.

Zerpa, R., Uchima, N. and Huicho, L. (1995). *Cyclospora cayetanensis* associated with water diarrhoea in Peruvian patients. *J Trop Med Hyg*, **98**: 325–329.

# 20

# *Entamoeba histolytica*

## Basic microbiology

Amoebic protozoa of the genus *Entamoeba* are members of the Phylum *Sarcodina*, Order *Amoebida*, Family *Endamoebidae*. Although many amoebae inhabit the human gastrointestinal tract (including *Entamoeba histolytica*, *Entamoeba coli*, *Entamoeba dispar*, *Entamoeba hartmanni*, *Entamoeba polecki*, *Chilmastix mesnili*, *Endolimax nana*, and *Iodamoeba buetschlii*), most are not pathogenic. *Entamoeba gingivalis* inhabits the mouth and has been associated with periodontal disease (Lyons *et al.*, 1983), while *E. polecki* is of uncertain pathogenicity (Levine and Armstrong, 1970). However, *E. histolytica* is the predominant pathogenic amoeba and causes amoebic dysentery (amoebiasis) (Brumpt, 1925; Sargeaunt *et al.*, 1978).

   *E. histolytica* is indistinguishable by microscopy from the non-pathogenic *E. dispar* and the differentiation is rarely made in routine clinical microbiological diagnoses. Historically, since these organisms could not be differentiated, both were referred to as *E. histolytica*, and it has been estimated that about 10% of the world's population are infected (Walsh, 1988). However, 90% of these infections are asymptomatic and probably with the non-pathogenic *E. dispar*. Despite this there remains a considerable burden of disease and more than 100 000 deaths occur annually from invasive amoebiasis, making it the third leading parasitic cause of death in developing countries (Reed, 1992).

The life cycle of *E. histolytica* has been described by Dobell (1928) and comprises an infective cyst form, metacyst, metacystic trophozoite, motile feeding trophozoite and precyst stages. The cyst form (10–16 μm), which develops only in the intestinal tract, is shed in the faeces and is capable of survival in food and water. The mature cysts, which contain four nuclei, are transmitted to humans by the ingestion of faecally-contaminated food, water or from body contact. The amoebae within the mature cyst are activated by the neutral or alkaline environment in the small intestine, and separate from the cyst wall which is digested by enzymes within the gut lumen. Rapid nuclear and cytoplasmic division results in eight uninucleate trophozoites. The trophozoites (20–40 μm) migrate to the large intestine where they multiply by binary fission and feed on the bacteria of the intestinal flora and on cell debris. Encystation is probably stimulated by the dehydrating luminal conditions and cysts develop, each with one to four nuclei, which are passed in the faeces. Although trophzoites may be shed during acute colitis, they are not responsible for the spread of infection because they do not survive outside the body and are destroyed by the low gastric pH. The robust cysts transmit infection.

Trophozoites of non-pathogenic *E. dispar* colonize the gut lumen and the infected host sheds cysts asymptomatically. *E. histolytica* can also cause asymptomatic infection but trophozoites can invade the intestinal mucosa, causing intestinal disease (dysentery), or travel via the blood stream to extraintestinal sites including the liver, brain and lungs (Ravdin, 1995).

Although the distribution of *E. histolytica* is world-wide, there is a higher prevalence of amoebiasis in developing countries and this appears to depend on sanitation, age, crowding and socioeconomic status (Ravdin, 1988). Although natural infections with indistinguishable organisms have been reported in macaque monkeys and pigs (Hoare, 1962), and primates, dogs and cats also shed morphologically similar amoebae, humans are the main reservoir of *E. histolytica*. Transmission is usually from a chronically ill or asymptomatic cyst shedder. Those at high risk in developed countries are travellers returning from developing countries, immigrants, migrant workers, immunocompromised individuals and sexually active male homosexuals (Ravdin, 1988). Waterborne transmission is commonly associated with faecally-contaminated water supplies in developing countries.

A number of biological features of *E. histolytica* affect its transmission and epidemiology. The robust cysts are resistant to gastric acid and once shed in the faeces are immediately infective. Indeed, contact with an asymptomatic carrier is regarded as the cause of most cases. However, environmental contamination can lead to waterborne transmission, which is common in developing countries where drinking water is untreated.

## Origin of the organism

A Russian clinical assistant first discovered organisms now classified as *E. histolytica* in 1873, by observing large numbers of amoebic trophozoites in the stools of a patient with bloody dysentery (Lösch, 1875). However, it was several decades before the concept that intestinal amoebae could cause disease was generally accepted, principally since, while large numbers of amoebae were often seen, disease was present in the minority of cases. Councilman and LaFleur made descriptions of the clinical and pathological outcomes of infection in 1891. Genus names *Amoeba*, *Endamoeba* and *Entamoeba* have been used in the past, and confusing and regularly changing taxonomy at the genus and species level persisted throughout the 20th century (Clark, 1998). Unclear pathogenicity was noted and differences between morphologically distinct cysts of *E. hartmanni*, *E. poleki* and *E. histolytica* were observed (Burrows, 1959). While pathogenic and non-pathogenic 'variants' of *E. histolytica* have been proposed in the past, particularly by Brumpt in the 1920s on the basis of clinical and experimental observations (Brumpt, 1925, 1928), the lack of morphological differences hampered acceptance of this proposal.

Isoenzyme, antigenic and genetic studies demonstrate differences between *E. histolytica* isolates (Sargeaunt *et al.*, 1978; Strachan *et al.*, 1988; Tannich *et al.*, 1989; Clark and Diamond, 1991). While changing zymodeme patterns suggested that genetic transfer occurred between pathogenic and non-pathogenic strains, this is not supported by genetic evidence and two stable, separate species were described by Diamond and Clark in 1993, proposing that the pathogenic species was to retain the name *E. histolytica* and the non-pathogenic species *E. dispar*. In 1997, the WHO recommended the acceptance of these two species (Anon, 1997). However, lack of distinction between the two species has certainly hampered the understanding, clinical management and epidemiology of *E. histolytica* infections.

## Clinical features

Infection with *E. histolytica* can range from asymptomatic infection and cyst passage, to acute amoebic rectocolitis, chronic non-dysenteric colitis and amoeboma (Reed, 1992; Ravdin, 1995). Severe invasive disease affects up to 20% of patients. Patients with acute amoebic colitis (dysentery) usually present with a 1–2 week history of watery stools with blood or mucus, abdominal pain, tenesmus and fever. Fulminant colitis mainly occurs in children, pregnant women and patients on corticosteroids and is characterized by abdominal pain, profuse bloody diarrhoea and fever: mortality is over 50% (Li and Stanley, 1996). Chronic amoebic colitis may be indistinguishable from irritable bowel disease (IBD) and must be ruled out before treatment with corticosteroids for IBD commences. Amoeboma (asymptomatic lesion or symptomatic dysentery with a tender mass) is a localized chronic infection that occurs in the caecum or ascending colon and can be differentiated from carcinoma by biopsy. Invasive extraintestinal amoebiasis depends on the site infected and includes liver and brain abscesses, peritonitis, pleuropulmonary abscess, cutaneous and genital amoebic lesions.

## Pathogenicity and virulence

As its name suggests, *E. histolytica* has a lytic effect upon host tissue. Invasion of the intestinal mucosa of the caecum and colon by trophozoites is initiated by depletion of the protective mucus blanket and proteolytic disruption of tissue, causing lysis and necrosis of host cells and characteristic flask-shaped lesions (Tse and Chadee, 1991). Virulence factors, such as the galactose-specific binding lectin cause lesions and permit spread through the blood stream. Trophozoites of invasive *E. histolytica* are resistant to complement-mediated lysis which may facilitate extraintestinal invasion (Ravdin, 1990a). Invasion of the liver occurs when trophozoites ascend the portal venous system and cause hepatic necrosis and amoebic liver abscesses which contain proteinaceous debris. Amoebic lysis of neutrophils releases toxic non-oxidative products that contribute to the destruction of host tissue. Periportal inflammation can cause liver enzyme abnormalities in the absence of demonstrable trophozoites.

Adherence of trophozoites to the mucosa, epithelial cells and host inflammatory cells is important in disease pathogenesis. *In vitro* studies have shown that this is mediated by an adherence lectin, which acts in the presence of extracellular calcium ions (Ravdin *et al.*, 1988). Cytolytic activity is augmented by phorbol esters and protein kinase activators (Weikel *et al.*, 1988). An ionophore-like protein has been identified in *E. histolytica* which induces leaking of sodium, potassium and calcium ions and acts as a parasite defence mechanism against ingested bacteria (Leippe *et al.*, 1994) and the bacterial gut flora probably plays a role in virulence and ability to colonize a host. Several strains of *E. histolytica* have been identified and cytotoxic haemolysins encoded by plasmid DNA have been identified in the most pathogenic strains (Jansson *et al.*, 1994).

## Causation

Disease severity is increased in children, particularly neonates, pregnancy and post-partum states, with the use of corticosteroids, and during malignancy and malnutrition (Ravdin, 1988). Initial invasion of the mucosa is probably not related to immunity but the severity of disease. This is indicated by the exacerbation of intestinal amoebiasis by corticosteroid therapy, in infants and pregnant women. Acquired immunity appears to have a protective effect since recurrences of amoebic colitis or liver abscesses in endemic areas are rare (DeLeon, 1970) and the presence of serum antibodies is associated with a lower rate of intestinal infection (Choudhuri *et al.*, 1991), although asymptomatic infection is recurrent and serum antibodies are only present at low levels when infection is with non-invasive amoebae. The role of secretory immunoglobulins may be limited by the organism's ability to shed anti-amoebic antibodies (Arhets *et al.*, 1995), but cell-mediated immunity plays an important role in limiting the extent of invasive amoebiasis and in protection of recurrence. It is not known whether infection with a non-pathogenic species confers any protection from pathogenic species.

## Treatment

Since primary testing laboratories rarely differentiate *E. histolytica* from *E. dispar*, and research has shown that the vast majority of infections in Europe and North America are in fact *E. dispar* (Sargeaunt, 1987), there is generally no need for the administration of antiparasitic agents in patients without symptoms of amoebiasis. Even in parts of the world where invasive amoebiasis is prevalent, such as Central America, southern Africa and India, *E. dispar* infections still outnumber *E. histolytica* by 10:1 (Petri, 1996). WHO recommends that unequivocal differential diagnosis is made (see below), or that there is strong reason to suspect amoebiasis, before treatment (Anon, 1997). If asymptomatic *E. histolyica* infection is detected by discriminatory tests, this should also be treated because of the risk of progression to symptomatic infection and risk of transmission.

Luminal amoebicides, such as iodoquinol, paromomycin and diloxanide furoate are effective against organisms in the intestinal lumen, but are not highly effective against invasive disease. For intestinal disease a tissue amoebicide such as metronizadole or tinidazole should be followed by a luminal amoebicide because luminal parasites are not otherwise eliminated. For severe or refactory disease dehydroemetine followed by iodoquinol, paromomycin or diloxanide furoate are suitable. Non-surgical aspiration may be necessary for patients with liver abscesses if they continue to be febrile. Treatment of laboratory confirmed asymptomatic *E. histolytica* with luminal amoebicides is important in the control of the spread of infection, but chemoprophylaxis is never appropriate.

## Survival in the environment and in water

Although trophozoites may be passed from the body in stools, they are rapidly destroyed and, furthermore, if ingested would not survive the gastric environment. The protective cyst wall, however, ensures cyst survival, and the use of human faeces as fertilizer is an important source of infection. In a study of 107 'zir' stored drinking water and 11 tap water samples in a village in the Nile Delta, 55% zir stored waters and 63% tap water samples contained *E. histolytica*, although differentiation was not made between *E. dispar* (Khairy *et al.*, 1982)

Methods for evaluating the survival of *Entamoeba* cysts have been based on chemical staining, *in vitro* culture (measuring total population kill) or on counting excysted cysts from populations, on the assumption that unexcysted (intact) cysts were not viable and therefore did not excyst (Stringer, 1972). However, while there are other reasons for which excystation may not occur, this was considered unlikely significantly to alter results. Cysts can remain viable for as long as 3 months depending on conditions, and survive at 4°C for 12 weeks from egg slant cultures (Neal, 1974). Early studies showed that treatment at 68°C for 5 minutes and boiling for 10 seconds killed cysts (Boek, 1921; Mills *et al.*, 1925). Time/temperature studies by Myjak (1967) showed that cysts from culture were killed at 47°C for 25 minutes, 49°C for 11 minutes, 51°C for 3 minutes and 53°C for 1 minute. Cysts in faeces from infected carriers showed a decrease in the time required to kill cysts with increasing temperature, although higher temperatures (57°C for 1 minute) were required to achieve kill from all five isolates tested. Cysts are destroyed by hyperchlorination or iodination (Kahn and Visscher, 1975; Markell *et al.*, 1986) but are resistant to chlorine at levels used in public water supplies.

## Methods of detection

Since all *E. dispar* and the majority of *E. histolytica* infections are asymptomatic, cysts may be detected in faeces without consequence for the patient. Diagnosis of intestinal amoebiasis is by both signs and symptoms of the patient and microscopical examination of stool or mucosal biopsy. If stools cannot be examined fresh they should be preserved in polyvinyl alcohol prior to examination. Occult blood is usually present in faeces, but presence or absence of faecal leucocytes is non-contributory to the diagnosis due to the lytic effect of the organism. The formalin-ethyl acetate concentration method is most commonly used to concentrate stools and, in addition to examination of wet preparations and stool concentrate, permanent stains of fresh or preserved faecal specimens should be examined. Two or three repeat stools taken over a 10-day period may be required for diagnosis. The presence of haematophagus trophozoites in faeces and other specimens, or the detection of trophozoites in biopsy material, is highly predictive of invasive *E. histolytica* (Gonzales-Ruiz *et al.*, 1994). Identification can be difficult since confusion between *Entamoeba* and other intestinal protozoa, leucocytes or macrophages in stool samples may occur: a pseudo-outbreak was reported in California when one laboratory identified 38 cases in 3 months against a background about 1 per month (Garcia *et al.*, 1985). This highlights the need for laboratory proficiency schemes. The issue of differentiation of *E. histolytica* cysts from *E. dispar* is outwith the scope of most primary testing laboratories, and when diagnosis is made by light microscopy, results should be reported as '*E. histolytica/E. dispar*' (Anon, 1997).

Differentiation of *E. histolytica* cysts from the non-pathogenic *E. dispar* is necessary when symptoms of amoebiasis are present, unless there are other reasons to suspect *E. histolytica*. Differentiation can only be undertaken by isoenzymatic, immunological or molecular analysis of isolates. Zymodemes, representing isoelectrophoretic patterns of various enzymes (Sargeunt, 1987), differentiate *E. histolytica* and *E. dispar* but this method requires high numbers of organisms that can often only be generated by culture, which may present problems of competition if both species or other species are present. Cultivation can be in association with bacteria (Robinson, 1968) or axenically (Diamond, 1968). Optimal growth is at 35–37°C, pH 7.0 with reduced oxygen tension. However, culture can never exclude the presence of *E. histolytica*.

Serological tests can be helpful for cyst-positive, symptomatic patients, where a positive result indicates current or remote *E. histolytica* infection, since an antibody response is not usually elicited by *E. dispar* (Ravdin, 1990b; Caballero-Salcedo, 1994). However, usefulness is limited for diagnosis of acute infection in endemic areas where seroprevalence can be high (Caballero-Salcedo, 1994). However, IgM does not persist in serum and may provide a good target during the acute phase (Abd-Alla *et al.*, 1998). Antibody detection is also useful in patients with extraintestinal disease where parasites are generally not found on stool examination.

Detection in environmental samples requires recovery from the sample matrix by filtration, followed by differentiation from other protozoan cysts. Membrane filtration using 1.2 μm, 47 mm diameter filters, followed by elution of the captured organisms from the filter, concentration by centrifugation and examination of the sediment is a long-established technique (Chang and Kabler, 1956). Recovery is probably erratic since cysts may be lost in the process.

Genetic markers confirm the separation of the pathogenic species. Techniques such as hybridization of genomic DNA with various gene probes and PCR-based tests targeting various gene loci, including small subunit rRNA genes, have been described (Clark and Diamond, 1991; Farthing *et al.*, 1996; Troll *et al.*, 1997). However, these are only undertaken in specialist testing laboratories. Monoclonal antibodies directed at the galactose-specific adherence lectin showed some epitopes present on pathogenic isolates and absent in non-pathogenic isolates (Petri *et al.*, 1990) and an ELISA kit has been developed and marketed commercially that distinguishes between *E. histolytica* and *E. dispar* on the basis of antigens detected directly in stools (Haque *et al.*, 1998). However, PCR offers greater sensitivity than ELISA (Mirelman *et al.*, 1997) and is therefore probably more appropriate for epidemiological studies. In-house immunological methods have also been developed (Bhaskar *et al.*, 1996).

Differentiation between the two species has assisted greatly in understanding the epidemiology of amoebiasis, and is essential for subsequent control of transmission, but there remains a need for simple, cheap differential diagnostic tests to reduce unnecessary treatment. It is important also for prevalence and epidemiological studies for risk factors for invasive disease.

## Critical review of the epidemiology

Greater understanding of the epidemiology of amoebiasis has been achieved by the differentiation of pathogenic from non-pathogenic isolates, since *E. dispar* accounts for 90% of infections. Accurate prevalence data for *E. histolyitca* are rare and improved methods for specific detection, including those appropriate for use in developing countries are required. In a study in the Philippines using PCR, 7% of 1872 individuals carried *E. dispar* and 1% were infected by *E. histolytica* (Rivera *et al.*, 1998). Similar to serological studies in other endemic areas, peak prevalence was detected in children aged 5–14 years. In a study in Bangladesh, age-specific seroprevalence rose sharply in the 1–2 year olds, peaked at 14 years and showed an age-related decline thereafter (Hossain *et al.*, 1983). Detection in stool specimens in the same study showed the lowest rate in children <1 year, increasing with age. A high proportion of urban adults were infected, although differentiation between *E. histolytica* and *E. dispar* was not made.

Risk factors have been identified epidemiologically for both increased prevalence and severity of disease (Ravidin, 1995). The groups at risk of increased prevalence of amoebiasis in developed countries include promiscuous male homosexuals, travellers and recent immigrants, institutionalized populations and communal living. In areas of endemicity, increased prevalence is linked to lower socioeconomic status including factors such as crowding and lack of indoor plumbing. Increased severity of disease is linked to neonates, pregnancy and post-partum, use of corticosteroids, malignancy and malnutrition. Waterborne transmission is considered common in developing countries due to faecal contamination and lack of water treatment. Other factors include the use of human faeces and wastewater in agriculture (Bruckner, 1992). In a study of the health effects of the agricultural reuse of urban wastewater in Morocco, the combined risk of infection rate for *Giardia* and *Entamoeba* was in the region of 41% (Amhmid and Bouhoum, 2000).

Amoebiasis, although more prevalent, is not restricted to developing countries. Outbreaks have been reported in long-term care facilities in the USA (Nicolle *et al.*, 1996) and one waterborne outbreak was reported in the UK in the 1950s among service personnel. Transmission was thought to be via the sewage system from personnel who probably acquired the infection while serving overseas (Galbraith *et al.*, 1987). Outbreaks have occurred worldwide, including in the USA (Lippy and Waltrip, 1984), Scandinavia, Taiwan (Lai *et al.*, 2000) and Tblisi, Georgia (Kreidl *et al.*, 2000). The cause of the Taiwanese outbreak was suspected to be contamination of the water supply by patients who had visited an endemic area. In Tblisi, the drinking water supply was also suspected.

Food-borne amoebiasis is possible, from directly or indirectly contaminated produce. Protozoan parasites were detected in 52% of 500 fresh clinical stool specimens collected over a 1-year period in Nigeria (Nzeako, 1992). Of those infected, six (1%) were from the University of Nigeria community, 89 (18%) were urban dwellers and 166 (33%) from rural areas. The highest incidence of *Entamoeba histolytica* (28%) was found among the rural community. Parasitic infections were seasonal, and started in April of each year (onset of rainy season), peaked between July and August, and were the lowest between November and March (dry season). The green vegetable *Amaranthus viridans*, which gets polluted by the sewage oxidation pond at the locality, was identified as the main vehicle of infection.

In a survey of vegetables in Costa Rica, *Entamoeba histolytica* cysts were found in 6.2% (5/80) of cilantro leaves, in 2.5% (2/80) cilantro roots, in 3.8% (3/80) lettuce, in 2.5% (2/80) radish samples and at least a 2% incidence of this amoeba was found in other vegetable samples (carrot, cucumber, cabbage and tomatoes) (Monge and Arias, 1996). In a follow-up study, more protozoan parasite-positive samples (including *E. histolytica*) were found during the dry season, although the association was only significant ($P < 0.05$) in radish (*Raphanus sativus*) and cilantro leaves (Monge *et al.*, 1996).

## Risk assessment

Differentiation between pathogenic *Entamoeba* and non-pathogenic species is fundamental to the epidemiology and control of infection. The use of human waste has been identified as an important source of human infection. Person-to-person transmission is common, as are family clusters of cases, with asymptomatic carriers responsible for disease spread and transmission since exposure of susceptible hosts to asymptomatic pathogen-excreters places them at risk of infection (Bruckner, 1982). If the proportion of the former in the community is high, detection and thus control of the spread of infection becomes difficult. However, if the majority of asymptomatic infections is with a non-pathogenic species, and symptomatic and thus reported infection due to a pathogenic species rare, then control is possible and transmission of the pathogenic species can be controlled. Differentiation between pathogenic and non-pathogenic species is clearly essential to risk assessment.

Although no differentiation was made in their study, Amhmid and Bouhoum (2000) highlighted the potential health effects of the agricultural reuse of urban wastewater in Morocco, and in a survey of sewage effluent, parasites including *E. histolytica* were detected at high levels discharging into the La Palta river, which is used for recreation and drinking water abstraction, both with and without treatment.

Additional to risk assessment from exposure is the predisposition of the host to infection and increased severity of disease.

## Overall risk assessment

*Health effects*: occurrence of illness, degree of morbidity and mortality, probability of illness based on infection:

- Infection with *Entamoeba histolytica dispar* is common – up to 10% of the world's population becomes infected. It is most common in developing countries with poor hygiene. Outbreaks, however, have occurred world-wide.
- Infection with *E. histolytica* causes a disease called amoebiasis. Infection can be asymptomatic or cause relatively mild intestinal upset and diarrhoea. Amoebic dysentery is a severe form of amoebiasis that causes stomach pain, bloody stools and fever. Rarely, *E. histolytica* can invade other parts of the body such as the liver, lungs, or brain.
- It is estimated that 1 in 10 people who are infected with *E. histolytica* develop disease. Severe invasive disease affects up to 20% of cases.
- Disease severity is increased in children, particularly neonates, pregnancy and post-partum states, with the use of corticosteroids and during malignancy and malnutrition.

*Exposure assessment*: routes of exposure and transmission, occurrence in source water, environmental fate:

- Infected people shed cysts in their faeces. The infection is transmitted by the ingestion of faecally-contaminated food, water, or from person-to-person contact.
- Waterborne transmission is commonly associated with faecally-contaminated water supplies in developing countries.
- Like other cysts, *E. histolytica* cysts can survive for a long time in the environment, though they are susceptible to heat (about 60°C for 1 minute or boiling for 10 seconds).

*Risk mitigation*: drinking-water treatment, medical treatment:

- Cysts are destroyed by hyperchlorination or iodination but are resistant to chlorine at levels used in public water supplies. Waterborne outbreaks have occurred when potable water was contaminated with sewage. Because of their size, standard sedimentation, flocculation and filtration should handle cysts in treated drinking water.
- Amoebicides, such as iodoquinol, paromomycin, and diloxanide furoate are effective against organisms in the intestinal lumen, but are not highly effective against invasive disease. For intestinal disease, a tissue amoebicide such as metronidazole or tinidazole should be followed by a luminal amoebicide because luminal parasites are not otherwise eliminated.

# Future implications

WHO has identified *E. histolytica* as an organism to be controlled by vaccination (Anon, 1997), which would reduce the incidence of disease and control the source of human infection.

From their study of the parasitological quality of zir stored water and tap water in a village in the Nile Delta, Khairy and colleagues (1982) concluded that simply supplying water via taps was not enough to ensure a safe drinking water supply: sufficient taps are required, supplying water protected from pollution, which in turn requires wastewater and sewage disposal. Safe storage of water in rural areas also needs to be addressed. Similar conclusions were reached by Feachem and colleagues (1983) who investigated the excreta disposal facilities and intestinal parasitism in sub-Saharan Africa: the provision of improved water supply methods and sanitation in individual houses or small clusters of houses did not necessarily protect from infection where the overall faecal contamination of the environment was high.

# References

Abd-Alla, M.D., Jackson, T.G. and Ravdin, J.I. (1998). Serum IgM antibody response to the galactose-inhibitable adherence lectin of *Entamoeba histolytica*. *Am J Trop Med Hyg*, **59**: 431–434.

Amhmid, O. and Bouhoum, K. (2000). Health effect of urban wastewater reuse in a peri-urban area in Morocco. *Environ Mgmt Hlth*, **11**: 263–269.

Anon. (1997). WHO/PAHO/UNESCO Report of a consultation of Experts on Amoebiasis, Mexico City, Mexico, January 1997.

Arhets, P., Gounon, P., Sansonetti, P. *et al.* (1995). Myosin II is involved in capping and uroid formation in the human pathogen *Entamoeba histolytica*. *Infect Immun*, **63**: 4358–4367.

Bhaskar, S., Singh, S. and Sharma, M. (1996). A single-step immunochromatographic test for the detection of *Entamoeba histolytica* antigen in stool samples. *J Immunol Mthds*, **196**: 193–198.

Boeck, W.C. (1921). The thermal-death point of the human intestinal protozoan cysts. *Am J Hyg*, **1**: 365–387.

Bruckner, D.A. (1992). Amebiasis. *Clin Microbiol Rev*, **5**: 356–369.

Brumpt, M.E. (1925). Etude sommaire de l'*Entomoebae dispar* n. sp. *Bull Acad Méd (Paris)*, **94**: 942–952.

Brumpt, M.E. (1928). Differentiation of human intestinal amoebae with four-nucleated cysts. *Trans Roy Soc Trop Med Hyg*, **22**: 101–114.

Burrows, R.B. (1959). Morphological differentiation of *Entamoeba hartmanni* and *E. polecki* from *E. histolytica*. *Am J Trop Med Hyg*, **8**: 583–589.

Caballero-Salcedo, A. *et al.* (1994). Seroepidemiology of amebiasis in Mexico. *Am J Trop Med Hyg*, **50**: 412–419.

Chang, S.L. and Kabler, P.W. (1956). Detection of cysts of *Entamoeba histolytica* in tap water by the use of membrane filters. *Am J Hyg*, **64**: 170–180.

Choudhuri, G., Prakash, V., Kumar, A. *et al.* (1991). Protective immunity to *Entamoeba histolytica* infection in subjects with antiamoebic antibodies residing in a hyperendemic zone. *Scand J Infect Dis*, **23**: 771–776.

Clark, C.G. (1998). *Entamoeba dispar*, an organism reborn. *Trans Roy Soc Trop Med Hyg*, **92**: 361–364.

Clark, C.G. and Diamond, L.S. (1991). Ribosomal RNA genes of 'pathogenic' and 'non-pathogenic' *Entamoeba histolytica* are distinct. *Mol Biochem Parasitol*, **49**: 297–302.

Clark, C.G. and Diamond, L.S. (1993). *Entamoeba histolytica*: an explanation for the reported conversion of 'non-pathogenic' amebae to the 'pathogenic' form. *Exp Parasitol*, **77**: 456–460.

Councilman, W.T. and LaFleur, H.A. (1891). Amoebic dystentery. *Johns Hopkins Hosp Rep*, **2**: 395–548.

DeLeon, A. (1970). Prognostico tardio en el absceso hepatico amibiano. *Arch Invest Med (Mexico)*, **1**(1): 205–206.

Diamond, L.S. (1968). Techniques of axenic cultivation of *Entamoeba histolytica* Schaudinn, 1903 and *Entamoeba*-like amoeba. *J Parasitol*, **54**: 1047–1056.

Diamond, L.S. and Clark, C.D. (1993). A redescription of *Entamoeba histolytica* Schaudinn, 1903 (emended Walker, 1911) separating it from *Entamoeba dispar* Brumpt, 1925. *J Eukaryotic Microbiol*, **40**: 340–344.

Dobell, C. (1928). Research on the intestinal protozoa of monkeys and man. *Parasitology*, **20**: 357–412.

Farthing, M.J.G., Cavellos, A.M. and Kelly, P. (1996). Intestinal protozoa. In *Manson's Tropical Diseases*, Farthing, M.J.G. (ed.)

Feachem, R.G., Guy, M.W., Harrison, S. *et al.* (1983). Excreta disposal facilities and intestinal parasitism in urban Africa: preliminary studies in Botswana, Ghana and Zambia. *Trans Roy Soc Trop Med Hyg*, **77**: 515–521.

Galbraith, N.S., Barrett, N.J. and Stanwell-Smith, R. (1987). Water and disease after Croydon. *J Inst Water Environ Mngmt*, **1**: 7–21.

Garcia, L., Sorvillo, F., Epstein, M. *et al.* (1985). Epidemiological notes and reports. Pseudo-outbreak of intestinal amoebiasis. California. *MMWR*, **34**: 125–126.

Gonzalez-Ruiz, A., Haque, R., Aguirre, A. *et al.* (1994). Value of microscopy in the diagnosis of dysentery associated with invasive *Entamoeba histolytica*. *J Clin Microbiol*, **47**: 236–239.

Haque, R., Ali, I.K.M., Alkther, S. *et al.* (1998). Comparison of PCR, isoenzyme analysis, and antigen detection for diagnosis of *Entamoeba histolytica* infection. *J Clin Microbiol*, **36**: 449–452.

Hoare, C.A. (1962). Reservoir hosts and natural foci of human protozoal infection. *Acta Trop*, **19**: 281–317.

Hossain, M.M., Ljungstrom, I., Glass, R.I. *et al.* (1983). Amoebiasis and giardiasis in Bangladesh: parasitological and serological studies. *Trans Roy Soc Trop Med Hyg*, **77**: 552–554.

Jansson, A., Gillin, F., Kagardt, U. *et al.* (1994). Coding of hemolysins within the ribosomal RNA repeat on a plasmid in *Entamoeba histolytica*. *Science*, **263**: 1440–1443.

Kahn, F.H. and Visscher, B.R. (1975). Water disinfection in the wilderness – a simple method of iodination. *West J Med*, **122**: 450–453.

Khairy, A.E.M., El Sebaie, O.E., Gawad, A.A. *et al.* (1982). The sanitary condition of rural drinking water in a Nile Delta village. *J Hyg*, **88**: 57–61.

Leippe, M., Andra, J. and Muller-Eberhard, H.J. (1994). Cytolytic and antibacterial activity of synthetic peptides derived from amoebapore, the pore-forming peptide of *Entamoeba histolytica*. *Proc Natl Acad Sci USA*, **91**: 2602–2060.

Levine, R.L. and Armstrong, D.E. (1970). Human infection with *Entamoeba polecki*. *Am J Clin Pathol*, **54**: 611–614.

Li, E. and Stanley, S.L. Jr (1996). Protozoa: Amebiasis. *Gastroenterol Clin North Am*, **25**: 471–492.

Lippy, E.C. and Waltrip, S.C. (1984). Waterborne disease outbreaks – 1946–1980: a thirty year perspective. *J Am Waterworks Assoc*, **76**: 60–67.

Lösch, F. (1975). MassenhafteEntwickelung von Amöben im Dickdarm. *Arch Pathol Anat Physiol Klin Med Rudolf Virchow*, **65**: 196–211.

Lyons, T., Scholten, T., Palmer, J.C. *et al.* (1983). Oral amoebiasis: the role of *Entamoeba gingivalis* in periodontal disease. *Quint Int*, **14**: 1245–1248.

Markell, E.K., Voge, M. and John, D.T. (1986). *Medical Parasitology*. Philadelphia: WB Saunders Co.

Mills, R.G., Bartlett, C.L. and Kessel, J.F. (1925). *Am J Hyg*, 5: 559–563.

Mirelman, D., Nuchamowitz, Y. and Stolarsky, T. (1997). Comparison of use of enzyme-linked immunosorbent assay-based kits and PCR amplification of rRNA genes for simultaneous detection of *Entamoeba histolytica* and *E. dispar*. *J Clin Microbiol*, 35: 2405–2407.

Monge, R. and Arias, M.L. (1996). Presence of various pathogenic microorganisms in fresh vegetables in Costa Rica. *Arch Latinoam Nutr*, 46: 292–294.

Monge, R., Chinchilla, M. and Reyes, L. (1996). Seasonality of parasites and intestinal bacteria in vegetables that are consumed raw in Costa Rica. *Rev Biol Trop*, 44: 369–375.

Myjak, P. (1967). The effect of temperature on the survival rate of *Entamoeba histolytica* (Schaudinn, 1903) cysts in water. 18: 35–42.

Neal, R.A. (1974). Survival of Entamoeba and related Amoebae at low temperature. I. Viability of Entamoeba cysts at 4 degrees C. *Int J Parasitol*, 4: 227–229.

Nicolle, L.E., Strausbaugh, L.J. and Garibaldi, R.A. (1996). Infections and antibiotic resistance in nursing homes. *Clin Microbiol Rev*, 9: 1–17.

Nzeako, B.C. (1992). Seasonal prevalence of protozoan parasites in Nsukka, Nigeria. *J Commun Dis*, 24: 224–230.

Petri, W.A. (1996). Recent advances in amoebiasis. *Crit Rev Clin Lab Sci*, 33: 1–37.

Petri, W.A., Jackson, T.F., Gathiram, V. *et al.* (1990). Pathogenic and non-pathogenic strains of *Entamoeba histolytica* can be differentiated by monoclonal antibody to the galactose specific adherence lectin. *Infect Immun*, 58: 1802–1806.

Ravdin, J.I. (1988). Intestinal disease caused by *Entamoeba histolytica*. In *Amebiasis: Human Infection by* Entamoeba histolytica, Ravdin, J.I. *et al.* (eds). New York: Churchill Livingstone, pp. 495–509.

Ravdin, J.L. (1990a). Pathogenic mechanisms, human immune response, and vaccine development. *Clin Enatamoeba histolytica Res*, 38: 215–225.

Ravdin, J.L. (1990b). Association of serum antibodies to adherence lectin with invasive amebiasis and asymptomatic infection with pathogenic *Entomoeba histolytica*. *J Infect Dis*, 162: 768–772.

Ravdin, J.I. (1995). Amebiasis. *Clin Infect Dis*, 20: 1453–1466.

Ravdin, J.I., Moreau, F., Sullivan, J.A. *et al.* (1988). The relationship of free intracellular calcium ions to the cytolytic activity of *Entamoeba histolytica*. *Infect Immun*, 56: 1505–1512.

Reed, S.L. (1992). Amebiasis: an update. *Clin Infect Dis*, 14: 385–393.

Rivera, W.L., Tachibana, H. and Kanbora, H. (1998). Field study on the distribution of *Entamoeba histolytica* and *Entamoeba dispar* in the Northern Philippines as detected by the PCR. *Am J Trop Med Hyg*, 59: 916–921.

Robinson, G.L. (1968). The laboratory diagnosis of human parasitic amoebae. *Trans Roy Soc Trop Med Hyg*, 62: 285–294.

Sargeaunt, P.G. (1987). The reliability of *Entamoeba histolytica* zymodemes in clinical diagnosis. *Parasitol Today*, 3: 40–43.

Sargeaunt, P.G., Williams, J.E. and Grene, J.D. (1978). The differentiation of invasive and non-invasive *Entamoeba histolytica* by isoenzyme electrophoresis. *Trans Roy Soc Trop Med Hyg*, 72: 519–521.

Strachan, W.D., Spice, W.M., Chiodini, P.L. *et al.* (1988). Immunological differentiation of pathogenic and non-pathogenic isolates of *Entamoeba histolytica*. *Lancet*, i: 561–563.

Stringer, R.P. (1972). New bioassay system for evaluating per cent survival of *Entamoeba histolytica* cysts. *J Parasitol*, 58: 306–310.

Tannich, E., Horstmann, R.D., Knoblock, J. *et al.* (1989). Genomic DNA differs between pathogenic and non-pathogenic *Entamoeba histolytica*. *Proc Natl Acad Sci USA*, 86: 5118–5122.

Troll, H., Marti, H. and Weiss, N. (1997). Simple differential detection of *Entamoeba hisolytica* and *Entamoeba dispar* in fresh stool specimens by sodium acetate-acetic acid-formalin concentration and PCR. *J Clin Microbiol*, 35: 1701–1705.

Tse, S.K. and Chadee, K. (1991). The interaction between intestinal mucus glycoproteins and enteric infections. *Parasitol Today*, 7: 163.

Walsh, J.A. (1998). Prevalence of *Entamoeba histolytica* infection. In *Amebiasis: Human Infection by* Entamoeba histolytica, Ravdin, J.I. (ed.). New York: Wiley, pp. 93–105.

Weikel, C.S., Murphy, C.F., Orozco, M.E. *et al.* (1988). Phorbol esters specifically enhance the cytolytic activity of *Entamoeba histolytica. Infect Immun*, 56: 1485–1491.

## 21

# *Giardia duodenalis*

## Basic microbiology

*Giardia* are bi-nucleate flagellated protozoa (Phylum *Sarcomastigophora*, Order *Diplomonadida*, Family *Hexamitidae*). Although the taxonomy of the genus is uncertain, three morphological types have been identified: *Giardia duodenalis* (*syn. lamblia*, *syn. intestinalis*) which is found in humans and many other mammals, birds and reptiles, *G. muris* which is found in rodents, birds and reptiles and *G. agilis* found in amphibians (Filice, 1952). *G. duodenalis* is the only type known to infect humans and indeed most other domesticated animals. Differentiation has been historically on the basis of size and morphology of cysts and trophozoites, size of the ventral adhesive disc relative to the length of the cell, and the shape of median bodies (Kulda and Nohýnková, 1996). Two further types have been described on the basis of ultrastructural features: *Giardia ardeae* (Erlandsen *et al.*, 1990) and *Giardia psittaci* (Erlandsen and Bemrick, 1987), and a further type proposed on the basis of cyst morphology and ss rRNA sequence analysis (Feely, 1988; van Keulen *et al.*, 1998). DNA analyses have confirmed differences between types (van Keulen *et al.*, 1991): *G. muris* is distant from *G. duodenalis*, *G. ardeae* is closer to *G. duodenalis* than to *G. muris* and *G. microti* is similar to *G. duodenalis* genotypes. *G. duodenalis* appears to be a clade with multiple

**Table 21.1** *Giardia duodenalis* genotypes

| Genotype | Subgroup | Host range |
|---|---|---|
| Assemblage A (syn. 'Polish' or 'Group 1/2') | All | Humans, livestock, dogs, cats, beavers, guinea-pigs, slow loris |
| | All | Humans |
| Assemblage B (syn. 'Belgian' or 'Group 3') | | Humans, slow loris, chinchillas, dogs, beavers, rats, siamang |
| Dog | | Dogs |
| Cat | | Cats |
| Hoofed animal | | Cattle, goats, pigs, sheep, alpaca |
| Rat | | Domestic rats |
| Wild rodents | | Muskrats, voles |

Adapted from Thompson, 2000

genotypes, variation having been demonstrated by isoenzyme and DNA analyses (Meloni *et al.*, 1995; Monis *et al.*, 1999). Genetic groups or 'assemblages' of *G. duodenalis* appear to have varying host-specificity (Thompson *et al.*, 2000) (Table 21.1), providing valuable information to further understanding of the epidemiology and sources of infection and provide better risk assessment (see below). However, *G. duodenalis* assemblages are highly divergent, possibly representing distinct species (Monis *et al.*, 1999) and, increasingly, genetic analyses illustrate the inadequacy of the current taxonomy within the genus.

*Giardia* has a two-stage life cycle, consisting of flagellate trophozoites (12–18 μm by 5–7 μm) and ellipsoid-shaped cysts (9–12 μm). Infection is non-invasive and follows the ingestion of cysts in contaminated water, food or by direct faecal-oral transmission. Excystation occurs in the small intestine when two trophozoites are released from each cyst. Excystation is probably stimulated by the acid environment of the stomach and can be induced *in vitro* by exposure to low pH (Bingham and Meyer, 1979), but has also been observed in pH neutral aqueous solutions and in patients with achlorhydria (Boucher and Gillin, 1990). Excystation has also been induced by 0.3 M sodium bicarbonate, suggesting that pancreatic secretions also have a role (Feely *et al.*, 1991). Emergence of the trophozoites is stimulated by chymotrypsin, trypsin and pancreatic fluid (Boucher and Gillin, 1990). Trophozoites are either free within the lumen of the duodenum or jejunum, or attached to the surface of mucosal enterocytes by a ventral attachment disc (Inge *et al.*, 1986). Adherence of trophozoites is optimal at 37°C and dependent on structures including the lateral crest and ventrolateral flange and a combination of forces within the gut lumen (Céu Sousa Gonçalves *et al.*, 2001). The trophozoites multiply by longitudinal binary fission in the lumen of the proximal small intestine. As the parasites pass from the small intestine through to the colon they mature to form resistant cysts, which are shed in the faeces, and are often detected in non-diarrhoeic stools. Encystation has also been induced *in vitro* under optimal conditions of bile salts and glycocholate with myristic acid at pH 7.8 (Gillin *et al.*, 1988). The cysts contain four nuclei and cytokinesis results in the development of two binucleated trophozoites immediately or shortly after being passed in the stool cysts can be transmitted either directly or via a vehicle such as contaminated food or water. Although trophozoites can also be shed in the faeces, they are not robust and are non-infectious (Meyer and Jarroll, 1980).

*Giardia*, along with *Entamoeba*, is one of the few eukaryotes that relies on fermentative metabolism of carbohydrates. Glucose is incompletely oxidized to ethanol, acetate and $CO_2$, and energy is obtained by substrate-level phosphorylation (Lindmark, 1980; Jarroll *et al.*, 1981). Trophozoite metabolism is affected by oxygen concentration and although minimal amounts of oxygen alter the metabolic pathway, trophozoites metabolize under anaerobic or microaerophilic conditions.

*Giardia* has a world-wide distribution but infection is more prevalent in warm climates. In developed countries, detection rates of 2–5% have been reported (Kappus *et al.*, 1994). In developing countries, carriage rates of 20–30% have been detected and is particularly prevalent in children (Bryan *et al.*, 1994). WHO figures show that some 200 million people in Asia, Africa and Latin America have giardiasis and the annual incidence is 500 000 cases. Many of the biological features of *Giardia* affect its transmission and epidemiology. Although transmission can occur between hosts, the life cycle is simple and dual or multiple hosts are not required for completion. The cysts are infective immediately or soon after being shed in the faeces, and direct host-to-host transmission can occur. The infective dose in experimental infections was 10–25 cysts (Rendtorff, 1954).

Recognized routes of transmission to humans include faecal-oral spread (including in child day-care centres and male homosexual contact) (Meyers *et al.*, 1977; Rauch *et al.*, 1990), drinking and recreational waters (Hunter, 1997; Marshall *et al.*, 1997), and consumption of contaminated food (Slifko *et al.*, 2000). Cysts are resistant to some degree to many environmental pressures, enabling the survival and environmental transmission of the parasite.

## Origin of the organism

Organisms now recognized as *G. duodenalis* were first described by Leeuwenhoek in 1681, who, on inspection of his own faeces, described their size, shape and morphology, noting their presence and link with his own diarrhoeal symptoms and dietary habits. The organism was formally described by Lambl in 1859, who named it *Cercomonas intestinalis*. While a number of genus and species names were proposed over the next 40 or so years, *G. lamblia* and *G. enterica* were proposed in 1915 and 1920 respectively (Kofoid and Christensen, 1915, 1920). However, *G. duodenalis* and *G. intestinalis* are more accurate taxonomically, although the former may be preferred since it conforms with Filice's morphologically-based nomenclature. The classification of *Giardia* remains controversial and difficult, however, since there are no well-defined criteria for species-designation for clonal organisms in the same clade. Early descriptions assigning species-status according to host overestimated the number of species, while those based solely on morphological differences probably underestimate the number. It is unfortunate that some medical literature has advocated the use of the term *G. lamblia* for *Giardia* isolated solely from humans, while genetic studies are demonstrating some of the genotypes present in humans are also found in animals (Thompson *et al.*, 2000).

Although the cyst and trophozoite flagella and nuclei were described in the 19th century, the most complete description of *Giardia* morphology was made by Filice in 1952. Dobell had identified the intestinal tract as the natural habitat for growth and reproduction in 1932, but universal acceptance of *Giardia* as a human pathogen took some time because many human infections are asymptomatic and the parasite is non-invasive (Erlandsen and Meyer, 1984). Clinical studies during the 1970s described the pathology of human infection (Kulda and Nohýnková, 1978) and in 1981 *Giardia* was included in the WHO list of parasitic pathogens (Anon, 1981). The recognition of waterborne outbreaks of giardiasis in Europe and the USA during the 1970s and 1980s (Craun, 1986) stimulated research into the epidemiology, pathogenesis and treatment of the disease.

## Clinical features

Clinical features vary widely in severity from asymptomatic carriage to severe diarrhoea and malabsorption (Meyer and Jarroll, 1980). This variability may be due to differences in pathogenicity of different isolates as well as differences in host responses (see below). In human volunteer studies, the prepatent period was 6–10 days when fresh cysts were used and 13 days using stored cysts (Rendtorff, 1954). In natural infections the pre-patent period can be considerably longer, and acute giardiasis usually lasts 1–3 weeks. Symptoms include diarrhoea, abdominal pain, bloating, flatulence, malaise, sulphurous belching, nausea and vomiting. The acute phase is usually resolved spontaneously in immune-competent individuals, but in some cases can develop into a chronic phase during which symptoms relapse in short recurrent bouts (Wolfe, 1990). Immunodeficient patients, particularly those with hypogammaglobulinaemia, often suffer chronic disease (see below).

In chronic cases the symptoms are recurrent and, if untreated, malabsorption and debilitation may occur. Untreated acute giardiasis lasts for at least 10 days and can last for 3 months or even years. *Cryptosporidium* and *Giardia* have been associated with persistent diarrhoea in children in developing countries (Lima *et al.*, 2000), where a cycle of malnourishment and continuing diarrhoea can be established and lead to poor growth rates and a substantial risk of death (Guerrant *et al.*, 1992; Fraser *et al.*, 2000). Other chronic sequelae of *Giardia* infection also include reactive arthritis (Tupchong *et al.*, 2001).

Asymptomatic cases can shed cysts in their faeces and present an important reservoir of infection, particularly in families, child-care settings and institutions, and may impact on cases in the community (Polis *et al.*, 1986).

## Pathogenicity and virulence

Enteropathy appears, in animal models, to involve enterocyte and micro-villous damage, villous atrophy and crypt hyperplasia, although normal appearance and function is restored on clearance of the infection (Ferguson *et al.*, 1990). Subsequent to brush border damage, reduction of the enzyme disaccharidase may occur. Adherence of trophozoites to the surface of mucosal enterocytes may provide sufficient suction to damage the microvilli, causing these histopathological changes and interfering with nutrient absorption (Inge *et al.*, 1988). Trophozoites multiply rapidly and can become so numerous as to create a physical barrier between the epithelial cells and the lumen, causing brush border enzyme deficiencies and impaired absorption. A relationship has been demonstrated between the amount of villous damage and malabsorption in animal infection (Wright and Tomkins, 1978). Cytopathic substances such as thiol proteinases and lectins may be released from the parasite and stimulation of the host immune response causes release of cytokines and mucosal inflammation, which have a pathological effect (Farthing, 1997). The role of inflammatory cells in giardial diarrhoea is not understood, since polymorphonuclear leucocytes and eosinophils have not been consistently detected in patients (Yardley, 1964; Brandborg *et al.*, 1967).

Although rarely invasive, human infection with *Giardia* has been reported invading tissues of the gallbladder and urinary tract (Meyers *et al.*, 1977; Goldstein *et al.*, 1978) and has been demonstrated in a mouse model (Owen *et al.*, 1979). However, approximately 50% of *Giardia* infections are asymptomatic, and risk factors for disease involve both host and parasite factors.

Genotypic and antigenic variation is marked between *Giardia* isolates (Nash *et al.*, 1987) and antigenic variation is important not only in defining immune response but also in the development of serological tests. Antigenic variation occurs as a result of expression of different surface proteins in populations of trophozoites in axenic culture, animal models and human volunteers. Variation may arise from immune responses in the host, adaptation to different environments or induced shift during excystation. The advantage for the parasite is that it promotes evasion of the host immune defences and enables survival in different environments. Ingestion of as few as 10–25 cysts can cause human disease (Rendtorff, 1954), but strain variation has been identified in variable infectivity results when two human isolates were used in human volunteer studies (Nash *et al.*, 1987). Thus it is thought that genetic differences may confer virulence and variation within *G. duodenalis* and may even explain some of the discrepant results obtained in infectivity/viability studies.

## Causation

Variation in manifestation of clinical symptoms includes the virulence of the *Giardia* strain, the number of cysts ingested, the age and immunocompetence of the host. Both antibodies and T cells appear to be required to control *Giardia* infections and giardiasis has been associated with nodular lymphoid hyperplasia, in which B and T cell functions are decreased. Interestingly, while the incidence of giardiasis is no different in AIDS patients than in people with an intact immune system, some such individuals may experience severe illness and prolonged or combination therapy may be required to clear the infection. More cases are, however, reported among hypogammaglobulinaemic patients (Wright *et al.*, 1977), who also appear to experience greater villous damage (Ferguson *et al.*, 1990). Their infections can be difficult to treat and may require prolonged courses of therapy.

## Treatment

Supportive treatment by fluid and electrolyte replacement is important in the treatment of giardiasis, but therapeutic agents are available to clear the infection and have been extensively reviewed by Gardner and Hill (2001). The nitroimidazoles (metronidazole, tinidazole, ornidazole and secnidazole) are the main agents used, while the nitrofuran compound furazolidone, the aminoglycoside paromomycin and benzimidazoles (i.e. albendazole) are also prescribed. Furazolidone is available in a liquid suspension and has therefore been advocated for paediatric treatment. Mepacrine is an alternative when nitroimidazole drugs are contraindicated or have failed. Treatment failure has been reported with all the drugs commonly used to treat giardiasis and may occur because of drug resistance, non-compliance with the treatment regimen, re-infection or immunological deficiency. Post-*Giardia* lactose intolerance may cause an apparent recurrence of symptoms (Duncombe *et al.*, 1978). Treatment of asymptomatic carriers may be suggested for the prevention of onward transmission, for example in households or in outbreak settings such as institutions. In endemic areas this has to be balanced with the risk of re-infection, although it may be desirable since clearance of the infection may allow catch-up growth in children of marginal nutritional status.

## Survival in the environment

Although the trophozoites are not robust, cysts survive in aqueous environments and show greater resistance to UV, chlorine and ozone than bacteria or viruses but, with the exception of UV, generally less resistance than *Cryptosporidium*.

*Giardia* cysts have been detected in a variety of environmental matrices, including dairy farm runoff, sewage plant effluent and sewer outflows (States *et al.*, 1997). Removal can be achieved by clarification and filtration (Sykora *et al.*, 1991), although recycling of filter back wash water has been identified as a means of reintroducing cysts into water supplies (States *et al.*, 1997). *Giardia* cysts have been detected in sewage and sewage-contaminated drinking water in La Palta, Argentina (Basualdo *et al.*, 2000) and studies in Scotland have shown that, although cysts are removed by both activated sludge and trickle filter treatment, they may occur in treated sewage (Robertson *et al.*, 2000). Payment *et al.* (2001) reported 75% removal from urban wastewater treatment by physicochemical treatment.

While the studies described above are for detection of cysts, methods for assessing viability are yet to be standardized. Cyst survival can either be measured as infectivity in a suitable host (usually mouse infectivity) or a surrogate marker for viability (using vital or fluorogenic dyes, or *in vitro* excystation). Although vital dye exclusion tests, particularly eosin, have been historically applied, comparison tests found that eosin exclusion consistently underestimated cyst viability compared with *in vitro* excystation (Bingham *et al.*, 1979). Comparison between eosin exclusion, *in vitro* excystation and experimental infection has shown lack of correlation (Kasprzak and Majewska, 1983). However, there may be additional influences that may inhibit *in vitro* culture that do not affect excystation, and workers recognize that the latter may at least provide a simple method for viability determination (Hoff *et al.*, 1985). Propidium iodide (PI) has more recently been used to assess *Giardia* viability, since the uptake of this vital dye indicates cell death or damage. However, while good positive correlation was found between PI staining and lack of excystation in the evaluation of cysts exposed to heat and quaternary ammonium compounds, there was no correlation when cysts were exposed to chlorine or monochloramines (Sauch *et al.*, 1991). Similarly fluorogenic dyes were found to overestimate viability compared to mouse infectivity when cysts were treated with ozone, although good correlation was observed between mouse infectivity and *in vitro* excystation (Labatiuk *et al.*, 1991). *In vitro* excystation was also found to have a good correlation with mouse infectivity for assessing the viability of cysts treated with chlorine (Hoff *et al.*, 1985). Although Smith and Smith (1989) also found that PI consistently overestimated viability by underestimating dead cysts, a suitable combination of PI and cyst morphology was suggested as a good marker for viability. Bukhari *et al.* (1999) identified problems in the use of surrogate indicators of viability compared with animal models and suggested that the latter be used as the gold standard.

Detection of respiratory activity has been used to measure the effects of drug susceptibility of trophozoites (Sousa and Poiares-Da-Silva, 1999), and has been applied as a test to determine enzyme redox activity of intact cysts in environmental isolates (Iturriaga *et al.*, 2001). Electrorotation has also been used to determine *Giardia* cyst viability, but application to environmental samples has yet to be undertaken (Dalton *et al.*, 2001). Amplification of total

RNA has been used as an indicator of *Giardia* viability (Mahbubani *et al.*, 1991). In a further refinement, test conditions have been used to induce the synthesis of stress-induced proteins, such as heat-shock proteins, and PCR to amplify the induced mRNA (Abbaszadegan *et al.*, 1997). This method correlated with PI staining, and since the half-life of mRNA is relatively short it is unlikely that residual mRNA is detected.

## Survival in water

*Giardia* cysts have been detected in surface waters in both urban and rural areas. LeChavallier *et al.* (1991a) studied waters in 14 American states and one Canadian province and found a greater proportion positive and higher concentration of cysts in urban areas. The same authors also found that 17% filtered drinking water supplies in the same study areas also contained *Giardia* cysts (LeChavallier *et al.*, 1991b). A Canadian study found *Giardia* in one-third of surface waters in remote rural areas and also found cysts in 17% drinking water supplies to one city, and in sewage (Roach *et al.*, 1993). Other Canadian studies have also found 21% raw waters and 18% finished waters (Wallis *et al.*, 1996) in 72 municipalities across Canada contained cysts but with wide geographical (watershed/catchment) variation. A detailed study in British Columbia, Canada found a high sample prevalence of *Giardia* oocysts (Isaac-Renton *et al.*, 1996), and was followed up by a longitudinal study of the effects of watershed management on parasite concentrations (Ong *et al.*, 1996), which showed that similarly high prevalence of *Giardia* cysts were detected, with a distinct seasonal variation, peaking in the winter months. Sampling in relation to agricultural activity in the catchments showed a significant increase in the number of cysts detected from sample points below cattle ranching than above. However, genotypic and phenotypic variation was observed between cattle, water and human isolates within the catchment indicating a variety of sources. Groundwaters have been found to contain *Giardia* cysts, and are vulnerable where they are under the influence of surface water or other sources of contamination (Moulton-Hancock *et al.*, 2000).

Studies have also been undertaken of *Giardia* cysts in surface waters and finished waters elsewhere. For example, in Russia, raw river water from the Sheksna River was found to contain *Giardia* cysts at a mean of 2 oocysts/100 l while finished water contained a mean of 1.6 per 10 000 litres (Egorov *et al.*, 2002). In the same study 26/87 (30%) source surface waters throughout the region contained *Giardia* cysts while they were detected in 5/70 (7%) finished water samples. In Japan, *Giardia* cysts were detected in 12/13 raw water samples (92%) compared with 3/26 (12%) treated water samples (Hashimoto *et al.*, 2002).

While some studies have shown a lack of detection in finished waters (Rose *et al.*, 1991), others have consistently detected cysts in the region of 2–5% samples (Hancock *et al.*, 1996; Rosen *et al.*, 1996). Observation of naturally occurring cysts has shown a log removal of *Giardia* cysts. For example, 1.54 log removal based on low initial concentrations of cysts in the raw water was reported by States *et al.* (1997). In contrast, a 3.26 log removal of *Giardia* cysts was achieved following experimental seeding of a full-scale conventional water treatment plant (Nieminski and Ongerth, 1995). The US EPA Surface Water Treatment Rule, 1989, requires that a 3 log reduction (99.9%) is achieved in *Giardia* cysts by a combination of removal and disinfection. Numbers of cysts detected may vary not only due to natural fluctuations but also according to the methods used for sampling and detection. Extreme runoff conditions can result in increased parasite contamination and need to be accounted for (Kistemann *et al.*, 2002).

*Giardia* cysts have been shown to maintain viability for up to 3 months in cold raw water sources and in tap water, with a range of 75–99.9% natural die-off (deReigner *et al.*, 1989). Although *Giardia* cysts can be inactivated by chlorine they do show some resistance (Jarroll *et al.*, 1981), and can survive water treatment, particularly where chemical dosing levels are unreliable. *Giardia* cysts have been shown to be sensitive to UV light at low doses under laboratory experimental conditions since a 2 log inactivation was observed at a $3\,\text{mJ/cm}^2$ dose (Modifi *et al.*, 2002) and 3 log inactivation at $20–40\,\text{mJ/cm}^2$ (Campbell and Wallis, 2002), both studies using an animal model for viability/infectivity assessment. In bench-scale trials, ozone has been evaluated but it was shown that to achieve a 2 log inactivation a Ct value 2.4 times that recommended under the SWTR was required (Finch *et al.*, 1993). Electroporation has been applied to both *Cryptosporidium* and *Giardia* to increase the permeability of the (oo)cyst wall and reduce the parasite's resistance to disinfection. Haas and Aturaliye (1999) reported that electroporation enhanced the effect of free chlorine and permanganate on *Giardia* in phosphate buffer. However, further evaluation needs to be undertaken in low conductivity waters and to optimize conditions for disinfection.

The effectiveness of water disinfection methods used to treat small volumes of drinking water was assessed by excystation by Jarroll and colleagues (1980), who found that, while saturated iodine, bleach, Globaline, tincture of iodine were all effective in both cloudy and clear water at 20°C, in cloudy water at 3°C saturated iodine failed to inactivate cysts and in clear water at the same temperature all methods failed. Commercial phenol-based disinfectants (Pine Glo, ammonia and Dettol) have been shown to inactivate 99% *Giardia* cysts following a 1-minute contact time while Pine Sol and vinegar were less effective (Lee, 1992). Sufficient residual and/or contact time are critical to the disinfection process, as was demonstrated by Ongerth *et al.* (1989) who found that iodine-based disinfectants were more effective than chlorine-based disinfectants. Although all failed to achieve 99.9% reduction in viability (measured by animal infectivity) after 30 minutes, iodine-based disinfection was effective after 8 hours. Heating water to 70°C for 10 minutes was effective. However, Gerba *et al.* (1997) reported that iodine tablets were effective in 'general case water' at pH 9 but relatively ineffective at pH 5.

## Methods of detection

Clinical diagnosis is by the identification of cysts or, during heavy infections, trophozoites in faeces either by direct examination of wet mounts or following concentration (Isaac-Renton, 1991). Organisms may also be detected in duodenal biopsy material or aspirates in difficult cases. Histological or immunofluorescent stains are also used to visualize the organisms and enzyme immunoassays are available for the detection of antigens. Cysts may be shed intermittently and so more than one stool may need to be examined for diagnosis and screening of more than one stool can significantly increase the chance of making a diagnosis (Cartwright, 1999). Screening of asymptomatic contacts of cases is desirable to control the spread of infection in some settings, for example during outbreaks in child-care centres and other institutions.

Low numbers of cysts in stools and inexperience of laboratory technicians can also contribute to poor sensitivity in microscopical detection (Wolfe, 1990). While some enzyme immunoassay kits are more sensitive than others, many show a greater sensitivity than microscopy on examination of a single stool specimen (Hanson and Cartwright, 2001) or when treatment is already underway (Nash *et al.*, 1987; Knisley *et al.*, 1989). However, antigenic variation may result in negative results from some antibody-based methods (Moss *et al.*, 1990).

Methods for detection in environmental samples require recovery of the cysts from the sample matrix (usually by filtration), concentration (by centrifugation and immunomagnetic separation) and staining with immunofluorescent antibodies prior to inspection by UV microscopy. Large volumes of samples are required for representative sampling and to recover low concentrations of parasites. Methods are detailed in the USEPA method 1623 (Anon, 1999). Development of alternative methods includes continuous flow centrifugation which has improved recovery of waterborne pathogens from water samples (Borchardt and Spencer, 2002) with recoveries over 90% reported, and membrane dissolution procedures (McCuin *et al.*, 2000).

The development of IMS and IFAs has improved the isolation and detection of organisms such as *Giardia* and *Cryptosporidium* from water and environmental samples, including food and enhanced our evaluation of water- and food-borne risks. Using IMS/IFA and improved washing procedures, Robertson and Gjerde (2000) reported mean recovery efficiencies for *Giardia* of 71% from iceberg lettuce, 65% from green lollo lettuce, 69% from Chinese leaves, 66% from an autumn salad mix, 62% strawberries and 4–42% from bean sprouts, largely depending on the age of the sprouts tested. Fewer cysts were recovered from older sprouts and therefore they recommend that fruit and vegetables should be tested as fresh as possible.

Molecular methods have been applied for both the detection and characterization of *Giardia* in both clinical and environmental samples. The application of electrophoresis-based and PCR-based methods for the investigation of

G. *duodenalis* isolates has demonstrated the phenotypic and genotypic heterogeneity of the species and contributed much towards our understanding of the epidemiology of the parasite. Methods have included pulsed field gel electrophoresis, isoenzyme typing, sequence-based analyses of the glutamate dehydrogenase, triosphosphate isomerase (tpi), elongation factor 1α and 18S rRNA genes (Monis *et al.*, 1999) and PCR/RFLP of the tpi gene locus (McIntyre *et al.*, 2000). The sensitivity and specificity of the PCR/RFLP of the tpi gene locus has been improved by Amar and colleagues (2002) to distinguish G. *duodenalis* assemblages A and B and the subgroups of assemblage A in human clinical isolates. PCR-based methods eliminate the need for *in vitro* or *in vivo* amplification of the parasite, and thus broaden the range of isolates that can be analysed. Improved tools for molecular analyses and the investigation of large, representative panels of isolates provide better data to identify genetic divisions within the G. *duodenalis* group. These need to be complemented by investigation of biotypic characteristics, for example, host range and pathogenicity, virulence markers and drug sensitivity.

Two viruses have been identified in G. *duodenalis*. One is a 6.2 kb non-enveloped double-stranded RNA virus (Wang and Wang, 1986) with a 100 kDa major capsid protein (Yu *et al.*, 1995) and the other also has 6.2 kb double-stranded RNA genome but encodes a smaller (95 kDA) protein (Tai *et al.*, 1996). These may serve as markers for strain variation.

# Epidemiology

Human cases of giardiasis are reported world-wide in both adults and children, but are more common where hygiene is poor and water treatment inadequate. For example, on examination of stools submitted for ova and parasites, 2–5% prevalence is observed in developed countries and 20–30% in developing countries. Seasonality has been observed with peak incidence in late summer in the UK, USA and Mexico. Person-to-person transmission occurs and is frequent where hygiene is poor, largely because *Giardia* is infective immediately or shortly after being passed in the stools. Giardiasis is of emerging importance, particularly since the number of outbreaks associated with day-care centres has increased, and the organism poses a risk in long-term care settings (Nicolle, 2001). In developed countries, many patients report foreign travel, probably reflecting exposure to poor hygiene, poorly treated water and unfamiliar strains of the parasite. Water is increasingly recognized as a vehicle for transmission and has caused concern among water providers in developed countries. CDC began surveillance for waterborne disease in 1971, and in the USA *Giardia* is the most frequently implicated organism in outbreaks of waterborne disease, although in the UK that accolade goes to *Cryptosporidium*. Waterborne outbreaks have been associated with the treated drinking water supply, indicating failure to remove/inactivate cysts.

In North America giardiasis is commonly referred to as 'beaver fever' from the perception that drinking untreated surface water contaminated by beavers is a cause of illness. Indeed, strains isolated from human cases during an outbreak were shown to be indistinguishable from those isolated from beavers (Isaac-Renton *et al.*, 1993). However, meta analysis of studies in the USA has shown that there was minimal evidence for an association between drinking untreated water and a high incidence of giardiasis among back country recreationalists (Welch, 2000), and that other recreational exposures such as contamination of hands and hand-to-mouth contact while camping should be explored and therefore proper hand washing prior to eating and drinking would be an effective control measure.

Much of the information about waterborne giardiasis has arisen from the study of outbreaks. While the relative contribution of the multiple possible routes of transmission of the parasite is unknown, it has been estimated that 60% of all cases of giardiasis in the USA are water-related (Bennett *et al.*, 1987).

By contrast to the predominance of waterborne outbreaks in the USA caused by *Giardia*, in the UK relatively few waterborne outbreaks of giardiasis have been reported. In the summer of 1985, 108 laboratory-confirmed cases were linked to consumption of the municipal water supply in the South West (Jephcott *et al.*, 1986). Cases were almost exclusively adults. The water supply had been treated by filtration and chlorination. A case control study showed a significant association with water consumption coinciding with the dates the water main supplying the community was opened for routine work. However, the symptoms of some cases anteceded the main outbreak and could have represented other aetiologies or the source of the outbreak itself. Reservoir samples showed no indication of pollution and ingress into the system was suspected, although environmental samples were negative for *Giardia*. The importance of laboratory screening of all stools for *Giardia* was illustrated by this outbreak.

In 1991/2 a private water supply in England supplying 260 people was implicated in an outbreak involving 31 cases, five of which were laboratory confirmed. Three *Giardia* cysts were detected per litre and faecal streptococci and coliforms were also detected. As a control measure a boil water notice was imposed. Inspection of the supply showed the spring collection chamber, although covered, had animals grazing round it. The reservoir was chlorine dosed twice weekly but filtration was installed only after the outbreak. The identification of private water supplies as a route of transmission of giardiasis is because of their general lack of appropriate treatment and their vulnerability to faecal contamination.

A case control study was undertaken in Avon and Somerset, England over a 12-month period from July 1992 to May 1993, to identify risk factors for sporadic giardiasis, in particular the risk associated with recreational water use and associations with foreign travel (Gray *et al.*, 1994). Swimming, travel to developing countries and accommodation type (camping/caravanning/staying in holiday chalets) were identified as independent risk factors for giardiasis. A larger study encompassing the same region of England was carried out for

12 months from April 1998 and excluded people who had travelled outside the UK in the 3 weeks prior to the onset of illness in order to investigate risk factors for endemic sporadic giardiasis (Stuart *et al.*, 2003). Swallowing water while swimming, recreational fresh water contact, drinking treated tap water, and eating lettuce were all associated with giardiasis.

Food-borne outbreaks of giardiasis are frequently related to infected food handlers or their contact with an infected person. Interestingly, while cysts may have been detected in subsequent screening, the food handlers and their contacts are frequently asymptomatic (Rose and Slifco, 1999). For example, the first report of food-borne transmission was of an outbreak in December 1979 when 29 cases occurred at a school in Minnesota, in which salmon was identified as the vehicle of infection. The food, including the fish, had been prepared by someone caring for a 1-year-old child who was excreting cysts. Other vehicles of infection have included noodle salad, fruit salad, sandwiches, various salad items or raw vegetables (Rose and Slifco, 1999). Transmission may also occur via fresh produce as a result of the use of uncomposted manure or human waste. *Giardia* has been detected in clams (*Macoma balthica* and *Macoma mitchelli*) from Rhode River, a subestuary of Chesapeake Bay (Graczyk *et al.*, 1999). Not only may this pose a risk through the consumption of undercooked shellfish, but it is possible that filter feeders could serve as biological indicators and contribute to the sanitary assessment of water.

Recreational waters clearly have a role in the epidemiology of giardiasis and outbreaks have occurred linked to the use of both chlorinated swimming pools and natural fresh recreational waters.

Molecular typing of isolates has indicated that an assumed potentially large animal reservoir of the organism may have lower zoonotic potential than previously thought, although this was previously difficult to determine because morphological differences between strains are lacking. Studies of albeit limited numbers of human clinical isolates from sporadic cases in the UK using PCR-RFLP of the tpi gene show that 21/33 (64%) were assemblage B, 9/33 (27%) were assemblage AII, 3/33 (9%) contained both B and AII and none of the 33 were assemblage AI (Amar *et al.*, 2002). These results therefore question the role of livestock animals as a major reservoir of human giardiasis in the UK. Sequence analysis of the 18S rDNA gene showed that human infections in an endemic area of Western Australia, previously assumed to be zoonotic, were different from dog isolates (Hopkins *et al.*, 1997), although there are some common genotypes in both humans, dogs and cats within assemblages A and B (Thompson, 2000). Case control studies have generated conflicting evidence for the role of both household pets and contact with farmed animals: a study in the South East of England showed an association with pigs, dogs and cats (Warburton *et al.*, 1994), while other studies found no such associations (Mathias *et al.*, 1992; Dennis *et al.*, 1993). Thus the role of pets is not fully established, nor is the pathogenicity of this parasite for cats and dogs. Interestingly, in the USA a vaccine is commercially available for the prevention of clinical signs and reduction of cyst shedding in cats and dogs.

Further molecular epidemiological studies are required to understand fully the reservoirs of human infection and their relative importance for public health.

## Risk assessment

It is likely, from epidemiological and genetic evidence, that two cycles of infection with *Giardia* probably exist: human-human and zoonotic (Thompson, 2000). Control of direct transmission between humans can be made by implementation of normal hygiene practices and there is a higher risk of transmission where faecal-oral contamination occurs (e.g. child-care centres; male homosexual practices). Sources of zoonotic *Giardia* need to be better defined by more extensive molecular epidemiological studies and sample surveys, but there is evidence that some species, such as the beaver, act as amplification hosts for strains infective for humans (Isaac-Renton *et al.*, 1993). Outbreak investigations have demonstrated the importance of the animal reservoir but the contribution to sporadic cases is unknown. Molecular epidemiology provides a powerful predictive tool for evidence of transmission and for determining sources of contamination and infection and investigations of the genomic characteristics of *G. duodenalis* have contributed greatly to our understanding of the epidemiology of giardiasis and contributes to risk assessment.

In a risk assessment model for Giardia, Rose and colleagues (1991) used *a priori* prevalence data showing average numbers of cysts in surface waters ranging from 0.33 to 104 cysts/100 l compared with 0.6–5 cysts/100 l in pristine waters. Assuming a source water contaminated with 0.7–70 cysts/100 l they demonstrated that treatment achieving 3 to 5 log reduction in cysts would be required to meet a risk target of less than one case giardiasis per 10 000 population. However, such models have yet to consider the infectivity of cysts for humans and their significance for public health. Although filtration is more effective for removal of *Giardia* cysts than *Cryptosporidium* oocysts, not all water supplies are filtered and both organisms have a low infectious dose.

## Overall risk assessment

*Health effects*: occurrence of illness, degree of morbidity and mortality, probability of illness based on infection:

- Human cases of giardiasis are reported world-wide in both adults and children, but are more common where hygiene is poor and water treatment inadequate. In developed countries, detection rates of 2–5% have been reported and between 2 and 20% of those infected will have symptoms. It has been estimated that 60% of all cases of giardiasis in the USA are water-related. In developing countries, carriage rates of 20–30% have been detected and are particularly prevalent in children.

- Clinical features vary from asymptomatic carriage to severe diarrhoea and malabsorption. Symptoms include diarrhoea, abdominal pain, bloating, flatulence, malaise, sulphurous belching, nausea and vomiting. The acute phase is usually resolved spontaneously in immune-competent individuals, but in some cases can develop into a chronic phase during which symptoms relapse in short recurrent bouts. Untreated acute giardiasis lasts for at least 10 days and can last for 3 months or even years. Immunocompromised patients, particularly those with hypogammaglobulinemia, often suffer chronic disease.
- Variation in clinical symptoms includes the virulence of the *Giardia* strain, the number of cysts ingested and the age and immunocompetence of the host.

*Exposure assessment*: routes of exposure and transmission, occurrence in source water, environmental fate:

- Asymptomatic cases can shed cysts in their faeces and provide an important reservoir of infection, particularly in families, child-care settings and institutions.
- Recognized routes of transmission to humans include faecal-oral spread through person-to-person contact, ingestion of drinking and recreational waters, consumption of contaminated food and zoonotic transmission.
- Many animals carry the pathogenic species of *Giardia* and a number of water-borne outbreaks have been associated with animal faecal contamination.
- The infective dose in experimental infections was 10–25 cysts, which makes *Giardia* easily spread from person to person and of significance even if detected in 'low' numbers.
- *Giardia* cysts have been found in surface waters in urban and rural areas, groundwater, sewage and treated drinking water.
- Cysts are resistant to many environmental pressures, enabling the survival and environmental transmission of the parasite. Cysts survive in water environments and show greater resistance to UV, chlorine and ozone than bacteria or viruses, but with the exception of UV, generally less resistance than *Cryptosporidium*.
- *Giardia* cysts have been shown to maintain viability for up to 3 months in cold raw water sources and in tap water, with a range of 75–99.9% natural die-off.

*Risk mitigation*: drinking-water treatment, medical treatment:

- Cyst removal can be achieved by clarification and filtration, although recycling of filter back wash water has been identified as a means of reintroducing cysts into water supplies. Proper filtration can remove cysts at levels of 99% or higher.
- With appropriate concentration and contact time, chemical disinfectants can effectively inactivate *Giardia* cysts.
- Supportive treatment by fluid and electrolyte replacement is important in the treatment of giardiasis. The nitroimidazoles (metronidazole, tinidazole, ornidazole and secnidazole) are the main agents used, while the nitrofuran

compound furazolidone, the aminoglycoside paromomycin and benzimidazoles (i.e. albendazole) are also prescribed.

## Future implications

Untreated water sources are vulnerable and treatment plants, particularly those operating under conditions of, for example, low temperature, may not be capable of consistently meeting annual risk targets of 1/10 000 for giardiasis set under the Surface Water Treatment Rule. Work by LeChevallier *et al.* (1991a) suggests not. Therefore, additional treatment may be required to prevent outbreaks occurring. Better identification of sources and routes of transmission would help prevent waterborne incidents.

## References

Abbaszadegan, M., Huber, M.S., Gerba, C.P. *et al.* (1997). Detection of viable *Giardia* cysts by amplification of heat shock-induced mRNA. *Appl Environ Microbiol*, **63**: 324–328.

Amar, C.F.L., Dear, P.H., Pedraza-Díaz, S. *et al.* (2002). Sensitive PCR-restriction fragment length polymorphism assay for detection and genotyping of *Giardia duodenalis* in human faeces. *J Clin Microbiol*, **40**: 446–452.

Anon. (1981). Intestinal protozoan and helminthic infections. *WHO Technical Report Ser*, **58**: 666–671.

Anon. (1999). Method 1623: *Cryptosporidium* and *Giardia* in water by filtration/IMS/FA. United States Environmental Protection Agency.

Basualdo, J., Pezzani, B., de Luca, M. *et al.* (2000). Screening of the municipal water system of La Palta, Argentina, for human intestinal parasites. *Int J Hyg Environ Hlth*, **203**: 177–182.

Bennett, J.V., Homberg, S.D. and Rogers, M.F. (1987). Infectious and parasitic diseases. *Am J Prevent Med*, **3**: 102–114.

Bingham, A.K. and Meyer, E.A. (1979). *Giardia* excystation can be induced *in vitro* in acidic solutions. *Nature*, **277**: 301–302.

Bingham, A.K., Jarroll, E.L. and Meyer, E.A. (1979). *Giardia* sp.: Physical factors of excystation *in vitro*, and excystation vs eosin exclusion as determinants of viability. *Exp Parasitol*, **47**: 284–291.

Borchardt, M.A. and Spencer, S.K. (2002). Concentration of *Cryptosporidium*, microsporidia and other water-borne pathogens by continuous separation channel centrifugation. *J Appl Microbiol*, **92**: 649–652.

Boucher, S.E.M. and Gillin, F.D. (1990). Excystation of in vitro-derived *Giardia lamblia* cysts. *Infect Immun*, **58**: 3516–3522.

Brabdborg, L.L., Tankersley, C.B., Gottlieb, S. *et al.* (1967). Histological demonstration of mucosal invasion by *Giardia lamblia* in man. *Gastroenterology*, **52**: 143–150.

Bryan, R.T., Pinner, R.W. and Berkelman, R.L. (1994). Emerging infectious diseases in the United States. *Ann NY Acad Sci*, **740**: 346–361.

Bukhari, Z., Hargy, T., Bolton, J. *et al.* (1999). Medium-pressure UV for oocyst inactivation. *JAWWA*, **91**: 86.

Campbell, A.T. and Wallis, P. (2002). The effect of UV irradiation on human-derived *Giardia lamblia* cysts. *Water Res*, **36**: 936–969.

Cartwright, C.P. (1999). Utility of multiple-stool-specimen ova and parasite examinations in a high-prevalence setting. *J Clin Microbiol*, **37**: 2408–2411.

Céu Sousa Gonçalves, C.A., Bairos, V.A. and Poiares-da-Silva, J. (2001). Adherence of *Giardia lamblia* trophozoites to Int-407 human intestinal cells. *Clin Diag Lab Immunol*, **8**: 258–265.

Craun, G.F. (1986). Waterborne giardiasis in the United States 1965–1984. *Lancet*, **ii**: 513–514.

Dalton, C., Goater, A.D., Pethig, R. *et al.* (2001). Viability of *Giardia intestinalis* cysts and viability and sporulation state of *Cyclospora cayetanensis* oocysts determined by electrorotation. *Appl Environ Microbiol*, **67**: 586–590.

Dennis, D.T., Smith, R.P., Welch, J.J. *et al.* (1993). Endemic giardiasis in New Hampshire: a case control study of environmental risks. *J Infect Dis*, **167**: 1391–1395.

deReigner, D.P., Cole, L., Schupp, D.G. *et al.* (1989). Viability of *Giardia* cysts suspended in lake, river and tap water. *Appl Environ Microbiol*, **55**: 1223–1229.

Duncombe, V.M., Bolin, T.D., Davies, A.E. *et al.* (1978). Histopathology in giardiasis: a correlation with diarrhoea. *Aust NZ J Med*, **8**: 392–396.

Egorov, A., Paulauskis, J., Petrova, L. *et al.* (2002). Contamination of water supplies with *Cryptospordiium parvum* and *Giardia lamblia* and diarrhoeal illness in selected Russian Cities. *Int J Hyg Environ Hlth*, **205**: 281–289.

Erlandsen, S.L. and Meyer, E.A. (eds) (1984). *Giardia and Giardiasis Biology, Pathogenesis, and Epidemiology*. New York: Plenum Press.

Erlandsen, S.L. and Bemrick, W.J. (1987). SEM evidence for a new species, *Giardia psittaci*. *J Parasitol*, **73**: 623–629.

Erlandsen, S.L., Bemrick, W.J., Wells, C.L. *et al.* (1990). Axenic culture and characterisation of *Giardia ardeae* from the great blue heron (*Ardea herodias*). *J Parasitol*, **76**: 717–724.

Farthing, M.J.G. (1997). The molecular pathogenesis of giardiasis. *J Paediatr Gastroenterol Nutr*, **24**: 79–88.

Feely, D.E. (1988). Morphology of the cyst of *Giardia microi* by light and electron microscopy. *J Protozool*, **35**: 52–54.

Feely, D.E., Gardner, M.D. and Hardin, E.L. (1991). Excystation of *Giardia muris* induced by a phosphate-bicarbonate medium: localisation of acid phosphatase. *J Parasitol*, **77**: 441–448.

Ferguson, A., Gillon, J. and Munro, G. (1990). Pathology and pathogenesis of the intestinal mucosal damage in giardiasis. In *Giardiasis*, Meyer, E.A. (ed.). New York: Elsevier Publishing Co, pp. 155–173.

Filice, F.P. (1952). Studies on the cytology and life history of a *Giardia* from the laboratory rat. *Univ California Publ Zool*, **57**: 53–146.

Finch, G.R., Black, E.K., Labatiuk, C.W. *et al.* (1993). Comparison of *Giardia lamblia* and *Giardia muris* cyst inactivation by ozone. *Appl Environ Microbiol*, **59**: 3674–3680.

Fraser, D., Bilenko, N., Deckelbaum, R.J. *et al.* (2000). *Giardia lamblia* carriage in Israeli Bedouin infants: risk factors and consequences. *Clin Infect Dis*, **30**: 419–424.

Gardener, T.B. and Hill, D.R. (2001). Treatment of Giardiasis. *Clin Microbiol Rev*, **14**: 114–128.

Gerber, C.P., Johnson, D.C. and Hasan, M.N. (1997). Efficacy of iodine water purification tablets against *Cryptospordium* oocysts and *Giardia* cysts. *Wilderness Environ Med*, **8**: 96–100.

Gillin, F.D., Reiner, D.S. and Boucher, S.E. (1988). Small intestinal factors promote excystation of *Giardia lamblia* in vitro. *Infect Immun*, **56**: 705–707.

Goldstein, F.J., Thornton, J.J. and Szyldlowski, T. (1978). Biliary tract dysfunction in giardiasis. *Am J Dig Dis*, **23**: 559–560.

Graczyk, T.K., Thompson, R.C., Fayer, R. *et al.* (1999). *Giardia duodenalis* cysts of genotype A recovered from clams in the Chesapeake Bay subestuary, Rhond River. *Am J Trop Med Hyg*, **61**: 526–529.

Gray, S.F., Gunnell, D.J. and Peters, T.J. (1994). Risk factors for giardiasis: a case control study in Avon and Somerset. *Epidemiol Infect*, **113**: 95–102.

Guerrant, R.L., Schorling, J.B., McAuliffe, J.F. *et al.* (1992). Diarrhoea as a cause and effect of malnutrition: diarrhoea prevents catch-up growth and malnutrition increases diarrhoea frequency and duration. *Am J Trop Med Hyg*, **47**: 28–35.

Haas, C.N. and Aturaliye, D. (1999). Semi-quantitative characterisation of electroporation-assisted disinfection processes for inactivation of *Giardia* and *Cryptosporidium*. *J Appl Microbiol*, **86**: 899–905.

Hancock, C.M. *et al.* (1996). Assessing plant performance using MPA. *JAWWA*, **88**: 24–30.

Hashimoto, A., Kunikane, S. and Hirata, T. (2002). Prevalence of *Cryptosporidium* oocysts and *Giardia* cysts in the drinking water supply in Japan. *Water Res*, **36**: 519–526.

Hoff, J.C., Rice, E.W. and Shaeffer, F.W. (1985). Comparison of animal infectivity and excystation as measures of *Giardia muris* cyst inactivation by chlorine. *Appl Environ Microbiol*, **50**: 115–117.

Hopkins, R.M., Meloni, B.P., Groth, D.M. *et al.* (1997). Ribosomal RNA sequencing reveals differences between the genotypes of *Giardia* isolates recovered from humans and dogs living in the same locality. *J Parasitol*, **83**: 44–51.

Hunter, P.R. (1997). *Waterborne Disease Epidemiology and Ecology*. Chichester: Wiley.

Inge, P.M.G., Edson, C.M. and Farthing, M.J.G. (1986). Attachment of *Giardia lamblia* to rat intestinal epithelial cells. *Gut*, **29**: 795–801.

Isaac-Renton, J., Cordeiro, C., Sarafis, K. *et al.* (1993). Characterisation of *Giardia duodenalis* isolates from a waterborne outbreak. *J Infect Dis*, **167**: 431–440.

Isaac-Renton, J., Moorehead, W. and Ross, A. (1996). Longitudinal studies of *Giardia* contamination in two community drinking water supplies: cyst levels, parasite viability and health impact. *Appl Environ Microbiol*, **62**: 47–54.

Iturriaga, R., Zhang, S., Sonek, G.J. *et al.* (2001). Detection of respiratory enzyme activity on *Giardia* cysts and *Cryptosporidium* oocysts using redox dyes and immunofluorescence techniques. *J Microbiol Meth*, **46**: 19–28.

Jarroll, E.L., Bingham, A.K. and Meyer, E.A. (1980). *Giardia* cyst destruction: effectiveness of six small-quantity water disinfection methods. *Am J Trop Med Hyg*, **29**: 8–11.

Jarroll, E.L., Bingham, A.K. and Meyer, E.A. (1981). Effect of chlorine on *Gardia lamblia* cyst viability. *Appl Environ Microbiol*, **41**: 483–487.

Jarroll, E.L., Muller, P.J., Meyer, E.A. *et al.* (1981). Lipid and carbohydrate metabolism of *Giardia lamblia*. *Mol Biochem Parasitol*, **2**: 187–196.

Jephcott, A.E., Begg, N.T. and Baker, I.A. (1986). Outbreak of giardiasis associated with mains water in the United Kingdom. *Lancet*, i: 730–732.

Kappus, K.D., Lundgren, R.G., Juranek, D.D. *et al.* (1994). Intestinal parasitism in the United States: update on a continuing problem. *Am J Trop Med Hyg*, **50**: 705–713.

Kasprzak, W. and Majewska, A.C. (1983). Infectivity of *Giardia* sp. cysts in relation to eosin exclusion and excystation *in vitro*. *Tropmed Parasitol*, **34**: 70–72.

Kistemann, T., Classen, T., Koch, C. *et al.* (2002). Microbial load of drinking water reservoir tributaries during extreme rainfall and runoff. *Appl Environ Microbiol*, **68**: 2188–2197.

Knisley, C.V., Engelkirk, P.G., Pickering, L.K. *et al.* (1989). Rapid detection of *Giardia* antigen in stool with the use of enzyme immunoassays. *Am J Clin Pathol*, **91**: 704–708.

Kofoid, C.A. and Christensen, E.B. (1915). On binary and multiple fission in *Giardia muris* (Grassi). *Univ California Publ Zool*, **16**: 30–54.

Kofoid, C.A. and Christensen, E.B. (1920). A critical review of nomenclature of human intestinal flagellates. *Univ California Publ Zool*, **20**: 160.

Kulda, J. and Nohýnková, E. (1978). Flagellates of the human intestine and of intestines of other species. In *Protozoa of Veterinary and Medical Interest*, Kreier, J.P. (ed.). New York: Academic Press, pp. 69–104.

Labatiuk, C.W., Schaefer, F.W., Finch, G.R. *et al.* (1991). Comparison of animal infectivity, excystation and fluorogenic dye as measures of *Giardia muris* cyst inactivation by ozone. *Appl Environ Microbiol*, **57**: 3187–3192.

Lambl, W. (1859). Mikroskopische untersuchungen der Darmexcrete. *Vierteljahrsschr Prakst Heikunde*, **61**: 1–58.

LeChavallier, M.W., Norton, W.D. and Lee, R.G. (1991a). Occurrence of *Giardia* and *Cryptosporidium* spp. in surface water supplies. *Appl Environ Microbiol*, **57**: 2610–2616.

LeChevallier, M.W., Norton, W.D. and Lee, R.G. (1991b). *Giardia* and *Cryptsporidium* spp. in filtered drinking water supplies. *Appl Environ Microbiol*, **57**: 2617–2621.

Lee, M.B. (1992). The effectiveness of commercially available disinfectants upon *Giardia lamblia* cysts. *Can J Public Hlth*, **83**: 171–172.

Lindmark, D.G. (1980). Energy metabolism of the anaerobic protozoon, *Giardia lamblia*. *Mol Biochem Parasitol*, **1**: 1–12.

Mahbubani, M.H., Bej, A.K., Perlin, M. *et al.* (1991). Detection of *Giardia* cysts by using the polymerase chain reaction and distinguishing live from dead cysts. *Appl Environ Microbiol*, **57**: 597–600.

Marshall, M.M., Naumovitz, D., Ortega, Y. *et al.* (1997). Waterborne protozoan pathogens. *Clin Microbiol Rev*, **10**: 67–85.

Mathias, R.G., Riben, P.D. and Osei, W.D. (1992). Lack of an association between endemic giardiasis and a drinking water source. *Can J Public Hlth*, **83**: 382–384.

McCuin, R.M., Bukhari, Z. and Clancy, J.L. (2000). Recovery and viability of *Cryptosporidium parvum* oocysts and *Giardia intestinalis* cysts using the membrane dissolution procedure. *Can J Microbiol*, **46**: 700–707.

McIntyre, L., Hoang, L., Ong, C.S.L. *et al.* (2000). Evaluation of molecular techniques to biotype *Giardia duodenalis* collected during an outbreak. *J Parasitol*, **86**: 172–177.

Meyer, E.A. and Jarroll, E.L. (1980). Giardiasis. *Am J Epidemiol*, **111**: 1–12.

Meyers, J.D., Kuharic, H.A. and Holmes, K.K. (1977). *Giardia lamblia* infection in homosexual men. *Br J Vener Dis*, **53**: 54–55.

Modifi, A.A., Meyer, E.A., Wallis, P.M. *et al.* (2002). The effect of UV light on the inactivation of *Giardia lamblia* and *Giardia muris* cysts as determined by animal infectivity assay (P-2951-01). *Water Res*, **36**: 2098–2108.

Monis, P.T., Andrews, R.H., Mayrhofer, G. *et al.* (1999). Molecular systematics of the parasitic protozoan *Giardia intestinalis*. *Mol Biol Evoln*, **16**: 1135–1144.

Moss, D.M., Mathews, H.M., Visvesvara, G.S. *et al.* (1990). Antigenic variation of *Giardia lamblia* in the faeces of Mongolian gerbils. *J Clin Microbiol*, **28**: 254–257.

Moulton-Hancock, C., Rose, J.B., Vasconcelos, G.J. *et al.* (2000). *Giardia* and *Cryptosporidium* occurrence in groundwater. *JAWWA*, **92**: 117–123.

Nanson, K.L. and Cartwright, C.P. (2001). Use of an enzyme immunoassay does not eliminate the need to analyse multiple stool specimens for sensitive detection of *Giardia lamblia*. *J Clin Microbiol*, **39**: 474–477.

Nash, T.E., Harrington, G.A., Losonsky, G.A. *et al.* (1987). Experimental human infections with *Giardia lamblia*. *J Infect Dis*, **156**: 974–984.

Nicolle, L.E. (2001). Preventing infections in non-hospital settings: long term care. *Emerging Infect Dis*, **7**: 205–207.

Ong, C., Moorehead, W., Ross, A. *et al.* (1996). Studies of *Giardia* spp. and *Cryptosporidium* spp. in two adjacent watersheds. *Appl Environ Microbiol*, **62**: 2798–2805.

Ortega, Y.R. and Adam, R.D. (1997). *Giardia*: overview and update. *Clin Infect Dis*, **25**: 545–550.

Owen, R.L., Nemanic, P.D. and Stevens, D.P. (1979). Ultrastructural observations of giardiasis in a murine model. *Gastroenterology*, **76**: 757–769.

Payment, P., Plante, R. and Cejka, P. (2001). Removal of indicator bacteria, human enteric viruses, *Giardia* cysts, and *Cryptosporidium* oocysts at a large wastewater primary treatment facility. *Can J Microbiol*, **47**: 188–193.

Polis, M.A., Tuazon, C.U., Alling, D.W. *et al.* (1986). Transmission of *Giardia lamblia* from a day care center to the community. *Am J Public Hlth*, **76**: 1142–1144.

Rauch, A.M., Van, R., Bartlett, A.V. *et al.* (1990). Longitudinal study of *Giardia lamblia* in a day care center population. *Pediatr Infect Dis J*, **9**: 186–189.

Rendtorff, R.C. (1954). The experimental transmission of human intestinal protozoan parasites. II. *Giardia lamblia* cysts given in capsules. *Am J Hyg*, **59**: 209–220.

Robertson, L.J. and Gjerde, B. (2000). Isolation and enumeration of *Giardia* cysts, *Cryptosporidium* oocysts, and Ascaris eggs from fruits and vegetables. *J Food Protect*, **63**: 775–778.

Robertson, L.J., Paton, C.A., Campbell, A.T. *et al.* (2000). *Giardia* cysts and *Cryptosporidium* oocysts at sewage treatment works in Scotland, UK. *Water Res*, **34**: 2310–2322.

Rose, J.B., Gerba, C.P. and Jakubowski, W. (1991). Survey of potable water supplies for *Cryptosporidium* and *Giardia*. *Environ Sci Technol*, **258**: 1393–1401.

Rose, J.B., Haas, C.N. and Regli, S. (1991). Risk assessment and control of waterborne giardiasis. *Am J Public Hlth Assoc*, **81**: 709–713.

Rose, J.B. and Slifco, T.R. (1999). *Giardia*, *Cryptosporidium* and *Cyclospora* and their impact on foods: a review. *J Food Protect*, **62**: 1059–1070.

Rosen, J.S. *et al.* (1996). Development and Analysis of a National Protozoan Database. Proceedings of the AWWA WQTC Boston, 1996.

Slifko, T.R., Smith, H.V. and Rose, J.B. (2000). Emerging parasite zoonoses associated with food and water. *Int J Parasitol*, **30**: 1379–1393.

Smith, A.L. and Smith, H.V. (1989). A comparison of fluorescein diacetate and propidium iodide staining and *in vitro* cultivation for determining *Giardia intestinalis* cyst viability. *Parasitology*, **99**: 329–331.

Sousa, M.A. and Poiares-Da-Silva, J. (1999). A new method for assessing metronizadole susceptibility of *Giardia lamblia* trophozoites. *Antimicrob Agents Chemother*, **43**: 2939–2942.

States, S., Stadterman, K., Ammon, L. *et al.* (1997). Protozoa in river water: sources, occurrence, and treatment. *JAWWA*, **89**: 74–83.

Sykora, J.L. *et al.* (1991). Distribution of *Giardia* cysts in wastewater. *Water Sci Technol*, **24**: 2187–2193.

Tai, J.H., Chang, S.C., Chou, C.F. *et al.* (1996). Separation and characterisation of two related giardiaviruses in the parasitic protozoan *Giardia lamblia*. *Virology*, **216**: 124–132.

Thompson, R.C.A. (2000). Giardiasis as a re-emerging infectious disease and its zoonotic potential. *Int J Parasitol*, **30**: 1259–1267.

Thompson, R.C.A., Hopkins, R.M. and Homan, W.L. (2000). Nomenclature and genetic groupings of *Giardia* infecting mammals. *Parasitol Today*, **16**: 210–213.

Tupchong, M., Simor, A. and Dewar, C. (2001). Beaver fever – a rare cause of reactive arthritis. *J Rheumatol*, **28**: 683.

van Keulan, H., Campbell, S.R., Erlandsen, S.L. *et al.* (1991). Cloning and restriction enzyme mapping of ribosomal DNA of *Giardia duodenalis*, *Giardia ardeae* and *Giardia muris*. *Mol Biochem Parasitol*, **46**: 275–284.

van Keulan, H., Feely, D., Macechko, M. *et al.* (1998). The sequence of *Giardia* small subunit rRNA shows that voles and muskrats are parasitised by a unique species *Giardia microti*. *J Parasitol*, **84**: 294–300.

Wallis, P.M., Erlandsen, S.L., Isaac-Renton, J.L. *et al.* (1996). Prevalence of *Giardia* cysts and *Cryptosporidium* oocysts and characterisation of *Giardia* spp. isolated from drinking water in Canada. *Appl Environ Microbiol*, **62**: 2789–2797.

Wang, A.L. and Wang, C.C. (1986). Discovery of a specific double-stranded RNA virus in *Giardia lamblia*. *Mol Chem Parasitol*, **21**: 269–276

Warburton, A.R.E., Jones, P.H. and Bruce, J. (1994). Zoonotic transmission of giardiasis: a case control study. *Commun Dis Rep*, **4**: R32–35.

Welch, T.P. (2000). Risk of giardiasis from consumption of wilderness water in North America: a systematic review of epidemiological data. *Int J Infect Dis*, **4**: 100–103.

Wolfe, M.S. (1990). Clinical symptoms and diagnosis by traditional methods. In *Giardiasis*, Meyer, E.A. (ed.). New York: Elsevier Publishing Co, pp. 175–185.

Wright, S.G. and Tomkins, A.M. (1978). Quantitative histology in giardiasis. *J Clin Pathol*, **31**: 712–716.

Wright, S.G., Tomkins, A.M. and Ridley, D.S. (1977). Giardiasis: clinical and therapeutic aspects. *Gut*, **18**: 343–350.

Yardley, J.H., Yakano, J. and Hendrix, T.R. (1964). Epithelial and other mucosal lesions of the jejunum in giardiasis: jejunal biopsy studies. *Bull Johns Hopkins Hosp*, **115**: 389–406.

Yu, D., Wang, C.C. and Wang, A.L. (1995). Maturation of giardiavirus capsid protein involves posttranslational proteolytic processing by a cysteine protease. *J Virol*, **69**: 2825–2830.

# 22

# *Naegleria fowleri*

## Basic microbiology

*Naegleria fowleri* is an amoeboflagellate of the Phylum *Sarcomastigophora*, Order *Schizopyrenida*, Family *Vahlkampfiidae*. This is a free-living protozoan, found in aquatic and soil habitats, and the life cycle consists of a motile, feeding trophozoite stage, non-feeding, non-replicating biflagellate stage and a resistant cyst stage (Bottone, 1993). Trophozoites (10–15 μm) excyst from the cysts via pores in the cyst wall and are elongate in shape with rounded processes called lobopodia. They feed off bacteria, including *Escherichia coli*. Trophozoites differentiate to pear-shaped, motile biflagellates, which can revert to trophozoites. The biflagellates encyst in adverse environmental conditions. The spherical cysts are approximately 10 μm in diameter, with a smooth double wall. Infection in humans occurs following exposure to biflagellates in recreational waters or cysts in soil or dust. Of the six species currently identified in the genus *Naegleria*, *N. fowleri* is the primary human pathogen. Following inhalation of biflagellates or cysts, trophozoites invade the nasopharyngeal mucosa, migrating through the olfactory nerve to invade the brain via the cribriform plate, typically causing primary amoebic meningoencephalitis (PAM) which is invariably fatal (Duma *et al.*, 1969). Other species including *N. australiensis* may also be pathogenic, but have not been identified to the same extent as *N. fowleri* (Bottone, 1993).

Although cases of PAM are rare, they have occurred throughout the world and are usually linked to swimming in contaminated water. The epidemiology of PAM and transmission of N. *fowleri* is driven by its survival in the aquatic environment and activity. Antibodies to *Naegleria* spp. are detected in human sera (Cursons *et al.*, 1980), indicating widespread exposure may occur. However, it is unknown why disease occurs when others undertaking similar activities are unaffected. Elucidation of risk factors for exposure and disease still need clarification.

## Origin of the organism

PAM was first recognized in Orange County, Florida, USA in 1962 (Butt, 1966) since when over 250 cases have been reported world-wide.

## Clinical features

PAM occurs in otherwise healthly, often active individuals, particularly young adults and children (Duma *et al.*, 1969). Symptoms occur within 7 days of exposure and are indistinguishable from fulminant bacterial meningitis, and include severe frontal headache, sore throat, high fever, anorexia, vomiting, signs of meningeal inflammation, altered mental status, occasional olfactory hallucinations, nucal rigidity, somnolence and coma (Lubor, 1981). Signs of brain stem compression and seizures follow. Disease is characterized by inflammation of the olfactory bulbs, cerebral hemispheres, brain stem, posterior fossa and spinal cord. Death typically occurs within 10 days of onset of symptoms, typically on days 5 or 6.

## Pathogenicity, virulence and causation

Infection occurs when the nasal mucosa is exposed to trophozoites following swimming in contaminated warm waters. Trophozoites adhere to and penetrate the nasopharyngeal mucosa and migrate along the olfactory nerve. N. *fowleri* exhibits rapid locomotion at 37°C, whereas non-pathogenic species are more active at 28°C (Thong and Ferrante, 1986). Trophozoites invade the brain through the cribiform plate and can reside in the subarachnoid and perivascular spaces. Tissue invasion is achieved by the production of cytopathic enzymes (Ferrante and Bates, 1988) and pathogenicity and virulence are assisted by phagocytosis of host neutrophils, possibly by sucker-like structures on the parasite surface (John *et al.*, 1984). Despite the abundance of

*N. fowleri* in the environment (see above), relatively few cases of PAM occur and it is likely that a multitude of host defence mechanisms are effective.

## Treatment

*N. fowleri* is sensitive to amphotericin B (Fungizone) but just a handful of survivors of PAM have been documented (Schmidt and Roberts, 1995). In these cases there was early diagnosis and administration of intravenous and intrathecal or intraventricular amphotericin B with intensive supportive care. One survivor received miconazole intravenously and intrathecally and rifampicin orally.

## Survival in water and the environment

Although cysts have been identified in soil and dust, *N. fowleri* appears to prefer warm moist environments, including warm water and soil (John, 1982). In water, most trophozoites transform into free-living biflagellates. *N. fowleri* tolerates temperatures of 40–45°C and has been isolated from bodies of temperate and warm fresh water, either natural (such as lakes, rivers and hot springs) or man-made bodies of water such as swimming pools, thermal effluent, sewage sludge and drinking water (Martinez, 1993; Fewtrell *et al.*, 1994). The organism has also been detected in water-cooling circuits from power stations (Hurzianga *et al.*, 1990).

Surveys of lakes in Florida have shown that 12/26 (46%) harbour pathogenic *Naegleria* spp. (Wellings *et al.*, 1979) and that they are common in lake bottom sediment and at the sediment/lake water interface (Wellings *et al.*, 1977). The source of the organism is probably soil, with heavy rain causing runoff into bodies of water (Marciano-Cabral, 1988). Occurrence is also linked to food and nutrient sources and large numbers have been detected where coliforms, filamentous cyanobacteria, eubacteria, increased iron and manganese concentrations are present and there is possibly an interaction between *Naegleria* spp. and *Legionella* spp. (Ma *et al.*, 1990). While chlorination of pools and other artificial waters inactivates *Naegleria* cysts and trophozoites, control measures have yet to be established for natural waters (De Jonckheere *et al.*, 1976).

## Methods of detection

PAM is usually diagnosed at autopsy, but in suspected cases diagnosis is made by examination of CSF which shows predominantly polymorphonuclear leucocytosis, increased protein and decreased glucose concentrations, mimicking bacterial meningitis (Bottone, 1993). Occasionally amoebae may be seen on Gram-stained smears. For diagnosis during life, there must be a clinical suspicion based on exposure history. If a previously healthy patient has swum in warm, fresh water within 7 days of onset of symptoms and displays symptoms of bacterial meningitis with predominantly basilar distribution of exudates by head CT, *N. fowleri* infection should be suspected. Examination of un-refrigerated CSF should be undertaken and care taken to examine further any atypical mononuclear cells, which may actually be amoebae. Transformation from the amoeboid to the biflagellate form can be induced within 1–20 hours to aid identification and is undertaken by dilution of 1 drop CSF with 1 ml distilled water.

*N. fowleri* is readily cultured from environmental samples on non-nutrient agar plates seeded with *Escherichia coli* using prior concentration by membrane filtration (Anon, 1989). However, differentiation from other closely related but non-pathogenic organisms is difficult, but essential for complete risk assessment. Phenotypic characterization, including morphology, thermotolerance, pathogenicity, detection of antigens by monoclonal antibodies and isoenzyme electrophoretic profiles, and genotypic characterization by PCR/PFLP have been used to differentiate the pathogenic *N. fowleri* from non-pathogenic thermophilic species (Ma *et al.*, 1990; Sparagano *et al.*, 1993; Kilvington and Beeching, 1995).

# Epidemiology

*N. fowleri* is acquired through exposure of the nasal passages to contaminated water. Most cases of PAM have occurred in children or in young adults who have been undertaking water-related activities, particularly during the summer. Clustering of cases has occurred pointing to a common or single source of exposure. For example, 16 cases were associated with a public swimming pool in Czechoslovakia (Cerva and Novak, 1968). The pool was supplied by river water. A similarly large outbreak of 16 cases occurred in Virginia, USA among people who had swam in three lakes within a 5-mile radius (Duma *et al.*, 1971). Although *N. fowleri* had been isolated from many of the numerous lakes in Virginia, cases of illness were not associated with other lakes and the reason for the clustering around these three lakes is unclear. An outbreak of five fatal cases of PAM occurred in Mexico in August and September 1990 (Lares-Villa *et al.*, 1993). All had swum in a canal from which *N. fowleri* was isolated, showing identical isoenzyme patterns to all the cases. Single cases also occur. For example, in the UK, a girl died of PAM after swimming in one of the thermal pools at Bath Spa (Cain *et al.*, 1981).

In Western Australia, cases have been associated with the reticulated water supply. Here remote localities are supplied with water via over-ground steel pipes in which solar heating can be substantial allowing proliferation of *N. fowleri* and resulting in about 20 cases of PAM (Robinson *et al.*, 1996). *N. fowleri* was detected in the household water supply of one case, demonstrating the link between drinking water and PAM and broadening the risk factors from recreational activities such as swimming to include drinking water (Marciano-Cabral, 1988).

## Risk assessment

*Health effects*: occurrence of illness, degree of morbidity and mortality, probability of illness based on infection:

- Following inhalation of *N. fowleri*, trophozoites invade the brain through the nasal passages, typically causing primary amoebic meningoencephalitis (PAM), which is almost always fatal.
- Antibodies in blood suggest that more people are exposed to *Naegleria* than become ill, but the reasons are unknown. *Naegleria* cysts are plentiful in the environment, yet few cases occur annually.
- Symptoms occur within 7 days of exposure and are indistinguishable from fulminant bacterial meningitis and include severe frontal headache, sore throat, high fever, anorexia, vomiting, altered mental status, occasional olfactory hallucinations, nucal rigidity, somnolence and coma. Death almost always follows.

*Exposure asssessment*: routes of exposure and transmission, occurrence in source water, environmental fate:

- Infection in humans occurs following inhalational exposure to biflagellates in recreational waters or cysts in soil or dust. Most cases have been linked to swimming in contaminated water, though two cases in the USA in 2002 might have been caused through ingestion or inhalation of unchlorinated drinking water.
- *N. fowleri* tolerates temperatures of 40–45°C, and has been isolated from bodies of temperate and warm fresh water, either natural (such as lakes, rivers, hot springs) or man-made such as swimming pools, thermal effluent, sewage sludge and drinking water.
- Surveys of lakes in Florida have shown that almost half contain pathogenic *Naegleria* spp., which are common in lake bottom sediment.

*Risk mitigation*: drinking-water treatment, medical treatment:

- Chlorination inactivates *Naegleria* organisms.
- *N. fowleri* is sensitive to amphotericin B (Fungizone), but just a handful of survivors of PAM have been documented, usually because diagnosis comes too late.

# References

Anon. (1989). Isolation and identification of *Giardia* cysts, *Cryptosporidium* oocysts and freeliving pathogenic amoebae in water etc. In *Methods for the Examination of Waters and Associated Materials*. London: HMSO.

Bottone, E.J. (1993). Free-living amoebas of the genera *Acanthamoeba* and *Naegleria*: an overview and basic microbiologic correlates. *Mount Sinai J Med*, **60**: 260–270.

Butt, C.G. (1966). Primary amoebic meningoencephalitis. *New Engl J Med*, **274**: 1473–1476.

Cain, A.R.R., Wiley, P.F., Brownell, B. *et al.* (1981). Primary amoebic meningencephalitis. *Arch Dis Childh*, **56**: 140–143.

Cerva, L. and Novak, K. (1968). Amoebic meningoencephalitis: sixteen fatalities. *Science*, **160**: 92.

Cursons, R.T.M., Brown, T.J., Keyes, E.A. *et al.* (1980). Immunity to pathogenic free-living amoeba: role of humoral immunity. *Infect Immun*, **29**: 401–407.

De Jonckheere, J. *et al.* (1976). Differences in the destruction of cysts of pathogenic and non-pathogenic *Naegleria* and *Acanthamoeba* by chlorine. *Appl Environ Microbiol*, **31**: 294–297.

Duma, R.J. *et al.* (1969). Primary amebic meningoencephalitis. *New Engl J Med*, **281**: 1315–1323.

Duma, R.J., Shumaker, J.B. and Callicott, J.H. (1971). Primary amebic meningo-encephalitis: a survey in Virginia. *Arch Environ Hlth*, **23**: 43–47.

Ferrante, A. and Bates, E.J. (1988). Elastase in pathogenic free-living amoebae *Naegleria* and *Acanthamoeba* species. *Infect Immun*, **56**: 3320–3321.

Fewtrell, L., Godfree, A.F., Jones, F. *et al.* (1994). *Pathogenic Microorganisms in Temperate Environmental Waters*. Cardigan, Dyfed: Samara Publishing.

Hurzianga, H.W. and McLaughlin, G.L. (1990). Thermal ecology of *Naegleria fowleri* from a power plant cooling reservoir. *Appl Environ Microbiol*, **56**: 2200–2205.

John, D.T. (1982). Primary amebic encephalitis and the biology of *Naegleria fowleri*. *Ann Rev Microbiol*, **36**: 101–123.

John, D.T., Cole, T.B. and Merciano-Cabral, F.M. (1984). Sucker-like structures on the pathogenic amoeba *Naegleria fowleri*. *Appl Environ Microbiol*, **47**: 12–14.

Kilvington, S. and Beeching, J. (1995). Development of PCR for identification of *Naegleria fowleri* from the environment. *Appl Environ Microbiol*, **61**: 3764–3767.

Lares-Villa, F., de Jonckhere, J.F., de Moura, H. *et al.* (1993). Five cases of primary amebic meningoencephalitis in Mexicali, Mexico: study of isolates. *J Clin Microbiol*, **31**: 685–688.

Lubor, C. (1981). Amebic meningoencephalitis. In *Medical Microbiology of Infectious Diseases*, Brause, A.I., Davies, C.E. and Fierer, J. (eds). Philadelphia, PA: WB Saunders Co, pp. 1281–1284.

Ma, P., Visvesvara, G.S., Martinez, A.J. *et al.* (1990). *Naegleria* and *Acanthamoeba* infections. *Rev Infect Dis*, **12**: 490–513.

Marciano-Cabral, F. (1988). Biology of *Naegleria* spp. *Microbiol Rev*, **52**: 114–133.

Martinez, A.J. (1993). Free-living amebas: infection of the central nervous system. *Mount Sinai J Med*, **60**: 271–278.

Schmidt, G.D. and Roberts, L.S. (1995). *Foundations of Parasitology*. Chicago: William C Brown.

Sparagano, O., Drouet, E., Brebant, R. *et al.* (1993). Use of monoclonal antibodies to distinguish pathogenic *Naegleria fowleri* (cysts, trophozoites, or flagellate forms) from other *Naegleria* species. *J Clin Microbiol*, **31**: 2758–2763.

Thong, Y.H. and Ferrante, A. (1986). Migration patterns of pathogenic and non-pathogenic *Naegleria* species. *Infect Immun*, **51**: 177–180.

Wellings, F.M., Amuso, P.T. and Chang, S.L. *et al.* (1977). Isolation and identification of pathogenic *Naegleria* from Florida lakes. *Appl Environl Microbiol*, **34**: 661–667.

Wellings, F.M. *et al.* (1980). Pathogenic Naegleria: Distribution in nature. EPA research and development bulletin No. 600/1-79-018.

# 23

# *Toxoplasma gondii*

## Basic microbiology

*Toxoplasma gondii* is the only species in the genus (Phylum *Apicomplexa*, Order *Eucoccidiorida*, Family *Eimeriidae*). This tissue-cyst forming coccidium has a heterogeneous life cycle comprising an asexual phase in a variety of warm-blooded intermediate hosts and a sexual phase in the intestines of carnivorous definitive hosts (Frenkel, 1973). Felids are the only known definitive hosts and are the main reservoirs of infection. Cats become infected by eating meat or offal containing tissue cysts, which are lysed by digestive enzymes following ingestion, releasing bradyzoites which invade the epithelial cells of the small intestine (Munday, 1972). Asexual multiplication occurs, followed by the sexual cycle resulting in the formation of immature oocysts which are shed in the faeces. Cats can shed oocysts for 1–2 weeks in extremely large numbers, peaking at over a million a day, even after the consumption of just one tissue cyst (Dubey and Beattie, 1988). The oocysts mature and sporulate over 1–5 days under ambient conditions: sporulation does not usually occur below 4°C or above 37°C. Mature oocysts measure about $12 \times 11$ μm and are infective for a wide range of warm-blooded intermediate hosts, including humans. When oocysts are ingested, an asexual cycle is initiated forming crescent or bow-shaped tachyzoites (2 μm by 6 μm) which localize to form tissue cysts (ranging from <12 μm to over 100 μm) in muscle and neural tissue, including the myocardium and brain. The cysts may persist throughout the life of intermediate hosts.

Intermediate hosts are therefore infected by ingestion of oocysts in cat faeces or by the consumption of tissue cysts (for example in meat) containing bradyzoites. Human infection is widespread: in the USA and UK estimates vary between 16 and 40% of the population being infected, while in continental Europe, Central and South America the figure ranges from 50 to 80% (Dubey and Beattie, 1988). However, disease occurs more rarely and the sequelae depend on the clinical status of the patient. If infection is acquired during pregnancy in the absence of prior immunity, tachyzoites can infect the fetus via the placenta and cause abortion, or congenital disease resulting in mental retardation or blindness in the infant (Remington *et al.*, 2001). Previously acquired latent infection can be reactivated in immunocompromised patients resulting in severe disease, including encephalitis (Luft and Remington, 1992). Many of the biological features of *T. gondii* affect its transmission and epidemiology: robust, highly infectious oocysts can be shed in large numbers by felids; they are infectious for many hosts including man; humans can also acquire infection by the ingestion of tissue cysts in raw or undercooked meat or tachyzoites in milk, via blood transfusion or organ transplant, or transplacentally from an acutely infected mother.

## Origin of the organism

The genus *Toxoplasma* was first proposed in 1909 by Nicolle and Manceaux following the identification of asexual stages of similar parasites in the tissues of birds and mammals, and merozoites in the blood of north African rodents called Gondi. Although several species were named, during the 1930s it was shown that these were identical to the type species *T. gondii*. Identification of sexual stages by electron microscopy during the 1960s provided evidence for the coccidian nature of the parasite (Levine, 1997). During the 1960s and 1970s the heterogeneous life cycle was elucidated by the discovery of sexual stages in the small intestine of cats, which followed the induction of infection in intermediate hosts by inoculation with cat faeces (Dubey and Beattie, 1988).

The first recorded case of human toxoplasmosis was recognized retrospectively in an 11-month old infant with congential hydrocephalus and microphthalmia. The clinical importance of *T. gondii* infection was established during the 1930s with the recognition of *T. gondii* as the aetiolgical agent of encephalomyelitis in neonates, and the description of the classic triad of human congenital toxoplasmosis: retinochoroiditis, hydrocephalus and encephalitis with cerebral calcification (Wolf *et al.*, 1939). Acutely acquired human disease and vertical transmission in humans were both recognized in the 1940s when the classic tetrad of symptoms of congenital toxoplasmosis was described: retinochoroiditis, hydrocephalus or microcephalus, cerebral calcification and psychomotor disturbances (Sabin *et al.*,

1952). *T. gondii* was recognized as a causative agent of lymphadenopathy in the early 1950s (Sinn, 1952). The major sequelae of congenital toxoplasmosis and recognition of risks to patients with malignancies were described in the 1960s (Eichenwald, 1960; Vietzke *et al.*, 1968), with recognition of *T. gondii* as an opportunistic pathogen in AIDS patients in the 1980s (Luft *et al.*, 1984).

## Clinical features

*T. gondii* infection is relatively common but is generally asymptomatic in approximately 85% of immunocompetent individuals, for whom there is no significant health risk; 10–20% of acute infections may develop flu-like illness or cervical lymphadenopathy, with symptoms resolving within a year. However, in immunodeficient patients, including HIV/AIDS patients, organ transplant recipients and patients undergoing chemotherapy, central nervous system disease is common and chorioretinitis or pneumonitis may develop. This is often due to reactivation of chronic or latent infection during immunodeficit, the most significant being toxoplasmic encephalitis due to reactiviation of bradyzoites in brain tissue (Renold *et al.*, 1992). Thus, *T. gondii* is a significant cause of morbidity and mortality in the immunodeficient. Additionally, organ transplant poses a risk, particularly from bradyzoites in infected heart, which may reactivate causing disease with 50% mortality rate.

Infection can also have significant health effects if acquired during pregnancy. Acute primary infection at this time can result in congenital toxoplasmosis and can cause spontaneous abortion, particularly during the first trimester, and fetal abnormalities of the brain, eyes and internal organs (Dunn, 1999). The spectrum of congenital infection is broad, ranging from slightly impaired vision to retinochoroiditis, hydrocephalus, convulsions and intracerebral calcification, ocular disease being most common (Desmonts and Couvreur, 1974; Remington *et al.*, 1995). Accurate diagnosis is essential since treatment of the mother can reduce fetal infection (Desmonts and Courvreur, 1974). Infants with subclinical infection at birth will develop signs or symptoms, frequently chorioretinitis, later in life unless treated.

## Pathogenicity and virulence

Infection with *T. gondii* is common in humans, but clinical disease is largely restricted to risk groups. Although mild symptoms may occur in immunocompetent individuals, the most significant being lymphadenopathy, most infections are asymptomatic and severe manifestations are rare. Mothers infected during pregnancy have a temporary parasitaemia and congenital infections acquired during the first trimester are usually more severe than those acquired later (Desmonts and Courvreur, 1974; Remington *et al.*, 1995). General infection of the fetus occurs following the development of focal lesions in the placenta and subsequently infection is cleared from visceral tissues to localize in the central nervous system. Ocular disease is the most common manifestation of congenital toxoplasmosis. Some controversy has surrounded whether ocular toxoplasmosis developing in later life is a manifestation of prenatal infection or recently acquired primary infection (Holland, 1999). However, multiple cases and outbreaks have been documented with compelling evidence of recent acquisition from a variety of sources.

Infection of immunosuppressed patients may occur in any organ, but encephalitis is the most clinically important manifestation, the predominant lesion in the brain being necrosis, usually of the thalamus (Renold *et al.*, 1992). Initially bilateral severe, persistent headaches develop, responding poorly to analgesics, followed by confusion, lethargy and coma.

There is evidence of at least two clonal lineages within *T. gondii*, supported by genetic analyses. Strains within one lineage are virulent for mice and show vertical transmission while strains in the other lineage are avirulent in mice (Johnson, 1997).

## Causation

*T. gondii* is a widespread parasitic infection, usually causing no symptoms. Infection with *T. gondii* results in life-long immunity, and if acquired before conception presents no substantial risk of transmission to the fetus. The exception is in women with systemic lupus erythematosus or AIDS. Risk of congenital disease appears to be related to gestational age at infection: while risk of transmission is highest in the third trimester, disease is most severe if acquired in the first trimester (Hohlfeld *et al.*, 1989; Dunn *et al.*, 1999). Previously acquired latent infection can be reactivated in immunocompromised individuals and is life-threatening, usually manifesting as encephalititis, although any organ can become infected. It was estimated in the 1980s that 10% AIDS patients in the USA and 30% in Europe died of toxoplasmosis (Dubey and Beattie, 1988).

## Treatment

Treatment of healthy immunocompetent individuals is not generally indicated: neither is treatment of immunocompromised individuals unless symptoms are severe or persistent, when treatment is generally with pyrimethamine and sulphadiazine (Kasper, 1998). Treatment of pregnant women is problematic. While treatment of acute primary infection of the mother during pregnancy can reduce the incidence of congenital infection by approximately half (Desmonts and Couvreur, 1974), treatment depends on gestational age and whether the fetus is known to be infected.

# Survival in the environment

Tachyzoites are fragile outside a host and are sensitive to heat, while bradyzoites can survive refrigeration and some may survive freezing, maintaining infectivity after storage at $-5°C$. However, they do succumb to heat and are killed by heating to $67°C$. Oocysts are shed unsporulated in felid faeces and take 1–5 days to sporulate and thus become infective. They have been detected in soil naturally contaminated with cat faeces (Ruiz *et al.*, 1973) and in soil from gardens (Coutinho *et al.*, 1982) and can survive in soil for 18 months (Frenkel *et al.*, 1975). Maternal contact with soil has been epidemiologically linked to increased risk of congenital toxoplasmosis (Decavalas *et al.*, 1990). Oocysts can become distributed in the environment by mechanical spread by invertebrates.

## Survival in water

Survival of *T. gondii* oocysts is readily assessed by the mouse infectivity model and has shown long-term maintenance of oocyst viability in survival water. Cell culture models, while currently not as sensitive as the mouse model, offer an alternative assay, particularly since there is evidence for at least two clonal lineages within *T. gondii*, one of which is avirulent for mice (Johnson, 1997). Few studies have been published on the survival of oocysts in water, but laboratory tests have shown they can survive for up to 4.5 years at $4°C$ (Dubey, 1998). Survival for several months has been demonstrated in water, as has resistance to many disinfectants, freezing at $-10°C$, and drying, but they are killed by heating to $55–60°C$ (Kuticic and Wikerhauser, 1996).

Infection has been detected in aquatic mammals and may imply survival of oocysts contaminating seawater (Cole *et al.*, 2000).

# Methods of detection

Diagnosis of toxoplasmosis in man is primarily by serology. The methylene blue dye test for the detection of antibodies, introduced in 1948 (Sabin and

Feldman), is maintained as a gold standard for serology tests by reference laboratories, but is labour-intensive and requires a continual supply of live organisms. Since IgG can persist for decades, IgM, which typically persists for 6–9 months, is used as a marker of recent infection, although IgM antibodies have been detected for up to 18 months (Wilson and McAuley, 1999). Other immunological methods include complement fixation tests, direct agglutination tests, ELISA, enzyme-linked fluorescent assay, indirect haemagglutination tests, immunosorbent agglutination test and latex agglutination test. Diagnosis in critical clinical cases (pregnant women, HIV/AIDS patients, neonates etc.) requires specialist testing including enhanced IgA/IgM detection, measurement of IgG avidity and direct detection by PCR, undertaken by reference laboratories.

Although cats may shed high numbers of oocysts for a limited period, concentration methods using high-density sucrose solution may be required and oocysts should be definitively recognized following sporulation and bioassay in mice (Dubey and Beattie, 1988). For epidemiological surveys, however, oocyst detection is impractical and serological prevalence is a better marker of exposure to *T. gondii*. Similarly, the detection of tissue cysts in meat animals is difficult since the numbers present are low and may be as few as 1/100 g meat. Digestion of the sample to rupture the cyst wall and release hundreds of bradyzoites prior to bioassay in mice or application of PCR to detect DNA has been used to assess *T. gondii* in meat samples (Dubey, 1988; Jauregue *et al.*, 2001).

Direct detection of oocysts in environmental samples is possible but hampered by the probable low density of oocysts within the sample. A method for the detection of *T. gondii* in filtered water retentate using a rodent model has been validated experimentally by Isaac-Renton and colleagues (1998), although failed to detect *T. gondii* in water samples following an outbreak in Canada.

## Epidemiology

Horizontal transmission via tissue cysts in pork to man was hypothesized in the 1950s (Weinman and Chandler, 1956) and traditionally the consumption of raw or undercooked meat has been regarded as the main horizontal route of acquisition by humans, although the importance could vary with dietary habits. However, the prevalence of infection in meat-producing animals has decreased and other routes of infection should be considered, particularly environmental routes, given the robust nature of the oocyst stage. Although the epidemiological role of cats in human infection was recognized in the late 1960s, the epidemiological importance of different routes of infection has been little evaluated and it is currently not possible to discriminate between infection caused by oocyst ingestion and that caused by tissue cyst ingestion.

Potential routes of transmission of oocysts to humans include direct contact with infected cat faeces, indirect contact following faecal contamination by cats

of water sources, vegetables, from unwashed hands, contact with contaminated soil, for example by children during play or by adults during gardening, cat litter trays or sand pits used as cat latrines. Seroprevalence in wild scavengers, such as racoons which in the USA was 60% (Dubey and Odening, 2001), indicates widespread occurrence of oocysts in the environment. Vehicles of tissue cyst infection include the consumption of raw or undercooked meat. *T. gondii* is common in many food animals, including sheep, pigs and rabbits. Outbreaks of acute toxoplasmosis in humans have been associated with the consumption of undercooked pork. In one outbreak in Korea, three adults ate a meal of uncooked wild pig spleen and liver and in a second Korean outbreak five soldiers were infected following a meal of raw pig liver (Choi *et al.*, 1997).

Human seroprevalence studies have shown large geographical variation both between and within countries and among different ethnic groups within the same geographic area. However, methods have not always been standardized and may account for some variation. Broad estimates of seroprevalence in women of child-bearing age with no obstetric history have shown a range of 37–58% in Central Europe, Eastern Europe, Australia and Northern Africa; 51–77% in Latin-American and West African countries; 4–39% in Southeast Asia, China and Korea and 11–28% in Scandinavian countries (Tenter *et al.*, 2000). Although mass antenatal screening is undertaken in a limited number of countries, including Canada, Austria and France, elsewhere such screening is not undertaken due to issues of cost-benefit analysis, and reduction in congenital toxoplasmosis and surveillance may rely on reports of congenital disease or acute symptoms (Peyron *et al.*, 2002).

Community outbreaks of toxoplasmosis are infrequently recognized and very few waterborne outbreaks have been reported. In those that have been documented, the association has been made epidemiologically, since accepted microbiological methods for the detection of *Toxoplasma* oocysts in water are lacking. In addition, it is likely that the time periods involved in the detection of clinical cases and the probable episodic nature of the occurrence of oocysts in water would render water testing unreliable for this purpose. The first waterborne outbreak, defined by descriptive epidemiology, was among US army personnel who drank untreated water from a stream in Panama, suspected to have been contaminated with jaguar faeces (Beneson *et al.*, 1982). The first outbreak to be associated with an unfiltered surface municipal water supply occurred in British Colombia, Canada over a 9-month period in 1995, during which 110 acute cases were identified, including 42 pregnant women and 11 neonates identified through an antenatal screening programme (Bowie *et al.*, 1997). Statistical associations were made between acute infection and residence in one water supply distribution zone, with no other conventional source of *Toxoplasma* implicated. The suspected cause of the outbreak was the contamination by domestic, feral or wild cats (including cougar) of the feeder streams to the drinking water reservoir (Isaac-Renton *et al.*, 1998; Aramini *et al.*, 1999). Serological studies of the riparian environment demonstrated that *T. gondii* cycles endemically among the wild and domestic animals of the watershed (Aramini *et al.*, 1999).

Other evidence for a waterborne route has been suggested by Hall *et al.* (1999) following a case of congenital toxoplasmosis in a Jain Hindu. Serological screening of 251 women in India showed no significant difference in seroprevalence among Jians, vegetarian Hindus and non-vegetarian Hindus. Although the numbers were small, the sample suggests that Jain Hindus are exposed to *Toxoplasma*. However, Jains strictly follow dietary rules where fruit and leafy vegetables are thoroughly washed to remove insects, root vegetables are rarely eaten to prevent introduction of soil to the kitchen, milk is boiled and farming or keeping cats is not undertaken and it was suggested that exposure might be from drinking water.

In a toxoplasmosis-endemic area of Brazil, seropositivity was related to socioeconomic group with 84%, 62% and 23% of subjects in the lower, middle and upper socioeconomic groups seropositive respectively (Bahia-Oliveira *et al.*, 2003). In multivariate analyses, drinking unfiltered water increased risk of seropositivity in the lower and middle socioeconomic groups.

## Risk assessment

Advice on the prevention of congenital toxoplasmosis issued to pregnant women includes handwashing after gardening, contact with soil, uncooked meat and cat faeces or litter trays. Interestingly, epidemiological studies have shown a lack of association between seropositivity for *T. gondii* and cat ownership (Stray-Pedersen and Lorentzen-Styr, 1980; Wallace *et al.*, 1993; Kapperud *et al.*, 1996; Cook *et al.*, 2000; Bahia-Oliveira *et al.*, 2003), possibly because cats often defecate away from home. One difficulty, however, is that IgG antibodies to *T. gondii* are long-lived and may reflect past rather than the more recent exposures recalled in questionnaires. IgM persists for 6–9 months and is generally used as an indicator of recent or acute infection (Wilson and McAuley, 1999).

The detection of waterborne outbreaks of toxoplasmosis is more difficult than, say, *Cryptosporidium*, since the symptoms of acute cryptosporidiosis are more likely to be reported. Where antenatal screening is in place, or special studies have been undertaken, links have been made with drinking water and either seropositivity or symptoms (Bowie *et al.*, 1997; Bahia-Oliveira *et al.*, 2003). However, the risk from waterborne infection is generally low where the numbers of cats, particularly wild cats, are not large, since only cats shed environmentally robust oocysts. Thus in some countries the potential for felid faeces to enter the water systems is lower than in others. Tissue cysts may enter water through degrading or rotting animal carcasses but their survival is unknown. Waterpipes should be covered during repair and renovation to prevent host animals entering and flushed prior to entering service.

Survival of oocysts and thus risk of environmental transmission will depend on climatic conditions since oocysts are susceptible to freezing and desiccation (Kuticic and Wikerhauser, 1996) and survival is enhanced by a mild, damp climate. It is likely that full water treatment (coagulation and filtration) will remove *T. gondii* oocysts, although oocysts are probably resistant to chlorine as used to treat potable water supplies.

# Overall risk assessment

*Health effects*: occurrence of illness, degree of morbidity and mortality, probability of illness based on infection:

- Prevalence data show that toxoplasmosis is one of the most common human infections world-wide. In the USA and UK, estimates of infection vary between 16% and 40% of the population, while in continental Europe and Central and South America, the figure ranges from 50% to 80%. Infection is more common in warm climates and at lower altitudes than in cold, high-altitude regions.
- *Toxoplasma* infection in immunocompetent persons is generally asymptomatic. However, 10–20% of patients may develop cervical lymphadenopathy or a flu-like illness. Infections are self-limited, and symptoms usually resolve within a few months to a year.
- Previously acquired latent infection can be reactivated in immunocompromised patients resulting in severe central nervous system disease including encephalitis, although any organ can become infected. Immunocompromised patients are also at greater risk of disease from newly acquired infections.
- If infection is acquired during pregnancy, the fetus can become infected causing abortion or congenital disease resulting in severe effects such as mental retardation or blindness in the infant. Most infants who are infected while in the womb have no symptoms at birth but may develop symptoms later in life if left untreated. Only a small percentage of infected newborns have serious eye or brain damage at birth. Though a fetus is more likely to become infected during the third trimester, the outcome is usually more severe in infections acquired during the first trimester. An estimated one-half of untreated maternal infections are transmitted to the fetus and an estimated 400–4000 cases of congenital toxoplasmosis occur each year in the USA.

*Exposure assessment*: routes of exposure and transmission, occurrence in source water, environmental fate:

- Cats are the only known definitive hosts. Intermediate hosts, including humans, are therefore infected by ingesting oocysts from faecally-contaminated food, hands, or water; from organ transplantation or blood transfusion (rarely); or from transplacental transmission.
- Very few waterborne outbreaks have been reported. The risk from waterborne infection is generally low where the numbers of cats, particularly wild cats, are not large, since only cats shed environmentally robust oocysts. Thus, in some countries the potential for felid faeces to enter the water systems is

lower than in others. There is the potential for *T. gondii* oocysts to enter surface water through wastewater effluent discharge if people flush cat faeces into the sewage system.

- Oocysts can survive in the environment for several months and are remarkably resistant to disinfectants, freezing and drying, but are killed by heating to 70°C for 10 minutes. Laboratory tests have shown they can survive for up to 4.5 years at 4°C.

*Risk mitigation*: drinking-water treatment, medical treatment:

- *T. gondii* oocysts are resistant to chlorine, though the process of coagulation and flocculation will probably remove oocysts.
- Medical treatment is not needed for a healthy person who is not pregnant. If symptoms exist, they will usually resolve within a few weeks. Treatment with pyrimethamine plus sulphadiazine may be recommended for pregnant women or persons who have weakened immune systems.

## Future implications

It is likely that further studies will identify waterborne risks in areas of high endemnicity of *Toxoplasma* infection. Other routes of contamination may be identified, such as the practice of emptying cat litter trays into domestic toilets and thus introducing oocysts to the sewage system. Work on *Giardia* and *Cryptosporidium* has shown that variable removal of oocysts occurs during sewage treatment (Robertson *et al.*, 2000) and thus *T. gondii* oocysts could enter surface water through effluent discharge.

## References

Aramini, J.J., Stephen, C., Dubey, J.P. *et al.* (1999). Potential contamination of drinking water with *Toxoplasma gondii* oocysts. *Epidemiol Infect*, 122: 305–315.

Bahia-Oliveira, L.M.G., Jones, J.L., Azevedo-Sila, J. *et al.* (2003). Highly endemic, waterborne toxoplasmosis in North Rio de Janeiro State, Brazil. *Emerging Infect Dis*, in press.

Beneson, M.W., Takafuji, E.T., Lemon, S.M. *et al.* (1982). Oocyst-transmitted toxoplasmosis associated with ingestion of contaminated water. *New Engl J Med*, 300: 694–699.

Bowie, W.R., King, S.A., Werker, D.H. *et al.* (1997). Outbreak of toxoplasmosis associated with municipal drinking water. *Lancet*, 350: 173–177.

Choi, W.Y., Nam, H.W., Kwak, N.H. *et al.* (1997). Foodborne outbreaks of human toxoplasmosis. *J Infect Dis*, 175: 1280–1282.

Cole, R.A., Lindsay, D.S., Howe, D.K. *et al.* (2000). Biological and molecular characterisations of Toxoplasma gondii strains obtained from southern sea otters (*Enhydra lutris nereis*). *J Parasitol*, 86: 526–530.

Cook, A.J., Gilbert, R.E., Buffolano, W. *et al.* (2000). Source of Toxoplasma infection in pregnant women: European multicentre case-control study. European Research network on congenital toxoplasmosis. *Br Med J*, **321**: 142–147.

Coutinho, S.G., Lobo, R. and Dutra, G. (1982). Isolation of *Toxoplasma* from the soil during an outbreak of toxoplasmosis in a rural area of Brazil. *J Parasitol*, **68**: 866–868.

Decavalas, G., Papapetropoulou, M., Giannoulaki, E. *et al.* (1990). Prevalence of *Toxoplasma gondii* antibodies in gravidas and recently aborted women and study of risk factors. *Eur J Epidemiol*, **6**: 223–226.

Desmonts, G. and Couvreur, J. (1974). Congenital toxoplasmosis. A prospective study of 378 pregnancies. *New Engl J Med*, **290**: 1110–1116.

Dubey, J.P. (1988). Refinement of pepsin digestion method for isolation of *Toxoplasma gondii* from infected tissues. *Vet Parasitol*, **74**: 75–77.

Dubey, J.P. (1998). *Toxoplasma gondii* oocyst survival under defined temperatures. *J Parasitol*, **84**: 862–865.

Dubey, J.P. and Beattie, C.P. (1988). *Toxoplasmosis of Animals and Man*. Boca Raton: CRC Press.

Dubey, J.P. and Odening, K. (2001). Toxoplasmosis and related infections. In *Parasitic Diseases of Wild Animals*, Samuel, W.M., Pybus, M.J. and Kocan, A.A. (eds). Ames: Iowa State University Press, pp. 478–5190.

Dunn, D., Wallon, M., Peyron, F. *et al.* (1999). Mother to child transmission of toxoplasmosis: risk estimates for clinical counselling. *Lancet*, **353**: 1829–1833.

Eichenwald, H.A. (1960). A study of congenital toxoplasmosis. In *Human Toxoplasmosis*, Siim, J.C. (ed.). Copenhagen: Ejnar Munksgaard Forlag, pp. 41–49.

Frenkel, J.K. (1973). Toxoplasmosis: parasite life cycle, pathology and immunology. In *The coccidian. Eimeria, Isospora, Toxoplasma and related genera*, Hammond, D. and Lond, P.L. (eds). Baltimore: University Park Press, pp. 343–410.

Frenkel, J.K., Ruiz, A. and Chinchilla, M. (1975). Soil survival of *Toxoplasma* oocysts in Kansas and Costa Rica. *Am J Trop Med Hyg*, **24**: 439–443.

Hohlfeld, P., Daffos, F., Thulliez, P. *et al.* (1989). Fetal toxoplasmosis: outcome of pregnancy and infant follow-up after in utero treatment. *J Pediatr*, **115**: 765–769.

Holland, G.N. (1999). Reconsidering the pathogenesis of ocular toxoplasmosis. *Am J Ophthalmol*, **128**: 502–505.

Isaac-Renton, J., Bowie, W.R., King, A. *et al.* (1998). Detection of *Toxoplasma gondii* oocysts in drinking water. *Appl Environ Microbiol*, **64**: 2278–2280.

Jauregue, L.H., Higgins, J.A., Zarlenga, D.S. *et al.* (2001). Development of a real-time PCR assay for the detection of *Toxoplasma gondii* in pig and mouse tissues. *J Clin Microbiol*, **39**: 2065–2071.

Johnson, A.M. (1997). Speculation on possible life cycles for the clonal lineages in the genus *Toxoplasma*. *Parasitol Today*, **13**: 393–397.

Kapperud, G., Jenum, P.A., Stray-Pedersen, B. *et al.* (1996). Risk factors for *Toxoplasma gondii* infection in pregnancy. Results of a prospective case-control study in Norway. *Am J Epidemiol*, **144**: 405–412.

Kasper, L.H. (1998). Toxoplasma infection. In *Harrison's Principles of Internal Medicine*, 14th edn, Fauci, A.S., Isselbacher, K.J. and Wilson, J. (eds). New York: McGraw-Hill, Health Professions Division, pp. 1535–1539.

Kuticic, V. and Wikerhauser, T. (1996). Studies of the effect of various treatments on the viability of *Toxoplasma gondii* tissue cysts and oocysts. In *Toxoplasma gondii*, Gross, U. (ed.). Berlin: Springer-Verlag, pp. 261–265.

Levine, N.D. (1977). Taxonomy of *Toxoplasma*. *J Parasitol*, **24**: 36–41.

Luft, B.J., Brooks, R.G., Conley, F.K. *et al.* (1984). Toxoplasmic encephalitis in patients with acquired immune deficiency syndrome. *JAMA*, **252**: 913–917.

Luft, B.J. and Remington, J.S. (1992). Toxoplasmic encephalititis in AIDS. *Clin Infect Dis*, **15**: 211–222.

Nicolle, C. and Manceaux, L. (1909). Sur un protozoaire nouveau du gondii. *CR Hebd Séanes Acad Sci*, **148**: 369–372.

Remington, J.S., McLeod, R. and Desmonts, G. (2001). Toxoplasmosis. In *Infectious Diseases of the Fetus and Newborn*, 5th edn, Remington, J.S. and Klein, J.O. (eds). Philadelphia: WB Saunders, pp. 205–346.

Renold, C., Sugar, A., Chave, J.P. *et al.* (1992). *Toxoplasma* encephalitis in patients with the acquired immunodeficiency syndrome. *Medicine*, **71**: 224–239.

Robertson, L.J., Paton, C.A., Campbell, A.T. *et al.* (2000). *Giardia* cysts and *Cryptosporidium* oocysts at sewage treatment works in Scotland, UK. *Water Res*, **34**: 2310–2322.

Ruiz, A., Frenkel, J.K. and Cerdas, L. (1973). Isolation of *Toxoplasma* from soil. *J Parasitol*, **59**: 204–206.

Sabin, A.B. and Feldman, H.A. (1948). Dyes as microchemical indicators of a new immunity phenomenon affecting a protozoan parasite (Toxoplasma). *Science*, **108**: 660–663.

Sabin, A.B., Eichenwald, H., Feldman, H.A. *et al.* (1952). Present status of clinical manifestations of toxoplasmosis in man: indications and provisions for routine serologic diagnosis. *JAMA*, **150**: 1063–1069.

Siin, J.C. (1952). Studies on acquired toxoplasmosis. II. Report of a case with pathologiocal changes in lymph node removed at biopsy. *Acta Pathol Microbiol Scand*, **30**: 104–108.

Stray-Pedersen, B. and Lorentzen-Styr, A.M. (1980). Epidemiological aspects of *Toxoplasma* infections among women in Norway. *Acta Obstet Gynecol Scand*, **59**: 323–326.

Tenter, A.M., Heckeroth, A.R. and Weiss, L.M. (2000). *Toxoplasma gondii*: from animals to humans. *Int J Parasitol*, **30**: 1217–1258.

Vietzke, W.M., Gelderman, A.H., Grimley, P.M. *et al.* (1968). Toxoplasmosis complicating malignancy: experience at the National Cancer Institute. *Cancer*, **21**: 816–827.

Wallace, M.R., Rossetti, R.J. and Olson, P.E. (1993). Cats and toxoplasmosis risk in HIV-infected adults. *JAMA*, **269**: 76–77.

Weinman, D. and Chandler, A.H. (1956). Toxoplasmosis and man and swine – an investigation of the possible relationship. *JAMA*, **161**: 229–232.

Wilson, M. and McAuley, J.B. (1999). Toxoplasma. In *Manual of Clinical Microbiology*, 7th edn, Murray, P. (ed.). Washington, DC: ASM Press, pp. 1374–1451.

Wolf, A., Cowen, D. and Paige, B.H. (1939). Toxoplasmic encephalomyelitis. III. A new case of granulomatous encephalitis due to a protozoan. *Am J Pathol*, **15**: 657–694.

Part 4

# Viruses

# 24

# Common themes

## Introduction

The challenge of proving that outbreaks or sporadic instances of disease are attributable to waterborne viruses is much greater than that associated with bacterial or protozoan disease. Viruses are more difficult to detect in all aquatic matrices, they are often associated with non-specific infections and epidemiology is usually difficult because the triviality of the majority of virus infections means that very few are reported to medical authorities and therefore tracing causes of outbreaks is a difficult task.

Nevertheless, evidence has accumulated over the last 30 years which links the ingestion of water contaminated by faecally-derived viruses with disease and the numbers of viruses for which this evidence is forthcoming is increasing as technology for their detection improves. Very early studies done in the 1940s as polioviruses were beginning to be propagated in cell culture indicated that the virus could survive in sewage and be transmitted to receiving waters, from where it could be recovered in an infectious state (Melnick, 1947). The notion, since largely refuted, that swimming in domestic pools during the summer led to outbreaks of poliomyelitis was held strongly in the USA for some years since the virus was on occasion recovered from domestic swimming pools where chlorination was defective. The enormous outbreak of infectious hepatitis, now recognized as caused by hepatitis E virus (HEV) in Delhi in 1955/56 demonstrated the potential for infection of large numbers of people when sanitary conditions are poor.

Despite waterborne viruses being enteric, that is to say they establish their initial infection in the gastrointestinal tract, conclusive links between polluted water and viruses causing specific instances of enteric disease have been difficult to demonstrate, mainly due to the problems in detecting viruses in aquatic matrices. Many groups of viruses inhabit the gastrointestinal tract of humans and animals, only some of which cause sufficient local damage for gastroenteritis or other illnesses to result. Primarily the faecal-oral route transmits viruses adapted to replicate in the gut, but epidemiological patterns vary. Although they may be present in the same water, enteric viruses have a range of important attributes that distinguish them from other enteric microorganisms. The infectious dose is low; it has been accepted for many years that a single clump, or plaque forming unit (pfu), of virus may be capable of initiating a viral infection. Indeed, it was on that premise that Berg (1967) stipulated that any virus in water would constitute a hazard. It is probable that this influential statement was the basis for the enterovirus parameter of the EU Bathing Water Directive, 1976. In common with the majority of virus families under normal circumstances, the enteric viruses are host specific, so animal enteric viruses do not usually infect humans. Viruses are the smallest of the microorganisms and the most likely to be able to travel through the pore spaces within rocks (Wellings *et al.*, 1975).

The principal areas for consideration of waterborne viral disease are the nature of outbreaks, the aetiology of the different viruses causing waterborne infection, the methods for detecting waterborne viruses and the survival of the agents in the aquatic environment. Predictions of the risk of contracting viral disease through ingestion of polluted water is becoming a reality as confidence in detection techniques improves and future trends in water virology will undoubtedly focus on acquisition of data for risk assessment models. Readily identified viral outbreaks are those of gastroenteritis as the incubation period is a matter of days and the symptoms well defined; noroviruses are currently the most commonly recognized cause. Viral hepatitis outbreaks are recognizable due to the severity of the symptoms, although the incubation period of 1–2 months makes investigation difficult; hepatitis A (HAV) and HEV outbreaks are well known. Outbreaks due to specific viruses are discussed in the section on each virus but general issues are addressed below.

Viral waterborne disease may be transmitted by consumption of drinking water, by immersion in recreational water or by contact through skin or inhalation if contamination of the water with human sewage has occurred. Pollution by animal faecal material is unlikely to pose a risk of transmission of animal viruses. The health effects caused by waterborne pathogens generally have been reviewed extensively, e.g. by Galbraith (1987) and by Stanwell-Smith (1994) who documented drinking water-associated disease outbreaks in the UK, and Hunter (1997), Dadswell (1993) and Pruss (1998) who considered disease associated with recreational water contact. A system for attaching strong or possible association between disease and exposure to water was developed by Tillett *et al.* (1998) and is based on the epidemiological picture, laboratory diagnosis and water quality.

Drinking water-associated outbreaks of disease are usually the result of one of four events:

- inadequate removal of organisms during treatment
- failure in the treatment process
- failure in the chlorination or other disinfection system
- breaks in the integrity of the distribution infrastructure or of sewage removal.

These events may occur alone or in concert, for example in Bramham, Yorkshire in 1980 (Short, 1988) a large outbreak of gastroenteritis, involving over 3000 people, occurred when sewage from a broken pipe contaminated drinking water supply pipes. At the time of the breakage the chlorination plant was faulty, so drinking water contaminated with sewage was distributed without adequate disinfection.

Viruses have been identified as responsible for a number of drinking water-associated disease outbreaks world-wide, though much less frequently than bacteria or protozoa. Though most transmission is by person to person, noroviruses (members of the family *Caliciviridae*) are the most widely implicated in the context of waterborne viral disease, though the evidence has often been circumstantial. Greater confidence in the association between water exposure and disease is generated when the same organism is detected in patients and in the water to which they have been exposed and there is a temporal consistency in the detection.

It has been even more problematic to link the use of recreational water with a specific virological cause of illness than drinking water. The most readily identifiable risks are those associated with swimming pools, spa-pools in leisure centres and domestic whirlpool baths. Outbreaks of adenovirus conjunctivitis have been associated with swimming pool exposure (Martone *et al.*, 1980; Turner *et al.*, 1987). Swimming pools and the spread of poliovirus was of great concern in the USA during the 1940s and 1950s. It is not clear whether transmission did in fact occur via the water, although the potential for spread is illustrated by an outbreak of echovirus 30 in Northern Ireland, where an outdoor pool was contaminated with vomit on its day of opening, followed by an outbreak of headache, diarrhoea and vomiting in swimmers (Kee *et al.*, 1994). Chlorine levels were said to be satisfactory and no other pathogens were detected but it was not stated if faeces were tested for the presence of noroviruses.

Illness after recreational contact in seawater is particularly difficult to confirm and quantify. Cohort studies on bathers have involved recruitment of individuals or by using individuals already on a section of beach. It was shown in the UK prospective studies (Pike, 1991a,b, 1994) that an increase in mild gastroenteritis, eye and ear symptoms occurred after swimming in microbiologically poor seawater. No virus or other microorganism was shown to be associated with any of these symptoms. These broad conclusions have been confirmed in studies all over the world. The quality of the water is most closely linked with health effects if enterococci are used as indicators of faecal pollution (Cabelli *et al.*, 1983; Fleisher *et al.*, 1993; Kay *et al.*, 1994). Although sewage is the most likely source of faecal contamination the bathers are also potential polluters (Fattal *et al.*, 1991).

Freshwater recreation has been more closely linked with waterborne illness than contact with seawater. Canoeists using the River Trent developed gastro-enteritis caused by noroviruses (Gray *et al.*, 1997) and undiagnosed gastroen-teritis was found to be more common in canoeists using the River Trent than those using unpolluted lakes in North Wales (Fewtrell *et al.*, 1992). Studies in The Netherlands indicate that illness and norovirus infection are linked to swimming in river water (van Olphen *et al.*, 1991; Medema *et al.*, 1995; de Roda Husman, personal communication).

River water used for irrigation has contaminated crops and transmitted disease. In the UK, the most likely viral pathogen would be the noroviruses. Regulations to limit the use of such water to particular crops and certain periods should reduce the small risk of transmission of these pathogens. Use of digested sewage sludge as a fertilizer in agriculture is currently an important issue. Retailers, such as the large supermarkets, are concerned that crops may be at risk of contamination, although no evidence is currently available that condi-tions in the UK have resulted in consumer disease. Research is now underway to improve the detection methods for enteric viruses, *Salmonella*, *E. coli* O157 and protozoa. The digestion processes used in sewage treatment are being more precisely characterized and the regulations governing the use of sewage sludge have been reviewed. The ADAS matrix (1999) has recently been drawn up that more closely defines the use of sewage sludge.

Bivalve shellfish, such as oysters, cockles and mussels are grown commercially in estuaries and in shallow seawater all around the coast of Britain. They are filter feeders of particulate matter that may include faecal material and conse-quently viruses. Norovirus has been identified in shellfish tissue (Lees *et al.*, 1995), has been detected in stools of individuals with gastroenteritis who have eaten shellfish and has been strongly linked in epidemiological studies to out-breaks of gastroenteritis (Advisory Committee on the Microbiological Safety of Food, 1998). Oysters are the most frequent cause of outbreaks as they are often eaten raw or poorly cooked. Between 1992 and 1996, 81 HAV infec-tions transmitted by shellfish were reported to CDSC. An astrovirus outbreak was reported at a workshop on food-borne viral infections in 1987 (Kurtz and Lee, 1987) to be associated with shellfish consumption.

# References

Advisory Committee on the Microbiological Safety of Food. (1998). *Report on foodborne viral infections*. London: HMSO.
ADAS (1999). *The safe sludge matrix. Guidelines for the application of sewage sludge to agricultural land*. London: Agricultural Development and Advisory Service (Ministry Agriculture Food and Fisheries).

Berg, G. (1967). *Transmission of Viruses by the Water Route*. New York: Wiley InterScience Publishers, pp. 1–484.

Cabelli, V.J., Dufour, A.P., McCabe, L.J. *et al.* (1982). Swimming associated gastro-enteritis and water quality. *Am J Epidemiol*, **115**: 606–616.

Dadswell, J.V. (1993). Microbiological quality of coastal waters and its health effects. *Int J Environ Hlth Res*, **3**: 32–46.

Fattal, B., Peleg-Olevsky, E. and Cabelli, V.J. (1991). Bathers as a possible source of contamination for swimming-associated illness at marine bathing beaches. *Int J Environ Hlth Res*, **1**: 204–214.

Fewtrell, L., Godfree, A.F., Jones, F. *et al.* (1992). Health effects of white-water canoeing. *Lancet*, **339**: 1587–1589.

Fleisher, J.M., Jones, F., Kay, D. *et al.* (1993). Water and non-water related risk factors for gastro-enteritis among bathers exposed to sewage contaminated marine waters. *Int J Epidemiol*, **22**: 698–708.

Galbraith, N.S. (1987). Water and disease after Croydon: a review of waterborne and water associated disease in the UK 1937–86. *J Instit Water Environ Manag*, **1**: 7–21.

Gray, J.J., Green, J., Gallimore, C. *et al.* (1997). Mixed genotype SRSV infections among a party of canoeists exposed to contaminated recreational water. *J Med Virol*, **52**: 425–429.

Hunter, P.R. (1997). *Waterborne Disease: Epidemiology and Ecology*. Chichester: John Wiley & Sons.

Kay, D., Fleisher, J.M., Salmon, R.L. *et al.* (1994). Predicting likelihood of gastroenteritis from sea bathing: results from randomised exposure. *Lancet*, **344**: 905–909.

Kee, F., McElroy, G., Sewart, D. *et al.* (1994). A community outbreak of echovirus infection associated with an outdoor swimming pool. *J Public Hlth*, **16**: 145–148.

Kurtz, J.B. and Lee, T.W. (1987). Astroviruses: human and animal. In *Novel Diarrhoea Viruses. Ciba foundation symposium 128*, Bock, G. and Whelan, J. (eds). London: J. and A. Churchill, pp. 92–107.

Lees, D.N., Hensilwood, K., Green, J. *et al.* (1995). Detection of small round structured viruses in shellfish by reverse transcription-PCR. *Appl Environ Microbiol*, **61**: 4418–4424.

Martone, W.J., Hierholzer, J.C., Keenlyside, R.A. *et al.* (1980). An outbreak of adenovirus type 3 disease at a private recreation center swimming pool. *Am J Epidemiol*, **111**: 229–237.

Medema, G.J., van Asperen, I.A., Kokman-Houweling, J.M. *et al.* (1995). The relationship between health effects in triathletes and microbiological quality of freshwater. *Water Sci Technol*, **31**: 19–26.

Melnick, J.L. (1947). Poliomyelitis virus in urban sewage in epidemic and in non-epidemic times. *Am J Hyg*, **45**: 240–253.

Pike, E.B. (1991a). *Health effects of sea bathing Phase I – Pilot studies at Langland Bay 1989*. Medmenham: WRc.

Pike, E.B. (1991b). *Health effects of sea bathing Phase II – Studies at Ramsgate and Moreton*. Medmenham: WRc.

Pike, E.B. (1994). *Health effects of sea bathing (WM 9021) Phase III – Final report to the Department of the Environment*. Medmenham: WRc.

Pruss, A. (1998). Review of epidemiological studies on health effects from exposure to recreational water. *Int Epidemiol Assoc*, **27**: 1–9.

Short, C.S. (1988). The Bramham incident, 1980: an outbreak of waterborne infection. *J Inst Water Environ Manag*, **2**: 383–390.

Stanwell-Smith, R. (1994). Recent trends in the epidemiology of waterborne disease. In *Water & Public Health*, Golding, A.M.B. (ed.). London: Smith-Gordon, pp. 39–56.

Tillett, H.E., de Louvois, J. and Wall, P. (1998). Surveillance of outbreaks of waterborne infectious disease: categorizing levels of evidence. *Epidem Infect*, **120**: 37–42.

Turner, M., Istre, G.R., Beauchamp, H. *et al.* (1987). Community outbreak of adenovirus type 7a infections associated with a swimming pool. *South Med J*, **80**: 712–715.

van Olphen, M., de Bruin, H.A.M., Havelaar, A.H. *et al.* (1991). The virological quality of recreational waters in the Netherlands. *Water Sci Technol*, **24**: 209–212.

Wellings, F.M., Lewis, A.L., Mountain, C.W. *et al.* (1975). Demonstration of virus in groundwater after effluent discharge onto soil. *Appl Environ Microbiol*, **29**: 751–757.

## 25

# The survival and persistence of viruses in water

The term 'survival' in this context means the ability of an infectious unit of virus to remain infectious in a body of water over a defined time. 'Persistence' refers to the continued presence of a particular virus type in a body of water over a period of time. The rates of decay for viruses vary in respect of a variety of factors, including the methods by which samples are taken, processed and assayed. There continues to be a need to standardize methodology as far as possible in order to accommodate this aspect of the variation. Cautionary notes have been sounded by Morris and Irving (1998).

Since viruses are obligate intracellular parasites and unable to multiply outside the host, the number of infectious enteric virus particles can only decline after being shed into the environment. However, enteric virions are robust and may survive for long periods dependent on the environmental conditions. Data on the survival characteristics of the individual virus types are generally scanty. Virus particles, comprising a protein coat enclosing the genome of nucleic acid, are protected from degradation by faecal organic material. Particles are destroyed in the environment by desiccation, by UV light, heat above 56°C, digestion by microorganisms and by predation. Particles shed in faecal material in soil at low temperature or in sediment under water will survive the longest and may be detectable for months or years.

Sewage may contain any of the human enteric virus groups that circulate within a population. Those present will vary from season to season, from year to year and between different geographical locations. During sewage treatment the heavier solid material settles to form sludge and, as virus particles clump together and are often attached to organic debris, the virions are most likely to be present in sludge. Treatment by biological filtration or the activated sludge process reduces the number of virus particles as a result of microbiological predation by an estimated 50% and 90% respectively (Berg, 1973) and discharge of effluent into fresh or marine waters will reduce the number of viruses further by microbial activity, the action of light and dispersal.

Treatment processes to reduce the numbers of microorganisms in sludge may reduce the numbers of infectious viruses present but the mechanisms are poorly understood since the variable and toxic nature of sludge has made it very difficult to undertake reproducible studies. Berg *et al.* (1988) showed that enteroviruses may survive up to 38 days in aeration sludges at 5°C at pH 6–8, but also that they sometimes survive at pH 3.5. Consequently the numbers and types of viruses present in fully treated sludge are unknown and so the risk from viruses is not yet possible to quantify. Mesophilic anaerobic digestion is currently the most widespread method of sludge treatment in the UK. This takes place at about 42°C for approximately a month, during which time virus particles may be degraded or consumed by other microorganisms. Anaerobic digestion or composting are processes more likely to produce sludge that contains residual infectious particles than the more extreme liming at pH 9. Drying the sludge or the production of dry pellets are treatments increasing in use and should eliminate all infectious virus particles.

Among the enteric viruses the enteroviruses and hepatitis A virus (HAV) have been the most intensively studied with respect to their survival and persistence because, although not usually associated with intestinal disease enteroviruses they are relatively easily recovered from aquatic matrices and can be grown in cell culture and HAV is a well-established waterborne pathogen. Other types of enteric virus such as rotaviruses and noroviruses, which do cause intestinal disease, are more difficult to detect and infectivity of these agents cannot easily, if at all, be determined in the laboratory. Hence the enteroviruses have been used as a model for survival of other enteric virus types.

Viruses are generally less numerous in sewage than bacteria; whereas faeces may consistently contain $10^6$–$10^7$ coliforms/g, viruses may only be present occasionally and it is generally accepted that raw (inlet) sewage contains approximately $10^8$ coliforms/100 ml and $10^3$–$10^4$ pfu enteroviruses per litre (though other individual studies report widely ranging figures).

Attempts have been made for more than 25 years to determine virus survival in different matrices, though relatively few have been done in recent years (e.g. Enriquez *et al.*, 1995; Wait and Sobsey, 2001). Since bacterial indicator organisms have been used to determine microbial water pollution, comparisons of bacterial and viral survival are common. Comments on experimental conditions and apparatus for survival studies, given elsewhere in the literature pertaining to bacterial or protozoan studies, apply generally to virus survival studies too. The general consensus is that viruses are more robust than bacteria; Vasl *et al.* (1981) in a field study of sewage-polluted marine coastal waters showed that while total and faecal coliforms and enterococci were correlated with enterovirus levels, enteroviruses were more resistant than coliforms and of resistance equal to enterococci. Parameters in this, as in other studies, varied; volumes of water analysed for enteroviruses ranged from 35 to 85 litres per sample, water temperature varied between 15°C and 29°C and the conductivity also varied. Nasser *et al.* (1993) compared the survival of poliovirus, hepatitis A virus (HAV) and $F^+$ phage in wastewater. Reduction of poliovirus and HAV was influenced by temperature. At 10°C the titres of these agents was reduced by 1–2 $\log_{10}$ after 90 days and the phage titre remained unaffected at the same temperature. At higher temperatures reduction in titres were seen. Kadoi and Kadoi (2001) showed that feline calicivirus, sometimes used as a cell-culturable surrogate for norovirus, remained infectious in seawater at up to 10°C for up to 30 days. Survival comparisons in respect of viruses in seawater and on sediments were done by Tsai *et al.* (1993), who found that, in seawater at 25°C, $F^+$ phage was inactivated faster than poliovirus, HAV and rotaviruses; at lower temperatures and in the presence of sediments there was less difference in inactivation rates, though it protected poliovirus and HAV at 5°C and 25°C, whereas it accelerated the inactivation of rotavirus. Abad *et al.* (1997) found 2 $\log_{10}$ units reduction in astrovirus infectivity titre after 60 days at 4°C and 3.2 $\log_{10}$ units at 20°C. In all tests the $T_{99.99}$ parameter (4 $\log_{10}$ reduction) was determined and for the enteric viruses was found to range from 2 weeks to years. Callahan *et al.* (1995) looked at virus survival in different types of seawater and concluded that $F^+$ phages survived about the same as poliovirus, undergoing a 4 $\log_{10}$ reduction in about one week and Johnson *et al.* (1997) compared the survival of poliovirus under different environmental conditions with that of *Cryptosporidium*, *Giardia* and *Salm. typhimurium* in marine waters, with reference to enterococci as the determinand of microbial water quality. The order of survival in sunlight was *Cryptosporidium* (greatest), poliovirus, *Giardia* and *Salm. typhimurium*, though poliovirus survived less well in the dark. The $T_{90}$ (the time required to reduce the number of viruses by 90%) values for poliovirus ranged from 10 hours to 26 hours, much less than previous studies, though some of this difference was probably due to different experimental conditions.

That the $T_{90}$ values vary under different circumstances is evident, however, the extent of this variation needs to be determined. It is also clear that viruses are more resistant to environmental degradation than other microorganisms, especially bacteria. Morris and Irving (1998) cited a series of seven papers which reported $T_{90}$ values for enteroviruses ranging, under different conditions, from 14 to over 288 hours.

The issue of virus adsorption to sediments needs to be considered, both in the context of the natural state (e.g. above) and in planning experimental protocols for virus survival studies. Turbidity dramatically attenuates the penetration of light into the water column; the nature of the material causing the turbidity may also markedly affect the survival of microorganisms, particularly the viruses as these can be readily absorbed to some materials such as clay and this could be more important than light for these organisms (Gerba and Schaiberger, 1975).

# References

Abad, F.X., Pinto, R.M., Villena, C. *et al.* (1997). Astrovirus survival in drinking water. *Appl Environ Microbiol*, **63**: 3119–3122.

Berg, G. (1973). Removal of viruses from sewage, effluents and waters. *Bull Wld Hlth Org*, **49**: 451–460.

Berg, G., Sullivan, G. and Venosa, A.D. (1988). Optimum pH levels for eluting enteroviruses from sludges solids with beef extract. *Appl Environ Microbiol*, **54**: 83–841.

Callahan, K.M., Taylor, D.J. and Sobsey, M. D. (1995). Comparative survival of hepatitis A virus, poliovirus and indicator viruses in geographically diverse seawaters. *Water Sci Technol*, **31**: 189–193.

Enriquez, C.E. and Gerba, C.P. (1995). Concentration of enteric adenovirus 40 from tap, sea and waste water. *Water Res*, **29**: 2554–2560.

Gerba, C.P. and Schaiberger, G.E. (1975). Effect of particulates on virus survival in seawater. *J Water Pollution Control Fed*, **27**: 125–129.

Johnson, D.C., Enriquez, C.E., Pepper, I.L. *et al.* (1997). Survival of *Giardia*, *Cryptosporidium*, poliovirus, and *Salmonella* in marine waters. *Water Sci Technol*, **35**: 261–268.

Kadoi, K. and Kadoi, B.K. (2001). Stability of feline caliciviruses in marine water maintained at different temperatures. *New Microbiol*, **24**: 17–21.

Morris, R. and Irving, T. (1998). *Review of the Fate of Enteroviruses in the Environment.* Report WW-11D: United Kingdom Water Industry Research Ltd Research and Development – Bathing Water Policy.

Nasser, A. M., Tchorch, Y. and Fattal, B. (1993). Comparative survival of *E. coli*, F$^+$ bacteriophages, HAV and poliovirus 1 in wastewater and groundwater. *Water Sci Technol*, **27**: 401–407.

Tsai, Y.L., Sobsey, M.D., Sangermano, L.R., Palmer, C.J. (1993). Simple method of concentrating enteroviruses and hepatitis A virus from sewage and ocean water for rapid detection by reverse transcriptase-polymerase chain reaction. *Appl Environ Microbiol*, **58**: 3488–3491.

Vasl, R., Fattal, B., Katzenelson, E. *et al.* (1981). Survival of enteroviruses and bacterial indicator organisms in the sea. In *Viruses and Wastewater Treatment*, Goddard, M. and Butler, M. (eds). Oxford: Pergamon Press, pp. 113–116.

Wait, D.A. and Sobsey, M.D. (2001). Comparative survival of enteric viruses and bacteria in Atlantic Ocean seawater. *Water Sci Technol*, **43**: 139–142.

# 26

# Methods for the detection of waterborne viruses

Detection of waterborne viruses is more complex than detection of other microorganisms because of difficulties in concentrating the sample and then in detecting the virus by cell culture or molecular biological means. With the exception of sewage, where the quantity of virus present may be sufficient to permit detection without further concentration, viruses in water are too dilute to be detected by direct analysis. Groundwater and drinking water will contain very few viruses and 100 l or more will need to be processed, while recreational fresh or marine waters may contain many more viruses so processing 10 l samples will be sufficient. Water samples are therefore taken through a two-stage process, first to concentrate the virus into a smaller volume (usually 5–10 ml), then detection of virus in the concentrate. The concentrate may be inoculated into cell cultures to detect infectious virus; if this is done in a quantitative fashion any virus present can be enumerated, the count being reported as plaque-forming units (pfu), tissue culture infectious doses ($TCD_{50}$), or MPN units. Where necessary virus can be isolated and identified from the cell cultures. Viruses which do not produce an identifiable cytopathic effect in culture may nevertheless be detected by immunoperoxidase or immunofluorescence staining. Where there is no viral replication in culture, as is the case with

(a)　　　　　　　　　　　　　　　　　　　　　　　(b)

**Figure 26.1**　Membrane filtration.

most gastroenteritis viruses, then molecular biological detection procedures may be used.

## Concentration methods

Viruses exhibit polarity and can adsorb to a wide variety of charged matrices, which may be immobilized (such as membranes) or fluid (such as glass powder) (Figures 26.1 and 26.2). They may thus be concentrated by adsorption to such matrices. Considered as protein, virus particles have a high relative molecular mass ($M_r > 10^6$) and lend themselves to sedimentation from suspension by ultracentrifugation and to concentration by ultrafiltration. Based on these general properties, numerous methods have been devised for the concentration of viruses from water (Table 26.1). These have been reviewed extensively by Wyn-Jones and Sellwood (1998) in respect of enteroviruses and by Wyn-Jones and Sellwood (2001) for other virus groups.

To be of optimum practical use any method must fulfil the following criteria (after Block and Schwartzbrod, 1989):

- be technically easy to accomplish in a short time
- have a high virus recovery rate
- concentrate a range of viruses
- provide a small volume of concentrate
- not be costly
- be capable of processing large volumes of water
- be repeatable (within a laboratory) and be reproducible (between laboratories).

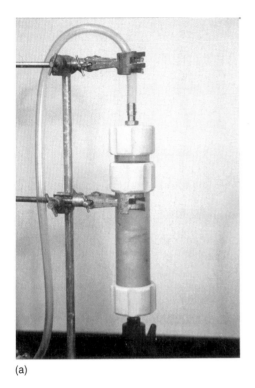 

(a)                                                    (b)

**Figure 26.2**   Filtration through glass wool. (a) column, (b) cartridge.

There is no single method that fulfils all these requirements.

There are four principal techniques, each based on a different property of the virus particle. Each technique has numerous variations. Most procedures can be used to concentrate viruses in sample volumes of 1–100 l. Adsorption/elution and some entrapment techniques comprise the first stage (primary concentration) in a two-stage concentration process which reduces the initial volume to between 100 and 400 ml. A secondary concentration stage concentrates the virus further by acid flocculation and low-speed centrifugation to deposit the virus-containing floc. This is dissolved in 5–10 ml neutral buffer.

## Adsorption/elution

The development of virus adsorption/elution (Viradel) methods suitable for the recovery of viruses from waters stems from the work of Melnick and his colleagues in Houston, Texas (e.g. Wallis and Melnick, 1967a,b,c; Wallis *et al.*, 1970). In general terms, the virus-containing sample is brought into contact with a solid matrix to which virus will adsorb under specific conditions of pH and ionic strength. Once virus is adsorbed, the water in which it was originally suspended is discarded. Virus is then eluted from the matrix into a smaller volume, though this is still too large to be inoculated directly on to cell cultures. Choice of adsorbing matrix, eluting fluid and processing conditions

**Table 26.1**  Summary of concentration techniques for viruses in water and water-related materials

| Technique | Method | Water quality | Initial volume | Relative virus content | Recovery | Capital cost | Revenue cost | 2ary concn required? | Comments |
|---|---|---|---|---|---|---|---|---|---|
| Adsorption/elution | Gauze pads | Sewage or effluent | Large | High | Low to medium | Nil | Very low | No | Not quantitative |
|  | Electronegative membranes | All waters | 1–1000 litres | Low to medium | 50–60% with practice | Medium | Medium | Yes | High vols require dosing pumps |
|  | Electropositive membranes | All waters | 1–1000 litres | Low to medium | 50–60% with practice | Medium | High | Yes | No preconditioning required |
|  | Electronegative cartridges | Any low turbidity | 1–50 litres | Low to medium | Variable: higher with clean waters | Low | Low | Yes | Clogs more quickly than membranes |
|  | Electropositive cartridges | All waters | 1–1000 litres | Low to medium | Variable | Medium | High | Yes | Wide range of viruses |
|  | Glass wool | All waters | 1–1000 litres | Low to medium | Variable | Low | Very low | Yes | No preconditioning required |
|  | Glass powder | All waters | <100 litres | Any | 20–60% | Medium | Low | If vol >100 litres | Special apparatus |
| Entrapment/ultrafiltration | Alginate membranes | Clean only | Low | High | Good | Low | Low | No | Very slow. Clogs rapidly if turbid |
|  | Single membranes | Clean | Low | Any | Variable | Medium | Low | No | Slow |
|  | Tangential (= cross) flow and hollow fibres | Treated effluents or better | High | Low | Variable | High | Medium | Sometime | Prefilter for turbid waters |
|  | Vortex flow | Treated effluents or better | High | Low | Unknown | High | Medium | Unknown | Undeveloped yet |
| Hydroextraction | PEG or sucrose | Any | Low | High | Variable (toxicity) | Negligible | Very low | No | High virus loss in waste waters |
| Ultra-centrifugation |  | Clean | Low | High | Medium | High | Medium | No | Wide range, but usu impractical |
| Other techniques | Fe oxide floc | All | Low | Any | Variable | Low | Low | No |  |
|  | Biph partition. Immunoaffinity | All | <7 litres | Any | Variable | Low | Low | No | Toxic to cells |
|  | and mag beads | Unknown | Low | Low | High | High | Low | No | New method |

will be influenced by the nature of the sample and by experience, but is commonly done using a solution containing beef extract or skimmed milk, both at high pH, which displaces the virus from the adsorbing matrix into the eluant. Glycine/NaOH solution may also be used. Following elution with proteinaceous fluids the virus is still too dilute to be inoculated directly into cell cultures and is therefore concentrated further; this is termed secondary concentration. Several methods are available, but that most commonly used is that of Katzenelson *et al.* (1976). The pH of the eluate is reduced to 3.5–4.5 which causes isoelectric coagulation (flocculation) of the protein. The virus is adsorbed to the floc which is deposited by centrifugation and dissolved in 5–10 ml neutral phosphate buffer. It may then be frozen or used for further analysis. Gilgen *et al.* (1997) developed a protocol for analysis of bathing waters and drinking water which used filtration through positively charged membranes followed by ultrafiltration as a secondary concentration step and Huang *et al.* (2000) used positively charged membranes followed by beef extract elution and PEG precipitation for the concentration of caliciviruses in water.

### a. Adsorption to electronegative membranes and cartridges

Concentration of viruses in water using negatively-charged microporous filters (e.g. Millipore, Sartorius) has been practised for many years and there are many variations of the technique, though little change in the basic process. The popularity of membranes, made of cellulose acetate or nitrate, is due to their availability in various pore sizes, configurations and compositions and, by judicious choice of pre-filters and adsorbing filters, it is possible to get good recoveries of virus accompanied by good flow rates and a minimum of filter clogging, even from turbid waters. In addition, many solids-associated virus can be recovered. Virus is bound to the filter by opposing electrostatic attractive forces and not by entrapment. In its simplest form, a virus-containing sample is passed under positive pressure or vacuum through a cellulose nitrate membrane 142 mm or 293 mm in diameter and of mean pore diameter 0.45 $\mu$m, 1.2 $\mu$m or 5 $\mu$m. For waters containing particulate material a pre-filter is used ahead of the membrane to prevent clogging. Virus is adsorbed and is eluted using beef extract or skimmed milk solution.

Since viruses and the filter materials are both negatively charged at neutral pH, the water sample is conditioned to allow electrostatic binding of virus to filter matrix. The water sample is adjusted to pH 3.5 and $Al^{3+}$ ions added in the form of $AlCl_3$ to a final concentration of $5 \times 10^{-4}$ M. Magnesium salts may also be used but most reports suggest better recoveries are obtained by using aluminium salts (e.g. Homma *et al.*, 1973; Metcalf *et al.*, 1974), though opinion is divided as to whether metal ions are needed at all when using cellulose nitrate membranes.

The choice of filter will depend partly on the nature of the sample; for sea-waters filters with pore diameters of 0.45 $\mu$m and 1.2 $\mu$m are commonly used and pre-filters are employed. Filters may be used singly or in series. This method is currently the preferred way of recovering viruses from effluent, diluted raw sewage and activated sludge samples, as well as recreational and surface waters in most water virology laboratories in the UK and is the recommended method (Standing Committee of Analysts, 1995). It is also a tentative standard method for the recovery of viruses from waters and wastewaters, as published by the American Public Health Association (APHA) (1980).

Negatively-charged filters may also be used in tube form. Balston filters are epoxy resin-bound glass fibre filters with an 8 μm nominal pore diameter. They were originally used for concentration of viruses from tap water (Jakubowski *et al.*, 1974) and have since been employed for concentration of viruses from river water (e.g. Morris and Waite, 1980) and other waters (e.g. Guttman-Bass and Nasser, 1984; United States Environmental Protection Agency, 1984). Their recoveries appear at least as good as membrane filters, they are less expensive and can be obtained complete in sterile cartridges in disposable form. Guttman-Bass and Nasser (1984) reported mean recoveries from seawater of 93, 75, 92 and 109% for poliovirus 1, echo 7, coxsackievirus B1 and coxsackievirus A9 respectively. They are, however, prone to clogging and cannot be used with even moderately turbid water and according to Gerba (1987) cannot be used at high flow rates. Because of problems of clogging of membrane or tube filters, the processing of seawater samples in this way is limited to a maximum of 20 l before filters have to be changed (Block and Schwartzbrod, 1989).

One way of overcoming the problem of clogging without having to change membranes or tubes frequently is to increase the surface area of filtration by the use of larger cartridge filters, where sheets of negatively-charged pleated filter material approximately 25 cm wide are rolled and used in 30 cm cartridge holders. These were evaluated by Farrah *et al.* (1976) who used fibreglass membrane material in a pleated format (Filterite Duo-Fine). The increased surface area allowed for higher flow rates and recoveries were better than given by membranes. Pre-filters were used in series to prevent further clogging where 0.45 μm pore diameter filters were used in processing marine waters. Seeded poliovirus was recovered from 378 l volumes of seawater with 53% efficiency. These filters are more expensive, but these authors reported that they could be regenerated up to five times by soaking for 5 minutes in 0.1 M NaOH.

Many modifications to the filtration process have been made but have had little effect on practicability, recovery, or other aspects. For example, Farrah *et al.* (1988) reported a higher adsorption of enteroviruses to Filterite (epoxy resin-bound glass fibre) filters after treatment with polyethyleneimine. Concentration by filtration through negatively-charged media has also been shown to be suitable for virus detection by molecular biological methods (e.g. Wyn-Jones *et al.*, 1995), though there are differences in the use of eluting fluids.

Generally, recovery rates are as variable with negatively-charged filter media as with any other kind. Block and Schwartzbrod (1989), citing Beytout *et al.* (1975), consider cellulose nitrate membranes relatively efficient insofar as they give 60% recovery of virus; the same authors recorded glass fibre filters giving a poor average yield on wastewater but 70% recovery with river water. Payment *et al.* (1979), using glass fibre filters, reported 38–58% recovery of $10^2$–$10^6$ pfu seeded in 100 ml to 1000 l volumes. Few studies have been done on recovery efficiencies from marine waters in a controlled way; however, controlled studies have been done to evaluate the recovery efficiency of the method using drinking water. Melnick *et al.* (1984) reported considerable variation in the quantity of virus recovered following processing of 100 l tap water samples containing poliovirus. Though the average recovery was 66% (of 350–860 pfu virus), values ranged from 8 to 20% in two laboratories, 49 to 63% in three laboratories and 198% in one laboratory. Recovery levels were higher and less variable where a higher input level of virus was used, but it must be noted that even the 'low' level of 350–680 pfu is more than is routinely found in bathing waters.

It is not usually possible to conduct studies where virus is deliberately added to water systems, however, Hovi *et al.* (2001), in assessing the feasibility of environmental poliovirus surveillance added poliovirus type 1 into the Helsinki sewers and recovered it over a period of 4 days by taking samples at downstream locations and concentrating 100-fold simply by polymer two-phase separation. If controlled studies are done then they should use virus input levels which would normally be expected in the waters to be monitored under actual conditions. In the UK External Quality Assessment Scheme for Water Virology the majority of participating laboratories use cellulose nitrate membranes for concentration; one laboratory uses Balston tube filters. The normal range of recovery for seeded clean water is up to 60%.

*b. Adsorption to electropositive membranes and cartridges*
Positively-charged filters adsorb virus from water and other materials without the need for prior conditioning of the sample with acid or cations. They will adsorb virus in the pH range 3–6; at pH values above 7 the adsorption falls off rapidly, so the pH still needs to be careful monitored. These properties make the use of positively-charged filters attractive, not only for the convenience of not having to condition the sample, but also because it makes possible the concentration of other viruses such as rotavirus and coliphages, which are sensitive to the low pH conditions needed for adsorption to negatively-charged media. Keswick *et al.* (1983) reported that type 1 poliovirus and rotavirus SA11 survived at least 5 weeks on electropositive filters at 4°C, so this may make them useful for extended surveys or transmission through postal systems. Other than not needing to condition the water sample, electropositive filters are used in the same way as electronegative materials. Virus is eluted from the filter and secondary concentration is carried out as for the electronegative types.

Recoveries from positively-charged filters are similar to those from negatively-charged filters; Sobsey and Jones (1979) reported 22.5% recovery using a two-stage procedure in the concentration of poliovirus from drinking water. The original positively-charged material, Zeta-plus Series S, is made of a cellulose/diatomaceous earth/ion-exchange resin mixture. It is commercially available as Virozorb 1MDS cartridges (AMF-Cuno). Sobsey and Glass (1980) compared these filters with Filterite (fibre glass) pleated cartridge filters for recovery of poliovirus from 1000 l tap water and obtained recoveries of about

30% with both types. The advantages of these filters lie in the large volumes they can handle without the need for conditioning the sample. Elution from the filter still needs to be carried out at pH 9 or above, which limits their use to viruses stable below that pH, though Bosch *et al.* (1988) successfully concentrated rotavirus in this way. Organic materials in the sample, especially fulvic acid, were reported to interfere more with virus recovery from Virozorb cartridges than from glass-fibre materials (Sobsey and Hickey, 1985; Guttman-Bass and Catalano-Sherman, 1986). A different electropositive material (MK) is cheaper but its recoveries were reported to be not as good as 1MDS in comparative tests (Ma *et al.*, 1994).

Advances in membrane technology resulted in charge-modified nylon membranes being available for concentration of viruses from water. Gilgen *et al.* (1995, 1997) described the use of positively-charged nylon membranes (Zetapor, AMF-Cuno) coupled with ultrafiltration for the concentration of a variety of enteric viruses prior to detection by RT-PCR. Other nylon membranes are also available; 'Biodyne B' and 'N66-Posidyne' (Pall), which are different grades of the same material, are made in pore sizes which would permit passage of virus (0.45 μm, 1.2 μm and 3 μm) and have a positive surface charge over the pH range 3 to 10, which would promote strong binding of negatively-charged particles. Posidyne filter material is also available in re-sterilizable cartridge form, which would increase the convenience of use. No studies have been done on this material in respect of recovering viruses from bathing waters, but its low cost and ease of use suggest that further evaluative research should be done. Triple-layered PVDF membranes and cartridges (Ultipor VF, Pall) have been used in industry for the removal of polio and influenza viruses from pharmaceutical products (AranhaCreado *et al.*, 1997), though whether the virus can be recovered from the filter is not known.

The need to determine the presence of *Cryptosporidium* and *Giardia* as well as viruses in water samples has led some workers to attempt the simultaneous concentration of both types of microorganism. Watt *et al.* (2002) compared the efficiencies of polypropylene spun fibre cartridges with Filterite or 1MDS microporous cartridges in concentrating *Giardia* cysts and *Cryptosporidium* oocysts and poliovirus from 400 l volumes of drinking water and treated wastewater. Filterite filters performed better than the 1MDS and were easier to process and the protozoa were trapped by both types of filter; the authors suggested that it is thus possible to concentrate virus and protozoa at the same time on a composite filter.

*c. Adsorption to glass wool*

Glass wool is an economic alternative to microporous filters. It is used in a column and, provided it is evenly packed to an adequate density, adsorption of enterovirus appears at least as efficient as with other filter types (see Figure 26.1). An advantage of the method is that virus will adsorb to the filter matrix at or near neutral pH, which makes it suitable for viruses sensitive to acid and without the addition of cations; elution still has to be done at high pH.

The technique has been pioneered in France principally by Vilaginès and co-workers (e.g. Vilaginès *et al.*, 1988), who applied it to the concentration of a range of viruses from surface, drinking and wastewaters. Glass wool packed into holders at a density of $0.5 \text{ g/cm}^3$ is washed through with HCl, water and NaOH and water to neutral pH before the sample is passed through the filter. Different sizes of filter can be prepared according to the type of water and flow rate.

In the French studies, sample sizes ranged from 100 to 1000 l for drinking waters, 30 l for surface waters and 10 l for wastewaters. The only pretreatment necessary was dechlorination of drinking waters. Surface water samples were filtered at 50 l/h in a 42 mm diameter filter holder. Virus was eluted from the filter with 0.5% beef extract solution and secondary concentration done by organic flocculation.

Recovery efficiency of approximately $10^2$ pfu poliovirus seeded into 400 l drinking water averaged 74% (s.d. 18.9%). For surface waters the recovery rate was 63% and 57% respectively. Lowering the flow rate to 50 l/h reduced clogging of the filters.

Since virus concentration on glass wool does not need the sample to be conditioned the technique lends itself to large sample monitoring (for surface waters) and to continuous monitoring (for drinking waters). Other viruses were also concentrated during field evaluation of the method; adenoviruses and reoviruses were also recovered, though as expected enteroviruses predominated. Vilaginès *et al.* (1993) also reported a survey of two rivers over a 44-month period and concluded that the technique was robust enough in physical and experimental terms to be used for routine monitoring of surface waters. Glass wool is also very cheap and thus the method is economic.

Glass wool has been used in other laboratories; Hugues *et al.* (1991) found it more sensitive than the glass powder method in terms of number of positives detected and in the level of virus when analysing biologically treated wastewaters; Wolfaardt *et al.* (1995) used glass wool to concentrate small round-structured viruses (SRSVs) from spiked sewage and polluted water samples prior to detection by RT-PCR.

*d. Adsorption to glass powder*

Powdered borosilicate glass with a bead size of 100–200 μm is a good adsorbent for viruses under conditions similar to those used for glass fibre microporous filters. However, glass beads constitute a fluidized bed and so have the advantage that the filter matrix cannot become clogged as with glass-fibre systems. Sarrette *et al.* (1977) first developed this technique, which was extended by Schwartzbrod and Lucena-Gutierrez (1978).

For low sample volumes (<100 l) the method gives a low eluate volume which may not need secondary concentration prior to further analysis. However, the recovery varies widely with the type of sample, from 60% with potable water to 20% with urban wastewater (Joret *et al.*, 1980). For sample volumes of less than 100 l containing relatively little organic material the powdered glass technique would appear useful in principle. A drawback is the complexity of the apparatus; specially constructed two-part mixing and elution chambers are required and this may render the method impracticable unless there were other strong reasons for its use.

## Entrapment

Entrapment refers to those techniques in which the virus in a sample is bound to a filter matrix principally by virtue of its molecular size rather than by any charges on the particle, though in practice electrostatic effects can influence binding to an ultrafilter.

### a. Ultrafiltration

Early ultrafiltration methods involved the filtering of the water sample under pressure through aluminium/lanthanum alginate filters (Poynter et al., 1975). While these had the unique advantage that they were soluble in sodium citrate, the flux obtained was too low for all but the cleanest waters unless they had been pre-filtered which precluded analysis of surface waters in volumes over 1 l.

More recent techniques involve passing the sample through capillaries (e.g. Rotem et al., 1979), membranes (e.g. Divizia et al., 1989a,b), hollow fibres (Belfort et al., 1982) with pore sizes that permit passage of water and low molecular mass solutes but exclude viruses and macromolecules, which are then concentrated on the membrane or fibre. Most laboratories now use membranes or fibre systems with cut-off levels of 30–100 kDa. In systems in which the fluid passes directly through the filter, non-filterable components quickly clog the filter or precipitate at the membrane surface, thus this type of filter is only useful for small volumes (1000 ml or less) of sample.

Recently developed ultrafilters employ tangential flow. The minimum 'dead' volume (e.g. 10–15 ml, Divizia et al., 1989a), which depends on the apparatus in use, is that beyond which it is impossible to reduce the retentate volume further; this is the final volume of concentrate. If this is small enough then it may be inoculated on to cell cultures or it may have to be further processed by secondary concentration.

Some workers have experienced binding of virus to the membrane rather than just prevention of its passage through it. In these cases the virus was eluted by backwashing with glycine buffer or beef extract and the eluate reconcentrated by organic flocculation. Some authors have even reported differences in binding between viruses that are related. Divizia et al. (1989b) for example noted that hepatitis A virus was recovered with 100% efficiency; poliovirus on the other hand was recovered very poorly under standard conditions but this improved if the membranes were pretreated with different buffers. Further, recovery was best if the virus was eluted with beef extract at neutral (not high) pH.

Ultrafiltration has also been used to reconcentrate viruses recovered from treated wastewater by adsorption/elution.

A variation in ultrafiltration is vortex flow filtration (VFF). The technique appears to offer a further step in the reduction of clogging while still retaining the ability of filters to process large volumes of sample, but there are few significant reports in the literature of its use in the field. One is that of Tsai *et al.* (1993), who used it for inshore waters in Southern California. Fifteen litres of each sample were concentrated to 100 ml using a 100 kDa cut-off membrane and the samples were further concentrated to 100 μml using Centriprep and Centricon units at 1000×*g*.

The advantages of ultrafiltration are principally that the sample requires no preconditioning and that a wide range of viruses can therefore be recovered, including bacteriophages (e.g. Urase *et al.*, 1993; Nupen *et al.*, 1981). Efficiency of recovery is usually good, though as with all methods it is variable. The main constraints upon its use are the high initial cost of the equipment and that, despite the advantages of tangential flow, turbid samples still tend to clog the membrane. Surface waters may take a long time to process if they are turbid; Nupen *et al.* (1981) were able to filter 50 l volumes but this took from 40 hours to 72 hours depending on the sample. In other studies workers have used different parameters (though generally all use 30–100 kDa membranes or hollow fibre cartridges) with differing results. The technique is generally seen as an advance on adsorption/elution (e.g. Grabow *et al.*, 1984; Muscillo *et al.*, 1997) and recovery efficiencies, though variable, appear to be higher than those obtained using adsorption/elution techniques.


*b. Ultracentrifugation*

Ultracentrifugation is a catch-all method since it is capable of concentrating all viruses in a sample provided sufficient g-force and time are used. Differential ultracentrifugation allows separation of different virus types. A number of studies have been reported using ultracentrifugation, including one in which virus from a polluted well was recovered (Mack *et al.*, 1972) and one where viral numbers in natural waters were as high as $2.5 \times 10^8$/ml, $10^3$ to $10^7$ times as high as had been found by plaque assay (Bergh *et al.*, 1989). However, the limited volumes that can be processed, even using continuous flow systems, together with the high capital costs and lack of portability of the equipment, limit its usefulness in concentrating viruses directly from natural waters. It does find a use as a secondary concentration method, however; Murphy *et al.* (1983) in an investigation of a gastroenteritis outbreak associated with polluted drinking water, concentrated 5 l samples of borehole water using an ultrafiltration hollow fibre device to 50 ml and followed this by ultracentrifugation to pellet the virus for electron microscopical examination. They were thus able rapidly to detect rotaviruses, adenoviruses and small round-structured viruses, as well as enteroviruses, which were confirmed by cell culture.

## Other methods

### Hydroextraction

Hydroextraction is the concentration of virus in a sample by the removal of water using a hygroscopic solid. Two solids are commonly used; polyethylene glycol (PEG, as the polymer in the range 6000–20 000) and sucrose. The technique involves filling a dialysis bag with the sample, immersing it in the solid and leaving it at 4°C for several hours. Water is drawn out of the sample which thus reduces in volume. Further dialysis against phosphate buffer is then required to remove the PEG/sucrose which has entered the bag.

Hydroextraction has been employed with some success, but its use is limited. Clearly, volume size is the principal constraint and the maximum volume that can be handled is about 1 l; sewage and wastewater were successfully concentrated in this way nearly 30 years ago (Wellings *et al.*, 1976). Further, dialysis membranes have a $M_r$ cut-off of approximately 12 000. Thus some organic compounds will be concentrated along with the virus. Many of these are cytotoxic and the virus cannot be assayed in cell culture. Hydroextraction has been used frequently as a secondary concentration step following microporous filtration or ultrafiltration (e.g. Ramia and Sattar, 1979).

### Iron oxide flocculation

Several reports in the literature describe the adsorption of enteroviruses to magnetic and non-magnetic iron oxides, either $Fe_2O_3$ or $Fe_3O_4$. Rao *et al.* (1968) described adsorption of a range of viruses to magnetic iron oxide and Bitton *et al.* (1976) carried out a detailed study of the adsorption of poliovirus to magnetite ($Fe_3O_4$). Adsorption occurred at pH 5–8 and elution could be effected with beef extract at pH 8–9. In the case of magnetic oxides the virus may be concentrated by removing the oxide with adsorbed virus from the sample water with a magnet, then eluting.

### Talc-celite adsorption

Talc (magnesium silicate) mixed with celite (diatomaceous earth) forms a good combined adsorbent which can be used as a fluid bed or sandwiched between two layers of filter paper. A range of viruses can be concentrated by this means and different waters, including tap water and wastewater can be processed (e.g. Sattar and Westwood, 1978; Sattar and Ramia, 1979; Ramia and Sattar, 1979).

### Adsorption to bituminous coal

Dahling *et al.* (1985) used powdered coal as an adsorbent with a view to transferring virus concentration technology to developing countries. Filters made this way were effective over the pH range 3 to 7 and recoveries did not differ significantly from those obtained with membranes filters when used with 100 l tap water samples. They could also be used for wastewater samples. Lakhe and Parhad (1988) described a similar system, and Chaudhiri and Sattar (1986) reported a system which could be used for the removal of enterovirus from water with a view to improving its quality. The same kind of matrix in a more refined state was used as granular activated carbon by Jothikumar et al. (1995) for the first stage concentration of enteroviruses, HEV and rotaviruses. Using RT-PCR as a detection method these authors claimed a 74% recovery of poliovirus 1.

*Two-phase separation*

Viruses and macromolecules can be partitioned between the two immiscible phases produced when two different organic polymers are dissolved in water. Lund and Hedstrom (1966) used sodium dextran sulphate and polyethylene glycol 6000 mixture for enterovirus recovery from sewage. By controlling the phases viruses can be partitioned into one of them. If the virus-containing phase is made small relative to the original volume of sample then concentration is achieved. Though effective, this method is limited by the occasional toxicity of the polymer for cell culture; it is also limited to about 7 l maximum volume.

*Immunoaffinity columns and magnetic beads*

These are relatively new techniques which have been used in a biochemical or molecular biological context. They are useful for small volumes but their application to virus concentration from larger volumes has yet to be demonstrated. Schwab et al. (1996) developed a broad-based antibody capture technique for a variety of viruses, in conjunction with primary concentration through 1MDS positively-charged filters and detection by RT-PCR. Myrmel et al. (2000) also described the separation of noroviruses in this way. Water samples seeded with genogroup I norovirus were brought in contact with magnetic beads coated with polyclonal antibodies raised against a recombinant capsid protein of the same virus. Virus was captured and detected by RT-PCR. An important attribute of this method is that it acts as a clean-up stage in that RT-PCR inhibitors are removed as well as the virus being concentrated.

*Drying/freeze-drying*

Forced removal of the water in samples has been used by several authors. The commonest approach seems to be the vacuum-drying of the sample; Bosch et al. (1988) concentrated rotavirus and astrovirus in this way and Kittigul et al. (2001) reported significantly higher rotavirus antigen recovery following SpeedVac concentration than when polyethylene glycol (PEG) precipitation was used as a secondary step, both after membrane filtration as a first stage.

It will be clear from the foregoing that there is a wide range of methods for concentration of viruses from different types of water. The UK preferred method is to concentrate the sample using negative polarity cellulose nitrate membranes in a 142 mm or 293 mm diameter filter holder and to elute the virus with beef extract solution. This is also the method recommended by the Standing Committee of Analysts (1995). Table 26.1 is intended as a quick reference to the methods for enterovirus concentration.

## Detection and enumeration of waterborne viruses

After sampling and concentration, the third major aspect of virological analysis of waters is the detection and enumeration of the viruses in the concentrate. Detection and enumeration are conveniently considered together since, for many viruses they are performed simultaneously, particularly where the virus multiplies in culture and infectivity assays are done. Broadly speaking, detection may be done by infectivity-based methods where the virus undergoes at least partial multiplication in cell culture, or it may be done by techniques based on properties other than infectivity. Most important in this latter category are the molecular biological techniques, especially the polymerase chain reaction which has found wide application in water virology as it has in other biological disciplines. Enumeration by molecular means may be semi-quantitative, such as end point dilution assays using the disappearance of a PCR band as the end point in a titration, or, increasingly, real-time PCR is becoming a realistic possibility for enumerating genome copies of a target virus, though the relationship between numbers of infectious units and genome copies depends on many variables and may not be achievable for all viruses.

Detection of virus infectivity is done by inoculating cell cultures with part or all of the concentrate and allowing the virus to multiply in the cells so that they are killed. Virus-specific cell killing (the cytopathic effect, CPE) of enteroviruses and some other types is visible to the naked eye. If a range of cell cultures is inoculated under liquid assay it should be possible to detect polio, coxsackie, echoviruses, as well as some adenoviruses and reoviruses. Hepatitis A virus may also be detected in this way but only after prolonged incubation of cultures and it is therefore not an approach used in routine waterborne HAV detection.

There are two approaches to the enumeration of virus infectivity. The *plaque assay*, where virus-mediated cell destruction is confined to a small area (the plaque) by incorporation of agar in the maintenance medium, may be done in cultures of cells growing in single layers (monolayers) or in cultures where the cells are suspended in the maintenance medium. In both cases, plaques develop following incubation and may be counted as they become visible, in the case of enteroviruses usually after about 3 days. The suspended cell assay, since it offers more adsorption sites to viruses, is three to four times more sensitive than the monolayer assay (UK Public Health Laboratory Service Water Virology External Quality Assurance (EQA) Scheme unpublished results), though the latter requires only a fifth of the cells. One plaque is taken as being the progeny of one infectious unit of virus; this may be the same as one virus particle, but this is unlikely given the association of virions with particulate matter, both organic and inorganic.

Virus infectivity may also be assayed in *liquid culture*, where virus concentrate is added to replicate cell cultures which are then observed for specific CPE. Computation of the positives allows a titre to be determined in terms of Most Probable Number units or Tissue Culture Dose$_{50}$ (TCD$_{50}$) units.

Virus infectivity may also be determined by immunofluorescence or immunoperoxidase techniques, which are particularly useful where limited replication occurs and a distinct CPE is not produced.

## Cell culture

Because viruses are obligate intracellular parasites they will only grow in living cells. They are also quite species specific, so that, for example, human viruses generally only grow in human cells and bovine viruses will usually only grow in bovine cells. As cultures and viruses become more adapted this division becomes blurred but, in general, human viruses will multiply only in cells of primate origin.

Cell cultures may be divided into three kinds; continuous cell lines, primary cell cultures and semi-continuous cell strains. The first are generally used in water virology because of their availability and susceptibility to virus strains normally encountered in water samples.

Continuous cell lines arise through spontaneous transformation of primary cultures or *in vivo* transformation as tumour tissue. They are usually heteroploid or even aneuploid. They tend to be sensitive to fewer viruses than primary cells but, being transformed, they will undergo extensive or indefinite serial passaging, though their properties change gradually as they do so. Provided the correct choice is made initially this type of cell culture is the most useful for a routine water virology laboratory.

There are many cell lines suitable for growing enteroviruses, including HEp-2, HeLa and VERO cells. The line most favoured for enumeration of waterborne enteroviruses is the Buffalo Green Monkey (BGM) line first described by Barron *et al.* (1970) and in a water context by Dahling *et al.* (1974). This was reported to give higher plaque assay titres of poliovirus, coxsackieviruses B, some echovirus and reoviruses than obtained in rhesus or grivet monkey kidney cells. It is interesting to note, however, that this apparent better sensitivity is not continued with isolation of enteroviruses from clinical specimens; Schmidt *et al.* (1976) found that BGM cells were less sensitive than primary RMK or human fetal kidney cells for the isolation of some echo and adenoviruses from clinical specimens and Pietri and Hugues (1985) found there was no difference between BGM and KB cells for quantification of poliovirus. Nevertheless, BGM cells are used almost exclusively for the detection and enumeration of waterborne enteroviruses. Morris (1985) examined ten cell lines for their ability to grow enteroviruses isolated from wastewater effluent. Eighty-two per cent of isolates were positive in BGM cells, 73% in RD cells and 64% in chimpanzee liver cells. BGM was also the most sensitive in the number of plaques counted.

Dahling and Wright (1986) carried out an extensive set of experiments to optimize the BGM line in respect of a number of assays for waterborne viruses and made recommendations in respect of many cell culture and assay parameters, as well as doing a comparative virus isolation study involving BGM cells and nine other cell lines. This work has become the accepted basis for many standard methods on detection of water-associated viruses.

Other cell lines, derived from intestinal tissue, have been investigated for their ability to support the growth of enteric viruses. Most of these studies have been directed at growing the more fastidious agents like rota- and astroviruses, but Patel *et al.* (1985) carried out a large survey on the susceptibility of a range of lines to different enteroviruses, including all 31 serotypes of echovirus; they found that two lines, HT-29 and SKCO-1, had a markedly wider sensitivity, with comparable or wider sensitivity for enteroviruses than PMK or rhabdomyosarcoma cell cultures. HT-29 and SKCO-1 (both of which supported the growth of all echoviruses) are derived from human colonic carcinoma tissue and so it is not surprising that they respond better to infection with enteric viruses. They require a high seed density and do not grow quickly, however and perhaps this is why they have not found greater favour, along with $CaCO_2$ cells (Fogh, 1977) which are of similar origin, in the detection of waterborne enteric viruses generally.

### The plaque assay

The plaque assay is the method most used for the estimation of waterborne enteroviruses; it is in plaque-forming units (pfu) that the levels of virus permitted per 10 l sample under the EU Bathing Water Directive are expressed. Although other methods are available, and have some advantages, this is the principal method in use. All the concentrate should be tested, but many laboratories test only a proportion of the concentrate and multiply the resulting count accordingly. This is unwise where there are likely to be small numbers of virus particles present and will not necessarily be randomly distributed. A subsample will therefore not be representative of the whole and an erroneous titre will be recorded. If all the sample is not assayed in one test adequate internal quality controls need to be included to verify the performance of the assay.

### Monolayer plaque assay

The virus concentrate is inoculated on to preformed monolayers in Petri dishes or flasks and the cells are reincubated under an agar overlay until a CPE is seen. Infection is confined to local areas of cell death (plaques) since the agar prevents the virus spreading over the whole of the monolayer and it can only infect adjacent cells. It has been recognized for many years that plaque morphology varies between enterovirus types (Hsiung and Melnick, 1957) and careful experienced examination of plaques in appropriate cell cultures may provide an indication of the enteroviruses present. Plaques are counted and the titre recorded as the number of plaque-forming units (pfu) in the sample. It is assumed that one plaque is the progeny of one virus particle or one infectious unit (but see below), hence 10 plaques counted in an assay means there were 10 infectious particles in the original sample, assuming the whole of the sample was assayed.

However, because viruses aggregate to different degrees under different conditions the one plaque = one particle assumption may be false and the actual numbers of particles may be greater than the plaque count would suggest. Whether this matters in practice is debatable; the decision in setting levels for compliance rests on the plaque count, not on the number of virus particles in the sample.

Plaques are counted daily starting at day 2. Since viruses multiply at different rates counting is continued after the first appearance of plaques. Echoviruses, for example, take longer to form plaques, if they do at all. The SCA (1995) method recommends counting plaques for 2–5 days; the US EPA (1984, revised 1987) method suggests counting should continue for 12 days or until no new plaques appear between counts; Block and Schwartzbrod (1989) recommend 6–14 days. The duration of the reading period will depend partly on the state of the negative control cultures; these are counted for up to 16 days, after which the test may be discarded. Visualization is enhanced either by incorporating neutral red into the agar overlay medium as the cells are inoculated, or by removing the agar and staining the monolayers with Giemsa or methylene blue (Figure 26.3). Neutral red stains only live cells, so plaques appear as clear areas against a pink background.

The monolayer plaque assay method is favoured in many European laboratories, often using Giemsa as a stain to reveal plaques.

*Suspended cell plaque assay*

The suspended cell assay (Cooper, 1967) increases the sensitivity of the ordinary plaque assay. More cells are used per vessel, suspended in the agar instead of being in a layer underneath it and thus offer many more adsorption sites to any virus present. It is also quicker in that it involves no prior establishment of monolayers or fluid changes since cells and concentrate are added to the culture vessels at the same time. The method is more sensitive than the monolayer assay, though more expensive in cells (by a factor of five to ten); Dahling and Wright (1988) reported a five to eight times greater sensitivity using the suspended cell assay compared with the monolayer assay and the UK Water Virology EQA Scheme records that the suspended cell assay has three times the sensitivity of the monolayer plaque assay. Plaques are easier to pick for subsequent identification and the system is independent of the surface area of the plaque assay vessel. However, the cultures only last a week (US EPA, 1984, revised 1987) to 10 days, so slow-plaquing viruses may not be detected. Plaques may also be more difficult to count compared with plaques in monolayers. Despite the constraints, it is a method well suited to the enumeration of enteroviruses in low numbers, such as are found in environmental samples. The USEPA method recommends that the suspended cell assay should be used where the level of indigenous virus is likely to be less than 5 pfu/ml.

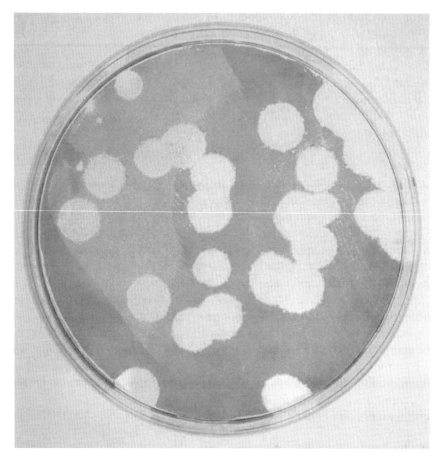

**Figure 26.3** Plaques in BGM cell monolayer caused by polioviruses. One plaque is the result of multiplication of one virus (or, more correctly, one virus aggregate) and so the technique can be used to enumerate plaque-producing viruses in concentrates.

Sometimes 'plaques' are produced by toxic components in the medium, or by an undissolved particle of agar. Recognition of these is clearly important and is a matter of experience. Where there is doubt as to the nature of a plaque the area is 'picked' and the medium plus cells in the vicinity is withdrawn and inoculated into a fresh tube culture to confirm its viral origin.

Calculation of the titre of a preparation assayed by the plaque method makes a number of assumptions. Further, the number that is calculated for the viruses in a sample will only at best be proportional to the number really present. The proportionality factors (usually unknown) depend on the virus type and the calculation method. The assumptions made are (a) that the presence of one virus is sufficient to cause cellular destruction and that the absence of cellular destruction signifies absence of virus; and (b) that viruses are randomly distributed in the sample.

Since it is impossible to predict the number of virus clumps that will produce a single plaque the titres are invariably expressed as plaque-forming units, or pfu. Four different ways of calculating this have been proposed by Block and Schwartzbrod (1989) in an attempt to allow for different ranges of counts in a plaque assay; calculation based on (a) fewer than 15 pfu counted at a level of dilution; (b) 15–100 pfu counted at one level of dilution; (c) more than 100 pfu counted at a level of dilution; and (d) counts from several successive dilutions are used. Different equations, all using 95% confidence intervals, are used in each case. Comparison of the results seems to indicate that the amplitude of the confidence intervals increases as the number of counted plaques decreases and the best results are obtained with a large number of plaques. However, the risks of confluence of plaques is then magnified and a truncation method of calculation (Beytout *et al.*, 1975) is recommended.

Most UK laboratories employ the suspended plaque assay method, though few EU laboratories appear to use it.

## Liquid assays

Cells under liquid media may support the growth of more enteric virus serotypes than cells growing in or under agar. Many enteroviruses, especially some echoviruses, do not form plaques and so will not be detected under agar; some viruses take a long time to produce a CPE and agar cultures may have deteriorated too far to be useful. In these cases cells growing under liquid medium are used. Virus multiplication produces cell degeneration and often a CPE characteristic of the infecting virus (though the CPE produced by all enteroviruses is consistent), so some idea may be gained of the agent at hand.

### Most probable number assay

In the most probable number of cytopathic units (MPNCU) assay the concentrate is divided into several fractions, each of which is inoculated into a separate cell culture. Dilutions of the concentrate may be used. The number of fractions and the size of cultures vary between laboratories but may range from five fractions each being inoculated into a 75 cm$^2$ flask of cells to over 40 fractions each being inoculated into a well in a microplate. Cultures are incubated for up to 21 days. Since plaques are not formed, the titre must be calculated in a different way. The MPNCU is calculated from a collection of positive results obtained for a series of dilutions, where a positive culture is one showing typical CPE. Probability tables (Chang *et al.*, 1958) are used for small numbers of replicates (e.g. three or five per dilution) and there are computer programs available for calculation of the MPNCU for larger numbers (e.g. Hurley and Roscoe, 1983).

This method is favoured by French workers and, until recently, by those in Austria.

### End point dilution assay (TCD$_{50}$)

Serial dilutions of the concentrate are inoculated into cell cultures and each culture is scored positive or negative after incubation. The titre is calculated (e.g. by the method of Reed and Muench, 1938) as the logarithm of the dilution of virus producing a CPE in 50% of the cultures. Though the method is simple and economic, its precision is difficult to evaluate. It is the least favoured of the three methods described.

*Choice of assay method*

Comparison of assay methods in agar and under liquid media provides the pointers indicated in Table 26.2.

It will be seen that there is no clear-cut best method. However, the plaque assay has a greater advantage of individualizing the plaque-forming units and providing entities (plaques) which are countable and directly related to the number of virus particles or to 'real' viruses. For many users this is an easier concept to grasp than the more abstract MPN or $TCD_{50}$. Despite its limitations, the plaque assay remains widely used as the unofficial standard method of assay. When enumeration methods are compared, a mathematical relationship can be deduced. Hugues (1981) showed that if 96 cultures per dilution were inoculated in an end point dilution assay, that 1 $TCD_{50}$ = 0.69 pfu, and that 1 pfu = 1.44 $TCD_{50}$. The MPN assay has advantages only when larger numbers of cultures are inoculated per dilution, since it is on this that the precision of the test depends. The three calculation techniques give very similar results if used properly and the 95% confidence limits overlap. The MPN is more reliable than the others provided the number of cultures inoculated per dilution exceeds 30 (Block and Schwartzbrod, 1989).

Several comparative studies have been done on methods for the detection of enteroviruses in water. Morris and Waite (1980), for example concluded that monolayers were the least sensitive system, tube cultures were of intermediate sensitivity (for MPN determination, though only four tubes were set up per dilution) and the suspended cell assay was the most sensitive. BGM cells gave the best recoveries and RD cells were variable. It is interesting to note that the

**Table 26.2** Comparison of assay methods

| Attribute | Liquid | Agar |
|---|---|---|
| Range of viruses detected | Wide range possible | Non-plaquing viruses not detected |
| Blind passage | Blind passage possible to increase titres to detectable levels | Faster-growing viruses in a mixture overgrow slower ones, which are not isolated |
| Sensitivity | Greater sensitivity (especially than monolayers) | Sensitivity improved using suspended cell assay |
| Subculture | Subculturing easy | Subculturing difficult (impossible without CPE) |
| Virus separation | Impossible to separate virus types | Separation of viruses possible by plaque picking |
| Statistical precision | Bad precision, large bias where few replicates used (as is usual) | Good, especially where all concentrate tested in one assay |

RD cells have been reported susceptible to coxsackievirus A strains (Block and Schwartzbrod, 1989) though they are less sensitive than suckling mice, which is the only other system that supports growth of this group of viruses; other cell cultures are refractory to these viruses which have to be assayed in suckling mice and are not therefore looked for routinely in waterborne virus detection. Toxicity problems can occur with both systems; experience from the UK EQA scheme suggests that toxicity occurs frequently in liquid cultures and less so in the plaque assay. Block and Schwartzbrod (1989) however suggest the reverse is true.

## Isolation and identification

The EU Bathing Water Directive requires that waters are analysed for enteroviruses, where circumstances demand; though this is a mixed group no requirement is made in respect of a specific identification, since the presence of any enterovirus in a sample will indicate the presence of human faecal contamination of the water and the potential presence of enteropathogenic viruses. However, it is essential to confirm that 'plaques' detected in an assay are indeed virus-specific and this is done by isolating the virus. Further, it may on occasion be necessary to do additional investigations on a sample, for example where repeated virological failures make it useful to see if the same organism is causing the failures.

### Isolation

Isolation of one specific virus type is done by picking plaques from the enumeration test. Dead cells constituting the plaque and the medium surrounding them are withdrawn and inoculated into a fresh cell culture. In 2–5 days any virus will have grown up sufficiently to be identified.

### Identification

Most laboratories identify isolated enteroviruses by the serum neutralization test (SNT). Aliquots of the isolate are incubated with different sera and each mixture inoculated into a cell culture. Absence of CPE indicates neutralization of the virus by the serum and thus indicates the identity of the agent.

The SNT, while flexible in that it can be used to identify groups or specific serotypes, is a time consuming and laborious test. It is difficult to standardize and sometimes gives equivocal results, necessitating a repeat of the procedure. Other serological procedures have been used to identify enteroviruses isolated from environmental sources, with varying degrees of success. Payment *et al.* (1982) described an immunoassay method for typing poliovirus isolates, and

Payment and Trudel (1985, 1987) described methods for detection and identification of enteric viruses, including enteroviruses, using immunoperoxidase techniques; they also suggested using this as an enumeration technique, based on scoring the number of peroxidase-positive cultures and calculating the titre by the MPN method. Pandya *et al.* (1988) used an immunosorbent assay system to identify coxsackievirus A isolated from sewage. Generally such methods show advantages over the SNT, though they do not indicate whether the virus is infectious, which is an important consideration. Further, cross-reactions between sera against different enterovirus types has been known, limiting the specificity of the tests. Polyclonal antibodies of sufficient specificity for enzyme immunoassays have not been available and refinement of serological tests has had to await detailed investigation of enterovirus structure.

A new approach to the identification of enteroviruses is the use of fluorescent monoclonal antibodies (S.J. Mabs, personal communication). These new reagents are available as blends or against individual serotypes, though not all are currently available. Infected cells, grown under liquid medium and showing a cytopathic effect, are pelleted, fixed and stained in wells on PTFE-coated slides and examined under a fluorescence microscope. The process is much quicker than the SNT, less ambiguous in its results and scanning of the test is also quicker. Testing in parallel with the SNT indicates a comparable level of sensitivity and specificity. Flow cytometry has been used by Abad *et al.* (1998) and Baradi *et al.* (1998) to sort automatically rotavirus infected $CaCO_2$ and MA-104 cells.


## Detection of viruses by molecular biology

Enteroviruses are the group that has been investigated most in the context of waterborne virus disease and is the viral determinant chosen for the EU Bathing Water Directive because they can be concentrated successfully and grown easily in cell culture, not because of their gastrointestinal pathogenesis. Detection of enteroviruses in a water sample implies that there may be other non-detectable agents present, which either do not survive the pH shifts in the concentration procedure or do not multiply in the available cell cultures (or both).

Enteric viruses other than polio, coxsackie and echo types are associated with gastrointestinal disease and therefore it is essential that methods are developed for their detection. Serological tests have been developed but these have not found favour generally, because they detect only viral protein and this as such does not represent a health hazard. The use of molecular biological tests offers a partial solution to the problem of the detection of enteropathogenic viruses in water. Based on knowledge of part or all of the sequence of a viral genome, complementary sequences can be synthesized and probes or PCR primers prepared which can be used to detect the relevant virus in concentrates of the sample.

Although the ultimate goal of such research is the direct (and rapid) detection of pathogenic viruses such as rotavirus and SRSVs in water, techniques first have to be validated against proven methods. This has led to the development of molecular biology-based detection methods for enteroviruses in environmental concentrates; these methods can be validated against traditional cell culture techniques and then taken forward in the development of methods for the detection of enteric pathogens. Richardson *et al.* (1991) reviewed the water industry application of gene probes.

Gene probes were the first approach made in the molecular biological detection of enteric viruses and have been widely used (Enriquez *et al.*, 1993; Dubrou *et al.*, 1991; Moore and Margolin, 1993; Margolin *et al.*, 1993). They are easy to prepare and, if used in conjunction with Southern blotting, they produce satisfactory results. However, they lack sensitivity and, even though some now incorporate digoxigenin instead of using radioisotopes, they have largely been superseded.

The PCR reaction (Saiki *et al.*, 1988) overcomes these shortcomings. The reaction has been used extensively in all branches of biology and has found a particular use in analysis of environmental materials. Detection of enteroviruses is a practical proposition since the picornavirus group contains well-conserved nucleotide sequences at the 5' end of the genome which are used to prepare pan-enterovirus primers which are the starting reagents in the PCR. Since they contain RNA, the nucleic acid of enteroviruses must first be reverse transcribed and cDNA prepared before the PCR proper can be done. The whole reaction is termed RT-PCR.

Methods and applications for detection of enteroviruses generally have been extensively described (e.g. Rotbart, 1990, 1991a,b; Chapman *et al.*, 1990; Tracy *et al.*, 1992). Numerous investigations have been done using RT-PCR to detect enteroviruses in different environmental samples, including river and marine recreational waters (e.g. Kopecka *et al.*, 1993; Wyn-Jones *et al.*, 1995; Gilgen *et al.*, 1995), ground waters (Abbaszadegan *et al.*, 1993; Regan and Margolin, 1997) and sludge-amended field soils (Straub *et al.*, 1995). The technique has been extended to cover other virus groups present in water including adenoviruses (Puig *et al.*, 1994), hepatitis A (Graff *et al.*, 1993), astrovirus (Marx *et al.*, 1995) and rotavirus (Gajardo *et al.*, 1995).

Though the technique is straightforward in principle it is not without practical problems. Chief among these is the presence of fulvic and humic acids in the concentrates which inhibit the RT and/or polymerase reactions. Different solutions have been found to remove these but most rely on adsorption of the extracted RNA to silica (e.g. Shieh *et al.*, 1995). Pallin *et al.* (1997) devised a method for recovering all the virus in a concentrate into a single PCR tube, which allowed direct comparisons of sensitivity with cell culture methods where the whole of the concentrate is tested at one time.

Refinement of the RT-PCR and restriction enzyme analysis of amplicons has permitted the differentiation of virus types within the enterovirus group. Hughes *et al.* (1993) compared the nucleotide sequences of six coxsackievirus B4 isolates from the aquatic environment with those of four CB4 isolates from clinical specimens and found that the isolates fell into two distinct groups not related to their origin. Wyn-Jones, Pallin and Lee (unpublished results) devised a method for identifying enterovirus concentrated from bathing waters based on restriction enzyme analysis which assigned any isolate to at least group level and many to serotype, using just four enzymes, the groupings correlating with those determined by immune electron microscopy and Sellwood *et al.* (1995) reported a system using RFLP analysis to discriminate between wild and vaccine-like strains of poliovirus. Egger *et al.* (1995) devised a multiplex PCR for the differentiation of polioviruses from non-polioviruses, which made an important step in the accumulation of public health information. None of these investigations would have been possible without RT-PCR.

RT-PCR detects viral RNA, not infectious virus and there is thus a theoretical objection to its use in a public health context. However, if virus-specific RNA is detected, even if it came from a non-infectious particle, that signifies there are likely to be infectious particles present also and that the water in which the RNA was detected is indeed polluted with enteroviruses. Further, it is well known that RNA does not survive in the environment so if a positive result is obtained by RT-PCR it indicates that it can only have come from intact, i.e. infectious virus particles. There is an important use for RT-PCR in the screening of samples for enteroviruses; negative ones can be discarded and positives investigated further for presence of infectious virus. The technique has been further employed by the combination of cell culture with RT-PCR. Reynolds *et al.* (1996) and Murrin and Slade (1997) inoculated BGM cultures with concentrates and tested the supernatants at intervals up to 10 days. Virus was detectable by RT-PCR as early as one day post-inoculation, instead of more than 3 days by normal visualization of CPE. This thus allows a more rapid analysis of waters to be carried out.

It will be obvious that RT-PCR detection of enteroviruses is a useful adjunct to the technology for analysing bathing waters for enteroviruses and its use should be considered when formulating strategies and monitoring schemes.

# References

Abad, F.X., Pintó, R.M. and Bosch, A. (1998). Flow cytometry detection of infectious rotaviruses in environment and clinical samples. *Appl Environ Microbiol*, **64**: 2392–2396.

Abbaszadegan, M., Huber, M.S., Gerba, C.P. *et al.* (1993). Detection of enteroviruses in ground water by PCR. *Appl Environ Microbiol*, **65**: 444–449.

American Public Health Association. (1980). *Standard methods for examination of water*, 15th edn.

AranhaCreado, H., Oshima, K., Jafari, S. *et al.* (1997). Virus retention by a hydrophilic triple-layer PVDF microporous membrane filter. *J Pharm Sci Technol*, **51**: 119–124.

Baradi, C.R.M., Emslie, K.R., Vesey, G. *et al.* (1998). Development of a rapid and sensitve quantitative assay for rotavirus based on flow cytometry. *J Virol Meth*, **74**: 31–38.

Barron, A.L., Olshevsky, C. and Cohen, M.M. (1970). Characteristics of the BGM line of cells from African green monkey kidney. *Archiv Gesamte Virusforschung*, **32**: 389–392.

Belfort, G., Paluszek, A. and Sturman, L.S. (1982). Enterovirus concentration using automated hollow fiber ultrafiltration. *Water Sci Technol*, **14**: 257–272.

Bergh, O., Borsheim, K.Y., Bratbak, G. *et al.* (1989). High abundance of viruses found in aquatic environment. *Appl Environ Microbiol*, **340**: 467–468.

Beytout, D., Laveran, H. and Reynaud, M.P. (1975). Mèthode practique d'evaluation numerique applicable aux techniques miniaturisèes de titrage en plages. *Ann Biol Clin*, **33**: 379–384.

Bitton, G., Pancorbo, O. and Gifford, G.E. (1976). Factors affecting the adsorption of poliovirus to magnetite in water and wastewater. *Water Res*, **10**: 973–980.

Block, J.C. and Schwartzbrod, L. (1989). *Viruses in Water Systems. Detection and Identification*. New York: VCH Publishers Inc.

Bosch, A., Pinto, R.M., Blanch, A.R. and Jofre, J.T. (1988). Detection of human rotavirus in sewage through two concentration procedures. *Water Res*, **22**: 343–348.

Chang, S.L., Berg, K.A., Busch, K. *et al.* (1958). Application of the 'most probable number' method for estimating concentrations of animal viruses by the tissue culture technique. *Virology*, **6**: 27–31.

Chapman, N.M., Tracy, S., Gauntt, C.J. *et al.* (1990). Molecular detection and identification of enteroviruses using enzymatic amplification and nucleic acid hybridization. *J Clin Microbiol*, **28**: 843–850.

Chaudhiri, M. and Sattar, S.A. (1986). Enteric virus removal from water by coal-based sorbents: development of low-cost water filters. *Water Sci Technol*, **18**: 77–82.

Cooper, P.D. (1967). The plaque assay of animal viruses. *Adv Virus Res*, **8**: 319–378.

Dahling, D.R. and Wright, B.A. (1986). Optimization of the BGM cell line and viral assay procedures for monitoring viruses in the environment. *Appl Environ Microbiol*, **51**: 790–812.

Dahling, D.R. and Wright, B.A. (1988). Optimisation of suspended cell method and comparison with cell monolayer technique for virus assays. *J Virol Meth*, **20**: 169–179.

Dahling, D.R., Berg, G. and Berman, D. (1974). BGM, a continuous cell line more sensitive than primary rhesus and African green kidney cells for the recovery of viruses from water. *Hlth Lab Sci*, **11**: 275–282.

Dahling, D.R., Phirke, P.M., Wright, B.A. *et al.* (1985). Use of bituminous coal as an alternative technique for the field concentration of waterborne viruses. *Appl Environ Microbiol*, **49**: 1222–1225.

Divizia, M., de Filippis, P., di Napoli, A. *et al.* (1989a). Isolation of wild-type hepatitis A virus from the environment. *Water Res*, **23**: 1155–1160.

Divizia, M., Santi, A.L. and Pana, A. (1989b). Ultrafiltration: an efficient second step for hepatitis A and poliovirus concentration. *J Virol Meth*, **23**: 55–62.

Dubrou, S., Kopecka, H., Lopez-Pila, J.M. *et al.* (1991). Detection of hepatitis A virus and other enteroviruses in wastewater and surface water samples by gene probe assay. *Water Sci Technol*, **24**: 267–272.

Egger, D., Pasamontes, L., Ostermayer, M. *et al.* (1995). Reverse transcription multiplex PCR for differentiation between polio- and enteroviruses from clinical and environmental samples. *J Clin Microbiol*, **33**: 1442–1447.

Enriquez, C.E., Abbaszadegan, M., Pepper, I.L. *et al.* (1993). Poliovirus detection in water by cell culture and nucleic acid hybridization. *Water Res*, **27**: 1113–1118.

Farrah, S.R., Gerba, C.P., Wallis, C. *et al.* (1976). Concentration of viruses from large volumes of tap water using pleated membrane filters. *Appl Environ Microbiol*, **31**: 221–226.

Farrah, S.R., Girard, M.A., Toranzos, G.A. *et al.* (1988). Adsorption of viruses to diatomaceous earth modified by in situ precipitation of metallic salts. *Zent Ges Hyg*, **34**: 520–521.

Fogh, J. (1977). Absence of Hela cell contamination in 169 cell lines derived from human tumours. *J Natl Cancer Inst*, **58**: 209–220.

Gajardo, R., Bouchrit, N., Pintó, R.M. *et al.* (1995). Genotyping of rotaviruses isolated from sewage. *Appl Environ Microbiol*, **61**: 3460–3462.

Gerba, C.P. (1987). Recovering viruses from sewage, effluents and water. In *Methods for Recovering Viruses from the Environment*, Berg, G. (ed.). Boca Raton, FL: CRC Press, pp. 1–23.

Gilgen, M., Germann, D., Luethy, J. *et al.* (1997). Three-step isolation method for sensitive detection of enterovirus, rotavirus, hepatitis A virus and small round-structured viruses in water samples. *Int J Food Microbiol*, **37**: 189–199.

Gilgen, M., Wegmuller, B., Burkhalter, P. *et al.* (1995). Reverse transcription PCR to detect enteroviruses in surface water. *Appl Environ Microbiol*, **61**: 1226–1231.

Grabow, W.O.K., Nupen, E.M. and Bateman, B.W. (1984). South African research on enteric viruses in drinking water. *Monogr Virol*, **15**: 146–155.

Graff, J., Ticehurst, J. and Flehmig, B. (1993). Detection of hepatitis A virus in sewage by antigen capture polymerase chain reaction. *Appl Environ Microbiol*, **59**: 3165–3170.

Guttman-Bass, N. and Nasser, A. (1984). Simultaneous concentration of four enteroviruses from tap, waste and natural water. *Appl Environ Microbiol*, **47**: 1311–1315.

Guttman-Bass, N. and Catalano-Sherman, J. (1986). Humic acid interference with virus recovery by electropositive microporous filters. *Appl Environ Microbiol*, **52**: 556–561.

Homma, A., Sobsey, M.D., Wallis, C. *et al.* (1973). Virus concentration from sewage. *Water Res*, 7: 945–952.

Hovi, T., Stenvik, M., Partanen, H. *et al.* (2001). Poliovirus surveillance by examining sewage specimens. Quantitative recovery of virus after introduction into sewerage at remote upstream location. *Epidemiol Infect*, **127**: 101–106.

Hsiung, G.D. and Melnick, J.L. (1957). Morphologic characteristics of plaques produced on monkey kidney monolayer culture by enteric viruses (poliomyelitis, coxsackie and ECHO groups). *Immunology*, **78**: 128–131.

Huang, P.W., Laborde, D., Land, V.R. *et al.* (2000). Concentration and detection of caliciviruses in water samples by reverse transcription-PCR. *Appl Environ Microbiol*, **66**: 4383–4388.

Hughes, M.S., Hoey, E.M. and Coyle, P.V. (1993). A nucleotide sequence comparison of Coxsackievirus B4 isolates from aquatic samples and clinical specimens. *Epidemiol Infect*, **110**: 389–398.

Hugues, B. (1981). Cited from Block, J.C. and Schwartzbrod, L. (1989). *Viruses in Water Systems; Detection and Identification*. New York: VCH Publishers Inc.

Hugues, B., André, M. and Champsaur, H. (1991). Virus concentration from waste water: glass wool versus glass powder. *Biomed Lett*, **46**: 103–107.

Hurley, M.A. and Roscoe, M.E. (1983). Automated statistical analysis of microbial enumeration by dilution series. *J Appl Bacteriol*, **55**: 159–164.

Jakubowski, W., Hoff, J.C., Anthony, N.C. *et al.* (1974). Epoxy-fiberglass adsorbent for concentrating viruses from large volumes of potable water. *Appl Microbiol*, **28**: 501.

Joret, J.C., Block, J.C., Lucena-Gutierrez, F. *et al.* (1980). Virus concentration from secondary wastewater: Comparative study between epoxy fibreglass and glass powder adsorbents. *Europ J Appl Microbiol Biotech*, **10**: 245–252.

Jothikumar, N., Khanna, P., Paulmurugan, R. *et al.* (1995). A simple device for the concentration and detection of enterovirus, hepatitis E virus and rotavirus from water samples by reverse transcription-polymerase chain reaction. *J Virol Meth*, **55**: 410–415.

Katzenelson, E., Fattal, B. and Hostovesky, T. (1976). Organic flocculation: an efficient second step concentration method for the detection of viruses in tap water. *Appl Environ Microbiol*, **32**: 638–639.

Keswick, B.H., Pickering, L.K., DuPont, H.L. *et al.* (1983). Organic flocculation: an efficient second step concentration method for the detection of viruses in tap water. *Appl Environ Microbiol*, **46**: 813–816.

Kittigul, L., Khamoun, P., Sujirarat, D. *et al.* (2001). An improved method for concentrating rotavirus from water samples. *Mem Inst Oswaldo Cruz*, **96**: 815–821.

Kopecka, H., Dubrou, S., Prèvot, J. *et al.* (1993). Detection of naturally-occurring enteroviruses in waters by reverse transcription, polymerase chain reaction, and hybridization. *Appl Environ Microbiol*, **59**: 1213–1219.

Lakhe, S.B. and Parhad, N.M. (1988). Concentration of viruses from water on bituminous coal. *Water Res*, **22**: 635–640.

Lund, E. and Hedstrom, C.E. (1966). The use of an aqueous polymer phase system for enterovirus isolations from sewage. *Am J Epidemiol*, **84**: 287–291.

Ma, J.-F., Naranjo, J. and Gerba, C.P. (1994). Evaluation of MK filters for recovery of enteroviruses from tap water. *Appl Environ Microbiol*, 60: 1974–1977.

Mack, W.N., Yue-Shoung, L. and Coohon, D.B. (1972). Isolation of poliomyelitis virus from a contaminated well. *Public Hlth Rep*, 87: 271–274.

Margolin, A.B., Gerba, C.P., Richardson, K.J. *et al.* (1993). Comparison of cell culture and a poliovirus gene probe assay for the detection of enteroviruses in environmental water samples. *Water Sci Technol*, 27: 311–314.

Marx, F.E., Taylor, M.B. and Grabow, W.O.K. (1995). Optimization of a PCR method for the detection of astrovirus type 1 in environmental samples. *Water Sci Technol*, 31: 359–362.

Melnick, J.L., Safferman, R., Rao, V.C. *et al.* (1984). Round robin investigations of methods for the recovery of poliovirus from drinking water. *Appl Environ Microbiol*, 47: 144–150.

Metcalf, T.G., Wallis, C. and Melnick, J.L. (1974). Environmental factors influencing isolation of enteroviruses from polluted surface waters. *Appl Environ Microbiol*, 27: 920.

Moore, N. and Margolin, A.B. (1993). Evaluation of radioactive and non-radioactive gene probes and cell culture for detection of poliovirus in water samples. *Appl Environ Microbiol*, 59: 3145–3146.

Morris, R. (1985). Detection of enteroviruses: an assessment of ten cell lines. *Water Sci Technol*, 17: 81–88.

Morris, R. and Waite, W.M. (1980). Evaluation of procedures for the recovery of viruses from water – I Concentration systems. *Water Res*, 14: 791–793.

Murphy, A.M., Grohmann, G.S. and Sexton, M.F.H. (1983). Infectious gastroenteritis in Norfolk Island and recovery of viruses from drinking water. *J Hyg*, 91: 139–146.

Murrin, K. and Slade, J. (1997). Rapid detection of viable enteroviruses in water by tissue culture and semi-nested polymerase chain reaction. *Water Sci Technol*, 35: 429–432.

Muscillo, M., Carducci, A., la Rosa, G. *et al.* (1997). Enteric virus detection in Adriatic seawater by cell culture, polymerase chain reaction and polyacrylamide gel electrophoresis. *Water Res*, 31: 1980–1984.

Myrmel, M., Rimstad, E. and Wasteson, Y. (2000). Immunomagnetic separation of a Norwalk-like virus (genogroup I) in artificially contaminated environmental water samples. *Int J Food Microbiol*, 62: 17–26.

Nupen, E.M., Basson, N.C. and Grabow, W.O.K. (1981). Efficiency of ultrafiltration for the isolation of enteric viruses and coliphages from large volumes of water in studies on wastewater reclamation. *Water Pollution Res*, 13: 851–863.

Pallin, R., Place, B.M., Lightfoot, N.F. *et al.* (1997). The detection of enteroviruses in large volume concentrates of recreational waters by the polymerase chain reaction. *J Virol Meth*, 57: 67–77.

Pandya, G., Jana, A.M., Tuteja, U. *et al.* (1988). Identification of Group A Coxsackieviruses from sewage samples by indirect enzyme-linked immunosorbent assay. *Water Res*, 22: 1055–1057.

Patel, J.R., Daniel, J. and Mathan, V.I. (1985). A comparison of the susceptibility of three human gut tumour-derived differentiated epithelial cell line, primary monkey kidney cells and human rhabdomyosarcoma cell line to 66-prototype strains of human enteroviruses. *J Virol Meth*, 12: 209–216.

Payment, P. and Trudel, M. (1979). Efficiency of several micro-fiber glass filters for recovery of poliovirus from tap water. *Appl Environ Microbiol*, 38: 365–368.

Payment, P. and Trudel, M. (1985). Immunoperoxidase method with human immune serum globulin for broad spectrum detection of cultivable viruses in environmental samples. *Appl Environ Microbiol*, 50: 1308–1310.

Payment, P. and Trudel, M. (1987). Detection and quantitation of human enteric viruses in wastewaters: increased sensitivity using a human immune serum immunoglobulin immunoperoxidase assay on MA-104 cells. *Can J Microbiol*, 33: 568–570.

Payment, P., Tremblay, C. and Trudel, M. (1982). Rapid identification and serotyping of poliovirus isolates by an immunoassay. *J Virol Meth*, 5: 301–308.

Pietri, C. and Hugues, B. (1985). The effect of the cell system on the quantification of viruses present in sewage eluates. *Microbios Lett*, 30: 67–72 (French).

Poynter, S.F.B., Jones, H.H. and Slade, J.S. (1975). Virus concentration by means of soluble ultrafilters. In *Methods for Microbiological Assay*, Board, R.G. and Lovelock, D.W. (eds). London: Academic Press, pp. 65–74.

Puig, M., Jofre, J., Lucena, F. (1994). Detection of adenoviruses and enteroviruses in polluted water by nested PCR amplification. *Appl Environ Microbiol*, **60**: 2963–2970.

Ramia, S. and Sattar, S.A. (1979). Second-step concentration of viruses in drinking and surface waters using polyethylene glycol extraction. *Can J Microbiol*, **25**: 587.

Rao, C., Sullivan, R., Read, R.B. *et al.* (1968). A simple method for concentrating and detecting viruses. *J Am Water Works Assoc*, **60**: 1288–1294.

Reed, L.J. and Muench, H. (1938). A simple method of estimating fifty percent endpoints. *Amer J Hyg*, **27**: 493–495.

Regan, P.M. and Margolin, A.B. (1997). Development of a nucleic acid capture probe with reverse transcriptase-polymerase chain reaction to detect poliovirus in groundwater. *J Virol Meth*, **64**: 65–72.

Reynolds, C.A., Gerba, C.P. and Pepper, I.L. (1996). Detection of infectious enterovirus by an integrated cell culture-PCR procedure. *Appl Environ Microbiol*, **62**: 1424–1427.

Richardson, K.J., Stewart, M.H. and Wolfe, R.L. (1991). Application of gene probe technology to the water industry. *JAWWA*, **83**: 71–81.

Rotbart, H.A. (1990). Enzymatic RNA amplification of the enteroviruses. *J Clin Microbiol*, **28**: 438–442.

Rotbart, H.A. (1991a). New methods for rapid enteroviral diagnosis. *Prog Med Virol*, **38**: 96–108.

Rotbart, H.A. (1991b). Nucleic acid detection systems for enteroviruses. *Clin Microbiol Rev*, **4**: 156–168.

Rotem, Y., Katzenelson, E. and Belfort, G. (1979). Virus concentration by capillary ultrafiltration. *J Environ Eng-ASCE*, **5**: 401–407.

Saiki, R.K., Gelfand, D.H., Stoffel, S. *et al.* (1988). Primer-directed enzymatic amplification of DNA with a thermostable DNA polymerase. *Science*, **239**: 487–491.

Sarrette, B., Danglot, B. and Vilagines, R. (1977). A new and simple method for the recuperation of enteroviruses from water. *Water Res*, **11**: 355–358.

Sattar, S.A. and Ramia, S. (1979). Use of talc-selite layers in the concentration of enteroviruses from large volumes of potable waters. *Water Res*, **13**: 1351–1353.

Sattar, S.A. and Westwood, J.C.N. (1978). Viral pollution of surface waters due to chlorinated primary effluents. *Appl Environ Microbiol*, **36**: 427–431.

Schimdt, N.J., Ho, H.H. and Lennette, E.H. (1976). Comparative sensitivity of the BGM cell line for isolation of enteric viruses. *Hlth Sci*, **13**: 115–117.

Schwab, K.J., Leon, R. and Sobsey, M.D. (1996). Immunoaffinity concentration and purification of waterborne enteric viruses for detection by reverse transcriptase PCR. *Appl Environ Microbiol*, **62**: 2086–2094.

Schwartzbrod, L. and Lucena-Gutierrez, F. (1978). Concentration des enterovirus dans les eaux par adsorption sur poudre de verre: proposition d'un appareillage simplifiè. *Microbia*, **4**: 55–58.

Sellwood, J., Litton, P.A., McDermott, J. *et al.* (1995). Studies on wild and vaccine strains of poliovirus isolated from water and sewage. *Water Sci Technol*, **31**: 317–321.

Shieh, Y.-S.C., Wait, D., Tai, L. *et al.* (1995). Methods to remove inhibitors in sewage and other faecal wastes for enterovirus detection by the polymerase chain reaction. *J Virol Meth*, **54**: 51–66.

Sobsey, M.D. and Glass, J.S. (1980). Poliovirus concentration from tap water with electropositive adsorbent filters. *Appl Environ Microbiol*, **40**: 201–210.

Sobsey, M.D. and Hickey, A.R. (1985). Effects of humic and fulvic acids on poliovirus concentration from water by microporous filtration. *Appl Environ Microbiol*, **49**: 259–264.

Sobsey, M.D. and Jones, B.L. (1979). Concentration of poliovirus from tapwater using positively-charged microporous filters. *Appl Environ Microbiol*, **37**: 588–595.

Standing Committee of Analysts (SCA). (1995). Methods for the isolation and identification of human enteric viruses from waters and associated materials. In *Methods for the examination of waters and associated materials*. London: HMSO.

Straub, T.M., Pepper, I.L. and Gerba, C.P. (1995). Comparison of PCR and cell culture for detection of enteroviruses in sludge-amended field soils and determination of their transport. *Appl Environ Microbiol*, **61**: 2066–2068.

Tracy, S., Chapman, N.M. and Pistillo, J.M. (1992). Detection of human enteroviruses using the polymerase chain reaction. In *Frontiers of Virology, 1: Diagnosis of Human Viruses by Polymerase Chain Reaction Technology*, Becker, Y. and Darai, G. (eds). New York: Springer-Verlag, pp. 331–344.

Tsai, Y.L., Sobsey, M.D., Sangermano, L.R. *et al.* (1993). Simple method of concentrating enteroviruses and hepatitis A virus from sewage and ocean water for rapid detection by reverse transcriptase-polymerase chain reaction. *Appl Environ Microbiol*, **59**: 3488–3491.

United States Environmental Protection Agency. (1984). *USEPA/APHA Standard Methods for the examination of water and wastewater.*

Urase, T., Yamamoto, K. and Ohgaki, S. (1993). Evaluation of virus removal in membrane separation processes using coliphage Q beta. In *Development and Water Pollution Control in Asia*, Bhamidimarri, R. *et al.* (eds), Oxford: Pergamon, pp. 9–15.

Vilaginès, P.H., Sarrette, B. and Vilaginès, R. (1988). Detection en continu du poliovirus dans des eaux de distribution publique. *CR Acad Sci Paris*, **307**, serie III: 171–176.

Vilaginès, P., Sarrette, B., Husson, G. *et al.* (1993). Glass wool for virus concentration at ambient water pH level. *Water Sci Technol*, **27**: 299–306.

Wallis, C. and Melnick, J.L. (1967a). Concentration of viruses from sewage by adsorption on to Millipore membranes. *Bull Wld Hlth Org*, **36**: 219–225.

Wallis, C. and Melnick, J.L. (1967b). Concentration of viruses on aluminium and calcium salts. *Am J Epidemiol*, **85**: 459–468.

Wallis, C. and Melnick, J.L. (1967c). Concentration of viruses on membrane filters. *J Virol*, **1**: 472–477.

Wallis, C., Melnick, J.L. and Fields, J.E. (1970). Detection of viruses in large volumes of natural waters by concentration on insoluble polyelectrolytes. *Water Res*, **4**: 787–796.

Watt, P.M., Johnson, D.C. and Gerba, C.P. (2002). Improved method for concentration of *Giardia, Cryptosporidium* and poliovirus from water. *J Environ Sci Hlth*, **37**: 321–330.

Wellings, F.M., Lewis, A.L. and Mountain, C.W. (1976). Demonstration of solids-associated virus in wastewater and sludge. *Appl Environ Microbiol*, **31**: 354–360.

Wolfaardt, M., Moe, C.L. and Grabow, W.O.K. (1995). Detection of small round-structured viruses in clinical and environmental samples by polymerase chain reaction. *Water Sci Technol*, **31**: 375–382.

Wyn-Jones, A.P. and Sellwood, J. (1998). Review of methods for the isolation, concentration, identification and enumeration of enteroviruses. Project WW-11B Research and Development – Bathing Water Policy, UK Water Industry Research Ltd.

Wyn-Jones, A.P. and Sellwood, J. (2001). Enteric viruses in the aquatic environment. invited review. *J Appl Microbiol*, **91**: 945–962.

Wyn-Jones, A.P., Pallin, R., Sellwood, J. *et al.* (1995). Use of the polymerase chain reaction for the detection of enteroviruses in river and marine recreational waters. *Water Sci Technol*, **31**: 337–344.

# 27

# Adenovirus

## Basic microbiology

One of the earliest reports of the isolation of adenoviruses was when secretions from patients with respiratory infections were inoculated on to human cell cultures resulting in cell death (Hilleman and Werner, 1954). The term 'adeno' was derived from the adenoid tissue which was found to be associated with persistent adenovirus infection (Rowe *et al.*, 1953). The human strains of adenoviruses are classified in the *Mastadenovirus* genus of the Adenovirus family, which also includes many other mammalian adenovirus strains. Forty-nine serotypes of human adenoviruses are recognized; these are based on neutralization studies using antisera raised against epitopes of the hexon protein and the knob portion of the fibre protein of the virion capsid. These serotypes form six subgroups, designated A to F, on the basis of their ability to agglutinate red blood cells.

The icosahedral structure of the virus particle is very distinct when visualized by electron microscopy (Figure 27.1). The virions are 70–100 nm in diameter with no membrane. The protein coat or capsid comprises 252 capsomeres of which 240 are hexons (surrounded by six other capsomeres) and 12 are pentons (surrounded by five capsomeres). The pentons form the angles of the icosaherdron and each has a gylcoprotein fibre extending from it. Four proteins and a genome of double-stranded DNA, approximately 36 kb in

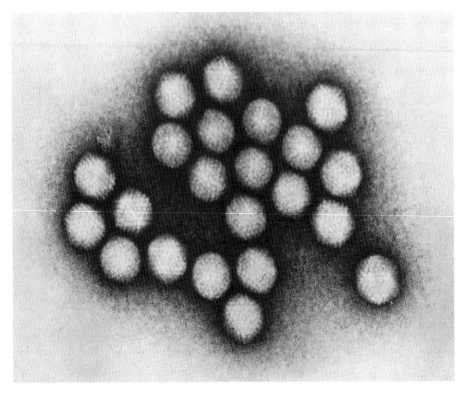

**Figure 27.1**  Transmission electron micrograph of adenovirus. (Courtesy of Dr G. William Gary, Jr, CDC, USA.)

length, form the core of each virion. These symmetrical particles can form extensive arrays or inclusions within the nucleolus of infected cells.

Adenoviruses are stable at pH 3, resistant to intestinal enzymes and replicate in the epithelial cells of the intestine. As adenoviruses do not possess a membrane, ether and chloroform do not inactivate the virus. The limited data that are available on the inactivation of adenoviruses by chlorine demonstrates that the virus is killed but it is not clear if the virus is more or less resistant than the enteroviruses. Adenoviruses are more resistant to the action of UV light than enteroviruses (Gerba *et al.*, 2002). Sewage seeded with adenovirus 40 and 41 serotypes had detectable virus after 60 days held at 15°C and 4°C which was longer than for enterovirus (Enriquez and Gerba, 1995; Enriquez *et al.*, 1995).

## Origin of the organism

The origin of human adenovirus strains in the environment will be human sewage. In common with all viruses the adenoviruses are obligate intracellular parasites, therefore no replication can occur in the environment.

## Clinical features and virulence

Adenoviruses infect the respiratory tract, the eye and the gastrointestinal tract. Asymptomatic infections are common and shedding of virus from pharynx and gut may be prolonged. The respiratory illness caused by adenoviruses is often associated with pharyngitis, fever, cough and cold-like symptoms. Tonsillitis may be involved as well as conjunctivitis (Wadell, 2000). As well as these common symptoms in children, an acute respiratory disease (ARD) has been recognized in young adults. It occurs particularly in young military recruits undergoing training and can be severe. Adenovirus infection has been associated with *Bordetella pertussis* infection and the symptoms of whooping cough.

Follicular conjunctivitis can occur as part of the general respiratory illness but also as an eye infection without generalized symptoms. It is usually a mild, short infection from which there is a full recovery. Epidemic keratoconjunctivitis is a more aggressive disease involving lymphadenopathy, pain and swelling around the eye. The residual effects may include corneal opacity and last for some years. It was easily spread in specific areas of work and became known as 'shipyard eye'.

Gastroenteritis in young children is caused by Group F adenoviruses, comprising serotypes 40 and 41. Several hundred reports are sent to the Health Protection Agency Communicable Disease Surveillance Centre throughout each year. Although many adenovirus serotypes may replicate in the gut and are detectable in faeces, it is now recognized that only serotypes 40/41 cause gastroenteritis (Krajden *et al.*, 1990; Lew *et al.*, 1991) unless part of a wider systemic illness. Several serotypes that replicate in the intestine have been linked to a syndrome termed 'intussusception'. Part of the lower bowel telescopes into itself thereby causing a blockage and may be caused by adenovirus replication in the associated mesenteric glands.

It is not understood why the severity of disease varies between different serotypes of adenovirus. Nor has the mechanism behind latency and long-term shedding of virus been elucidated.

## Pathogenicity

Intranuclear inclusions consisting of an array of virions may be found in some host cells such as alveolar cells when pneumonia has developed. The infection shuts off the expression of host cell messenger RNA which, in conjunction with the production of excess virus protein, leads to cell death. The different subgenera (A–F) may have a range of cell tropisms, e.g. adenovirus 8 usually infects the eye, while 40 and 41 infect the gastrointestinal tract. However, most strains gain entry to the host via the mouth or eye; initial infection is in non-ciliated epithelium cells and glandular tissue (adenoids and tonsils) in the nasopharynx. Incubation periods vary according to the serotypes and the site of infection; respiratory illness and gastrointestinal disease can result after a few days, while eye infections may have an incubation period of 2–6 days.

Immune responses include the production of 'complement-fixing antibodies' or immunofluorescent antibody which are group-specific (raised against the soluble hexon antigen) and of neutralizing antibodies which are type specific (raised against a mixture of hexon and fibre antigens) and are measured by a neutralizing or haemagglutination test. All the antibody levels wane in time but will rise after further infection. Immunity to group F viruses appears to be life-long as the majority of reported infections occur in very young children.

## Transmission and epidemiology

The transmission route for group F, serotypes 40/41 is faecal-oral spread as the virus is shed in faecal material in large numbers. Other serotypes are spread by aerosols of respiratory droplets and mechanically into the conjunctiva of the eyes. Swimming pool waters have been suggested to be the vehicle of transmission for serotypes 3 or 7 (Foy et al., 1968; Martone et al., 1980; Turner et al., 1987; Papapetropolou and Vantarakis, 1995; Harley et al., 2001) resulting in eye infections. Problems with the chlorination of these pools have usually been identified. Outbreaks of eye infections, often serotype 8, have also occurred in hospital eye departments after the use of contaminated equipment on a series of patients (Jernigan et al., 1993).

Infection with the adenoviruses that cause respiratory infections is widespread and common with and without obvious symptoms. Group F adenovirus infection causes diarrhoea in babies and immunity to these serotypes can reach 50% or more by the age of 4 years.

## Treatment

There is no specific treatment for enteric adenovirus infection and symptoms are rarely sufficiently severe to require rehydration.

## Distribution in the environment

Adenoviruses do not form plaques and thus have not been detected in the widely used BGM cell plaque assays but do grow well in human cell culture and monkey kidney cell culture under liquid medium. They form an easily recognizable cytopathic effect which can be confirmed as adenovirus by IF or ELISA. Many of the culturable serotypes were isolated from sewage by Sellwood *et al.* (1981) and Irving and Smith (1981), however, this method of detection is laborious and few reports have been published.

Molecular methods such as PCR have allowed all enteric adenoviruses to be detected with greater ease from sewage, other water matrices and shellfish. Primers based on the hexon gene detect both culturable and group F serotypes of human origin and other mammals (Allard *et al.*, 1990; Puig *et al.*, 1994), whereas primers based on the fibre gene can be directed at group F specifically (Tiemessen and Nel, 1996). An alternative method is to use RFLP to distinguish the different sized PCR products of the different subgenuses (Kidd *et al.*, 1996). Genthe *et al.* (1995) used not PCR but gene probes to identify adenoviruses in a high proportion of sewage and other waters in South Africa.

The Spanish group of Puig *et al.* (1994) and Pina *et al.* (1998) have detected adenoviruses in sewage, polluted water and shellfish throughout the year in most samples investigated. Sewage from the south of England also contains adenovirus in most samples all year round (Sellwood unpublished results). This frequency of detection has prompted several calls for the virus group to be used as a marker of human virus contamination in water. A European collaborative project comprising Spain, UK, Sweden and Greece identified adenoviruses in shellfish from each of the countries using hexon-based primers and a quantitative PCR (Formiga-Cruz *et al.*, 2002). Adenoviruses including group F serotypes were detected in sewage from San Paulo in Brazil (Santos *et al.*, 2002) and in the UK (unpublished results Sellwood and Wyn-Jones). Drinking water was shown to contain adenoviruses including some 'closely related' to adenovirus 40/41 in nearly 40% of Korean tap water samples by Lee *et al.* (2002). The disinfection system was found not to be effective and therefore level of chlorination of the water was low.

Integrated cell culture-PCR may be used to identify infectious virus by an initial incubation period in cell culture and then, to increase the sensitivity using PCR as the detection method. This was used by Chapron *et al.* (2000) to detect adenoviruses 40 and 41 (among others) in 29 surface water samples; 14 were adenovirus 40/41 positive, of which 11 contained infectious virus. Grabow *et al.* (2001) has detected adenovirus in 4% of drinking water supplies in South Africa using a similar approach.

Adenoviruses were detected in the air filters of a building's ventilation system where an outbreak of adenovirus respiratory illness had occurred (Echavarria *et al.*, 2000).

## Waterborne outbreaks

Kukkula *et al.* (1997) identified adenoviruses in the sewage contaminated water that had caused an outbreak of norovirus infection as well as the norovirus itself.

Most other reports have been swimming pool associated, as mentioned above, particularly when the chlorination system had failed (Harley *et al.*, 2001).

## Risk assessment

*Health effects*: occurrence of illness, degree of morbidity and mortality, probability of illness based on infection:

- Adenoviruses mostly cause respiratory illness; however, depending on the infecting serotype, they may also cause gastroenteritis (from group F), conjunctivitis, cystitis and rash illness. In healthy people, illness from adenovirus is generally mild.
- In a summary of studies of sources of diarrhoeal illness, enteric adenovirus was identified as a sole cause around 5–10% of the time, representing between 0 and 40% of all serotypes in adenovirus cases; generally, 0–20% prevalence of faecal shedding in healthy hosts.
- Asymptomatic infections are common and shedding of virus from pharynx and gut may be prolonged. About 50% of childhood enteric adenovirus infections result in illness; the probability of illness increases when the infection is respiratory. Immunity to group F (enteric) viruses appears to be life-long as the majority of reported infections occur in very young children.

*Exposure assessment*: routes of exposure and transmission, occurrence in source water, environmental fate:

- Adenovirus has not been positively associated with any outbreaks from drinking water. A number of outbreaks of conjunctivitis have occurred as a result of exposure to recreational water.
- Routes of exposure include direct (touching), aerosolization and ingestion. The main route of exposure for enteric adenovirus infection is most likely faecal-oral.
- There is limited evidence of secondary spread except among very young (preschool age) children.
- Water is contaminated by human faecal matter/sewage, so the occurrence in source water is dependent on level of human faecal contamination. In published reports, concentrations in river water have ranged from 0 to 25 pfu/l and from 70 to 3200 cpu/l.

*Risk mitigation*: drinking-water treatment, medical treatment:

- The limited data that are available suggest that adenoviruses are inactivated by free chlorine, but its level of resistance compared to other viruses is not clear. Adenoviruses are more resistant to the action of UV light than enteroviruses.
- The illness is self-limiting. Rehydration therapy may be necessary in some children with diarrhoea.

# References

Allard, A., Girones, R., Juto, P. *et al.* (1990). PCR for detection of adenovirus in stool samples. *J Clin Microbiol*, **28**: 2659–2667.

Chapron, C.D., Ballester, N.A., Fontaine, J.H. *et al.* (2000). Detection of astroviruses, enteroviruses and adenovirus types 40 and 41 in surface waters collected and evaluated by the Information Collection Rule and an Integrated Cell Culture-Nested PCR procedure. *Appl Environ Microbiol*, **66**: 2520–2525.

Echavarria, M., Kolavic, S.A., Cersovsky, S. *et al.* (2000). Detection of adenoviruses in culture negative environmental samples by PCR during an adenovirus associated respiratory disease outbreak. *J Clin Microbiol*, **38**: 2982–2984.

Enriquez, C.E. and Gerba, C.P. (1995). Concentration of enteric adenovirus 40 from tap, sea and waste water. *Water Res*, **29**: 2554–2560.

Enriquez, C.E., Hurst, C.J. and Gerba, C.P. (1995). Survival of the enteric adenoviruses 40 and 41 in tap, sea and waste water. *Water Res*, **29**: 2548–2553.

Formiga-Cruz, M., Tofino-Quesada, G., Bofill-Mas, S. *et al.* (2002). Distribution of human virus contamination in shellfish from different growing areas in Greece, Spain, Sweden and the UK. *Appl Environ Microbiol*, **68**: 5990–5998.

Foy, H.M., Cooney, M.K. and Hatlen, J.G. (1968). Adenovirus type 3 epidemic associated with intermittent chlorination of a swimming pool. *Arch Environ Hlth*, **17**: 795–802.

Genthe, B., Gericke, M., Bateman, B. *et al.* (1995). Detection of enteric adenoviruses in South African waters using gene probes. *Water Sci Technol*, **31**: 345–350.

Gerba, C.P., Ramos, D.M. and Nwachuka, N. (2002). Comparative inactivation of enteroviruses and adenovirus 2 by UV light. *Appl Environ Microbiol*, **68**: 5167–5169.

Grabow, W.O., Taylor, M.B. and Villiers, J.C. (2001). New methods for the detection of viruses: call for review of drinking water quality guidelines. *Water Sci Technol*, **43**: 1–8.

Harley, D., Harrower, B., Lyon, M. *et al.* (2001). A primary school outbreak of pharyngo-conjunctival fever caused by adenovirus type 3. *Commun Dis Intell*, **25**: 9–12.

Hilleman, M.R. and Werner, J.H. (1954). Recovery of new agents from patients with acute respiratory illness. *Proc Soc Exp Biol Med*, **85**: 183–188.

Irving, L.G. and Smith, F.A. (1981). One-year survey of enteroviruses, adenoviruses and reoviruses isolated from effluent at an activated sludge purification plant. *Appl Environ Microbiol*, **41**: 51–59.

Jernigan, J.A., Lowry, B.S. and Hayden, F.G. (1993). Adenovirus type 8 epidemic keratoconjunctivitis in an eye clinic: risk factors and control. *J Infect Dis*, **167**: 1307–1313.

Kidd, A.H., Jonsson, M., Garwicz, D. *et al.* (1996). Rapid subgenus identification of human adenovirus isolates by a general RT-PCR. *J Clin Microbiol*, **34**: 622–627.

Krajden, M., Brown, M. and Petrasek, A. (1990). Clinical features of adenovirus enteritis: a review of 127 cases. *Pediatr Infect Dis J*, **9**: 636–641.

Kukkula, M., Arstila, P., Klossner, M.L. *et al.* (1997). Waterborne outbreak of viral gastro-enteritis. *Scand J Infect Dis*, **29**: 415–418.

Lee, S.H. and Kim, S.J. (2002). Detection of infectious enteroviruses and adenoviruses in tap water in urban areas in Korea. *Water Res*, **36**: 248–256.

Lew, J.F., Moe, C.L. and Monroe, S.S. (1991). Astrovirus and adenovirus associated with diarrhoea in children in day care settings. *J Infect Dis*, **164**: 673–678.

Martone, W.J., Hierholzer, J.C., Keenlyside, R.A. *et al.* (1980). An outbreak of adenovirus type 3 disease at a private recreation center swimming pool. *Am J Epidemiol*, **111**: 229–237.

Papapetropoulou, M. and Vantarakis, A.C. (1995). Detection of adenovirus outbreak at a municipal swimming pool by nested PCR amplification. *J Infect*, **36**: 101–103.

Pina, S., Puig, M., Lucena, F. *et al.* (1998). Viral pollution in the environment and in shellfish: human adenovirus detection by PCR as an index of human viruses. *Appl Environ Microbiol*, **64**: 3376–3382.

Puig, M., Jofre, J., Lucena, F. *et al.* (1994). Detection of adenoviruses and enteroviruses in polluted water by nested PCR amplification. *Appl Environ Microbiol*, **60**: 2963–2970.

Rowe, W.P., Huebner, R.J., Gillmore, L.K. *et al.* (1953). Isolation of a cytopathogenic agent from human adenoids undergoing spontaneous degeneration in tissue culture. *Proc Soc Exp Biol Med*, **84**: 570–573.

Santos, F.M., Vieira, M.J., Monezi, T.A. *et al.* (2002). Discrimination of adenovirus types circulating in urban sewage and surface waters in San Paulo City, Brazil. Poster International Water Association. Health Related Microbiology Symposium. Cascaise Portugal.

Sellwood, J., Dadswell, J.V. and Slade, J.S. (1981). Viruses in sewage as an indicator of their presence in the community. *J Hyg Camb*, **86**: 217–225.

Tiemessen, C.T. and Nel, M.J. (1996). Detection and typing of subgroup F adenoviruses using PCR. *J Virol Meth*, **59**: 73–82.

Turner, M., Istre, G.R., Beauchamp, H. *et al.* (1987). Community outbreak of adenovirus type 7a infections associated with a swimming pool. *South Med J*, **80**: 712–715.

Wadell, G. (2000). Adenoviruses. In *Principles and Practice of Clinical Virology*, 4th edn, Zuckerman, A.J., Banatalava, J.E. and Pattison, J.R. (eds). Chichester: John Wiley and Sons, pp. 307–327.

# 28

# Astrovirus

## Introduction

Among the viruses recognized as agents of viral gastroenteritis, the role of astroviruses is perhaps the least well understood. First described in stool specimens from babies with gastroenteritis by Appleton and Higgins (1975) and since observed regularly, their significance in enteric disease has been less well defined than that of noroviruses, rotaviruses or the enteric adenoviruses. Generally, in healthy patients, astroviruses cause only a mild or symptomless infection, but several authors have reported incidences second only to rotaviruses as causing infantile gastroenteritis (e.g. Marx *et al.*, 1998a) and they are occasionally associated with more severe disease, especially in the immunocompromised individual. In addition to numerous reports of their involvement in diarrhoea in babies and young children (e.g. Madeley, 1979), astroviruses have also been associated with disease in the elderly (Wilson and Cubitt, 1988) and in a wide variety of young animals including lambs (Snodgrass and Gray, 1977), calves (Woode and Bridger, 1978), deer (Tzipori *et al.*, 1981), piglets (Bridger, 1980), kittens (Hoshino *et al.*, 1981), mice (Kjeldsberg and Hem, 1985) and puppies (Marshall *et al.*, 1984). In most animals they cause mild diarrhoeal disease, though in avian species they are more virulent (e.g. Gough *et al.*, 1984; Koci *et al.*, 2000; Imada *et al.*, 2000). Astrovirus infections appear to be species specific and there is currently no animal model for infection in humans. Given that the role of viruses in general in waterborne disease has yet to be fully understood, it will be clear that the less well defined role of human astrovirus infection will compound the difficulties in ascribing outbreaks of waterborne disease to this agent.

## Basic microbiology

Astroviruses were discovered in the UK by electron microscopical examination of faecal specimens (Appleton and Higgins, 1975; Madeley and Cosgrove, 1975). While most particles appear as round structures approximately 28 nm in diameter, about 15% of particles (in a fresh specimen) show a characteristic five- or six-pointed star-shaped motif on the particle surface. This structure is clearly different from that of other viruses and serves as a diagnostic feature. Buoyant densities range from $1.32 \, \text{g/cm}^3$ to $1.35 \, \text{g/cm}^3$ for human strains to $1.39 \, \text{g/cm}^3$ for ovine strains (Herring et al., 1981). Astroviruses are moderately resistant to chemical and physical agents; they are stable at pH 3, are resistant to chloroform, a variety of detergents (non-ionic, anionic and zwitterionic) and to other lipid solvents. Human astroviruses retain their infectivity after 5, but not 10, minutes at 60°C (Kurtz and Lee, 1987). They are stable to storage at low temperatures ($-70$°C) but, in common with most viruses and other micro-organisms, may be disrupted by repeated freezing and thawing. There is no evidence to suggest that astrovirus is not sensitive to chlorine. Ovine astrovirus capsids were shown by Herring et al. (1981) to contain at least two structural proteins of $M_r 30$ and $32 \, \text{kDa}$, an observation which initiated the classification of the group separate from the picornaviruses and caliciviruses.

Astroviruses contain a positive-sense single-stranded polyadenylated RNA genome of 6.8–7.2 kb encompassing three open reading frames (ORF1a, 1b and 2). Complete nucleotide sequences are available for two (HastV-1 and HastV-2) of the eight recognized astrovirus serotypes (Jiang et al., 1993; Willcocks et al., 1994). The subgenomic fragment of about 2500 nucleotides encodes a single open reading frame (ORF2) and hence the capsid precursor polypeptide. A database search revealed the amino acid sequence of this polypeptide to be unique and it was thus proposed that astroviruses belong to a separate family, the Astroviridae (Monroe et al., 1993).

In 1981, Lee and Kurtz (1981) reported the adaptation of astroviruses to grow in cell culture, first in human embryo kidney (HEK) monolayers and, after about six passages, to grow in the LLCMK$_2$ monkey kidney cell line. This is achieved by growing the virus in the presence of trypsin sufficient to cleave the major capsid protein but not high enough to damage the host cells. Viral infectivity may be monitored by immunofluorescence because the trypsin causes detachment of the cells and a cytopathic effect is difficult to determine, though with experience it is possible to see a deterioration in the quality of the cells compared with uninfected controls (Wyn-Jones, unpublished observations). The need for initial passage in HEK cells was circumvented by the observation that the virus could be grown directly from clinical specimens by inoculation into monolayers of the human colonic carcinoma cell line CaCo-2 (Willcocks *et al.*, 1990; Wyn-Jones and Herring, 1991). Following growth in cell culture further studies on the viral proteins were possible. Willcocks *et al.* (1990) determined that there are three virus specific polypeptides in human astroviruses of 33.5, 31.5 and 24 kDa, the latter being loosely associated with the particle. For astroviruses generally, it is agreed that the capsid consists of three major proteins, two in the range 29–33 kDa and one of more variable composition in the range 13–26.5 kDa (Willcocks *et al.*, 1992). The polypeptides are the specific cleavage products of a 90 kDa precursor, which is cleaved only in the presence of trypsin.

Human astroviruses are currently divided into eight serotypes on the basis of immune EM, immunofluorescence and enzyme immunoassays (Kurtz and Lee, 1984; Lee and Kurtz, 1994). High divergence in the capsid protein sequences of astroviruses infecting hosts of different species reflects human strains being serologically distinct from the animal astroviruses, though limited similarities between capsid sequences of human, feline and porcine astroviruses have suggested that zoonoses involving these species could occur (Jonassen *et al.*, 2001). There is no epidemiological evidence for this however. Nucleotide sequence analysis of a limited region of the ORF2 of human astrovirus strains by Noel *et al.* (1995) showed there is good correlation between antigenic and genomic types. Relatively few isolates of HAstV type 8 have been reported and there is little information available with regard to the antigenic and genetic relationships between HAstV type 8 (HAstV-8) and the other serotypes. Taylor *et al.* (2001) reported on the analysis of a human astrovirus isolated from a South African paediatric patient with diarrhoea. Immune electron microscopy and EIA, and genetic analysis of selected regions of the ORF1a, ORF1b and ORF2 characterized the virus as HAstV-8 and confirmed that HAstV-8 represents a distinct antigenic type and genotype.

The principal technique for detection of astroviruses in clinical specimens remains electron microscopy (Figure 28.1), done on faecal specimens from children of five years and under in sporadic cases and adults in outbreaks. This technique has the advantage of being a catch-all method, in that it will reveal the presence in a specimen of a range of enteric viruses provided they have a distinct morphology and are there in sufficient quantity, about $10^6$ particles/g. However, it suffers from the very real problem of missing astrovirus-positive specimens owing to fewer than 15% of particles usually showing the surface star structure in a fresh specimen. This will be less (5–10%) if the specimen is old or has been stored. Immune EM is not a good technique as the antibody tends to obscure the surface morphology and leads to false identification. Solid phase immune electron microscopy (SPIEM) has, however, been used with good effect and contributed to the separation of serotypes (Kurtz, 1994). Astrovirus EIA kits of high sensitivity, based on monoclonal antibodies, are available but are not widely used, mainly due to cost considerations.

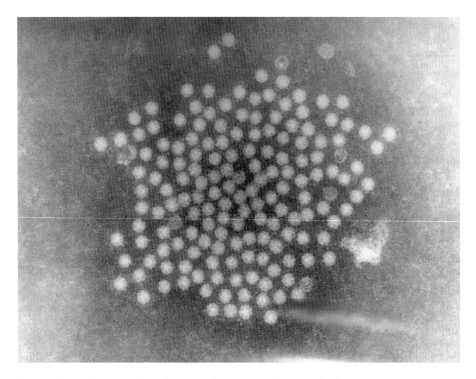

**Figure 28.1**    Transmission electron microscope micrograph of astrovirus.

Other diagnostic techniques involve inoculation of CaCo-2 cell cultures and performing immunofluorescence on the culture 18–24 hours later; this has the advantage of detecting infectious virus and is a sensitive technique. Detection can also be carried out using the reverse-transcriptase polymerase chain reaction (RT-PCR); reagents are available which will detect all the serotypes of astrovirus in one reaction (Jonassen *et al.*, 1995). These authors reported on 38 astrovirus EM-positive stool specimens. Using RT-PCR, 36 were positive by RT-PCR and the remaining two were considered to have been either EM false positives or had lost the RNA. Tests for other enteric viruses were negative. These authors and others have also described reagents for the detection of separate astrovirus serotypes by this technique. Authors have also described integrated cell culture RT-PCR for astrovirus detection in environmental sample concentrates; here, concentrate is inoculated into permissive or partially permissive cell cultures such as PLC/PRF/5 or CaCo-2 and allowed to develop for between 24 and 60 hours; the culture is then analysed by RT-PCR. This technique has been used with good effect by several groups, e.g. Pintó *et al.* (1996), Chapron *et al.* (2000) and Taylor *et al.* (2001) as well as for enteroviruses by Reynolds *et al.* (1998).

Plaque assays in cell culture have also been described (Hudson *et al.*, 1989), but the methods have not been generally adopted, either because they have not been found reproducible or because they depend on the use of primary human cell culture. Attempts to develop a plaque assay in CaCo-2 cells have been unsuccessful (Wyn-Jones, unpublished data; M. Carter, personal communication), though Tree (1997) reported a plaque assay in HT29 cells, a colonic carcinoma cell line different from CaCo-2.

Typing of astroviruses is carried out by immunofluorescence, immune electron microscopy (Lee and Kurtz, 1994), enzyme immunoassay and nucleotide sequencing (Noel *et al.*, 1995) and immunogold staining electron microscopy (Kjeldsberg, 1994).

## Origin of the organism

The *source* of astroviruses in the environment may be defined as the host animal (including man) from which they are derived. In common with all viruses multiplying in the gut, the source of human astrovirus in the environment is faecal material from infected individuals, whether they are symptomatic or not. No multiplication takes place outside the host. The *input* to the environment is the material which contains the virus before it enters a controlled water. Inputs include those materials, such as faeces, sewage, sludge, sediments and septic tanks which may contain human faeces and subsequently enter controlled waters.

## Pathogenesis and clinical features

Phillips *et al.* (1982) showed that astroviruses infect the mucosal epithelial cells of the lower parts of the duodenal villi and Snodgrass *et al.* (1979) reported transient villous atrophy in the small intestine of infected lambs. The virus is assembled in pseudo-crystalline arrays and is released from cells into the gut lumen; a characteristic of astrovirus infection is the occurrence of these arrays on EM grids. Excretion of virus may be followed by diarrhoea, though this is rarely as severe as rotavirus diarrhoea and asymptomatic excretion is common. Diarrhoea, where present, generally lasts 2–3 days but may continue for a week and is paralleled by virus shedding. During the acute stage, virus levels of $10^{10}$ particles/g have been observed. A good immunity follows infection so that repeated episodes of disease are rare in a healthy individual.

The incubation period of the disease, determined by human volunteer studies, is 3–4 days (Kurtz and Lee, 1987) though the virus is excreted for a day prior to symptoms appearing and Konno *et al.* (1982) calculated a 24–36 hours incubation period based on secondary spread characteristics. In addition to diarrhoea, which may be watery but is usually just soft stools and sometimes not even recorded as diarrhoea, clinical features of the infection include headache, malaise, nausea and occasionally vomiting. There is also a mild fever. The diarrhoea generally lasts 2–3 days but there are reports of it continuing for as long as 14 days. Generally, astrovirus diarrhoea is milder than that caused by rotavirus and does not lead to significant dehydration.

The early volunteer studies (Kurtz and Lee, 1987) showed that not all patients suffer diarrhoea. These studies were done on adults, which may reflect a different picture from children, but nevertheless showed fewer individuals affected with diarrhoea; of 17 volunteers only one had overt diarrhoea (2–6 loose motions per day). There is probably a dose-response mechanism operating; following the Marin County outbreak (see below; Midthun *et al.*, 1993) virus was given to 19 volunteers; none of the first 17 became ill so the dose was increased by 20-fold for the remaining two, of whom one became ill. These studies confirm the relatively low pathogenicity of astrovirus, at least for adults, compared with that of noroviruses.

## Transmission and epidemiology

Astroviruses are transmitted from person to person by the faecal-oral route. The PHLS Communicable Disease Surveillance Centre totals included 116 astrovirus laboratory reports for 2001 and 96 for 2002. This compares with 256 enteric adenovirus (serotypes 40/41) for 2001 and 56 for 2002; 28 'classic calicivirus' (now known as sappovirus) in 2001 and 32 in 2002. These three enteric viruses share a similar epidemiology. Though person-to-person spread is the most common mode of enteric virus transmission within closed communities (hospital wards, schools, families etc.) modifications of this route exist including transmission via water or sewage. However, unequivocal waterborne transmission of astroviruses has not been reported (see below).

The epidemiology of astroviruses was reviewed by Glass *et al.* (1996) and a 3-year study by Guix *et al.* (2002) described a molecular epidemiological approach. Person-to-person spread occurs in families, nurseries or paediatric wards, where endemic infections may occur (Caul, 1996). Astrovirus infections have been found world-wide, mainly in young children with diarrhoea. Outbreaks have also been recorded among the elderly (Gray *et al.*, 1987) and in military recruits (Belliot *et al.*, 1997). As with any virus, infection is more pronounced in immunocompromised individuals. Many infected individuals shed virus in the absence of marked symptoms, hence they may still partake in daily activities; in the case of children, they will still go to school and will mix with other children. The agent is thus likely to be transmitted efficiently through the community, though numbers of individuals reporting illness may be quite low and many cases will go unreported.

In Marin County, California, 51% of the residents in a home for the elderly developed gastroenteritis due to astrovirus type 5 (Herrmann *et al.*, 1990). Outbreaks have been reported where staff in day-care centres or on paediatric wards have become infected, showing that susceptibility into adult life does occur (Caul, 1996). Gray *et al.* (1987) reported an outbreak of gastroenteritis in a home for the elderly where 80% of the residents and 44% of the staff were affected. Calicivirus was involved as well as astrovirus.

Nevertheless a defined role for astroviruses in gastrointestinal disease is difficult to determine; reports indicate that their incidence and significance may be underestimated. On the one hand, they seem to be a minor cause of gastroenteritis in children (see above) and are only rarely associated with adult disease; surveys have shown that astroviruses are endemic in most communities and so it would seem that there is a reservoir of virus ready to be taken up by susceptible individuals. Conversely, studies in Guatemala, Thailand (Hermann *et al.*, 1991) and Australia (Palombo and Bishop, 1996) found that astroviruses were the second most frequently detected viral pathogen in children with diarrhoea, exceeding adenoviruses in frequency of detection and only rotaviruses being more prevalent. The high frequency of detection is therefore not confined to developing countries. Astrovirus outbreaks in adults are reported infrequently; Konno *et al.* (1982) described an outbreak in Japan involving four adults and 84 children aged 5–6 years and Oishi *et al.* (1994) reported a large outbreak in Osaka, Japan affecting both students and teachers; over 4700 people were affected. Astrovirus was detected by EM and confirmed by serology and molecular biology tests. Utagawa *et al.* (1994) reported on samples collected between 1982 and 1992 in Japan from sporadic and outbreak cases and used different tests to identify the agent.

## Distribution in the environment

Not all cases of astrovirus diarrhoea arise from direct person-to-person transmission. Food, water and fomites have been implicated in astrovirus disease, which suggests that astroviruses persist in the environment. The 1991 outbreak described by Oishi *et al.* (1994) was food-borne, canteens in 14 schools having received food prepared in three central kitchens which, in turn, had been supplied from a single source. Astroviruses have also been implicated in shellfish-related outbreaks, though not as commonly as noroviruses; Kurtz and Lee (1987) cited a report where a third of 39 naval officers attending a party suffered gastroenteritis 5 days after consuming oysters. Astrovirus was seen in specimens from five patients and two showed astro-specific IgM antibodies.

Astroviruses have been detected in most aquatic matrices. Generally speaking, however, there are no unequivocal accounts of the waterborne transmission of astrovirus disease and little circumstantial or anecdotal evidence to suggest that astroviruses constitute a significant waterborne public health risk. Cubitt (1991) reported one instance where children became ill 24–36 hours after drinking stream water which was the presumed source of the infection. Astrovirus particles were detected in stool samples from the children and high titres of IgG suggestive of recent infection were detected. In another incident in 1994, soldiers returning from a training exercise in the country drank treated effluent at a treatment works under the misconception that it was drinking water. Four developed gastroenteritis and astrovirus particles were seen by electron microscopy in specimens from two of them (CDR, 1996).

Detection of astrovirus in several types of environmental sample has been described by Marx *et al.* (1995, 1998b). These authors tested sewage sludge, hospital and abattoir effluents, source water (raw water, 40-litre samples), stream water (20-litre samples) and treated wastewater. The source water and stream water were concentrated by glass wool adsorption. No information was given on the sludge and wastewater processing. Concentrates were tested by a modification of the RT-PCR method of Jonassen *et al.* (1993) for astrovirus type 1 or using more broadly reactive primers described by Mitchell *et al.* (1995) and by Jonassen *et al.* (1995). Virus detection was achieved either directly following concentration or after a virus amplification step involving inoculating the concentrates in cell cultures. Positive results were obtained for the source water even without concentration and water from the suburban stream was positive following cell culture amplification. Effluent and sludge were also positive by RT-PCR. In contrast, no samples were positive by EM or IEM.

Taylor *et al.* (2001) analysed river and dam water in South Africa over a 12-month period for human astrovirus and hepatitis A virus (HAV) and found astrovirus in 11/51 river samples and 3/51 dam samples. The river was used by the local rural community for 'domestic purposes' and both sites were used for recreation; the dam water was also abstracted for drinking water treatment. The river was continually faecally polluted and the dam, while a large storage facility, nevertheless, received input from a number of sources of varying water quality. The authors also highlighted that virus was detected in dam water samples in the absence of bacterial and phage indicators and that there was no seasonal peak for astroviruses, although there was for HAV.

Pintó *et al.* (1996) reported detection of infectious astroviruses in 500-litre samples from a dam receiving sewage effluent in an area of Spain during a gastroenteritis outbreak. Astrovirus detection in one sample out of six was done by molecular hybridization with an astrovirus-specific cDNA probe following virus growth in CaCo-2 cell cultures. Stool samples from affected individuals were shown to contain astrovirus particles by electron microscopy. These authors determined that the limit of detection by this technique was 10 virus particles. Pintó *et al.* concluded that the water concentrates and unconcentrated samples would have contained 200 and 20 astrovirus particles per litre, respectively. The samples positive for astrovirus also contained enterovirus, rotavirus and adenovirus 40.

The same group later reported astrovirus detection (by RT-PCR) in samples from three sewage treatment plants in France and two in Barcelona, and found that peak environmental prevalence occurred in the winter months, coincident with the peak of clinical activity (Pinto *et al.*, 2001). Samples could be typed by restriction fragment length polymorphism (RFLP) analysis of a fragment of the ORF1a. By this means different patterns were established, three different patterns of isolates (corresponding to different viral genotypes) together accounting for about 87% of isolates.

Abad *et al.* (1997) studied the survival of astroviruses in different water types. Study of survival in dechlorinated tap water at 4°C and marine water at 20°C showed a reduction of around 2 $logs_{10}$ at 4°C and about 3.6 $logs_{10}$ at 20°C over a period of 60 days. After 90 days the reduction was 3.3 $logs_{10}$ and 4.3 $logs_{10}$ respectively. These figures would indicate that virus persists in water for extended periods of time but that inactivation does eventually occur. The virus displayed marked stability in dechlorinated drinking water at 4°C, with only a 1.2 $log_{10}$ reduction over a 45-day period.

Several groups have studied the presence of astroviruses in shellfish. Le Guyader *et al.* (2000) reported the detection, over a 3-year period, of HAV, noroviruses, enteroviruses, rotavirus and astrovirus by reverse transcription-PCR and hybridization in shellfish tissue. In respect of astroviruses, noroviruses and rotaviruses, mussel samples were more highly contaminated than oysters. Viral contamination was mainly observed during winter months, although there were some seasonal differences among the viruses.

Chlorine has an inactivating effect on astrovirus, as would be expected. In the presence of 0.5 mg/l free chlorine astrovirus showed a log titre reduction of 2.5 after a one-hour contact time. Increasing the free chlorine concentration to 1 mg/l resulted in a 3 log reduction in titre (Abad *et al.*, 1997). The survival characteristics found in these studies were comparable to those determined for rotavirus and enteric adenoviruses, though the astrovirus decay was more pronounced at the higher temperatures. Astroviruses will therefore be inactivated by chlorine to the same extent as other viral agents.

Quantification of astroviruses in environmental samples is difficult given the poor growth in cell culture. Enumeration by plaque assay (Hudson *et al.*, 1989) has not been repeated in other laboratories so, until recently, semiquantitative approaches such as end-point dilution methods have been the best available techniques. In 2002, El-Senousy *et al.* (2002) reported the detection and quantification of astroviruses in sewage and river water in the Cairo (Egypt) area by competitive RT-PCR. An unrelated internal control RNA which reverse transcribed and co-amplified with the astrovirus target was used to provide a quantitative standard. In eight raw sewage samples the number of RNA copies ranged from $3.4 \times 10^3/l$ to $2.3 \times 10^6/l$, and in two treated effluent samples the titres were $3.4 \times 10^3/l$ and $1.1 \times 10^4/l$. Two water samples (before and after treatment) were positive, both with titres of $2.3 \times 10^3$ copies astrovirus RNA per litre. Using an integrated cell culture RT-PCR technique estimates were also made of the viral infectious units in the samples and these ranged from zero to 33.3 'cell culture-RT-PCR units' per litre.

Increasing sophistication of techniques has promoted the study of astroviruses in the environment. One of the persistent problems associated with cause and effect in waterborne disease is to demonstrate that the virus detected in the patient is the same as that found in the water. Nadan *et al.* (2003) compared astrovirus cDNA sequences from contemporaneous clinical and environmental sources by sequence analysis and found that there was coincidence of types. Human astroviruses types 1, 3, 5, 6 and 8 were found in the clinical specimens and types 1, 2, 3, 4, 5, 7 and 8 were detected in the environmental samples. Phylogenetic analysis revealed that the types 1, 3, 5 and 8, present in stool specimens and environmental isolates, clustered together, thus demonstrating they were closely related. Whether this was coincidence or true cause and effect cannot be further determined but it is a significant step forward. The enigma was further unravelled in a study by Gofti-Laroche *et al.* (2003) which showed that the presence of astrovirus RNA in tap water samples was associated with a significant risk of acute digestive conditions (defined as an episode of abdominal pain, nausea, vomiting with or without diarrhoea; RR = 1.51 and a $P = 0.002$). Thus astroviruses may yet have a part to play in waterborne gastrointestinal disease.

## Risk assessment

*Health effects*: occurrence of illness, degree of morbidity and mortality, probability of illness based on infection:

- Astroviruses are endemic in most communities world-wide and mostly infect young children and babies, although they have been associated with disease in the elderly. They appear to have a very low pathogenicity for adults.
- Generally, astroviruses cause only a mild or symptomless infection, but reported incidences are second only to rotaviruses as causing infantile gastroenteritis.
- In addition to diarrhoea, which may be watery but is usually just soft stools, clinical features of astrovirus infection include headache, malaise, nausea, occasionally vomiting and a mild fever. The diarrhoea generally lasts 2–3 days but there are reports of it continuing for as long as 14 days. Diarrhoea is paralleled by virus shedding.
- A good immunity follows infection, so that repeated episodes of disease are rare in a healthy individual.

*Exposure assessment*: routes of exposure and transmission, occurrence in source water, environmental fate:

- Person-to-person spread by the faecal-oral route is thought to be the most common route of transmission. Food, water and fomites have been implicated in astrovirus disease, which suggests that astroviruses persist in the environment; however, there are no confirmed cases of waterborne transmission.
- Astroviruses have been found in sewage, wastewater and surface water (over 50% of samples were positive for astrovirus (Chapron *et al.*, 2000)). Also, 12% of tap water samples in France were positive and the presence of astrovirus RNA was associated with a significant increased risk of gastroenteritis (Gofti-Laroche *et al.*, 2003).
- Astroviruses are moderately resistant to chemical and physical agents. Human astroviruses retain their infectivity after 5, but not 10 minutes at 60°C.

*Risk mitigation*: drinking-water treatment, medical treatment:

- Chlorine inactivates astrovirus. In the presence of 0.5 mg/l free chlorine, astrovirus showed a log titre reduction of 2.5 after a one-hour contact time. Increasing the free chlorine concentration to 1 mg/l resulted in a 3-log reduction in titre. Astroviruses should be inactivated by chlorine to the same extent as other viral agents.
- No medical treatment is necessary as astrovirus diarrhoea is mild and self-limiting and does not lead to significant dehydration.

# References

Abad, F.X., Pintó, R.M., Villena, C. *et al.* (1997). Astrovirus survival in drinking water. *Appl Environ Microbiol*, **63**: 3119–3122.

Appleton, H. and Higgins, P.G. (1975). Viruses and gastroenteritis in infants. *Lancet*, **1**: 7919.

Belliot, G., Laveran, H. and Monroe, S.S. (1997). Outbreak of gastro-enteritis in military recruits associated with serotype 3 astrovirus infection. *J Med Virol*, **51**: 101–106.

Bridger, J.C. (1980). Detection by electron microscopy of caliciviruses, astroviruses and rotavirus-like particles in the faeces of piglets with diarrhoea. *Vet Rec*, **107**: 532–533.

Caul, E.O. (1996). Viral gastroenteritis: small round-structured viruses, caliciviruses and astroviruses. Part II – the epidemiological perspective. *J Clin Pathol*, **49**: 959–964.

CDR (1996). General outbreaks of infectious intestinal disease in England and Wales 1992–1994. *Commun Dis Rep*, **6**: R57–R63.

Chapron, C.D., Ballester, N.A., Fontaine, J.H. *et al.* (2000). Detection of astroviruses, enteroviruses, and adenovirus types 40 and 41 in surface waters collected and evaluated by the information collection rule and an integrated cell culture-nested PCR procedure. *Appl Environ Microbiol*, **66**: 2520–2525.

Cubitt, W.D. (1991). A review of the epidemiology and diagnosis of waterborne viral infections. *Water Sci Technol*, **24**: 197–203.

El-Senousy, W.M., Caballero, S., Pintó, R.M. *et al.* (2002). Detection and quantification of human astroviruses in sewage and river water from Cairo (Egypt) by a competitive RT-PCR. *Am Water Works Assoc*. International Symposium on Waterborne Pathogens. Poster.

Glass, R.I., Noel, J., Mitchell, D. *et al.* (1996). The changing epidemiology of astrovirus-associated gastroenteritis: a review. *Arch Virol*, **12**(Suppl.): 287–300.

Gofti-Laroche, L., Gratacap-Cavallier, B., Demannse, D. *et al.* (2003). Are waterborne astrovirus implicated in acute digestive morbidity (EMIRA study). *J Clin Virol*, **27**(1): 74–82.

Gough, R.E., Collins, M.S., Borland, E. *et al.* (1984). Astrovirus-like particles associated with hepatitis in ducklings. *Vet Rec*, **114**: 279.

Gray, J.J., Wreghitt, T.G., Cubitt, W.D. *et al.* (1987). An outbreak of gastroenteritis in a home for the elderly associated with astrovirus type 1 and human calicivirus. *J Med Virol*, **23**: 377–381.

Guix, S., Caballero, S., Villena, C. *et al.* (2002). Molecular epidemiology of astrovirus infection in Barcelona, Spain. *J Clin Microbiol*, **40**: 133–139.

Herring, A.J., Gray, E.W. and Snodgrass, D.R. (1981). Purification and characterization of ovine astrovirus. *J Gen Virol*, **53**: 47–55.

Herrmann, J.E., Cubitt, D.W., Hudson, R.W., Perron-Henry, D.M. *et al.* (1990). Immunological characterisation of the Marin county strain of astrovirus. *Arch Virol*, **110**: 213–220.

Herrmnan, J.E., Taylor, D.N., Echevarria, P. *et al.* (1991). Astroviruses as a cause of gastro-enteritis in children. *New Engl J Med*, **324**: 1757–1760.

Hoshino, Y., Zimmer, J.F., Moise, N.S. *et al.* (1981). Detection of astroviruses in faeces of a cat with diarrhoea. *Arch Virol*, **84**: 135–140.

Hudson, R.W., Herrmann, J.E. and Blacklow, N.R. (1989). Plaque quantitation and virus neutralization assays for human astroviruses. *Arch Virol*, **108**: 33–38.

Imada, T., Yamaguchi, S., Masaji, M. *et al.* (2000). Avian nephritis virus (ANV) as a new member of the family *Astroviridae* and construction of infectious ANV cDNA. *J Virol*, **74**: 8487–8493.

Jiang, B., Monroe, S.S., Koonin, E.V. *et al.* (1993). RNA sequence of astrovirus: distinctive genome organisation and a putative retrovirus-like ribosomal frameshifting signal that directs the viral replicase synthesis. *Proc Natl Acad Sci USA*, **90**: 10539–10543.

Jonassen, C.M., Jonassen, T.O., Saif, Y.M. *et al.* (2001). Comparison of capsid sequences from human and animal astroviruses. *J Gen Virol*, **82**: 1061–1067.

Jonassen, T.O., Kjeldsberg, E. and Grinde, B. (1993). Detection of human astrovirus serotype 1 by the polymerase chain reaction. *J Virol Meth*, **44**: 83–88.

Jonassen, T.O., Monceyron, C., Lee, T.W. *et al.* (1995). Detection of all types of human astrovirus by the polymerase chain reaction. *J Virol Meth*, **52**: 327–334.

Kjeldsberg, E. (1994). Serotyping of human astrovirus strains by immunogold staining electron microscopy. *J Virol Meth*, **50**: 137–144.

Kjeldsberg, E. and Hem, A. (1985). Detection of astroviruses in gut contents of nude and normal mice. *Arch Virol*, **84**: 135–140.

Koci, M.D., Seal, B.S. and Schultz-Cerry, S. (2000). Molecular characterization of an avian astrovirus. *J Virol*, **74**: 6173–6177.

Konno, T., Suzuki, H., Ishida, N. *et al.* (1982). Astrovirus-associated epidemic gastroenteritis in Japan. *J Med Virol*, **9**: 11–17.

Kurtz, J.B. (1994). Astroviruses. In *Viral Infections of the Gastrointestinal Tract*, 2nd edn, Kapikian, A.Z. (ed.). New York: Marcel Dekker.

Kurtz, J.B. and Lee, T.W. (1984). Human astrovirus serotypes. *Lancet*, **2**: 1405.

Kurtz, J.B. and Lee, T.W. (1987). Astroviruses: human and animal. In *Novel Diarrhoea Viruses, CIBA Foundation Symposium No. 128*. Chichester: Wiley Interscience.

Lee, T.W. and Kurtz, J.B. (1981). Serial propagation of astroviruses in tissue culture with the aid of trypsin. *J Gen Virol*, **57**: 421–424.

Lee, T.W. and Kurtz, J.B. (1994). Prevalence of human astrovirus serotypes in the Oxford region 1976–92, with evidence for two new serotypes. *Epidem Infect*, **112**: 187–193.

Le Guyader, F., Haugarreau, L., Miossec, L. *et al.* (2000). Three-year study to assess human enteric viruses in shellfish. *Appl Environ Microbiol*, **66**: 3241–3248.

Madeley, C.R. (1979). Viruses in the stools. *J Clin Pathol*, **32**: 1–10.

Madeley, C.R. and Cosgrove, B.P. (1975). 28 nm particles in faeces in infantile gastroenteritis. *Lancet*, **2**: 451–452.

Marshall, J.A., Healey, D.S., Studdert, M.J. *et al.* (1984). Viruses and virus-like particles in the faeces of dogs with and without diarrhoea. *Aust Vet J*, **61**: 33–38.

Marx, F.E., Taylor, M.B. and Grabow, W.O.K. (1995). Optimization of a PCR method for the detection of astrovirus type 1 in environmental samples. *Water Sci Technol*, **31**: 359–362.

Marx, F.E., Taylor, M.B. and Grabow, W.O.K. (1998a). The prevalence of human astrovirus and enteric adenovirus infection in South African patients with gastroenteritis. *S Afr J Epidem Infect*, **13**: 5–9.

Marx, F.E., Taylor, M.B. and Grabow, W.O.K. (1998b). The application of a reverse transcriptase-polymerase chain reaction-oligonucleotide probe assay for the detection of human astrovirus in environmental water. *Water Res*, **32**: 2147–2153.

Midthun, K., Greenberg, H.B., Kurtz, J. *et al.* (1993). Characterization and seroepidemiology of a type 5 astrovirus associated with an outbreak of gastroenteritis in Marin County, California. *J Clin Microbiol*, **31**: 955–962.

Mitchell, D.K., Monroe, S.S., Jiang, X. *et al.* (1995). Virologic features of an astrovirus diarrhea outbreak in a day care center revealed by reverse transcriptase polymerase chain reaction. *J Infect Dis*, **172**: 1437–1444.

Monroe, S.S., Jiang, B. and Stine, S.E. (1993). Subgenomic RNA sequence of human astrovirus supports classification of Astroviridae as a new family. *J Virol*, **65**: 641–648.

Nadan, S., Walter, J.E., Grabow, W.O.K. *et al.* (2003). Molecular characterization of astroviruses by reverse transcriptase PCR and sequence analysis: comparison of clinical and environmental isolates from South Africa. *Appl Environ Microbiol*, **69**: 747–753.

Noel, J.S., Lee, T.W., Kurtz, J.B. *et al.* (1995). Typing of human astroviruses from clinical isolates by enzyme immunoassay and nucleotide sequencing. *J Clin Microbiol*, **33**: 797–801.

Oishi, I., Yamazaki, K., Kimoto, T. *et al.* (1994). A large outbreak of acute gastroenteritis associated with astrovirus among students and teachers at schools in Osaka, Japan. *J Infect Dis*, **170**: 430–443.

Palombo, E.A. and Bishop, R.F. (1996). Annual incidence, serotype distribution, and genetic diversity of human astrovirus isolates from hospitalized children in Melbourne, Australia. *J Clin Microbiol*, **43**: 1750–1753.

Phillips, A.D., Rice, S.J. and Walker-Smith, J.A. (1982). Astrovirus within human small intestinal mucosa. *Gut*, **23**: A923–924.

Pintó, R.M., Abad, F.X., Gajardo, R. *et al.* (1996). Detection of infectious astroviruses in water. *Appl Environ Microbiol*, **62**: 1811–1813.

Pinto, R.M., Villena, C., Le Guyader, F. *et al.* (2001). Astrovirus detection in wastewater samples. *Water Sci Technol*, **43**: 73–76.

Reynolds, C.A., Gerba, C.P. and Pepper, I.L. (1998). Detection of infectious enterovirus by an integrated cell culture-PCR procedure. *Appl Environ Microbiol*, **62**: 1424–1427.

Snodgrass, D.R. and Gray, E.W. (1977). Detection and transmission of 30 nm particles (astroviruses) in the faeces of lambs with diarrhoea. *Arch Virol*, **55**: 287–291.

Snodgrass, D.R., Angus, K.W., Gray, E.W. *et al.* (1979). Pathogenesis of diarrhoea caused by astrovirus infections in lambs. *Arch Virol*, **60**: 217–226.

Taylor, M.B., Cox, N., Very, M.A. *et al.* (2001). The occurrence of hepatitis A and astroviruses in selected river and dam waters in South Africa. *Water Res*, **35**: 2653–2660.

Tree, J. (1997). PhD Thesis, University of Surrey, UK.

Tzipori, S., Menzies, J.D. and Gray, E.W. (1981). Detection of astroviruses in the faeces of red deer. *Vet Rec*, **108**: 286.

Utagawa, E.T., Nishizawa, S., Sekine, S. *et al* (1994). Astrovirus as a cause of gastroenteritis in Japan. *J Clin Microbiol*, **32**: 1841–1845.

Willcocks, M.M., Brown, T.D.K., Madeley, C.R. *et al.* (1994). The complete sequence of a human astrovirus. *J Gen Virol*, **75**: 1785–1788.

Willcocks, M.M., Carter, M.J. and Madeley, C.R. (1992). Astroviruses. *Rev Med Virol*, **2**: 97–106.

Willcocks, M.M., Carter, M.J., Laidler, F.R. *et al.* (1990). Growth and characterisation of human faecal astrovirus in a continuous cell line. *Arch Virol*, **113**: 73–81.

Wilson, S.A. and Cubitt, W.D. (1988). The development and evaluation of radioimmune assays for the detection of immunoglobulins M and G against astrovirus. *J Virol Meth*, **19**: 151–160.

Woode, G.N. and Bridger, J.C. (1978). Isolation of small viruses resembling astroviruses and caliciviruses from acute enteritis of calves. *J Med Microbiol*, **11**: 441–452.

Wyn-Jones, A.P. and Herring, A.J. (1991). Growth of clinical isolates of astrovirus in a cell line and preparation of viral RNA. *Water Sci Technol*, **24**: 285–290.

# 29

# Enterovirus
## (poliovirus, coxsackievirus, echovirus)

## Basic microbiology

Enteroviruses are small, icosahedral particles approximately 27 nm in diameter with no obvious surface structure, as seen by EM, and no envelope. The virus consists of a protein capsid enclosing a single-strand genome of positive sense RNA. The RNA is approximately 7.5 kb long and contains a single open reading frame (ORF) from which the various proteins are derived (Racaniello, 2001).

The Enterovirus genus is part of the *Picornaviridae* and is further classified by biomolecular, biochemical characterization and serological typing which also relates to the species in which the viruses are found. Many animal enteroviruses have been identified including bovine and porcine species, which have distinct genomic organizations from the human species. In common with all viruses they are obligate intracellular parasites and, as for most virus groups, under normal conditions they are species specific.

The capsid of the enteroviruses is an icosahedron with 20 faces and 12 apices, composed of 60 protomers, each containing a single copy of each of the proteins VP1, VP2, VP3 and VP4. The capsid outer surface undulates markedly with deep canyons between star-shaped protrusions. The canyons have been shown to be the receptor binding sites of poliovirus (Mendelsohn *et al.*, 1989). The virus attaches to the cell membrane receptor, a protein in the case of poliovirus, after which the virus gains entry into the cell cytoplasm by endocytosis. After the protein coat is lost the single strand of viral RNA is translated directly into virion proteins which are cleaved and processed by proteases into structural and non-structural units. Viral RNA synthesis takes place on small sections of cell membranes. Replication continues with the assembly of capsids and genome into infectious particles but also of empty capsids in the cellular cytoplasm. Cell lysis and death follows.

The enterovirus virions are stable within the pH range 3–10, resistant against quaternary ammonium compounds, 70% ethanol and stable in many detergents. As they have no lipid envelope, they are resistant to ether and chloroform. The virions are sensitive to chlorine, sodium hypochlorite, formaldehyde, gluteraldehyde and UV radiation (Porterfield, 1989).

Human enteroviruses were initially classified by their ability to cause disease in suckling mice and later by growth characteristics in monkey and human cell culture (Melnick, 1976). The strains have now been shown to have different genomic sequences by molecular typing including the use of RT-PCR and RFLP (Muir *et al.*, 1998). The International Committee for the Taxonomy of Viruses Picornavirus Study Group (Van Regenmortel *et al.*, 2000) has decided that the genus Enterovirus contains the human species of poliovirus, human enterovirus A, B, C, D (HEV A–D) as well as bovine and porcine species. In the traditional groupings shown below coxsackievirus B and many echoviruses are classed as HEV-B while coxsackievirus A are classed as mostly HEV-A and HEV-C:

- polioviruses (serotypes 1–3)
- coxsackievirus A (serotypes 1–22, A24)
- coxsackievirus B (serotypes 1–6)
- echoviruses (serotypes 1–9, 11–27, 29–33)
- enteroviruses 68–71 (within species HEV-A, B, D).

Coxsackievirus A strains do not grow well in cell culture so less is known of their occurrence than the other enteroviruses. The other three groups grow readily in human or monkey cell cultures and have been widely studied. Poliovirus and coxsackievirus B form plaques in monkey cells (such as BGM cells) under agar, but echoviruses vary in this respect.

Since 1968, newly identified enteroviruses have been given sequential numbers (enterovirus 68–71). Prior to 1991 hepatitis A virus (HAV) was included as enterovirus 72 but distinct differences in protein size, cell culture growth characteristics and temperature resistance compared with the majority of human enteroviruses determined that it should be placed in the separate genus 'Hepatovirus' (Minor, 1991) and is discussed in detail in Chapter 30.

## Virology of infectious poliovirus vaccine

Poliovirus vaccine was developed from the serial passage of wild-type virus strains in monkey kidney cells to produce an attenuated strain which causes asymptomatic infection (Sabin and Boulger, 1973; Minor, 1998). Attenuated virus replicates only in the small intestine and does not usually have a viraemic phase. The local immune response includes IgA as well as circulating IgG, both of which are protective against future infection. The molecular basis of this shift in virulence was reviewed by Minor (1992) and was identified as a small number of mutations. Attenuation of serotype 1 resulted from one major change at base 480 of the 5' non-coding region and some smaller changes elsewhere in the genome. Serotype 2 had changes at base 481 in the 5' non-coding region and another at 2903 that affected residue 143 of the capsid protein VP1. The major attenuating mutations for serotype 3 were changes at base 2034 which produced a change at residue 91 of the VP3 capsid protein and at base 472 in the 5' non-coding region. The viruses that are shed after vaccination have mutated and often reverted to virulent forms but the genomes remain distinguishable from wild-type. Despite these common changes the vaccine in practice is very safe and does not revert to the virulence of wild-type (Freidrich, 1998). Symptoms of paralytic polio occur only at approximately 1 in a million doses of vaccine.

This live vaccine strain of poliovirus will be present in sewage throughout the world wherever a programme of vaccination is in place. The USA, Finland and increasingly other countries have introduced inactivated poliovirus vaccine as the WHO Poliovirus Eradication Programme progresses. This killed vaccine, given by intramuscular injection will not be present in sewage. The elimination of wild and then the vaccine strain of poliovirus from the environment should be complete in the foreseeable future.

## Origin of the organism

In common with all viruses, the enteroviruses are obligate intracellular parasites and, as for most virus groups, under normal conditions they are species specific. The origin of human viruses in sewage will be human faecal material. No multiplication will occur outside the living host cell.

## Clinical features

Enterovirus infections may cause a wide range of symptoms but most commonly infection is asymptomatic (Grist et al., 1978). Approximately 1% of those infected show clinical illness. For many serotypes, when symptoms do occur they present as flu-like including fever, malaise, respiratory disease, headache and muscle ache. Occasionally the virus is more neurotropic and meningitis develops. Paralysis is a major feature of poliovirus symptomatic infections and a temporary feature of some coxsackievirus B infections. Coxsackievirus A infections may be associated in particular with hand, foot and mouth disease; coxsackievirus B infections with Bornholme disease, pericarditis and myocarditis; echoviruses with encephalitis and Guillain-Barré syndrome; enterovirus 70 with haemorrhagic conjunctivitis. None of the enterovirus symptomatic infections are associated with diarrhoea and vomiting other than as part of the wider disease spectrum.

The virulence of vaccine poliovirus was discussed previously. For other enteroviruses the virulence is dependent on genetic makeup but also on the initial virus dose and the host age and immune response.

## Virulence and pathogenicity

The alimentary canal and the respiratory system are the main sites of multiplication for the enteroviruses with entry through the mouth (Loria, 1988); replication may also take place in the lymphoid tissue of the pharynx. The incubation period from exposure to onset of symptoms may be between 2 and 30 days for different strains but is most commonly 5–14 days (Pallansch and Roos, 2001). Virus is shed, in asymptomatic and symptomatic infections, in faeces and in oral secretions usually from 7 days after infection and may be prolonged, especially in the faeces. It is via this route that enteroviruses reach the environment. A viraemia may occur during which specific target organs may be infected, determined by the serotype involved.

Of all the enteroviruses, wild poliovirus causes the most severe neurological disease. Virus replication in the lower motor neurons causes flaccid paralysis; when respiratory muscles are affected death will ensue. Viral meningitis, most commonly caused by enterovirus infection, is the inflammation of the meninges together with fever and headache. Recovery is usually without lasting effects. Serotypes coxsackievirus B5 and echovirus 4, 6, 9, 11, 13 and 30 are known to be more neurotropic than other strains. Encephalitis that involves the brain tissue and symptoms of confusion and seizures is more serious in outcome and may produce lasting damage.

## Treatment

No specific treatment is available for enterovirus infection but symptomatic treatment is used for serious infections. Normal human immunoglobulin comprising high titre antibody has been used to support neonates with echovirus 11 infection (Nagington, 1982).

## Environment

More information is available on the occurrence of enteroviruses in the environment than any other virus group because their identification in water samples has been, for many years, the most straightforward. However, even the type of enterovirus detected in water is in part dependent on the isolation method used. In general, if a range of human and monkey cell cultures under liquid medium are used, poliovirus, coxsackievirus B and echovirus serotypes will be detectable (Sellwood et al., 1981; Irving and Smith, 1981; Hovi et al., 1996). When an agar-based plaque assay with BGM cells is used then poliovirus and coxsackievirus B are the most likely serotypes to be found (Morris, 1985).

Many studies on the presence of enteric viruses in water and associated materials took place between 1975 and 1985 in the UK and the USA. The techniques in use at that time were primarily for detection of culturable enterovirus and were similar in efficiency to those used for current enterovirus surveys and monitoring projects and may, indeed have had greater resources available. The studies that included the use of liquid culture and a range of cell culture types were done during this early period. More recently, molecular methods, RT-PCR, real-time RT-PCR and integrated cell culture and RT-PCR (ICC-PCR), have been utilized thus increasing the sensitivity of detection. The latter assay incorporates an infectivity component which addresses a major disadvantage of molecular techniques. (See also section on Concentration and detection.)

Coxsackievirus B serotypes 1–6 have all been detected and reported, although B3, B4 and B5 are the most frequently identified (Irving and Smith, 1981; Lewis et al., 1986; Hughes et al., 1992). The predominating serotype has been shown to change over time and to be associated with the serotype most frequently found in samples from clinical material. Echoviruses are least frequently identified (Martins et al., 1983; Morris and Sharp, 1984; Krikelis et al., 1986; Hovi et al., 1996) except if liquid assay is used when the combined serotypes of echoviruses are the most numerous strain (Sellwood et al., 1981).

Enteroviruses are usually detectable all year round with some seasonal variation in the studies cited above. An increase in the number of isolates has been reported in summer and autumn for both northern and southern hemispheres in the above references. Poliovirus isolates also seem to follow this pattern. Sampling error, sample size, differences in water flow quantity, differences in techniques and recovery efficiency between laboratories makes a comparison of numbers difficult per standard volume at any single time.

Of the polioviruses, serotype 2 is usually predominant, type 3 being the least numerous across all water types (Sellwood *et al.*, 1981; Payment *et al.*, 1983; Morris and Sharp, 1984). Wild strains were identified (Sattar and Westwood, 1977; Payment *et al.*, 1983; Lucena *et al.*, 1985) in early studies, although in the UK and more recently in other parts of the world only vaccine strains have been reported (Sellwood *et al.*, 1981, 1995; Pöyry *et al.*, 1988). Wild strains were found in water associated with a poliovirus outbreak in Finland (Pöyry *et al.*, 1988).

Surveillance for poliovirus has been carried out in several countries during outbreaks of poliomyelitis. Sewage in Japan has been investigated by Matsuura *et al.* (2000) and Horie *et al.* (2002) for the presence of poliovirus. Only vaccine strains were identified but some of the strains had neurovirulent characteristics. Manor *et al.* (1999), in Israel, has developed a selective cell culture system for processing large numbers of water samples and plaques and found a small number of wild poliovirus isolates. Pöyry *et al.* (1988) and Böttinger *et al.* (1992), in Scandinavia, found poliovirus of vaccine and wild type during the previous ten years. During an outbreak in the Netherlands in 1992–93 van der Avoort *et al.* (1995) studied poliovirus in sewage and Oostvogel *et al.* (1994) analysed the isolates from patient samples. Wild type strains were identified in both and sewage collected early in the epidemic also contained wild type. An outbreak in Albania was investigated by Divizia *et al.* (1999) and wild type, vaccine related and some recombinant isolates were identified. As part of the WHO Poliovirus Eradication Programme, RD cells have been recommended for monitoring purposes.

The investigation of water and associated materials for the distribution of enteroviruses has been undertaken in many countries. The epidemiology of enteroviruses in communities and the presence of enteroviruses in sewage and water have been shown to follow a similar pattern in temperate countries across the world. UK reports include those of Sellwood *et al.* (1981), Edwards and Wyn-Jones (1981), Morris and Sharp (1984), Hughes *et al.* (1992), Murrin *et al.* (1997), Pallin *et al.* (1997).

Other European studies have been reported from Sweden (Böttinger, 1973; Böttinger and Herrström, 1992), Finland (Pöyry *et al.*, 1988; Hovi *et al.*, 1996), Denmark (Lund *et al.*, 1973), Italy (Carducci *et al.*, 1995), France (Agbalika *et al.*, 1983; Kopecka *et al.*, 1993), Greece (Kirkelis *et al.*, 1986), Romania (Nestor and Costin, 1976), Hungary (Palfi, 1971) and Russia (Drozdov and Kazantseva, 1977).

Early work by Melnick *et al.* (1954) in the USA has continued to the present day by researchers such as Gerba *et al.* (1978), Sobsey *et al.* (1973), Abbaszadegan *et al.* (1999), Griffin *et al.* (1999) and Reynolds *et al.* (2001). Chapron *et al.* (2000) investigated a range of enteric viruses, including enterovirus in source waters for drinking water treatment, as part of the monitoring required by the US Environmental Protection Agency Information Collection Rule. Payment *et al.* (1983, 1988) and Sattar and Ramia (1978) have continued to investigate enteroviruses over many years in Canada. Studies in Japan (Tani *et al.*, 1992; Katayama *et al.*, 2002), South Africa (Grabow and

Nupen, 1981; Geldenhuys and Pretorius, 1989; Vivier *et al.*, 2001), Australia (Irving and Smith, 1981; Grohmann *et al.*, 1993), New Zealand (Lewis *et al.*, 1986) and Israel (Fattal *et al.*, 1977; Guttman-Bass *et al.*, 1981) have all confirmed the presence of enteroviruses in a wide range of water types, such as surface freshwater and marine waters and throughout the seasons. High numbers of enteroviruses have been identified in tropical countries such as Puerto Rico (Dahling *et al.*, 1989) and Colombia (Tambini *et al.*, 1993).

A range of human enteric viruses are potentially present in any type of water contaminated by human faecal material or polluted by untreated or treated sewage. Viruses are likely to be present associated with debris and will be clumped together in faecal material. Therefore untreated sewage will contain the most viruses. Sewage that has been processed to break up clumps may have more evenly dispersed virus particles, but these should then be more accessible to factors that may destroy them. Any treatment which removes the solids of sewage before disposal to a receiving water will also remove a proportion of the viruses. The resulting effluent will contain fewer viruses.

Most rivers in the UK have treated sewage discharged into them and therefore are likely to contain enteroviruses. Sewage is discharged into in-shore marine waters all along the coasts of Britain. Designated bathing waters, the majority of which are marine sites in the UK, are more likely to have higher levels of enteroviruses if sewage is discharged nearby through short sea-outfalls than if long sea-outfalls are in place. Fresh water, i.e. rivers, streams, brooks and canals, contain enteroviruses if that water has had treated sewage, storm water overflows or untreated discharges added to it. The increasing use of UV as a tertiary treatment in the UK will reduce the number of infectious virions in sewage effluent. The vast majority of enteroviruses in controlled waters originate from un-disinfected, continuous point source sewage discharges. The particulate phase of river water was also found to contain enteroviruses by Payment *et al.* (1988). Figure 29.1 shows the range of enterovirus serotypes present in a UK river in 2002. Coxsackieviruses B2 and B4 were the most common serotypes identified throughout 2002 but numbers waned later in the year. During the summer coxsackievirus B5 numbers increased. Coxsackievirus B1 isolates became increasingly numerous after the middle of the year and coxsackievirus B3 was present in low numbers throughout the year.

Enteroviruses have been detected in sewage sampled at the point of entry to a treatment works and final effluent in studies by Sellwood *et al.* (1981), Irving and Smith (1981), Payment *et al.* (1983), Martins *et al.* (1983), Lewis *et al.* (1986), Pöyry *et al.* (1988), Dahling *et al.* (1989), Hovi *et al.* (1996) and Green and Lewis (1999). The enteroviruses of effluent were investigated by Carducci *et al.* (1995) and Rose *et al.* (1989); effluent and wastewater were studied in association with river waters by Morris and Sharp (1984), Lucena *et al.* (1985), Payment *et al.* (1988) and Geldenhuys and Pretorius (1989). Figure 29.2 shows the range of enterovirus serotypes identified in sewage and effluent from a sewage treatment works in the UK in 2001. Coxsackievirus B4 was the most common serotype identified throughout the year. During the autumn months coxsackieviruses B2 and B5 were found in similar numbers to that of B4.

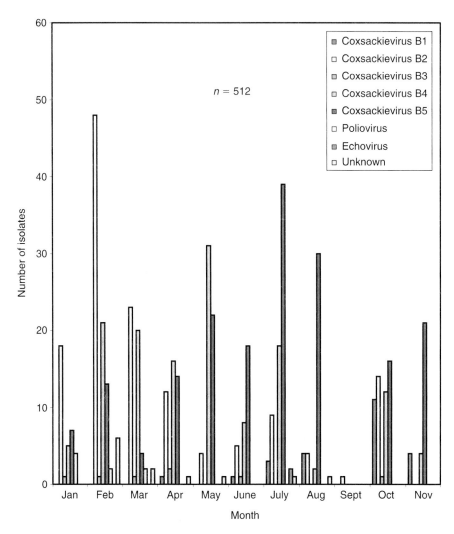

**Figure 29.1**  Enterovirus serotypes, UK recreational river water isolates, 2002.

Sewage sludge comprises a high proportion of solids to which enteroviruses are attached (Hamparian *et al.*, 1985; Hurst and Goyke, 1986). Anaerobic digestion of sludge reduces, but does not remove all enteroviruses. Re-suspension or introduction of this material into the water cycle by dumping at sea, even in mid-Atlantic (Goyal *et al.*, 1984) or spread onto pasture or soil (Straub *et al.*, 1992) and subsequent rain may release any viruses present. Enteroviruses in sludge spread to land will be denatured over time, at a rate dependent on the weather conditions. Lagoons and oxidation ponds for sewage sludge may also contain enteroviruses. Leachates from landfill may contain enteroviruses but it is unlikely to be in high numbers. Aerosols of sewage have occasionally been reported to contain low numbers of enteroviruses (Carducci *et al.*, 1995).

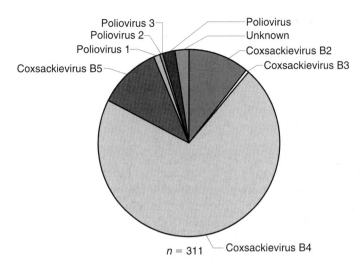

**Figure 29.2** Enterovirus serotypes, UK sewage isolates, 2001 percentage distributions.

Enteroviruses have been detected in bathing water, especially seawater in many parts of the world, including the UK (Merrett *et al.*, 1989; Hughes *et al.*, 1992; Sellwood, 1994), Spain (Lucena *et al.*, 1985), the Netherlands (Van Olphen *et al.*, 1991), France (Hugues *et al.*, 1979), Italy (Patti *et al.*, 1990; Pianetti *et al.*, 2000), Australia (Grohmann *et al.*, 1993), Hawaii (Reynolds *et al.*, 1998) and Florida Keys (Donaldson *et al.*, 2002). Bathers, especially children, may pollute recreational waters directly (Fattal *et al.*, 1991) and affect the virus count in localized areas.

Sediments from fresh and marine waters have been shown to be associated with enteroviruses (Lewis *et al.*, 1985, 1986), Rao *et al.* (1984) and Bosch *et al.* (1988). These sediments may be re-suspended by rainfall, strong winds, strong tides and currents thereby releasing their attached viruses back into the water column. The enteroviruses in sediments may persist for long periods and so be capable of affecting counts for some time after discharging of sewage or sludge may have stopped. Enteroviruses, rotaviruses and adenoviruses were found in sediments near Bondi beach up to 2 years after disposal of sewage had moved off-shore (G. Grohmann, personal communication).

Enteroviruses have occasionally been detected in groundwater (Slade, 1981; Abbaszadegan, 1993, 1999) and in wastewater-recharged groundwater (Vaughn *et al.*, 1978) and there has been concern that this may affect the quality of the source water for drinking water abstraction. An urban aquifer in the UK was shown to be contaminated with enteroviruses and noroviruses during different times of the year (Powell *et al.*, 2003). Drinking water in the UK after treatment contains little organic material and therefore a minimal chance of containing microorganisms. Before distribution as potable water it is also disinfected with chlorine which inactivates any viruses. Morris and Sharp (1985) found no evidence of viruses after treatment but Tyler (personal communication) detected occasional enteroviruses in other waters originating in smaller reservoir treatment works. This was thought to be because of failure of chlorination equipment. Using molecular assays Grabow *et al.* (2001) and Lee and Kim (2002) have detected enteroviruses in finished tap water in South Africa and Korea respectively.

Bird and animal waste is unlikely to contain human enteroviruses.

## Waterborne outbreaks

Swimming pools and the spread of poliovirus was of great concern in the USA during the 1940s and 1950s. It is not clear whether transmission did in fact occur via the water although the potential for spread is illustrated by an outbreak of echovirus 30 in Northern Ireland. An outdoor pool was contaminated with vomit on its day of opening followed by an outbreak of headache, diarrhoea and vomiting in swimmers (Kee *et al.*, 1994). Chlorine levels were said to be satisfactory and no other pathogens were detected, but it was not stated if faeces were tested for the presence of norovirus.

Although often investigated, little strong evidence of waterborne transmission of poliovirus or any enterovirus has been documented in developed countries (Metcalf *et al.*, 1995). As enterovirus infections are so common in the community and easily spread person to person and within families, any transmission by drinking water or recreational water contact will be difficult to confirm by cohort studies. In addition, only 1 in 100 infections results in symptomatic illness making it difficult to trace contacts. One case control study investigated children with febrile illness who presented at a hospital paediatric clinic (D'Alessio *et al.*, 1981). When an enterovirus infection was confirmed by laboratory isolation, a history of swimming in a local lake was significantly more common than in controls. Coxsackievirus infection was found to be more common in divers using a river in France than in a control population (Garin *et al.*, 1994). Community outbreaks of coxsackievirus B infections in Cyprus in 1998 (Papageorgiou, personal communication) and Belarus (Amvrosieva *et al.*, 1998) have again raised questions regarding the significance of enteroviruses in water and sewage as possible vehicles of transmission, but epidemiological or virological evidence remains scanty. Enterovirus will be readily detectable in sewage and associated polluted recreational water if it is circulating in a community, but to confirm transmission via water is problematic.

The relevance of enterovirus in recreational water as a pathogen is unclear. The EU Bathing Water Directive, 1976 included the parameter for enterovirus to be monitored if the quality of the water was suspected to have changed. The choice of this test was a practical one, no other virus group could at that time, or can still be concentrated and detected reliably. Although the enteroviruses should be regarded as pathogens and a hazard potentially transmissible in water, the degree of risk via water, remains a matter of debate.

In areas of the world where full poliovirus vaccination cover is yet to be achieved the virulent virus continues to cause serious outbreaks. Gaspar *et al.* (2000) reported that between January and June 1999 there were 1100 suspected cases of poliomyelitis in Angola, where poor sanitation and water supply and low herd immunity contributed to the rapid spread of the virus and a high incidence of paralytic disease. In Albania in 1996 an outbreak occurred with a high (12%) mortality rate. Samples of water and patient stools were analysed by RT-PCR and significant homology found between sequences of environmental isolates and faecal isolates (Divizia *et al.*, 1999). Further characteristics of the genomes of isolates from the two sources were found in common. Polluted drinking water supplies may have facilitated the rapid spread of poliovirus within these communities once the infection was established and therefore significant amounts of virus were being shed into water supplies.

Drinking water that is efficiently chlorinated or disinfected should not contain infectious enteroviruses.

## Epidemiology

Transmission of virus is primarily person to person by the faecal-oral route and by secretory droplets from the nasopharynx. Faecal material contains large amounts of virus which is subsequently incorporated into sewage. There is no known animal reservoir for human strains. Spread of the viruses within families and between close contacts, such as in schools, is rapid and takes place with ease. Poor hygiene and lack of hand washing facilitate viral spread. Adults are more likely to demonstrate symptoms. A clear seasonal pattern shows a peak of reported illness during the summer months of the general and neurological manifestations of enterovirus infection (Maguire *et al.*, 1999). This pattern has been reported world-wide in temperate climates. During any one season a single serotype may predominate clinical reports but other serotypes will also be present. In the summer of 2000, echovirus 13 was a common cause of meningitis in the UK (PHLS Communicable Disease Report August 2000) but a range of coxsackievirus B serotypes were found in sewage (Figure 29.3) reflecting the range of serotypes present in the community.

Several studies have attempted to determine the infectious dose for enteroviruses using poliovirus (Minor *et al.*, 1981) and echovirus 12 (Schiff *et al.*, 1984) but the results are difficult to assess. The lowest dose was 17 plaque forming units (pfu) of echovirus 12, which would infect 1% of those inoculated. Electron microscopy of the virus preparation suggested this was equivalent to 700 virus particles. Racaniello (2001) states that poliovirus and other picornaviruses have a particle to pfu range between 30 and 1000. Such studies are extremely difficult to standardize and it is generally assumed that for most viruses one infectious particle is capable of infecting a susceptible host (Schiff *et al.*, 1984). It is difficult to distinguish experimentally between one virion (virus particle) and one 'infectious unit', which may comprise an unknown and variable number of virions clumped together. Risk assessment for water contact assumes that one virus infectious unit is capable of initiating an infection in a susceptible host (Berg, 1967).

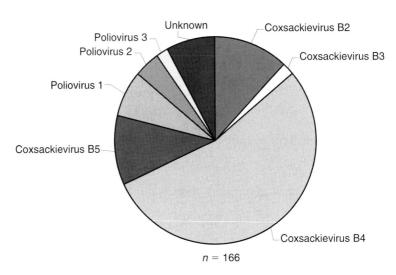

**Figure 29.3**   Enterovirus serotypes, UK sewage isolates, 2000 percentage distributions.

## Risk assessment

*Health effects*: occurrence of illness, degree of morbidity and mortality, probability of illness based on infection:

- Occurrence of enterovirus infection and illness varies by serotype, geography, season and the host's age and antibody status.
- For many serotypes, when symptoms occur they present as fever, malaise, respiratory disease, headache and muscle ache. Coxsackievirus primarily causes meningitis, unspecified febrile illness, hand-foot-and-mouth disease and conjunctivitis. Echovirus primarily causes meningitis, unspecified febrile illness, and respiratory illness. Enterovirus 70 is associated with haemorrhagic conjunctivitis. Wild poliovirus causes the most severe neurological disease. Most enterovirus disease resolves on its own within a week of infection.
- Susceptibles, such as newborns, can have a fatality rate of up to 10%, but otherwise, fatality is uncommon.
- The probability of illness based on infection varies widely by serotype and the host's age and antibody status. Overall, approximately 1% of those infected show clinical illness.

*Exposure assessment*: routes of exposure and transmission, occurrence in source water, environmental fate:

- Transmission is faecal-oral and by secretory droplets from the nasopharynx: primarily person to person directly or via fomites. Theoretically, transmission is possible through water or food ingestion or aerosolization.
- A volunteer study on infectious doses of enteroviruses was performed with echovirus 12. The study of 149 volunteers concluded that 919 pfu are

necessary to infect 50% of an exposed population and estimated that 17 pfu are necessary to infect 1% of an exposed population. Such studies are extremely difficult to standardize and it is generally assumed that for most viruses one infectious particle is capable of infecting a susceptible host.

- Attack rates of infection have been reported variously, depending on the serotype and host characteristics.
- Secondary spread of enterovirus infections is common. Viral shedding generally lasts from a week to a month. This relatively long period increases the probability of secondary spread. Generally, illness in secondary cases is lessened from that of the index case; frequently secondary infection is asymptomatic.
- Enteroviruses are found in all types of water that can be polluted by human sewage: lakes, rivers, sediments, marine environments, groundwater, wastewater, and drinking water. As enterovirus infections are so common in the community and easily spread person to person and within families, any transmission by drinking water or recreational water contact will be difficult to confirm. Little evidence supports a waterborne route of infection. Most outbreaks appear to result from person-to-person spread. The predominating serotype in water sources has been shown to change over time and to be associated with the serotype most frequently found in samples from clinical material.

*Risk mitigation*: drinking-water treatment, medical treatment

- The enterovirus virions are sensitive to chlorine, sodium hypochlorite, formaldehyde, gluteraldehyde and UV radiation. Standard water treatment with chlorine disinfection should efficiently chlorinate or control enteroviruses.
- Until recently, no antiviral drugs effectively prevented or treated echovirus infections. Gammaglobulin therapy and exchange transfusions have been used with limited success in the critically ill. However, an antipicornaviral agent, pleconaril, has shown promise in clinical trials for echovirus meningitis. It is expected to receive approval from the FDA (Rotbart and Webster, 2001).

# References

Abbaszadegan, M., Huber, M.S., Gerba, C.P. *et al.* (1993). Detection of enteroviruses in ground water with the polymerase chain reaction. *Appl Environ Microbiol*, **59**: 1319–1324.

Abbaszadegan, M., Stewart, P. and LeChevallier, M. (1999). A strategy for detection of viruses in groundwater by PCR. *Appl Environ Microbiol*, **65**: 444–449.

Agbalika, F., Hartemann, P., Briigaud, M. *et al.* (1983). Enterovirus circulation in the wastewaters of the Lorraine region. *Rev Epidemiol Sante Publique*, **31**: 209–221.

Amvrosieva, T.V., Titov, L.P., Mulders, M., Hovi, T. *et al.* (2001). Viral contamination as the cause of aseptic meningitis outbreak in Belarus. *Cent Eur J Public Hlth*, **9**: 154–157.

Bell, E.J. and Grist, N.R. (1967). Viruses in diarrhoeal disease. *Br Med J*, **582**: 741–742.

Berg, G. (1967). *Transmission of Viruses by the Water Route*. New York: Wiley InterScience Publishers, pp. 1–484.

Bosch, A., Lucena, F., Girones, R. *et al.* (1988). Occurrence of enteroviruses in marine sediment along the coast of Barcelona, Spain. *Can J Microbiol*, **34**: 921–924.

Böttinger, M. (1973). Experiences from investigations of virus isolations from sewage over a two year period with special regard to poliovirus. *Arch Gesamte Virusforsch*, **41**: 80–85.

Böttinger, M. and Herrström, E. (1992). Isolation of poliovirus from sewage and their characteristics: experience over two decades in Sweden. *Scand J Infect Dis*, **24**: 151–155.

Carducci, A., Arrighi, S. and Ruschi, A. (1995). Detection of coliphages and enteroviruses in sewage and aerosol from an activated sludge wastewater treatment plant. *Lett Appl Microbiol*, **21**: 207–209.

Chapron, C.D., Ballester, N.A., Fontaine, J.H. *et al.* (2000). Detection of astroviruses, enteroviruses and adenovirus types 40 and 41 in surface waters collected and evaluated by the Information Collection Rule and an Integrated Cell Culture-Nested PCR procedure. *Appl Environ Microbiol*, **66**: 2520–2525.

D'Alessio, D.J., Minor, T.E., Allen, C.I. *et al.* (1981). A study of the proportions of swimmers among well controls and children with enterovirus-like illness shedding or not shedding an enterovirus. *Am J Epidemiol*, **113**: 533–541.

Dahling, D.R., Safferman, R.S. and Wright, B.A. (1989). Isolation of enterovirus and reovirus from sewage and treated effluents in selected Puerto Rican communities. *Appl Environ Microbiol*, **55**: 503–506.

Divizia, M., Palombi, L., Buonomo, E. *et al.* (1999). Genomic characterization of human and environmental polioviruses isolated in Albania. *Appl Environ Microbiol*, **65**: 3534–3539.

Donaldson, K.A., Griffin, D.W. and Paul, J.H. (2002). Detection, quantitation and identification of enteroviruses from surface waters and sponge tissue from the Florida Keys using real-time RT-PCR. *Water Res*, **36**: 2505–2514.

Drozdov, S.G. and Kazantseva, V.A. (1977). Importance of the results of a virological examination of sewage. *Vopr Virusol*, **5**: 597–602.

Edwards, E. and Wyn-Jones, A.P. (1981). Incidence of human enteroviruses in water and wastewater systems associated with the River Wear. In *Viruses and Wastewater Treatment*, Goddard, M. and Butler, M. (eds). Oxford: Pergamon Press, pp. 109–112.

Fattal, B. and Nishmi, M. (1977) Enterovirus types in Israel sewage. *Water Res*, **11**: 393–396.

Fattal, B., Peleg-Olevsky, E. and Cabelli, V.J. (1991). Bathers as a possible source of contamination for swimming-associated illness at marine bathing beaches. *Int J Environ Hlth Res*, **1**: 204–214.

Freidrich, F. (1998). Neurological complications associated with oral poliovirus vaccine and genomic variability of the vaccine strains after multiplication in humans. *Acta Virol*, **42**: 187–195.

Garin, D., Fuchs, F., Crance, J.M. *et al.* (1994). Exposure to enteroviruses and hepatitis A virus among divers in environmental waters in France, first biological and serological survey of a controlled cohort. *Epidemiol Infect*, **113**: 541–549.

Gaspar, M., Morais, A., Brumana, L. and Stella, A. (2000). Outbreak of poliomyelitis in Angola. *J Infect Dis*, **181**: 1777–1779.

Geldenhuys, J.C. and Pretorius, P.D. (1989). The occurrence of enteric viruses in polluted water, correlation to indicator organisms and factors influencing their numbers. *Water Sci Technol*, **21**: 105–109.

Gerba, C.P., Farrah, S.R., Goyal, S.M. *et al.* (1978). Concentration of enteroviruses from large volumes of tap water, treated sewage and seawater. *Appl Environ Microbiol*, **35**: 540–548.

Goyal, S.M., Adams, W.N., O'Malley, M.L. *et al.* (1984). Human pathogenic viruses at sewage sludge disposal sites in the middle Atlantic region. *Appl Environ Microbiol*, **48**: 758–763.

Grabow, W.O.K. and Nupen, E.M. (1981). Comparison of primary kidney cells with the BGM cell line for the enumeration of enteric viruses in water by means of a tube dilution technique. In *Viruses and Wastewater Treatment*, Goddard, M. and Butler, M. (eds). Oxford: Pergamon Press, pp. 253–256.

Grabow, W.O., Taylor, M.B. and de Villiers, J.C. (2001). New methods for the detection of viruses: call for review of drinking water quality guidelines. *Water Sci Technol*, **43**: 1–8.

Green, D.H. and Lewis, G.D. (1999). Comparative detection of enteric viruses in waste-waters, sediments and oysters by RT-PCR and cell culture. *Water Res*, **33**: 1195–1200.

Griffin, D.W., Gibson, C.J. and Lipp, E.K. (1999). Detection of viral pathogens by RT-PCR and of microbial indicators by standard methods in the canals of the Florida Keys. *Appl Environ Microbiol*, **65**: 4118–4125.

Grist, N.R., Bell, E.J. and Assaad, F. (1978). Enteroviruses in Human Disease. *Prog Med Virol*, **24**: 114–157.

Grohmann, G., Ashbolt, N.J., Genova, M.S. *et al.* (1993). Detection of viruses in coastal and river water systems in Sydney, Australia. *Water Sci Technol*, **27**: 457–461.

Guttman-Bass, N., Tchorsh, Y., Nasser, A. *et al.* (1981). Rapid detection of enteroviruses in water by a quantitative fluorescent technique. In *Viruses and Wastewater Treatment*, Goddard, M. and Butler, M. (eds). Oxford: Pergamon Press, pp. 247–251.

Hamparian, V.V., Ottolenghi, A.C. and Hughes, J.H. (1985). Enteroviruses in sludge: mul-tiyear experience with four wastewater treatment plants. *Appl Environ Microbiol*, **50**: 280–286.

Horie, H., Yoshida, H., Matsuura, K. *et al.* (2002). Neurovirulence of type 1 polioviruses isolated from sewage in Japan. *Appl Environ Microbiol*, **68**: 138–142.

Hovi, T., Stenvik, M. and Rosenlew, M. (1996). Relative abundance of enterovirus serotypes in sewage differs from that in patients: clinical and epidemiological implications. *Epidemiol Infect*, **116**: 91–97.

Hughes, M.S., Coyle, P.V. and Connolly, J.H. (1992). Enteroviruses in recreational waters of Northern Ireland. *Epidemiol Infect*, **108**: 529–536.

Hugues, B., Bougis, M.A., Plissier, M. *et al.* (1979). Evaluation of the viral contamination of the sea water after the emission of an effluent into the sea. *Zentralbl Bakteriol*, **169**: 253–264.

Hurst, C.J. and Goyke, T. (1986). Stability of viruses in wastewater sludge eluates. *Can J Microbiol*, **32**: 649–653.

Irving, L.G. and Smith, F.A. (1981) One-year survey of enteroviruses, adenoviruses and reoviruses isolated from effluent at an activated sludge purification plant. *Appl Environ Microbiol*, **41**: 51–59.

Katayama, H., Shimasaki, A. and Ohgaki, S. (2002). Development of a virus concentration method and its application to detection of enterovirus and Norwalk virus from coastal seawater. *Appl Environ Microbiol*, **68**: 1003–1039.

Kee, F., McElroy, G., Sewart, D. *et al.* (1994). A community outbreak of echovirus infection associated with an outdoor swimming pool. *J Public Hlth*, **16**: 145–148.

Kopecka, H., Dubrou, S., Prévot, J. *et al.* (1993). Detection of naturally-occurring enteroviruses in waters by reverse transcription, polymerase chain reaction, and hybridization. *Appl Environ Microbiol*, **59**: 1213–1219.

Krikelis, V., Markoulatos, P. and Spyrou, N. (1986). Viral pollution of coastal waters resulting from disposal of untreated sewage effluents. *Water Sci Technol*, **18**: 43–48.

Lee, S.H. and Kim, S.J. (2002). Detection of infectious enteroviruses and adenoviruses in tap water in urban areas in Korea. *Water Res*, **36**: 248–256.

Lewis, G.D., Austin, M.W. and Loutit, M.W. (1986). Enteroviruses of human origin and faecal coliforms in river water and sediments down stream from a sewage outfall in the Taieri River, Otago. *NZ J Marine Freshwater Res*, **20**: 101–105.

Lewis, G.D., Loutit, M.W. and Austin, F.J. (1985). A method for detecting human enteroviruses in aquatic sediments. *J Virol Meth*, **10**: 153–162.

Loria, R.M. (1988). Host conditions affecting the course of Coxsackievirus infections. In *Coxsackieviruses: A General Update*, Beninelli, M. and Friedman, H. (eds). New York: Plenum Press, pp. 125–150.

Lucena, F., Bosch, A., Jofre, J. *et al.* (1985). Identification of viruses isolated from sewage, river water and coastal seawater in Barcelona. *Water Res*, **19**: 1237–1239.

Lund, E. and Rönne, V. (1973). On the isolation of viruses from sewage treatment plant sludges. *Water Res*, 7: 863–866.

Maguire, H.C., Atkinson, P., Sharland, M. *et al.* (1999). Enterovirus infections in England and Wales: laboratory surveillance data: 1975 to 1994. *Commun Dis Public Hlth*, **2**: 122–125.

Manor, Y., Handsher, R., Halmut, T. *et al.* (1999). Detection of poliovirus circulation by environmental surveillance in the absence of clinical cases in Israel and the Palestinian authority. *J Clin Microbiol*, **37**: 1670–1675.

Martins, M.T., Soares, L.A., Marques, E. *et al.* (1983). Human enteric viruses isolated from influents of sewage treatment plants in S Paulo Brazil. *Water Sci Technol*, **15**: 69–73.

Matsuura, K., Ishikura, M., Yoshida, H. *et al.* (2000). Assessment of poliovirus eradication in Japan: genomic analysis of polioviruses isolated from river water and sewage in Toyama Prefecture. *Appl Environ Microbiol*, **66**: 5087–5091.

Melnick, J.L. (1976). Enteroviruses. In *Viral Infections of Humans: Epidemiology and Control*, Evans, A.S. (ed.). New York: J Wiley & Sons, pp. 163–207.

Melnick, J.L., Emmons, J., Opton, E.M. *et al.* (1954). Coxsackieviruses from sewage. *Am J Hyg*, **59**: 183–195.

Mendelsohn, C.L., Wimmer, E., Racaniello, V.R. *et al.* (1989). Cellular receptor for poliovirus: molecular cloning, nucleotide sequence, and expression of a new member of the immunoglobulin superfamily. *Cell*, **56**: 855–865.

Merrett, H., Pattinson, C., Stackhouse, C. *et al.* (1989). The incidence of enteroviruses around the Welsh coast – a three year intensive survey. In *Watershed 89. The Future of Water Quality in Europe*. Oxford: Pergamon Press, pp. 345–351.

Metcalf, T.G., Melnick, J.L. and Estes, M.K. (1995). Environmental virology: from detection of virus in sewage and water by isolation to identification by molecular biology – a trip of over 50 years. *Ann Rev Microbiol*, **49**: 461–487.

Minor, P. (1991). Picornaviridae. In *Classification and Nomenclature of Viruses (Arch Virol Suppl 2)*, Francki, R.I.B., Fauquet, C.M., Knudson, D.L. *et al.* (eds). Wien: Springer-Verlag, pp. 320–326.

Minor, P. (1992). The molecular biology of poliovaccines. *J Gen Virol*, **72**: 3065–3077.

Minor, P. (1998). Picornaviruses. In *Topley and Wilson's Microbiology and Microbial Infections*, 9th edn, Mahy, B.W.J. and Collier, L. (eds). London: Arnold, pp. 485–509.

Minor, T.E., Allen, C.I., Tsiatis, A.A. *et al.* (1981). Human infective dose determination for oral poliovirus type 1 vaccine in infants. *J Clin Microbiol*, **13**: 388–389.

Morris, R. (1985). Detection of enteroviruses: an assessment of ten cell lines. *Water Sci Technol*, **17**: 81–88.

Morris, R. and Sharp, D.N. (1984). Enteric viruses levels in wastewater effluents and surface waters in the Severn Trent Water Authority 1979–81. *Water Res*, **18**: 935–939.

Morris, R. and Sharp, D.N. (1985). Failure to detect cytopathic enteroviruses in drinking water. *Water Sci Technol*, **17**: 105–109.

Muir, P., Kämmerer, U., Korn, K. *et al.* (1998). Molecular typing of enteroviruses: current status and future requirements. *Clin Microbiol Rev*, **11**: 202–227.

Murrin, K. and Slade, J. (1997). Rapid detection of viable enteroviruses in water by tissue culture and semi-nested polymerase chain reaction. *Water Sci Technol*, **35**: 429–432.

Nagington, J. (1982). Echovirus 11 infection and prophylactic antiserum. *Lancet*, **1**: 446.

Nestor, I. and Costin, L. (1976). Presence of certain enteroviruses (Coxsackievirus) in sewage effluents and in river waters of Romania. *J Hyg Epidemiol Microbiol Immunol*, **20**: 137–149.

Oostvogel, P.M., van Wijngaarden, J.K., van der Avoort, H.G.A.M. *et al.* (1994). Poliomyelitis outbreak in an unvaccinated community in the Netherlands, 1992–93. *Lancet*, **344**: 665–669.

Palfi, A.B. (1971). Virus content of sewage in different seasons in Hungary. *Acta Microbiol Acad Sci Hung*, **18**: 231–237.

Pallansch, M.A. and Roos, R.P. (2001). Enteroviruses: Polioviruses, Coxsackieviruses, Echoviruses, and newer Enteroviruses. In *Fields Virology*, 4th edn, Knipe, D.M. and Howley, P.M. (eds). Philadelphia, PA: Lippincott Williams & Wilkins, pp. 723–775.

Pallin, R., Place, B.M., Lightfoot, N.F. *et al.* (1997). The detection of enteroviruses in large volume concentrates of recreational waters by the polymerase chain reaction. *J Virol Meth*, **57**: 67–77.

Patti, A.M., Aulicino, F.A., de Filippis, P. *et al.* (1990). Identification of enteroviruses isolated from seawater: indirect immunofluorescence. *Boll Soc Ital Biol Sper*, **66**: 595–600.

Payment, P., Ayache, R. and Trudel, M. (1983). A survey of enteric viruses in domestic sewage. *Can J Microbiol*, **29**: 111–119.

Payment, P., Affoyon, F. and Trudel, M. (1988). Detection of animal and human enteric viruses in water from the Assomption River and its tributaries. *Can J Microbiol*, **34**: 967–973.

Pianetti, A., Baffone, W., Citterio, B. *et al.* (2000). Presence of enteroviruses in the waters of the Italian coast of the Adriatic Sea. *Epidemiol Infect*, **125**: 455–462.

Porterfield, J.S. (1989). Picornaviridae. In *Andrewes' Viruses of Vertebrates*. London: Bailliere Tindall, pp. 120–145.

Powell, K.L., Taylor, R.G., Cronin, A.A. *et al.* (2003). Microbial contamination of two urban sandstone aquifers in the UK. *Water Res*, **37**: 339–352.

Pöyry, T. and Stenvik Hovi, T. (1988). Viruses in sewage waters during and after a poliomyelitis outbreak and subsequent nationwide oral poliovirus vaccination campaign in Finland. *Appl Environ Microbiol*, **54**: 371–374.

Rao, V.C., Seidel, K.M., Goyal, S.M. *et al.* (1984). Isolation of enteroviruses from water, suspended solids and sediments from Galveston Bay: survival of poliovirus and rotavirus adsorbed to sediments. *Appl Environ Microbiol*, **48**: 404–409.

Ravcaniello, V.R. (2001). *Picornaviridae*: The viruses and their replication. In *Fields Virology*, 4th edn, Knipe, D.M. and Howley, P.M. (eds). Lippincott Williams & Wilkins, pp. 685–722.

Reynolds, K.A., Roll, K., Fujioka, R.S. *et al.* (1998). Incidence of enteroviruses in Mamala Bay, Hawaii using cell culture and direct PCR methodologies. *Can J Microbiol*, **44**: 598–604.

Reynolds, K.A., Gerba, C.P., Abbaszadegan, M. *et al.* (2001). ICC/PCR detection of enteroviruses and hepatitis A virus in environmental samples. *Can J Microbiol*, **47**: 153–157.

Rose, J.B., de Leon, R. and Gerba, C.P. (1989). *Giardia* and virus monitoring of sewage effluent in the state of Arizona. *Water Sci Technol*, **21**: 43–47.

Rotbart, H.A. and Webster, A.D. (2001). Treatment of potentially life-threatening enterovirus infections with pleconaril. *Clin Infect Dis*, **32**: 228–235.

Sabin, A.B. and Boulger, L.R. (1973). History of Sabin attenuated poliovirus oral live vaccine strains. *J Biol Standard*, **1**: 115–118.

Sattar, S.A and Westwood, J.C. (1977). Isolation of apparently wild strains of poliovirus type 1 from sewage in the Ottawa area. *Can Med Assoc J*, **116**: 25–27.

Sattar, S.A. and Ramia, S. (1978). Viruses in sewage: effect of phosphate removal with calcium hydroxide (lime). *Can J Microbiol*, **24**: 1004–1008.

Schiff, G.M., Stefanovic, G.M., Young, E.C. *et al.* (1984). Studies of echovirus 12 in volunteers: determination of minimal infectious dose and the effect of previous infection on infectious dose. *J Infect Dis*, **150**: 858–866.

Sellwood, J. (1994). Viruses and the EC Bathing Water Directive. *PHLS Microbiol Digest*, **11**: 81–82.

Sellwood, J., Dadswell, J.V. and Slade, J.S. (1981). Viruses in sewage as an indicator of their presence in the community. *J Hyg Camb*, **86**: 217–225.

Sellwood, J., Litton, P.A., McDermott, J. and Clewley, J.P. (1995). Studies on wild and vaccine poliovirus isolated from water and sewage. *Water Sci Technol*, **31**: 317–321.

Slade, J.S. (1981). Viruses and bacteria in a chalk well. *Water Sci Technol*, **17**: 111–126.

Sobsey, M.D., Wallis, C., Henderson, M. *et al.* (1973). Concentration of enteroviruses from large volumes of water. *Appl Environ Microbiol*, **26**: 529–531.

Straub, T.M., Pepper, I.L. and Gerba, C.P. (1992). Persistence of viruses in desert soils amended with anaerobically digested sewage sludge. *Appl Environ Microbiol*, **58**: 636–641.

Tambini, G., Andrus, J.K., Marques, E. *et al.* (1993). Direct detection of wild poliovirus circulation by stool surveys of healthy children and analysis of community wastewater. *J Infect Dis*, **168**: 1510–1514.

Tani, N., Shimamoto, K., Ichimura, K. *et al.* (1992). Enteric virus levels in river water. *Water Res*, **26**: 45–48.

van der Avoort Reimerink, J.H.J., Ras, A., Mulders, M.N. *et al.* (1995). Isolation of epidemic poliovirus from sewage during the 1992–93 type 3 outbreak in the Netherlands. *Epidemiol Infect*, **114**: 481–491.

Van Regenmortel, M.H.V., Faugeuet, C.M., Bishop, E.B. *et al.* (2000). *Virus taxonomy: The classification and nomenclature of viruses. Seventh Report of the International Committee on Taxonomy of Viruses*. San Diego, CA: Academic Press.

Van Olphen, M., de Bruin, H.A.M., Havelaar, A.H. *et al.* (1991). The virological quality of recreational waters in the Netherlands. *Water Sci Technol*, **24**: 209–212.

Vaughn, J.M., Landry, E.F., Baranosky, L.J. *et al.* (1978). Survey of human virus occurrence in wastewater recharged groundwater on Long Island. *Appl Environ Microbiol*, **36**: 47–51.

Vivier, J.C., Clay, C.G. and Grabow, W.O. (2001). Detection and rapid differentiation of human enteroviruses in water sources by restriction enzyme analysis. *Water Sci Technol*, **43**: 209–212.

# 30

# Hepatitis A virus (HAV)

## Basic microbiology

Mild jaundice as a clinical entity has been recognized for thousands of years and by the beginning of the last century epidemic hepatitis was associated with a short incubation period and transmission by the faecal-oral route. By the 1960s, Krugman differentiated between clinical cases of the blood-borne hepatitis B virus and 'infectious hepatitis', hepatitis A (Krugman *et al.*, 1962).

The virion of HAV is 25–28 nm in diameter without obvious morphology by EM and has no envelope. The RNA is a single strand of 7478 bases with three sections including a large open reading frame. This organization is typical of picornaviruses, but the HAV organization is sufficiently distinct to be classified in its own genus: the Hepatovirus. A range of strains of the single HAV serotype are found around the world but the antibodies produced after infection are cross-protective.

The virus is stable at pH 3, resistant to intestinal enzymes and a temperature of 60°C for 10 hours, although it is inactivated at 100°C after a few minutes (Provost *et al.*, 1975). As the virions have no membrane ether, freon and chloroform have no effect. The virus remains viable for months after storage at room temperature and in water, sewage and shellfish (Sobsey *et al.*, 1988). The virion is inactivated by free residual chlorine concentration at 2 mg/l after 15 minutes, by iodine, by UV radiation and by 70% ethanol.

## Origin of the organism

The origin of HAV strains in the environment will be human sewage. Other than man, only primates are known to be susceptible to HAV infection. In common with all viruses HAV is an obligate intracellular parasite, therefore no replication can occur in the environment.

## Clinical features and virulence

A prodromal stage of malaise and fever is followed in a few days by nausea, vomiting and abdominal pain, then dark urine and jaundice develop. Jaundice may persist for 1–2 weeks and malaise for 2–3 months. There is rarely progression to fulminant liver disease and no chronic or carrier state. Severity of the symptoms is age related with infection more severe in adulthood. The icteric stage is short or absent in children and asymptomatic infections are common.

## Pathogenicity

The virus replicates in the hepatocytes of the liver which causes local necrosis and a marked response by lymphocytes (inflammatory cell infiltration). Blockage of the bile canaliculi may result in cholestasis, although the basic structure of the liver remains unchanged. Repair of damaged tissue is usually complete in 3–4 months. Subsequent to infection, immunity is life long as neutralizing antibody will prevent re-infection.

## Transmission and epidemiology

Spread is by the faecal-oral route as large numbers of virus particles are shed in faecal material for approximately a week before and a few days after jaundice is apparent. There is a short period of low titre viraemia which has resulted in rare cases of blood-borne transmission in men having sex with men, in post-transfusion hepatitis and in the past as a result of blood products such as factor VIII. Current inactivation procedures used on blood products now eliminate this latter route.

The transmission of HAV via the faecal-oral route is most common within families and close contacts (Maguire *et al.*, 1995). If hand hygiene is poor, water is scarce or sanitation lacking, the virus will be easily spread. The population of many countries with poor sanitation have asymptomatic infections early in life resulting in a high level of immunity in adults. HAV infection is not common in the UK (Gay *et al.*, 1994) and many infections are due to imports from travellers returning from endemic areas. Localized outbreaks have occurred in the UK involving drug users living in poor housing conditions. The poor standard of hygiene and the contacts between groups in different cities lead to localized but linked increased numbers of cases (Crowcroft, 2003). The virus spreads readily between children within schools and day-care facilities, but infection is often only recognized when adult family members develop symptomatic disease.

Diagnosis is by detection of circulating antibody HAV IgM which is detectable in serum for 3 months after the onset of symptoms. Virions are present in faeces for only a few days after the onset of symptoms. Although the virus may eventually grow in a very limited range of cell culture and has been detected using EM and the more sensitive PCR, they are not appropriate for diagnosis.

Faecal contamination of food and water with HAV has led to many point source outbreaks and is discussed below. HAV is endemic in many developing countries of the world where children become infected early in life and a high level of immunity is present in the adult population.

## Treatment

There is no specific treatment for HAV infection and symptoms are rarely severe enough to warrant hospitalization. Bed rest and adequate diet during the periods of anorexia provide the required support. There is an effective vaccine which is recommended for travellers to endemic areas and to close contacts of cases. High titre immunoglobulin is also available for close contacts of cases and may prevent the spread of the virus if given within 10 days of contact.

## Distribution in the environment

As HAV is very difficult and slow to grow in cell culture, the detection of virus particles in water and associated materials has relied mainly on molecular means. Only during the last 10 years has virus been reported in water and shellfish. HAV antigen capture PCR was used by Graff *et al.* (1993) in Germany for sewage samples and by Deng *et al.* (1994) in the USA on septic tank effluent.

Shieh *et al.* (1991) used gene probes to detect HAV in Danube River water, in well water in Maryland during an outbreak and from a drinking water supply in Mexico. Gajardo *et al.* (1991), also using gene probes, found HAV in sewage samples collected in Barcelona.

RT-PCR was used to detect HAV in sewage and occasionally in treated effluent from Pune in India, an endemic area throughout the year (Vaidya *et al.*, 2002). Drinking water tested by Grabow *et al.* (2001) in South Africa had HAV in 3% of samples. In contrast, the sewage and effluent from New Zealand was not shown to contain HAV, although marine sediments from near the sewage treatment works outfall did have detectable virions. It was not clear if this was as a result of continuing shedding in the community or the persistence of virus (Green and Lewis, 1999). Only three domestic wells out of 50, sampled in Wisconsin four times a year, were found to contain HAV by the use of PCR on one occasion (Borchardt *et al.*, 2003). HAV, however, was the enteric virus most frequently found but did not correlate with the presence of any of the bacterial indicator organisms. The wells were of generally good bacterial quality. Groundwater in the eastern states also contained HAV in nearly 9% of samples (Abbaszadegan *et al.*, 1999). Jothikumar *et al.* (1998), also in the USA, investigated the distribution of HAV in sewage from California using PCR linked to an immunocapture bead system. Although problems of inhibition were encountered, samples from May and November were positive. Sewage contaminated canal water was found to contain HAV in the Florida Keys in 63% of samples tested (Griffin *et al.*, 1999).

Shellfish have been implicated in the transmission of HAV and some of the outbreaks are discussed below. Shellfish material is toxic to cell culture and inhibitory to some molecular systems, consequently detection of virus is not often reported. Enriquez *et al.* (1992) were able to detect HAV by cell culture in mussels that had been kept in faecally-contaminated river water. Pietri *et al.* (1988), Le Guyader *et al.* (1994) and Chung *et al.* (1996) detected HAV, with considerable difficulty, from oysters raised in seawater. A survey of shellfish in Europe (Formiga-Cruz *et al.*, 2002) found HAV in shellfish from all the countries' samples except Sweden. However, the number of positive samples was small compared to other enteric viruses found. Romalde *et al.* (2002a) detected HAV in 27% of sampled bivalves grown in north-west Spain and remarked on the inadequacy of EU standards of microbiological quality. The same group (Romalde *et al.*, 2002b) investigated imported shellfish from South America after an outbreak of HAV and found four of 17 batches to be contaminated, including the batch implicated in a Spanish outbreak of hepatitis. Green and Lewis (1999) did not detect HAV by PCR in New Zealand oysters.

HAV will only be present in sewage if the virus is circulating within the community and therefore is less likely to be found in developed countries where the level of endemic disease is low. The virus has not been identified in UK sewage but has been detected in shellfish that was influenced by sewage associated with a local outbreak (Lees, personal communication). Thornton *et al.* (1995) in Ireland found no evidence of HAV infection in a follow-up epidemiological study after a large outbreak of gastroenteritis resulting from a sewage contaminated borehole supplying drinking water to a town. Garin *et al.* (1994) did not find any evidence that divers using surface water in France had a higher prevalence of antibody than military personnel in the same age range. These results indicate that the level of HAV present in sewage in northern Europe should be low.

## Waterborne outbreaks

Hepatitis virus A may be spread by drinking polluted water, by contact with recreational water and by consuming infected shellfish. The detection of virus in the implicated environmental matrix has been reported in relatively few episodes. Divizia *et al.* (1993) linked an outbreak in an Italian school with the pollution of well water that supplied the drinking water for the school. HAV was detected from the water after concentration using ultrafiltration and detection by RT-PCR. Epidemiological evidence linking polluted water with outbreaks has been successively reinforced by the use of probes (e.g. Shieh *et al.*, 1991) and RT-PCR (Jothikumar *et al.*, 1998). HAV was detected in well water, a cesspool and specimens from patients associated with a community outbreak in Quebec, Canada (De Serres *et al.*, 1999). Prior to the use of molecular techniques, cell culture was able to detect infectious virus after prolonged incubation of samples (Bloch *et al.*, 1990). This was also described by Hejkal *et al.* (1982) using radioimmunoassay on sewage and well water in Texas.

Conaty *et al.* (2000) undertook an epidemiological investigation that implicated HAV infection with the consumption of oysters in New South Wales, Australia. HAV was detected in samples of shellfish grown in the lake linked to the outbreak. During the large outbreak of hepatitis in Shanghai, China, during 1988, HAV was grown in cell culture and experimentally infected marmosets after inoculation with a suspension of clam material (Xu *et al.*, 1992).

More commonly it is only epidemiological evidence that links HAV outbreaks with water or consumption of shellfish. Contaminated spring water was responsible for an outbreak in Kentucky (Bergeisen *et al.*, 1985). An outbreak in India was associated with use of a drinking water reservoir in a college cafeteria which was shown to be heavily contaminated with indicator bacteria (Poonawagul *et al.*, 1995). A recent outbreak associated with tap water in Albania was described in a ProMED website posting (January 2003). A swimming pool in a campsite in Louisiana was implicated in an outbreak of HAV (Mahoney *et al.*, 1992). Direct contact with sewage after storm flooding was linked to a community outbreak in Florida (Vonstille *et al.*, 1993). Contaminated well water was responsible for a hardware store outbreak (Bowen and McCarthy, 1983).

A large outbreak of hepatitis was attributed to the consumption of clams in China (Halliday *et al.*, 1991), to clams in the UK (Anon, 1982) and to oysters (Desenclos *et al.*, 1991). Many other waterborne outbreaks have been reported, especially from developing countries (Hunter, 1997).

Transmission may occur by other food routes not directly associated with water such as raw fruits (Niu *et al.*, 1992; Hutin *et al.*, 1999) and by direct faecal-oral route if an infected person with poor hygiene is involved in food preparation (CDC, 1993). Glasses used to serve drinks by a barman in a pub were linked by Sundkvist *et al.* (2000) to customers who developed hepatitis.

## Risk assessment

*Health effects*: occurrence of illness, degree of morbidity and mortality, probability of illness based on infection:

- Many hepatitis A viral infections are asymptomatic. The severity of the symptoms is age related; infection in adults is more severe.
- If symptoms are present, a prodromal stage of malaise and fever is followed in a few days by nausea, vomiting and abdominal pain, then dark urine and jaundice. Jaundice may persist for 1–2 weeks and malaise for 2–3 months. A few patients are ill for as long as 6 months. There is rarely progression to fulminant liver disease and no chronic or carrier state.
- The population of many countries with poor sanitation have asymptomatic infections early in life resulting in a high level of immunity in adults.

*Exposure assessment*: routes of exposure and transmission, occurrence in source water, environmental fate:

- Since the virus is difficult to detect in the environment using cell culture, information on occurrence has been reported in the past 10 years using molecular means. HAV has been found in sewage, surface water sources, groundwater sources, and in food, especially shellfish. The occurrence in the environment is dependent on the amount of infection circulating in a community; therefore, developing countries will have higher levels of HAV in sewage than developed countries.
- Hepatitis A is spread through faecal-oral contact. Therefore, it may be spread by drinking polluted water, by contact with recreational water or by consuming infected shellfish. It is also transmitted by close person-to-person contact.
- Bloodborne cases of transmission have been rarely reported.
- Many waterborne outbreaks have been linked to drinking water and recreational water contact.

*Risk mitigation*: drinking-water treatment, medical treatment:

- There is no specific treatment for HAV infection, though symptoms are rarely severe enough to warrant hospitalization.
- There is an effective vaccine which is recommended for travellers to endemic areas and other high-risk populations, such as intravenous drug users and people with liver disease.
- Immunoglobulin is a preparation of antibodies that can be given before exposure for short-term protection against hepatitis A and for persons who have already been exposed to hepatitis A virus.

- HAV is susceptible to chlorine disinfection, so adequate residuals in drinking water should be sufficient to inactivate it.

# References

Abbaszadegan, M., Stewart, P. and LeChevallier, M. (1999). A strategy for detection of viruses in groundwater by PCR. *Appl Environ Microbiol*, **65**: 444–449.

Anon. (1982). Hepatitis A: frozen cockles. *Commun Dis Rep*, **82**: 18.

Bergeisen, G.H., Hinds, M.W. and Skaggs, J.W. (1985). A waterborne outbreak of hepatitis A in Meade County, Kentucky. *Am J Public Hlth*, **75**: 161–164.

Bloch, A.B., Stramer, S.L., Smith, J.D. *et al.* (1990). Recovery of hepatitis A virus from a water supply responsible for a common source outbreak of hepatitis A. *Am J Public Hlth*, **80**: 428–430.

Borchardt, M.A., Bertz, P.D., Spencer, S.K. *et al.* (2003). Incidence of enteric viruses in groundwater from household wells in Wisconsin. *Appl Environ Microbiol*, **69**: 1172–1180.

Bowen, G.S. and McCarthy, M.A. (1983). Hepatitis A associated with a hardware store water fountain and a contaminated well in Lancaster County, Pennsylvania, 1980. *Am J Epidemiol*, **117**: 695–705.

CDC (1993). Foodborne Hepatitis A – Missouri, Wisconsin, Alaska, 1990–1992. *MMWR*, **1993**: 526–529.

Chung, H., Jaykus, L. and Sobsey, M.D. (1996). Detection of human viruses in oysters by *in vivo* and *in vitro* amplification of nucleic acids. *Appl Environ Microbiol*, **62**: 3772–3778.

Conaty, S., Bird, P., Bell, G. *et al.* (2000). Hepatitis A in New South Wales, Australia from consumption of oysters: the first reported outbreak. *Epidemiol Infect*, **124**: 121–130.

Crowcroft, N.S. (2003). Hepatitis A infections in injecting drug users. *CDPH*, **6**: 82–85.

De Serres, G., Cromeans, T.L., Levesque, B. *et al.* (1999). Molecular confirmation of Hepatitis A virus from well water: epidemiology and public health implications. *J Infect Dis*, **179**: 37–43.

Deng, M.Y., Day, S.P. and Cliver, D.O. (1994). Detection of HAV in environmental samples by antigen-capture PCR. *Appl Environ Microbiol*, **60**: 1927–1933.

Desenclos, J.-C.A., Klontz, K.C., Wilder, M.H. *et al.* (1991). A multi-state outbreak of hepatitis A caused by the consumption of oysters. *Am J Public Hlth*, **81**: 1268–1272.

Divizia, M., Gnesivo, C., Amore Bonapasta, R., Morace, G. *et al.* (1993). Hepatitis A virus identification in an outbreak by enzymatic amplification. *Eur J Epidemiol*, **9**(2): 203–208.

Enriquez, R., Frosner, G.G., Hochstein-Mintzel, V. *et al.* (1992). Accumulation and persistence of HAV in mussels. *J Med Virol*, **37**: 174–179.

Formiga-Cruz, M., Tofino-Quesada, G., Bofill-Mas, S. *et al.* (2002). Distribution of human virus contamination in shellfish from different growing areas in Greece, Spain, Sweden and the UK. *Appl Environ Microbiol*, **68**: 5990–5998.

Gajardo, R., Diez, J.M., Jofre, J. *et al.* (1991). Adsoption-elution with negatively and positively-charged glass powder for the concentration of hepatitis A virus from water. *J Virol Meth*, **31**: 345–352.

Garin, D., Fuchs, F., Crance, J.M. *et al.* (1994). Exposure to enteroviruses and HAV among divers in environmental waters in France, first biological and serological survey of a controlled cohort. *Epidemiol Infect*, **113**: 541–549.

Gay, N.J., Morgan-Capner, P., Wright, J. *et al.* (1994). Age-specific antibody prevalence to hepatitis A in England: implications for disease control. *Epidemiol Infect*, **113**: 113–120.

Grabow, W.O., Taylor, M.B. and de Villiers, J.C. (2001). New methods for the detection of viruses: call for review of drinking water quality guidelines. *Water Sci Technol*, **43**: 1–8.

Graff, J., Ticehurst, J. and Flehmig, B. (1993). Detection of hepatitis A virus in sewage by antigen capture polymerase chain reaction. *Appl Environ Microbiol*, **59**: 3165–3170.

Green, D.H. and Lewis, G.D. (1999). Comparative detection of enteric viruses in waste-waters, sediments and oysters by RT-PCR and cell culture. *Water Res*, **33**: 1195–1200.

Griffin, D.W., Gibson, C.J., Lipp, E.K. *et al.* (1999). Detection of viral pathogens by RT-PCR and of microbial indicators by standard methods in the canals of the Florida Keys. *Appl Environ Microbiol*, **65**: 4118–4125.

Halliday, M.L., Kang, L.Y., Zhou, T.K. *et al.* (1991). An epidemic of Hepatitis A attributable to the ingestion of raw clams in Shanghai China. *J Infect Dis*, **164**: 852–859.

Hejkal, T.W., Keswick, B., LaBelle, R.L. *et al.* (1982). Viruses in a community water supply associated with an outbreak of gastro-enteritis and infectious hepatitis. *JAWA*, **74**: 318–321.

Hunter, P.R. (1997). Viral hepatitis. In *Waterborne Disease Epidemiology and Ecology.* Chichester: John Wiley & Sons, pp. 206–221.

Hutin, Y.J., Pool, V., Crammer, E.H. *et al.* (1999). A multistate, foodborne outbreak of HAV. *New Engl J Med*, **340**: 595–602.

Jothikumar, N., Cliver, D.O. and Mariam, T.W. (1998). Immunomagnetic capture RT-PCR for rapid concentration and detection of HAV from environmental samples. *Appl Environ Microbiol*, **64**: 504–508.

Krugman, S., Ward, R. and Giles, J.P. (1962). The natural history of infectious hepatitis. *Am J Med*, **32**: 717–728.

Le Guyader, F., Dubois, E., Menard, D. *et al.* (1994). Detection of HAV, rotavirus, and enterovirus in naturally contaminated shellfish and sediment by RT-PCR. *Appl Environ Microbiol*, **60**: 3665–3671.

Maguire, H.C., Handford, S., Perry, K.R. *et al.* (1995). A collaborative case control study of sporadic hepatitis A in England. *Commun Dis Rep CDR Rev*, **5**: R33–R40.

Mahoney, F.J., Farley, T.A., Kelso, K.Y. *et al.* (1992). An outbreak of HAV associated with swimming in a public pool. *J Infect Dis*, **165**: 613–618.

Niu, M.T., Polish, L.B., Robertson, B.H. *et al.* (1992). Multistate outbreak of HAV associated with frozen strawberries. *J Infect Dis*, **166**: 518–524.

Pietri, Ch., Huges, B., Crance, J.M. *et al.* (1988). HAV levels in shellfish exposed in a natural marine environment to effluent from a treated sewage outfall. *Water Sci Technol*, **20**: 229–234.

Poonawagul, U., Warintrawat, S., Snitbhan, R. *et al.* (1995). Outbreak of hepatitis A in a college traced to contaminated water reservoir in cafeteria. *SE Asian J Trop Med Public Hlth*, **26**: 705–708.

Provost, P.J., Wolanski, W., Miller, W.J. *et al.* (1975). Physical, chemical and morphological dimensions of human HAV strain CR326. *Proc Soc Exp Biol Med*, **148**: 532–539.

Romalde, J.L., Area, E., Sanchez, G. *et al.* (2002a). Prevalence of enterovirus and HAV in bivalve molluscs from Galicia (NW Spain): inadequacy of the EU standards of microbiological quality. *Int J Food Micro*, **74**: 119–130.

Romalde, J.L., Torrado, I. and Barja, J.L. (2002b). Global market: shellfish imports as a source of re-emerging food-borne HAV infections in Spain. *Int Microbiol*, **4**: 223–226.

Shieh, Y.S., Baric, R.S., Sobsey, M.D., Ticehurst, J. *et al.* (1991). Detection of hepatitis A virus and other enteroviruses in water by ssRNA probes. *J Virol Methods*, **31**(1): 119–136.

Sobsey, M.D., Shields, P.A., Haunchman, F.S. *et al.* (1988). Survival and persistence of hepatitis A virus in environmental samples. In *Viral hepatitis and liver disease*, Zuckerman, A.J. (ed.). New York: Alan R Liss, pp. 121–124.

Sundkvist, T., Hamilton, G.R., Hourihan, B.M. *et al.* (2000). Outbreak of HAV spread by contaminated glasses in a public house. *Commun Dis Public Hlth*, **3**: 60–62.

Thornton, I., Fogarty, J., Hayes, C. *et al.* (1995). The risk of HAV from sewage contamination of a water supply. *Commun Dis Rep*, **5**: R1–R4.

Vaidya, S.R., Chitambar, S.D. and Arankalle, V.A. (2002). PCR based prevalence of HAV, HEV and TT viruses in sewage from an endemic area. *J Hepatol*, **37**: 131–136.

Vonstille, W.T., Stille, W.T. and Sharer, R.C. (1993). Hepatitis A epidemics from utility sewage in Ocoee, Florida. *Arch Environ Hlth*, **48**: 120–124.

Xu, Z.Y., Li, Z.H., Wang, J.X. *et al.* (1992). Ecology and prevention of a shellfish associated hepatitis A epidemic in Shanghai, China. *Vaccine*, **10**(Suppl.): 67–68.

# 31

# Hepatitis E virus (HEV)

## Basic microbiology

The most recently recognized viral cause of hepatitis is now known as hepatitis E virus (Wong *et al.*, 1980), but its classification is currently unspecified. Although initially classified in the *Caliciviridae* family, the similarities between HEV, beet necrotic yellow vein virus (a furovirus) and rubella virus (a togavirus) suggest that they should be linked in related families. The genome is single-stranded, positive sense RNA approximately 7.5 kb long. It is a non-enveloped particle about 30 nm in diameter with little surface structure visible by EM. Several strains have been identified, but all seem to belong to a single serotype. A virus closely resembling human HEV has been shown to infect pigs in Europe and in the USA, but it is unclear what the exact relationship is (Meng *et al.*, 1999).

As the virus successfully passes through the stomach, the virions must be acid stable and similarly as no envelope is present, virions should be resistant to ether and chloroform. Further studies on sensitivity ranges to chlorine and other chemicals have yet to be done.

## Origin of the organism

The origin of HEV strains in the environment will be human sewage, although the precise distribution of the pig HEV-like virus remains to be investigated.

In common with all viruses HEV is an obligate intracellular parasite, therefore no replication can occur in the environment.

## Clinical features and virulence

The acute clinical features are indistinguishable from HAV and similar to a mild case of HBV. Anorexia, abdominal pain and jaundice are the main features but are usually mild and the illness is self-limiting; there is no chronic or carrier state. The infection can, however, be severe in pregnant women with a significant mortality of up to 20% (Khuroo, 1991). It is not known what causes this increase in virulence. Symptoms are most apparent in young adults. Prevalence studies in Vietnam demonstrated an immunity rate of only 9%, indicating a significant potential for outbreaks to occur (Hau *et al.*, 2000).

## Pathogenicity

It is likely that after replication in the intestine the virus reaches the liver through the portal vein and then replicates in the hepatocytes, as is the case for HAV. Progeny virus particles are likely to be shed back into the intestine in bile. Cell-mediated factors may contribute to liver damage and symptoms.

## Treatment

There is no specific treatment for HEV infection. Neither specific immunoglobulin nor vaccine is available.

## Transmission and epidemiology

Spread of this virus is by the faecal-oral route but water and also food seem to be the major routes rather than direct person-to-person contact (Aggarwal and Naik, 1994). Major epidemics of waterborne disease have been linked with HEV and are discussed below. The main diagnostic tool is circulating antibody HEV IgM which is detectable in serum for 3 months after onset of symptoms. Virions are present in faeces for a week or more after the onset of symptoms, which is longer than is the case for HAV. The virus has been detected using EM and PCR which is more sensitive. Cell culture is not appropriate.

Outbreaks of HEV hepatitis have been recognized in India and most of Asia, the Middle East, Africa and Mexico. Antibody studies suggest some circulation of the virus in the USA and in Europe, but again the relationship to the pig virus is currently unclear.

# Distribution in the environment

Methods have been developed to detect HEV in water matrices (Grimm and Fout, 2002) but there are few reports on natural environmental samples. Jothikumar *et al.* (1993) were the first to publish a study on the detection of HEV in sewage in Asia. Vaidya *et al.* (2002) found 10% of sewage samples collected in India over a period of one year assayed using PCR were positive for HEV compared to 24% positive for HAV.

## Waterborne outbreaks

A large outbreak occurred in Somalia during 1988 and 1989 which was linked to consumption of river water, particularly after heavy rain. Communities that drank well water were less likely to suffer disease (Bile *et al.*, 1994). As seen in other outbreaks the mortality rate among pregnant women was significant at 13%. This was similar to the outbreak reported by Naik *et al.* (1992) in India. Singh *et al.* (1995) reported an outbreak in a small, educated Indian community during a time of water scarcity and subsequent contamination of a piped water supply. The attack rate was nearly 2% in both children and adults. The first recognized waterborne outbreak of HEV in Indo China was reported by Corwin *et al.* (1996) and linked to consumption of river water for drinking. The same group undertook epidemiological studies which indicated that boiling river water used for drinking was a significant protective measure (Corwin *et al.*, 1999). A mixed outbreak of HAV and HEV was identified in Djibouti which was reported as waterborne (Coursaget *et al.*, 1998). French soldiers were infected with HAV and the local residents with HEV. A large waterborne outbreak of HEV occurred in Islamabad when the drinking water treatment plant malfunctioned (Rab *et al.*, 1997). HEV was identified as the causative agent of an outbreak in Kashmir in 1978 (Skidmore *et al.*, 1992).

The potential for food-borne HEV infection spreading from endemic areas to non-endemic areas has been reviewed by Smith (2001). As evidence of person-to-person spread of HEV is not strong and waterborne transmission well documented in endemic areas, it seems likely that the drinking water treatment infrastructure in developed countries should safeguard the population from HEV infection if the virus is introduced.

## Risk assessment

*Health effects*: occurrence of illness, degree of morbidity and mortality, probability of illness based on infection:

- Many hepatitis E viral infections are asymptomatic. The severity of the symptoms is age related; infection in young to middle-aged adults is more severe.
- Typical clinical signs and symptoms of acute hepatitis E are similar to those of other types of viral hepatitis and include abdominal pain, anorexia, dark urine, fever, hepatomegaly, jaundice, malaise, nausea and vomiting.
- No evidence of chronic infection has been detected in long-term follow up of patients with hepatitis E.
- The case fatality rate for hepatitis E is 1–3%; however, the fatality rate is up to 20% in pregnant women.
- Death of the mother and fetus, abortion, premature delivery, or death of a live-born baby soon after birth are common complications of hepatitis E infection during pregnancy.

*Exposure assessment*: routes of exposure and transmission, occurrence in source water, environmental fate:

- HEV is transmitted primarily by the faecal-oral route and faecally-contaminated drinking water is the most commonly documented vehicle of transmission.
- Unlike hepatitis A infection, person-to-person transmission of HEV appears to be uncommon.
- Hepatitis E may be a zoonotic disease, with pigs possibly serving as reservoirs for human infection.
- In developed countries, most cases that occur are in people who have travelled to areas where hepatitis E is endemic.
- There are few studies looking at the occurrence of hepatitis E in source water; however, it is probably common in endemic areas, because of the number of waterborne infections that occur.

*Risk mitigation*: drinking-water treatment, medical treatment:

- There is no specific treatment for HEV.
- No products are available to prevent hepatitis E, though a vaccine is currently in development.
- Hepatitis E is susceptible to chlorination. Water that has been adequately disinfected should be free of infectious hepatitis E.

## References

Aggarwal, R. and Naik, S.R. (1994). Hepatitis E: Intrafamilial transmission versus waterborne spread. *J Hepatol*, **21**: 718–723.

Bile, K., Isse, A., Mohamud, O., Allebeck, P. *et al.* (1994). Contrasting roles of rivers and wells as sources of drinking water on attack and fatality rates in a hepatitis E epidemic in Somalia. *Am J Trop Med Hyg*, **51**: 466–474.

Corwin, A.L., Khiem, H.B., Clayson, E.T. *et al.* (1996). A waterborne outbreak of HEV transmission in southwestern Vietnam. *Am J Trop Med Hyg*, **54**: 559–562.

Corwin, A.L., Tien, N.T., Bounlu, K. *et al.* (1999). The unique riverine ecology of HEV transmission in South-East Asia. *Trans Roy Soc Trop Med Hyg*, **93**: 255–260.

Coursaget, P., Buisson, Y., Enogat, N. *et al.* (1998). Outbreak of enterically-transmitted hepatitis due to HAV and HEV. *J Hepatol*, **28**: 745–750.

Grimm, A.C. and Fout, G.S. (2002). Development of a molecular method to identify hepatitis E virus in water. *J Virol Meth*, **101**: 175–188.

Hau, C.H., Hien, T.T., Khiem, H.B. *et al.* (2000). Prevalence of enteric hepatitis A and E viruses in the Mekong River delta region of Vietnam. *Am J Trop Med Hyg*, **60**: 277–280.

Jothikumar, N., Aparna, K., Kamatchiammal, S. *et al.* (1993). Detection of hepatitis E virus in raw and treated wastewater with the polymerase chain reaction. *Appl Environ Microbiol*, **59**: 2558–2562.

Khuroo, M.S. (1991). Hepatitis E: the enterically transmitted non-A, non-B hepatitis. *Indian J Gastroenterol*, **10**: 96–100.

Meng, X.J., Dea, S., Engle, R.E. *et al.* (1999). Prevalence of antibodies to the hepatitis E virus (HEV) in pigs from countries where hepatitis E is common or rare in the human population. *J Med Virol*, **59**: 297–302.

Naik, S.R., Aggarwal, R., Salunke, P.N. *et al.* (1992). A large waterborne viral hepatitis E epidemic in Kanpur, India. *Bull WHO*, **70**: 597–604.

Rab, M.A., Bile, M.K., Mubarik, M.M. *et al.* (1997). Waterborne HEV epidemic in Islamabad, Pakistan: a common source outbreak traced to the malfunction of a modern water treatment plant. *Am J Trop Med*, **57**: 151–157.

Singh, J., Aggarwal, N.R., Bhattacharjee, J. *et al.* (1995). An outbreak of viral hepatitis E: role of community practices. *J Commun Dis*, **27**: 92–96.

Skidmore, S.J., Yarbough, P.O., Gabor, K.A. *et al.* (1992). Hepatitis E virus: The cause of a waterborne hepatitis outbreak. *J Med Virol*, **37**: 58–60.

Smith, J.L. (2001). A review of HEV. *J Food Protect*, **64**: 572–586.

Vaidya, S.R., Chitambar, S.D. and Arankalle, V.A. (2002). PCR based prevalence of HAV, HEV and TT viruses in sewage from an endemic area. *J Hepatol*, **37**: 131–136.

Wong, D.C., Purcell, R.H. and Sreenivasan, M.A. (1980). Epidemic and endemic hepatitis in India: evidence for nonA/nonB hepatitis virus aetiology. *Lancet*, **2**: 876–878.

# 32

# Norovirus and sapovirus

## Basic microbiology

Norwalk virus was identified in faecal material by immune electron microscopy during an outbreak of winter vomiting disease in Ohio (Kapikian *et al.*, 1972) after which many similar viruses causing similar outbreaks were identified around the world. New strains have been named after the place in which they were found. A name for this collection of viruses has been more problematic, but recently, the International Committee for the Taxonomy of Viruses has designated 'norovirus' for the name of the genus, to replace the terms Norwalk virus, Norwalk-like virus, NLV and the UK term 'small round-structured virus' (SRSV) (Caul and Appleton, 1982). Comparisons of the RNA polymerase and capsid sequences of the genome have shown that noroviruses are a genus within the *Caliciviridae* and have at least two distinct genogroups. Genogroup I viruses include Norwalk virus and Southampton virus, genogroup II includes Hawaii, Bristol and Lordsdale viruses (Green *et al.*, 2000b, 2001). Genomic relationships of newly identified viruses are determined by sequence analysis of the capsid proteins (Green *et al.*, 1995, 2000a). Clusters (genotypes) of strains with approximately 80% or more similarity in amino acid sequence are grouped together and linked, with the first reported strain as reference virus. Bovine and porcine noroviruses have also been identified.

Three other genera within the *Caliciviridae* are currently recognized. The International Committee for the Taxonomy of Viruses has designated the term 'sapovirus' as the name of the genus, to replace Saporo virus, classic calicivirus, human calicivirus or Saporo-like virus (SLV). The genus includes Saporo virus and Manchester virus. Virus particles have a very distinct cup-shape pattern on the surface and ragged outline visible by EM and genetically are significantly different to the noroviruses. The porcine enteric calicivirus is a sapovirus. The other two genera do not contain human strains of virus, but of note is the feline calicivirus which, although a respiratory virus, has been used as a surrogate for norovirus in some experimental studies due to its ability to replicate in cell culture.

Calicivirus particles are 27–30 nm in diameter and have a ragged outline. The capsid surface is a series of arches and cup-like depressions which, using electron microscopy, is seen as an amorphous surface with no discernible features in noroviruses. In contrast, sapovirus virions have well-defined cup shapes on the surface. The capsid is made of a single protein arranged in icosahedral symmetry and the genome is single-stranded RNA (Clarke *et al.*, 1998) approximately 7.5 kb in length with three open reading frames (ORFs). As the virus cannot yet be grown in cell culture and there are few animal models, the characterization of the virus components has been difficult and little is known of the replication mechanisms *in vivo*.

Studies on Norwalk virus indicate that the virus is not inactivated by pH 3, or ether or chloroform or by 30 minutes at 60°C and is only partially inactivated by 70% alcohol (Clarke *et al.*, 1998). However, it is inactivated by chlorine at >10 mg/l, although not by free residual chlorine at 0.5–1.0 mg/l (Keswick *et al.*, 1985). Studies on the disinfection of noroviruses are hampered by the lack of an infectivity assay, but current limited data suggest that they may be more resistant than enteroviruses. Further studies are needed to confirm this.

## Origin of the organism

In common with all viruses the noroviruses and sapoviruses are obligate intracellular parasites and, as for most virus groups, under normal conditions they are species specific. The origin of human viruses in sewage will be human faecal material. No multiplication will occur outside the living host cell.

## Clinical features and virulence

Diarrhoea and vomiting are the predominant symptoms of norovirus infection with accompanying headache and myalgia. Projectile vomiting is often present in more than 50% of adult cases and is one of the features that indicate an outbreak may be due to norovirus (Kaplan *et al.*, 1982a). Asymptomatic infection does occur although may not be common. Symptoms can be incapacitating but usually resolve within 2–3 days with no complications.

Sapovirus causes gastroenteritis in which diarrhoea is the predominate feature and vomiting is less likely to occur.

Nothing is known of the factors that influence virus virulence.

## Pathogenicity

Virus particles enter the host through the mouth; the virions are unaffected by the acidic pH as they pass through the stomach. Virus replication probably takes place in the mucosal epithelium of the small intestine resulting in damage to the epithelium (enterocytes) and flattening of the villi. The incubation period is commonly 24 hours followed by the sudden onset of nausea and vomiting. The projectile vomiting that is such a marker feature of norovirus symptoms may be due to delayed emptying of the stomach caused by abnormal gastric motor function (Meeroff *et al.*, 1980). Virus shedding in faecal material continues for a week or sometimes two but rarely longer.

Immunity appears to be short lived and different strains of noroviruses are not cross-protective. Children become infected early in life but symptomatic re-infections occur throughout adult life. Results from volunteer infection studies are complex but suggest that immunity to any one strain may only last 9 months in some individuals (Green *et al.*, 2001). Factors other than circulating antibody may be important in determining the host response to infection.

Sapovirus immunity has been less studied but as infection is most common in babies and young children, it may be more long lasting than norovirus.

## Epidemiology

Transmission of noroviruses is mainly direct person to person by the faecal-oral route or by the ingestion of particles of vomitus. Projectile vomiting has been shown to contaminate the environs from which re-suspended, aerosolized particles may be swallowed and infection transmitted (Chadwick and McCann, 1994; Cheesbrough *et al.*, 2000). Marks *et al.* (2000) demonstrated transmission of the virus most frequently in persons situated near to an index case of vomiting and less frequently when further away. In a report by Evans *et al.* (2002) transmission occurred over time when an index case contaminated an area of theatre seating and successive clients using the seating became ill. Transmission by food and water is discussed in detail below. Norovirus gastroenteritis, with vomiting a particular feature, occurs in all ages and occurs in sporadic and epidemic disease patterns (Caul, 1996a,b). Secondary cases are common among close contacts in the family, in schools, in residential settings, in hospitals and on cruise ships.

Initially called 'winter vomiting disease' the infection is now recognized throughout the year in temperate climates. The Infectious Intestinal Disease Study in the UK 1993–1996 investigated gastrointestinal disease in the community and found that norovirus gastroenteritis was significantly more frequent than previously recognized (Wheeler *et al.*, 1999). Noroviruses were the most common cause of gastroenteritis in adults and have an epidemic or outbreak pattern overlying the endemic disease. The PHLS Communicable Disease Report of December 19, 2002 stated that 3029 cases of noroviruses in the UK had been reported in the first 10 months of that year. This was the highest level yet recognized and substantially more than any other gastroenteritis-causing pathogen. There had not been the usual decline in the summer months. Patients aged over 65 years of age had accounted for 68% of cases. As noroviruses are readily transmitted, outbreaks are common within enclosed and residential areas such as schools, nursing homes, hospital wards, hotels and cruise ships (Caul *et al.*, 1979; Cubitt *et al.*, 1981; Dolin *et al.*, 1982; Ho *et al.*, 1989).

More outbreaks of gastroenteritis are caused by noroviruses than any other infectious agent in the UK. In 2000, over 250 outbreaks were reported as caused by noroviruses compared to approximately 50 due to *Salmonella* (Public Health Laboratory Service unpublished data). Similar levels of infection have been reported world-wide but as laboratory diagnosis is difficult much is based on clinical presentation and outbreak features (Kaplan *et al.*, 1982a).

Oliver *et al.* (2003) investigated the risk of infection to humans by bovine noroviruses and concluded that there was no evidence of transmission.

Detection of human noroviruses in faecal material may be done by electron microscopy (EM), but this requires specialist equipment and staff. It is also relatively insensitive as more than $10^6$ particles/ml must be present for the virus to be identified. However, much of the initial investigation of this virus group was based on EM (Caul and Appleton, 1982). It remains the only method of detecting any virus that may be present in faecal material. More recent advances have been possible using molecular techniques directed specifically at noroviruses. Polymerase chain reaction (PCR) is discussed fully under 'Detection'; it is much more sensitive than EM which has allowed virus identification for a wider time range after the onset of symptoms. This has in turn lead to a more accurate picture of norovirus infections. ELISA kits are now available for clinical specimens which will mean that diagnosis will not be limited to specialist centres.

Reported clinical infections of noroviruses are usually linked to genogroup II viruses but the predominant genotype or genetic cluster can vary from year to year (Hale *et al.*, 2000). Paradoxically the same strain can cause large outbreaks year after year.

The transmission and epidemiology of sapoviruses are less well understood. Direct person-to-person spread by the faecal-oral route is assumed to be the most common method of transmission. Symptomatic illness in the UK occurs mainly in babies throughout the year (Public Health Laboratory Service Communicable Disease Report, 2002). Outbreaks have been identified mainly in nurseries but occasionally in schools.

## Treatment

There is no specific treatment and the symptoms are rarely sufficiently severe to require rehydration.

## Distribution in the environment

Noroviruses have only recently been identified in aquatic samples. As with enterovirus investigation, sewage, river and marine samples must usually be processed to concentrate the virus. Processing methods that have been used for enterovirus concentration have been effective for noroviruses. Techniques, particularly adaptation of molecular methods used to detect virus in clinical samples, have now been used successfully to identify the norovirus genome in a range of water matrices. It is not possible to grow the virus in animal models or in cell culture so it cannot be proven that the molecular material originated in an infectious particle.

In the UK, noroviruses have been identified in sewage, effluent, river and sea-water samples (Wyn-Jones et al., 2000; Sellwood, unpublished results) (Table 32.1). Both genogroups and a range of strains of virus have been found throughout the year using the primer set that is also used in the UK for many clinical investigations. Lodder et al. (1999) detected both genogroups of virus in sewage in the Netherlands. Griffin et al. (1999) identified noroviruses in 10% of canal water sites in the Florida Keys using one of two primer sets. Interestingly, only a primer set that detected Norwalk strain gave positive results. Waters most influenced by sewage discharge were the most often found to contain virus. Katayama et al. (2002) detected noroviruses in Tokyo Bay coastal seawater during the winter.

Less polluted water has been shown to contain noroviruses; in the UK, urban groundwater samples were positive during the winter months (Powell et al., 2003). Mineral waters were investigated by Beuret et al. (2002) in Switzerland and several brands were positive by their tests in litre size samples; this study remains to be repeated by others. Borchardt et al. (2003) found evidence of transient norovirus contamination in a small number of private household wells in the USA.

Environmental contamination in hospitals, hotels and restaurants has been reported as the transmission route for noroviruses during outbreaks (Green et al., 1998b; Cheeseborough et al., 1997; Marks et al., 2000). Contamination of

**Table 32.1**  Norovirus detected in UK sewage, 2000

| Month | Crude sewage[a] unprocessed | Crude sewage[b] 100 ml processed | Final effluent[c] 1 litre processed |
|---|---|---|---|
| January | D | D | Nd |
| February | Nd | D | Nd |
| March | D | D | D |
| April | D | D | Nd |
| May | D | D | Nd |
| June | D | D | Nd |
| July | D | D | Nd |
| August | D | D | Nd |
| September | D | D | Nd |
| October | Nd | D | Nd |
| November | Nd | Nd | D |
| December | Nd | D | Nd |

[a] 140 μl aliquot was used for all RT-PCR (Wyn-Jones *et al.*, 2000).
[b] 100 ml sewage sample concentrated by beef extract protein precipitation to 10 ml final volume.
[c] 1 l final sewage effluent sample concentrated by beef extract protein precipitation to 10 ml final volume.
D: detected.
Nd: not detected.

carpets was detected in a hotel where a prolonged outbreak of norovirus gastroenteritis occurred (Cheesbrough *et al.*, 2000). Swabs from areas where vomiting had contaminated carpets and also from curtain material at high level were positive, thus supporting the suggestion that such contamination is a significant factor in the transmission of virus. Virus has also been detected from surfaces used in food preparation (Taku *et al.*, 2002).

Bivalve shellfish such as oysters and mussels have been found to contain noroviruses by groups in the UK (Lees *et al.*, 1995; Henshilwood *et al.*, 1998), other parts of Europe (Le Guyader *et al.*, 2000; Hernroth *et al.*, 2002; Formiga-Cruz *et al.*, 2002) and the USA (Le Guyader *et al.*, 1996; Kingsley *et al.*, 2002).

## Waterborne outbreaks

Outbreaks on record with a greater or lesser degree of association with noroviruses include one at a tourist resort hotel, where sewage had seeped through faults in the rock strata into the borehole water used to supply the hotel (Lawson, 1991) and another at a mobile home park, which was the result of sewage gaining access to the well (McNulty, 1993). Sewage also gained access to a borehole supply in Naas, Ireland in 1991, which resulted in a reported 6000 cases of gastroenteritis (Fogarty, 1995). Nearly 50 cases of gastroenteritis occurred in 1992 after water from the River Thames, being used for irrigation purposes, entered the drinking-water distribution system as a result of a backflow from a farm installation (Gutteridge *et al.*, 1994) and a similar incident occurred in Riding Mill, Northumberland, in 1992 (Stanwell-Smith, 1994). In all of these early outbreaks it was epidemiological assessment and clinical symptoms that suggested noroviruses as the infective cause.

Epidemiological evidence combined with the detection of norovirus in stool specimens supported the assertion that contaminated ice produced on a cruise ship caused the outbreak of gastroenteritis reported by Khan *et al.* (1994). An outbreak, in which 45 out of 70 hikers who drank water from a general store on the Appalachian Trail in the USA became ill (Peipins *et al.*, 2002), was confirmed by epidemiology, clinical symptoms and the detection of norovirus in stool specimens and somewhat strengthened by the finding of faecal coliform organisms in the water. Contaminated water that was used in the production of cakes in a bakery is likely to have been the cause of an outbreak of noroviruses among employees and customers in South Wales (Brugha *et al.*, 1999); the strength of association here was the detection and typing of the virus in stool samples and again poor bacterial quality of a drinking water supply.

A well in Finland, polluted by river water during spring floods, transmitted several groups of pathogens to the inhabitants of a village who then developed gastroenteritis (Kukkula *et al.*, 1997). The drinking water of an Italian resort (Boccia *et al.*, 2002) during the summer of 2000 and drinking water in Andorra during the skiing season (Pedalino *et al.*, 2003) were strongly associated with widespread outbreaks of noroviruses and poor quality water. Kukkula *et al.* (1999) reported another outbreak of norovirus gastroenteritis in Heinavesi, Finland where by epidemiological estimate 1700–3000 cases occurred. Here, not only was the virus detected in the stool samples of those affected, but it was also detected in the municipal water supply and in consumers' taps. The water supply to the village had poor or non-existent chlorination. The virus was typed and genogroups I and II found in the stools and genogroup II virus found in the water, so the cause and effect were well reconciled. Maurer and Sturchler (2000) reported an outbreak of gastroenteritis of mixed aetiology in La Neuveville, Switzerland involving over 1600 individuals. *Campylobacter jejuni*, *Shigella sonnei* and noroviruses were detected in stool samples and a norovirus isolate showed identical sequence homology with one isolated from the water supply. Another instance of the value of molecular techniques was demonstrated by Brown *et al.* (2001) in linking the water supply to an outbreak of gastroenteritis involving 448 individuals in a resort hotel in Bermuda where 18 out of 19 stool specimens were positive for genogroup II noroviruses which were also detected in a 3-litre water sample. Epidemiology and clinical features of a norovirus outbreak combined with noroviruses identified in stool samples and noroviruses in water with identical nucleotide sequences in an outbreak was reported by Anderson *et al.* (2003) in snowmobilers at a winter resort.

Many of these outbreaks are associated with small drinking-water systems with limited supply that were in challenging weather conditions: dry and hot or frozen/thawing or flood. In these circumstances the integrity of the infrastructure can be compromised and drinking-water supply becomes contaminated with sewage. If disinfection is not available or is not adequate, transmission may occur.

Freshwater recreation has been more closely linked with waterborne illness than contact with seawater. Canoeists using the River Trent developed gastro-enteritis caused by noroviruses (Gray *et al.*, 1997) and undiagnosed gastro-enteritis was found to be more common in canoeists using the River Trent than those using unpolluted lakes in North Wales (Fewtrell, 1992). Studies in the Netherlands indicate that illness and norovirus infection are linked to swimming in river water (Medema *et al.*, 1995; van Olphen *et al.*, 1991; de Roda Husman personal communication).

Bivalve shellfish, such as oysters and mussels, are a major transmission route of norovirus gastroenteritis and have been reviewed by Lees (2000). The large outbreak in the USA in 1993 included at least 25 clusters of gastroenteritis spread over seven states and infected over 100 people (Kohn *et al.*, 1995). Linking outbreaks in restaurants to batches of shellfish and obtaining stool specimens is difficult in practice so much of the evidence for shellfish transmission is epidemiological.

A common feature of waterborne and shellfish-associated norovirus outbreaks is the finding that mixed genogroups of virus are present in the stool. Many of the outbreaks described above noted this finding. A single genogroup is usually found in faecal specimens from patients infected by the major route of norovirus transmission, i.e. person-to-person transmission as seen in sporadic cases, hospitals, residential and commercial settings. Transmission by aerosolized vomitus and environmental contamination is being increasingly recognized and has been reported in cruise ships, theatre seating, hospital wards and restaurants.

Surface contamination of food by food-handlers is another well-recognized transmission route of norovirus gastroenteritis (Advisory Committee on the Microbiological Safety of Food, 1998). Food which is served uncooked or handled after cooking is a potential hazard. Many types of food have been associated, on epidemiological evidence, with transmission: sandwiches, salad, raspberry gateaux, watercress, frosting mix, carrots, lobster tail, melon cocktail and even hamburger and fries. As yet detection of the virus from the food is difficult and under development. Good hygiene and information for food handlers is crucial for prevention of transmission.

## Risk assessment

*Health effects*: occurrence of illness, degree of morbidity and mortality, probability of illness based on infection:

- Seroprevalence of identified strains is usually high – approaching 100% – by adulthood. Seropositivity does not affect the susceptibility of re-infection, except perhaps in sapoviruses. Noroviruses are estimated to be the number one cause of viral gastroenteritis outbreaks in the USA and the UK.

- Clinical symptoms of norovirus are generally mild and self-limiting: nausea, vomiting, diarrhoea and fever that last for 1–3 days. Mortality is rare and has occurred mainly in people with pre-existing conditions and the elderly. Sapovirus causes gastroenteritis in which diarrhoea is the more predominating feature.
- Attack rates in drinking water outbreaks have ranged from 31 to 87%. Three studies have shown 46%, 68% and 85% of people developing symptoms after becoming infected. The probability of developing illness may relate to genetic susceptibility in the host.

*Exposure assessment*: routes of exposure and transmission, occurrence in source water, environmental fate:

- Transmission is faecal-oral, either person to person or through a common source such as water or food. Aerosolization from vomitus is a source of infection as well.
- Secondary spread can be high in noroviruses but apparently uncommon in sapoviruses.
- An infectious dose of 10 PCR-detectable units has been shown through oral ingestion.
- Waterborne outbreaks of norovirus have been identified. Researchers estimated that 23% of waterborne outbreaks in the USA between 1975 and 1981 were due to Norwalk-like viruses. Freshwater recreation has been more closely linked with waterborne illness than contact with seawater. Food has been identified as a common outbreak source – especially shellfish.
- Noroviruses have only recently been identified in aquatic samples. In the UK and other countries, noroviruses have been identified in sewage, effluent, river and seawater. Noroviruses have been found in groundwater and bottled mineral water samples. Concentrations at intake are unknown, but dependent on level of human faecal contamination in the source water.

*Risk mitigation*: drinking-water treatment, medical treatment:

- Studies on the disinfection of noroviruses are hampered by the lack of an infectivity assay, but norovirus is inactivated by chlorine at $>10$ mg/l, although not by free residual chlorine at 0.5–1.0 mg/l.
- The illness is self-limiting. Rehydration therapy may be necessary in rare cases.

# References

Advisory Committee on the Microbiological Safety of Food. (1998). *Report on foodborne viral infections*. London: HMSO.

Anderson, A.D., Heryford, A.G., Sarisky, J.P. *et al.* (2003). A waterborne outbreak of Norwalk-like virus among snowmobilers – Wyoming, 2001. *J Infect Dis*, **15**: 303–306.

Beuret, C., Kohler, D., Baumgartner, A. *et al.* (2002). Norwalk-like virus sequences in mineral waters: one-year monitoring of three brands. *Appl Environ Microbiol*, **68**: 1925–1931.

Boccia, D., Tozzi, A.E., Cotter, B. *et al.* (2002). Waterborne outbreak of Norwalk-like virus gastro-enteritis at a tourist resort, Italy. *Emerg Infect Dis*, **8**: 563–568.

Borchardt, M.A., Bertz, D., Spencer, S.K. *et al.* (2003). Incidence of enteric viruses in groundwater from household wells in Wisconsin. *Appl Environ Microbiol*, **69**: 1172–1180.

Brown, C.M., Cann, J.W., Simons, G. *et al.* (2001). Outbreak of Norwalk virus in a Caribbean island resort: application of molecular diagnostics to ascertain the vehicle of infection. *Epidemiol Infect*, **126**: 425–432.

Brugha, R., Vipond, I.B., Evans, M.R. *et al.* (1999). A community outbreak of food-borne SRSV gastro-enteritis caused by a contaminated water supply. *Epidemiol Infect*, **122**: 145–154.

Caul, E.O. (1996a). Viral gastroenteritis: small round structured viruses, caliciviruses and astroviruses. Part I. The clinical and diagnostic perspective. *J Clin Pathol*, **49**: 874–880.

Caul, E.O. (1996b). Viral gastroenteritis: small round structured viruses, caliciviruses and astroviruses. Part II. The epidemiological perspective. *J Clin Pathol*, **49**: 959–964.

Caul, E.O. and Appleton, H. (1982). The electron microscopical and physical characteristics of SRSVs: an interim scheme for classification. *J Med Virol*, **9**: 257–265.

Caul, E.O., Ashley, C.R. and Pether, J.V.S. (1979). 'Norwalk'-like particles in epidemic gastro-enteritis in the UK. *Lancet*, **ii**: 1292.

Chadwick, P.R. and McCann, R. (1994). Transmission of SRSV by vomiting during a hospital outbreak of gastro-enteritis. *J Hosp Infect*, **26**: 251–259.

Cheesbrough, J.S., Barkess-Jones, L. and Brown, D.W.G. (1997). Possible prolonged environmental survival of SRSV. *J Hosp Infect*, **35**: 325–326.

Cheesbrough, J.S., Green, J. and Gallimore, C.I. (2000). Widespread environmental contamination with Norwalk-like viruses (Norovirus) detected in a prolonged hotel outbreak of gastro-enteritis. *Epidemiol Infect*, **125**: 93–98.

Clarke, I.N., Lambden, P.R. and Caul, E.O. (1998). Human enteric RNA viruses: Caliciviruses and astroviruses. In *Topley and Wilsons' Microbiology and Microbial Infections*, 9th edn, Collier, L., Ballows, A. and Sussman, M. (eds). London: Edward Arnold, pp. 511–535.

Cubitt, W.D., Paed, P.J. and Saeed, A.A. (1981). A new serotype of calicivirus associated with an outbreak of gastro-enteritis in a residential home for the elderly. *J Clin Pathol*, **34**: 924–926.

Dolin, R., Reichmann, R.C., Roessner, K.D. *et al.* (1982). Detection by immune electron microscopy of the Snow Mountain agent of acute viral gastro-enteritis. *J Infect Dis*, **146**: 184–189.

Evans, M.R., Meldrum, R., Lane, W., Gardner, D. *et al.* (2002). An outbreak of viral gastroenteritis following environmental contamination at a concert hall. *Epidemiol Infect*, Oct, **129**(2): 355–360.

Fewtrell, L., Godfree, A.F., Jones, F., Kay, D. *et al.* (1992). Health effects of white-water canooeing. *Lancet*, **339**: 1587–1589.

Fogarty, J., Thornton, L. and Hayes, C. (1995). Illness in a community associated with an episode of water contamination with sewage. *Epidemiol Infect*, **114**: 289–295.

Formiga-Cruz, M., Tofino-Quesada, G., Bofill-Mas, S. *et al.* (2002). Distribution of human virus contamination in shellfish from different growing areas in Greece, Spain, Sweden and the UK. *Appl Environ Microbiol*, **68**: 5990–5998.

Gray, J.J., Green, J., Gallimore, C. *et al.* (1997). Mixed genotype SRSV infections among a party of canoeists exposed to contaminated recreational water. *J Med Virol*, **52**: 425–429.

Green, J., Hale, A.D. and Brown, D.W.G. (1995). Recent developments in the detection and characterisation of small round structured viruses. *PHLS Microbiol Digest*, **12**: 219–222.

Green, J., Wright, P.A., Gallimore, C.I. *et al.* (1998). The role of environmental contamination with small round structured viruses in a hospital outbreak investigated by reverse-transcriptase polymerase chain reaction assay. *J Hosp Infect*, **39**: 39–45.

Green, J., Vinje, J., Gallimore, C.I. *et al.* (2000a). Capsid diversity among Norwalk-like viruses. *Virus Genes*, **20**: 227–236.

Green, K.Y., Ando, T., Balayan, M.S. *et al.* (2000b). Taxonomy of the caliciviruses. *J Infect Dis*, **181**: S322–S330.

Green, K.Y., Chanock, R.M. and Kapikian, A.Z. (2001). *Human Caliciviruses*. In *Fields Virology*, Knipe, D.M. and Howley, P.M. (eds). Philadelphia, PA: Lippincott Williams & Wilkins, pp. 841–874.

Griffin, D.W., Gibson, C.J., Lipp, E.K. and Riley, K. (1999). Detection of viral pathogens by reverse transcriptase PCR and of microbial indicators by standard methods in the canals of the Florida Keys. *Appl Environ Microbiol*, **65**: 4118–4425. Erratum in: *Appl Environ Microbiol*, 2000 Feb, **66**(2): 876.

Gutteridge, W. and Haworth, E.A. (1994). An outbreak of gastrointestinal illness associated with contamination of the mains supply by river water. *PHLS Commun Dis Rep CDR Rev*, **4**: R50–R51.

Hale, A., Mattick, K., Lewis, D. *et al.* (2000). Distinct epidemiological patterns of Norwalk-like virus infections. *J Med Virol*, **62**: 99–103.

Henshilwood, K., Green, J., Gallimore, C.I. *et al.* (1998). The development of PCR assays for detection of SRSV and other human enteric viruses in molluscan shellfish. *J Shellfish Res*, **17**: 1675–1678.

Hernroth, B.E., Conden-Hansson, A.-C., Rehnstam-Holm, A.-S. *et al.* (2002). Environmental factors influencing human viral pathogens and their potential indicator organisms in the Blue Mussel, *Mytilus edulis*: the first Scandinavian report. *Appl Environ Microbiol*, **68**: 4523–4533.

Ho, M.S., Glass, R.I.M. and Monroe, S.S. (1989). Viral gastro-enteritis aboard a cruise ship. *Lancet*, **2**: 961–965.

Kapikian, A.Z., Wyatt, R.G., Dolin, R. *et al.* (1972). Visualisation by immune electron microscopy of a 27nm particle associated with infectious non-bacterial gastro-enteritis. *J Virol*, **10**: 1075–1081.

Kaplan, J.E., Feldman, R., Campbell, D.S. *et al.* (1982a). The frequency of a Norwalk-like pattern of illness in outbreaks of acute gastro-enteritis. *Am J Public Hlth*, **72**: 1329–1332.

Katayama, H., Shimasaki, A. and Ohgaki, S. (2002). Development of a virus concentration method and its application to detection of enterovirus and Norwalk virus from coastal seawater. *Appl Environ Microbiol*, **68**: 1033–1039.

Keswick, B.H., Satterwhite, T.K., Johnson, P.C. *et al.* (1985). Inactivation of Norwalk virus in drinking water by chlorine. *Appl Environ Microbiol*, **50**: 261–264.

Khan, A.S., Moe, C.L., Glass, R.I. *et al.* (1994). Norwalk virus associated gastro-enteritis traced to ice consumption aboard a cruise ship in Hawaii: comparison and application of molecular method-based assays. *Appl Environ Microbiol*, **32**: 318–322.

Kingsley, D.H., Meade, G.K. and Richards, G.P. (2002). Detection of both HAV and Norwalk-like viruses in imported clams associated with food-borne illness. *Appl Environ Microbiol*, **68**: 3914–3918.

Kohn, M., Farley, T.A., Ando, M. *et al.* (1995). An outbreak of Norwalk virus gastro-enteritis associated with eating raw oysters. *JAMA*, **273**: 466–471.

Kukkula, M., Arstila, P., Klossner, M.-L. *et al.* (1997). Waterborne outbreak of viral gastro-enteritis. *Scand J Infect Dis*, **29**: 415–418.

Kukkula, M., Maunula, L., Silvennoinen, E. *et al.* (1999). Outbreak of viral gastroenteritis due to drinking water contaminated by Norwalk-like viruses. *J Infect Dis*, **180**: 1771–1776.

Lawson, H.W., Braun, M.M., Glass, R.I.M. *et al.* (1991). Waterborne outbreak of Norwalk virus gastro-enteritis at a southwest US resort: role of geological formations in contamination of well water. *Lancet*, **337**: 1200–1204.

Le Guyader, F., Haugarreau, L., Miossec, L. *et al.* (2000). Three year study to assess human enteric viruses in shellfish. *Appl Environ Microbiol*, **66**: 3241–3248.

Le Guyader, F., Neill, F.H., Estes, M.K. *et al.* (1996). Detection and analysis of a SRSV strain in oysters implicated in an outbreak of acute gastro-enteritis. *Appl Environ Microbiol*, **62**: 4268–4272.

Lees, D.N. (2000). Viruses and bivalve shellfish. *Int J Food Microbiol*, **59**: 81–116.

Lees, D.N., Hensilwood, K., Green, J. *et al.* (1995). Detection of small round structured viruses in shellfish by reverse transcription-PCR. *Appl Environ Microbiol*, **61**: 4418–4424.

Lodder, W.J., Vinje, J., van de Heide de Roda Husman, A.M. *et al.* (1999). Molecular detection of Norwalk-like caliciviruses in sewage. *Appl Environ Microbiol*, **65**: 5624–5627.

Maurer, A.M. and Sturchler, D. (2000). A waterborne outbreak of small round structured virus, *Campylobacter* and *Shigella* co-infection in La Neuveville, Switzerland, 1998. *Epidemiol Infect*, **125**: 325–332.

Marks, P.J., Vipond, I.B., Carlisle, D. *et al.* (2000). Evidence for Norwalk-like virus (NLV) in a hotel restaurant. *Epidemiol Infect*, **124**: 481–487.

McAnulty, J.M., Rubin, G.L., Carvan, C.T., Huntley, E.J., Grohmann, G. and Hunter, R. (1993). An outbreak of Norwalk-like gastroenteritis associated with contaminated drinking water at a caravan park. *Aust J Public Hlth*, **17**: 36–41.

Medema, G.J., van Asperen, I.A., Kokman-Houweling, J.M., Nooitgedagt, A. *et al.* (1995). The relationship between health effects in triathletes and microbiological quality of freshwater. *Water Sci Technol*, **31**: 19–26.

Meeroff, J.C., Schreiber, D.S., Trier, J.S. *et al.* (1980). Abnormal gastric motor function in viral gastro-enteritis. *Ann Intern Med*, **92**: 370–373.

Oliver, S.L., Dastjerdi, A.M., Wong, S. *et al.* (2003). Molecular characterisation of bovine enteric caliciviruses: a distinct third genogroup of noroviruses (NLV) unlikely to be of risk to humans. *J Virol*, **77**: 2789–2798.

Pedalino, B., Feely, E., McKeown, B. *et al.* (2003). An outbreak of Norwalk-like viral gastro-enteritis in holidaymakers travelling to Andorra, January–February 2002. *Eur Comm Dis Bull*, **8**: 1–8.

Peipins, L.A., Highfill, K.A., Barrett, E. *et al.* (2002). A Norwalk-like virus outbreak on the Appalachian Trail. *J Environ Hlth*, **64**: 18–23.

Powell, K.L., Taylor, R.G., Cronin, A.A. *et al.* (2003). Microbial contamination of two urban sandstone aquifers in the UK. *Water Res*, **37**: 330–352.

Public Health Laboratory Service. (2002). *Communicable Disease Report* 2002. Annual figures.

Stanwell-Smith, R. (1994). Recent trends in the epidemiology of waterborne disease. In *Water & Public Health*, Golding, A.M.B. (ed.). London: Smith-Gordon, pp. 39–56.

Taku, A., Gulati, B.P., Allwood, P.B. *et al.* (2002). Concentration of caliciviruses from food contact surfaces. *J Food Protect*, **65**: 999–1004.

Van Olphen, M., de Bruin, H.A.M., Havelaar, A.H. and Schijven, J.F. (1991). The virological quality of recreational waters in the Netherlands. *Water Sci Technol*, **24**: 209–212.

Wheeler, J.G., Sethi, D., Cowden, J.M. *et al.* (1999). Study of infectious intestinal disease in England: rates in the community, presenting to general practice, and reported to national surveillance. *Br Med J*, **318**: 1046–1050.

Wyn-Jones, A.P., Pallin, R., Dedoussis, C. *et al.* (2000). The detection of small round-structured viruses in water and environmental materials. *J Virol Meth*, **87**: 99–107.

# 33

# Rotavirus

## Basic microbiology

Rotaviruses are classified on the basis of serogroups, serotypes and most recently, genogroups. Six serogroups, termed groups A–F, are recognized and between each of these groups no cross-reaction occurs using assays based on the inner capsid antigen, VP6. Group A is the most common human rotavirus infection; members of groups B and C are the only other groups known to infect humans. Serotypes are based on neutralization assays for VP7 and to a lesser extent VP4. Genogroups are based on the structure of specific genes; G types (G1–G14) refer to the VP7 protein (gene 9) and P types (P1A[8]-P1B[4]) refer to VP4 protein (gene 4) and are differentiated by RT-PCR. Rotaviruses can therefore be typed and strains circulating in the community can be monitored (Iturriza *et al.*, 2000). The strains change in frequency of circulation over time and occasionally new reassortment strains are introduced to a community.

Rotaviruses, at 75 nm in diameter, are relatively large icosahedral virus particles with a triple layer of protein (Estes, 2001). The virion consists of a core and a shell of intermediate and outer layers from which 60 spikes protrude. The

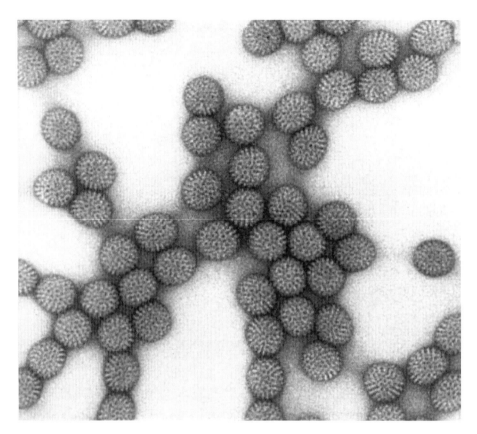

**Figure 33.1**  Transmission electron micrograph of rotavirus. (Courtesy of Dr Erskine Palmer, CDC, USA.)

genome consists of 11 segments of double-stranded RNA (dsRNA) that are capable of genetic reassortment. The complete, infectious particle has a smooth appearance by EM which becomes a rough outline when the outer shell is lost (Figure 33.1). It was the wheel-like appearance of the complete particle that was the basis for the name 'rotavirus' (Flewett *et al.*, 1974). The core comprises nucleic acid and three structural proteins (VP1, VP2 and VP3). The intermediate layer of the rough, non-infectious particle is made up of protein VP6 and the outer shell of VP4 (the spikes) and VP7. One hundred and thirty two channels link the outer capsid with the inner core (Desselberger, 1998).

Virus replication *in vitro* may be enhanced by pre-treatment with a proteolytic enzyme such as trypsin to cleave the outer capsid spike protein VP4 into two products, designated VP5* and VP8* (Beards, 1992). Infectivity is lost when the outer shell is disrupted by the action of a chelating agent such as EDTA. Infectivity and particle integrity are usually resistant to ether and chloroform. Infectivity is stable from pH 3–9 and at 4°C especially when stabilized in $CaCl_2$. Phenol, formalin, chlorine and ethanol inactivate the virus.

Replication takes place in the cytoplasm of infected cells, involves the endoplasmic reticulum and release is by cell lysis after a 12-hour replication cycle.

Binding of the virion to the cell receptor site is by the VP4 spike and may depend on sialic acid presence in the cell membrane. The method by which the virus penetrates the membrane is not yet fully understood (Estes, 2001). Two modes of entry seem to be available *in vitro* depending on whether the virus has been treated with a proteolytic enzyme such as trypsin or not (Suzuki *et al.*, 1985). Trypsin-treated particles may enter directly through the cell membrane within moments of attachment, whereas non-trypsin-treated particles are internalized more slowly by phagocytosis and then released. Accumulations of proteins that assemble into the incomplete particles occur in cytoplasmic inclusions called viroplasms. Virus replication *in vivo* results in the production of large numbers of incomplete, non-infectious particles as well as infectious particles. Rotavirus is fastidious *in vitro* and will grow only in a limited range of cell culture, often only in an abortive replication cycle producing only incomplete particles.

The rotavirus was initially identified by Bishop in biopsy material (Bishop *et al.*, 1973) and then was seen in faeces (Flewett *et al.*, 1973) and was quickly associated with diarrhoea in children.

## Origin of the organism

In common with all viruses rotavirus is an obligate intracellular parasite and, as for most other groups, under normal conditions they are species specific. The origin of human viruses in sewage will be human faecal material. No replication will occur outside the living host cell.

## Clinical features and virulence

Diarrhoea is the predominant symptom of rotavirus infection. This can become a life-threatening illness if severe dehydration develops. It has been estimated that 1 in 8 rotavirus infections in developing countries is severe and that 1 in 160 infected children will die (Kapikian *et al.*, 2001) resulting in an annual death toll of nearly 900 000. Although not as severe an illness in developed countries, in the under 5-year age group admitted to hospital with gastroenteritis, rotavirus is the most common pathogen found. Immunity studies have shown that by age 3 years 90% of children have had at least one rotavirus infection.

## Pathogenicity

Rotavirus replication takes place in the small intestine, in the mature entero-cytes lining the villi leading to shortening and atrophy. Diarrhoea is the main symptom of primary rotavirus infection, although re-infections may be less severe or asymptomatic. The virus is eliminated and cellular recovery begins within 48 hours. The incubation period is also about 48 hours and transmission is by the faecal-oral route. First infections with rotavirus are usually at less than 1 year old resulting in virtually all children over 5 years being immune (Kapikian *et al.*, 2001). Infection confers long-lasting, cross-strain protection, to some degree based on antibody to VP7, although adults can be re-infected though usually without symptoms (Hrdy, 1987). The elderly in residential care may become more susceptible as their immunity wanes and the opportunity for transmission occurs. Symptoms in babies may include vomiting as well as diarrhoea and can be severe with resulting dehydration.

## Transmission and epidemiology

The rotaviruses are transmitted directly person to person by the faecal-oral route. The transfer of virus from surfaces in nurseries to susceptible children may occur as rotavirus have been detected from this environment (Sattar *et al.*, 1986; Wilde and Pickering, 1992). It has been suggested that the respiratory route may be involved but there is little evidence to support the theory. It would, however, help to explain the rapid transmission between young children in all settings. The usual age of the first episode of infection and hence symptoms is between 6 months and 2 years old.

Rotaviruses have been detected world-wide and throughout the year in non-temperate climates. In temperate countries a marked peak of incidence occurs in the cooler, often winter months. No satisfactory explanation of this seasonality has been found.

Very large numbers of particles, many of which are the rough non-infectious virions, are shed into the faeces. For diagnostic purposes this means that EM is a suitable tool for identification. An ELISA has been available for many years and is used widely for diagnosis. It is based on detection of the VP6 antigen of the capsid of the rough particle which is present at more than $10^6$/g in faeces. Rotavirus only grows in certain cell culture types and this is not suitable for diagnostic purposes. Other methods available are latex agglutination, reverse passive haemagglutination assay or immunoblot assay. None of these methods is useful or practical for investigation of environmental samples. Investigation by PCR of the genes for the outer capsid proteins (VP4 and VP7) is used to genogroup strains and hence describe the molecular epidemiology.

## Treatment

No specific antiviral treatment is available so the main aim is rehydration by the replacement of lost fluid and electrolytes. This may be by intravenous administration or by oral fluids. The latter, containing a mixture of sodium, chloride and potassium salts plus glucose, has been used very successfully in many developing countries resulting in a decrease in infant mortality.

## Distribution in the environment

There are only limited numbers of reports on the presence of rotavirus in the environment. The first report was by Steinmann (1981) of virus in sewage in Germany followed by Rao *et al.* (1986), Smith and Gerba (1982) in the USA, Oragui *et al.* (1995) and Mehnert and Stewien (1993) in Brazil. Hejkal *et al.* (1984) reported on the occurrence of rotavirus in sewage in Texas. The presence of rotavirus in seawater and river water was reported in South Africa by Genthe *et al.* (1991), in the UK by Merrett *et al.* (1991), in Canada by Raphael *et al.* (1985) and in Spain by Bosch *et al.* (1998). Abad *et al.* (1998) reported a semi-automated system of rotavirus detection from river water using MA104 cells and flow cytometry. The latter stage counted the cells labelled with rotavirus specific antibody conjugated to a fluorescein dye. Failure to detect rotavirus was reported by Guttman-Bass *et al.* (1987) and Tsai *et al.* (1994).

Dubois *et al.* (1997) and Le Guyader *et al.* (1994) reported the detection of rotavirus using PCR on treated effluent samples and shellfish and sediments from western France. Gajardo *et al.* (1995) detected and genotyped rotavirus using RT-PCR. A single well water sample was positive for the presence of rotavirus in a study based in Wisconsin (Borchardt *et al.*, 2003). Green and Lewis (1999) detected rotavirus from sewage, effluent and oysters in New Zealand using RT-PCR.

Difficulties with concentration and virus detection are the reason for the limited number of reported studies. Unlike most other viruses investigated, rotavirus cannot be concentrated effectively using standard membrane filtration methods. Methods of detection of infectious virus based on IF or immunoperoxidase are extremely tedious and very difficult to undertake for large numbers of samples. The stability of complete particles in the environment is uncertain as the outer shell, which is essential for infection, is readily lost but the resulting incomplete particle is robust and stable, consequently the detection of infectious particles is difficult. Of the rotavirus that reaches water it is not known how much is infectious and therefore a potential hazard. Methods based on RT-PCR will detect genome from infectious and non-infectious particles and have allowed greater numbers of samples to be assayed using a sensitive technique.

As the majority of symptomatic rotavirus infections in developed countries occur in babies in disposable nappies this may affect the quantity of virus reaching sewage. Information on the seasonal variation in sewage or its persistence in the environment is limited.

### Waterborne outbreaks

Rotavirus waterborne outbreaks are uncommon as childhood infection results in life-long immunity. Infections may recur throughout life but they are mainly asymptomatic. In a review for the American Water Works Association Research Foundation, Gerba *et al.* (1996) reported eight outbreaks, some of which were on Pacific islands with known water supply problems.

## Risk assessment

*Health effects*: occurrence of illness, degree of morbidity and mortality, probability of illness based on infection:

- Rotavirus is the most common cause of severe diarrhoea among children. This can become a life-threatening illness if severe dehydration develops. Other symptoms can include vomiting and fever, especially in young children.
- Immunization from infection is incomplete, but re-infections are less severe or asymptomatic.
- Immunity studies have shown that by the age of 3 years, 90% of children have had at least one rotavirus infection and almost all children over 5 years are immune.
- The elderly in residential care may become more susceptible as their immunity wanes and the opportunity for transmission occurs.
- It is estimated that 900 000 children die each year around the world from rotavirus infection – mostly in developing countries; however, in developed countries, it is still the primary cause of hospitalization in children under 5 years.

*Exposure assessment*: routes of exposure and transmission, occurrence in source water, environmental fate:

- Transmission is by the faecal-oral route either directly from person to person or via fomites. There is some evidence of a respiratory route, though that has not been proven.
- Rotavirus has been detected in sewage as well as in seawater and river water as well as shellfish and sediments, but because of difficulties in testing, there are few data on rotavirus occurrence in source waters.
- Rotavirus waterborne outbreaks are uncommon since childhood infection results in lifelong immunity.

*Risk mitigation*: drinking-water treatment, medical treatment:

- Rotaviruses are susceptible to chlorine, ozone and UV light disinfection, though they are more resistant to UV than enteroviruses.
- For persons with healthy immune systems, rotavirus gastroenteritis is a self-limited illness, lasting for only a few days. Rehydration to replace

fluid and electrolytes may be required; no specific antiviral treatment is available.

# References

Abad, F.X., Pinto, R.M. and Bosch, A. (1998). Flow cytometry detection of infectious rotaviruses in environment and clinical samples. *Appl Environ Microbiol*, **64**: 2392–2396.

Beards, G.M. (1982). Polymorphism of genomic RNAs within rotavirus serotypes and subgroups. *Arch Virol*, **74**: 65–70.

Bishop, R.F., Davidson, C.P., Holmes, J.H. *et al.* (1973). Virus particles in epithelial cells of duodenal mucosa from children with acute non-bacterial gastro-enteritis. *Lancet*, **2**: 1281–1283.

Borchardt, M.A., Bertz, P.D., Spencer, S.K. *et al.* (2003). Incidence of enteric viruses in groundwater from household wells in Wisconsin. *Appl Environ Microbiol*, **69**: 1172–1180.

Bosch, A., Pinto, R.M., Blanch, A.R. *et al.* (1998). Detection of human rotavirus in sewage through two concentration procedures. *Water Res*, **22**: 343–348.

Desselberger, U. (1998). Viruses associated with acute diarrhoeal disease. In *Principles and Practice of Clinical Virology*, 4th edn, Zuckerman, A.J., Banatvala, J.E. and Pattison, J.R. (eds). London: J Wiley and Sons.

Dubois, E., Le Guyader, F., Haugarreau, L. *et al.* (1997). Molecular epidemiological survey of rotaviruses in sewage by reverse transcriptase seminested PCR and restriction fragment length polymorphism assay. *Appl Environ Microbiol*, **63**: 1794–1800.

Estes, M. (2001). Rotavirus and their replication. In *Fields Virology*, 4th edn, Knipe, D.M. and Howley, P.M. (eds). Philadelphia, PA: Lippincott Williams and Wilkins, pp. 1747–1785.

Flewett, T.H., Bryden, A.S. and Davies, H. (1974). Virus particles in gastro-enteritis. *Lancet*, **2**: 1497.

Gajardo, R., Bouchriti, N., Pinto, R.M. *et al.* (1995). Genotyping of rotaviruses isolated from sewage. *Appl Environ Microbiol*, **61**: 3460–3462.

Genthe, B., Idema, G.K., Krif, R. *et al.* (1991). Detection of rotavirus in South African waters: a comparison of a cytoimmunolabelling technique with commercially available immunoassays. *Water Sci Technol*, **24**: 241–244.

Gerba, C.P., Rose, J.B., Haas, C.N. *et al.* (1996). Waterborne rotavirus: a risk assessment. *Water Res*, **30**: 2929–2940.

Green, D.H. and Lewis, G.D. (1999). Comparative detection of enteric viruses in wastewaters, sediments and oysters by RT-PCR and cell culture. *Water Res*, **33**: 1195–1200.

Guttman-Bass, N., Tchorsh, Y. and Marva, E. (1987). Comparison of methods for rotavirus detection in water and results of a survey of Jerusalem wastewater. *Appl Environ Microbiol*, **53**: 761–767.

Hejkal, T.W., Smith, E.M. and Gerba, C.P. (1984). Seasonal occurrence of rotavirus in sewage. *Appl Environ Microbiol*, **47**: 588–590.

Hrdy, D.B. (1987). Epidemiology of rotaviral infection in adults. *Rev Infect Dis*, **9**: 461–469.

Iturriza Gomara, M., Green, J., Brown, D.W.G. *et al.* (2000). *Seroepidemiological and molecular surveillance of human rotavirus infections in the UK*. London: Public Health Laboratory Service.

Kapikian, A.Z., Hoshino, Y. and Chanock, R.M. (2001). Rotavirus. In *Fields Virology*, 4th edn, Knipe, D.M. and Howley, P.M. (eds). Philadelphia, PA: Lippincott Williams & Wilkins, pp. 1787–1833.

Le Guyader, F., Dubois, E., Menard, D. *et al.* (1994). Detection of hepatitis A virus, rotavirus and enterovirus in naturally contaminated shellfish and sediment by RT-PCR. *Appl Environ Microbiol*, **60**: 3665–3671.

Mehnert, D.U. and Stewien, K.E. (1993). Detection and distribution of rotavirus in raw sewage and creeks in San Paulo, Brazil. *Appl Environ Microbiol*, **59**: 140–143.

Merrett, H., Stackhouse, C. and Cameron, S. (1991). The incidence of rotavirus in the marine environment: a two-year study. In *Proceedings of the First UK Symposium on Health-Related Water Microbiology, Glasgow, 3–5 Sept.,* Morris, R., Alexander, L., Wyn-Jones, A.P. *et al.* (eds), pp. 148–157.

Oragui, J.I., Arridge, H., Mara, D., Pearson, H.W. *et al.* (1995). Rotavirus removal in experimental waste stabilisation pond systems with different geometries and configurations. *Water Sci Technol*, **31**: 285–290.

Rao, V.C., Metcalf, T.G. and Melnick, J.L. (1986). Development of a method for concentration of rotavirus and its application to recovery of rotavirus from estuarine waters. *Appl Environ Microbiol*, **52**: 484–488.

Raphael, R.A., Sattar, S.A. and Springthorpe, V. (1985). Rotavirus concentration from raw water using positively charged filters. *J Virol Meth*, **11**: 131–140.

Sattar, S.A., Lloyd-Evans, N. and Springthorpe, V.S. (1986). Institutional outbreaks of rotavirus diarrhoea: potential role of fomites and environment surfaces as vehicles for virus transmission. *J Hyg (Camb)*, **96**: 277–289.

Smith, E.M. and Gerba, C.P. (1982). Development of a method for detection of human rotavirus in water and sewage. *Appl Environ Microbiol*, **43**: 1440–1450.

Steinmann, J. (1981). Detection of rotavirus in sewage. *Appl Environ Microbiol*, **41**: 1043–1045.

Suzuki, H., Kitoaka, S., Sato, T. *et al.* (1985). Further investigation on the mode of entry of human rotavirus into cells. *Arch Virol*, **91**: 135–144.

Tsai, Y.L., Tran, B., Sangermano, L.R. *et al.* (1994). Detection of poliovirus, hepatitis A virus, and rotavirus from sewage and ocean water by triplex reverse transcriptase PCR. *Appl Environ Microbiol*, **60**: 2400–2407.

Wilde, J. and Pickering, L. (1992). Detection of rotaviruses in the day care environment by RT-PCR. *J Infect Dis*, **166**: 507–511.

# Part 5

# Helminths

# 34

# Dracunculiasis

Dracunculiasis is the only helminth that has unequivocally been shown to be spread through drinking infected water. The disease is one of those that are targeted by the World Health Organization for global eradication and it had been hoped that by the year 2000, this campaign would have achieved its objective. However, the disease remains a problem in certain parts of the world. In 1986 there were an estimated 3.5 million cases of dracunculiasis world-wide, though by 1995 this had fallen to 129 825 cases. In the years since then the decline in reported cases has been much less marked with 63 717 cases reported in 2001 (Figure 34.1) (Anon, 2002). These cases were reported from 17 countries, though 49 471 (77.6%) were reported from Sudan. Other countries reporting more than 1000 cases were Burkino Faso (1032), Ghana (4739), Nigeria (5355) and Togo (1354).

## Basic parasitology

The agent of dracunculiasis, *Dracunculus medinensis*, is the only nematode worm to have been shown unequivocally to be transmitted through drinking water. The adult female worm measures up to 1 m long and 2 mm in diameter.

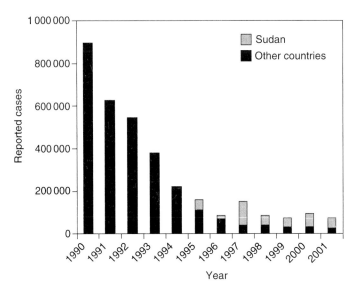

**Figure 34.1** Cases of dracunculiasis reported to the World Health Organization for the years 1990–2001. (Taken from Weekly Epidemiological Record, Anon, 2002.)

The head end of the worm is rounded with a triangular mouth. Most of the body of the worm is taken up with a double uterus. The adult males remain small, about 4 cm in length.

The worm lives in connective tissue where it does no harm until it migrates down (usually to the legs and feet). Here lytic secretions from glands in the head probably combined with the irritant effect of the larvae cause a blister that eventually ruptures to expose the head of the worm. When the head is doused in water the uterus is extruded through the mouth of the worm and larvae are expelled into the water.

## Origin of the organism

Dracunculiasis is a disease that has been known since antiquity and accurate descriptions can be found in texts dating to 1350 BC (Muller, 2001; Cox, 2002). The name dates back to Lineus in 1758. The worm is a nematode or roundworm belonging to the phylum Nemathelminthes, the class Nematoda and the superfamily Dracunculoidea.

## Life cycle

Once expelled from the uterus, the larvae can survive in water for about 6 days. In muddy water or moist earth, survival can be as long as 3 weeks. Larvae are then ingested by a species of microcrustaceans previously known as *Cyclops*, now divided into three genera, *Tropocyclops*, *Mesocyclops* and *Metacyclops*. Once ingested the larvae penetrate the gut wall into the body cavity where they develop further, requiring 14 days at above 21°C. The next stage of the life cycle happens when the water is drunk by the next human host. The infected *Cyclops* are dissolved in the stomach acid and the larvae released. The larvae then penetrate the gut wall and enter the abdominal cavity. About 3–4 months later, the males and females mate in the subcutaneous tissues of the thoracic region. After mating the males die, though the females continue to grow and eventually migrate downward, usually to the feet where the blister forms until the foot is placed in water. The blister then ruptures, the larvae are released and the life cycle starts again. The whole cycle of infection takes about 1 year.

## Clinical features

In the majority of cases the first clinical feature is the appearance of the blister, which grows over a few days to about 3 cm diameter (Cairncross *et al.*, 2002). The site of the blister is usually preceded by burning, intense itching and urticaria. The blister then ruptures and the worm becomes extruded, about 1 cm per day. Left to itself and assuming the site does not become secondarily infected, the worm track will resolve within about 6 weeks. Unfortunately, secondary infection is frequent, affecting more than 50% of cases and this can cause severe pain and disability and rarely death (Adeyeba, 1985). Tetanus, as a result of secondary infection, has also been described. If the worm breaks before complete removal, the remnant can cause severe inflammation and scarring. Dracunculiasis is rarely fatal, though it can be severely disabling. Given that agricultural labourers are among the people most frequently affected, this can have severe consequences for the family due to the reduced harvest.

There are no laboratory diagnostic tests for dracunculiasis and diagnosis is made clinically by a history of residence in an endemic area and examination of the ulcer.

## Treatment

No specific antiparasitic agents are effective against the disease and the traditional method of slow removal of the worm by winding around a small stick is still the method of choice. There is some evidence that niridazole reduces inflammation around the worm and aids removal. Surgical removal of the worm before emergence will reduce associated disability.

Care must be taken to avoid the risk of secondary infection; clean dressings and the prophylactic use of antibiotics are important. Any secondary infection should be treated aggressively with antimicrobial agents. Tetanus immunization is also recommended.

## Epidemiology

The epidemiology follows what can be predicted from a knowledge of the life cycle of this helminth. Cairncross *et al.* (2002) review the published studies on the epidemiology of dracunculiasis. They point out that there is a strong relationship in a number of studies that villages where cases occur take their drinking water from ponds or shallow step wells. Villages taking their water from deep wells or from running water tend not to be affected.

There are a number of different interventions that have been shown to reduce the risk of infection: provision of a safe water supply, filtering drinking water, active case ascertainment and adequate treatment, preventing cases from having contact with water supplies and killing cyclops in ponds (Cairncross *et al.*, 2002). Filtration is particularly easy as the cyclops are quite large and many simple fabrics are effective.

## Risk assessment

*Health effects*: occurrence of illness, degree of morbidity and mortality, probability of illness based on infection:

- There were 63 717 cases of *Dracunculus* reported in 2001, mostly in Africa.
- The worm lives harmlessly in connective tissue until it migrates down to the legs and feet. Secretions from glands in the worm's head probably combined with the irritant effect of the larvae cause a blister that grows to about 3 cm diameter and that eventually ruptures to expose the head of the worm. The blister ruptures and the worm extrudes about 1 cm per day. Left alone, the worm track will resolve within about 6 weeks. Secondary infection is frequent and affects more than 50% of cases. This can cause severe pain and disability and, rarely, death.
- If the worm breaks before complete removal, the remnant can cause severe inflammation and scarring.

*Exposure assessment*: routes of exposure and transmission, occurrence in source water, environmental fate:

- *Dracunculus* is the only helminth that has been shown to be spread through drinking infected water.
- Larvae can survive in water for about 6 days. In muddy water or moist earth, survival can be as long as 3 weeks.

*Risk mitigation*: drinking-water treatment, medical treatment:

- Interventions to reduce the risk of infection include the provision of a safe water supply, filtering drinking water and preventing infected people from having contact with water supplies. Filtration is particularly easy as the organism is large and simple fabrics are effective.
- No specific antiparasitic agents are effective against the disease. The traditional method of slow removal of the worm by winding around a small stick is still used. Any secondary infection should be treated aggressively with antimicrobial agents.

# References

Adeyeba, O.A. (1985). Secondary infection in dracunculiasis: bacteria and morbidity. *Int J Zoonoses*, **12**: 147–149.

Anon. (2002). Dracunculiasis eradication. Global surveillance summary, 2001. *Wkly Epidemiol Rec*, **77**: 143–152.

Cairncross, S., Muller, R. and Zagaria, N. (2002). Dracunculiasis (Guinea Worm Disease) and the eradication initiative. *Clin Microbiol Rev*, **15**: 223–246.

Cox, F.E.G. (2002). History of human parasitology. *Clin Microbiol Rev*, **15**: 595–612.

Muller, R. (2001). Dracunculiasis. In *Principles and Practice of Clinical Parasitology*, Gillespie, S.H. and Pearson, R.D. (eds). Wiley: Chichester, pp. 553–559.

Part 6

# Future

# 35

# Emerging waterborne infectious diseases

It could be argued that one of the more important philosophical developments in our understanding of the epidemiology of infectious diseases was the introduction of the concept of the emerging infectious disease. Some of the first authors to use the term were Morse and Schluederberg (1990). It was only some 5 years later in 1995 that the new journal *Emerging Infectious Diseases* was first published. This journal is now one of the most highly cited infectious disease journals.

Emerging infections can be defined as those infections that have newly appeared in the population, or have existed but are rapidly increasing in incidence or geographic range (Morse, 1995). Sometimes emerging infections are distinguished from re-emerging infections. The latter being those diseases that were once common in a community but then declined in incidence only to increase again a number of years later.

Many of the diseases that are covered in this book can be defined as emerging infections. Cryptosporidiosis, enterohaemorrhagic *E. coli*, noroviruses, as well as many other pathogens not known even 30 years ago. Others such as cholera and typhoid are better described as re-emerging.

In this chapter we shall consider the factors that are responsible for the emergence and re-emergence of infections, especially as they apply to waterborne

**Table 35.1** Factors responsible for the emergence or re-emergence of water-borne pathogens

| Factor | Examples |
| --- | --- |
| Microbial evolution | *E. coli* O157:H7, *Vibrio cholerae* O139 |
| Improved diagnostic technology | *Cryptosporidium*, hepatitis E virus |
| New technology | Legionnaires' disease and air conditioning systems |
| Ecological change | Schistosomiasis following building of dams |
| Demographic change | Increased pressure on water supplies |
| International travel and trade | Movement of *Vibrio cholerae* |
| Breakdown in public health systems | Re-emergence of cholera and typhoid after the collapse of the Soviet Union |

pathogens. Morse (1995) described a range of factors (Table 35.1). These factors will be discussed in turn.


## Microbial evolution

Some infections are emerging because they are due to new pathogens that simply did not exist a few decades ago. The two waterborne pathogens that have recently evolved are *Escherichia coli* O157:H7 and *Vibrio cholerae* O139 (Rubin *et al.*, 1998). *Escherichia coli* O157:H7 was first recognized as a human pathogen in 1982. *E. coli* O157:H7 appears to have developed by a series of steps whereby the organism evolved from a related strain by the acquisition of genes for the expression of additional virulence factors (Whittam *et al.*, 1998). *Vibrio cholerae* O139 is also thought to have evolved in a similar way (Rubin *et al.*, 1998).

Clearly for any newly evolved microbial pathogen to be thought of as emerging, the evolution must give the organism some advantage, either to transmit itself in the environment or to increase its capacity to cause disease. In the two examples quoted above, the likely explanation is the acquisition of additional virulence factors that gave organisms that were already present in the environment the enhanced ability to cause disease. In the case of *E. coli* O157:H7 the increased virulence probably also increased the ability to spread in the environment by allowing increased excretion into the environment from infected cattle and other mammals.


## Improved diagnostic technology

It is probably obvious when thinking about it that a disease will become emergent only after a diagnostic test has been developed or introduced. Perhaps the two most obvious emerging pathogens in this regard are *Cryptosporidium* and hepatitis E virus.

Cryptosporidiosis is a useful example in that the technology was not particularly complex. Indeed the veterinary pathologists had known about *Cryptosporidium* since 1907 when it was described in mice. It was not until 1976 that the pathogen was first described in humans (Meisel *et al.*, 1976; Nime *et al.*, 1976). Since then of course, microbiologists increasingly realized that *Cryptosporidium* is a cause of diarrhoea and more and more laboratories started to look for the organism in stool samples with the result that reports increased substantially over the following years. The technology was around to diagnose *Cryptosporidium* for decades but nobody thought to look. Hepatitis E, on the other hand, is a disease that had to await the development of molecular biological tools before it could become emergent. Although non-A, non-B hepatitis was known about for many years, hepatitis E could not be readily distinguished from other non-A, non-B hepatitis infections as no serological test was available. It was not until it was possible to synthesize synthetic peptides and recombinant antigens from the viral genome for use in ELISA or Western blot technologies that it was possible to diagnose the condition. Once commercial diagnostic tests became available, reports of infections increased and the disease became emergent.

## New technology

The classic example of an emergent disease developing because of new technology is Legionnaires' disease. The first outbreak to be diagnosed was in Philadelphia in 1976 (Fraser *et al.*, 1977). Although it was a little time before the causative agent could be identified and a diagnostic test became available, it soon thereafter became clear that a major risk factor for this infection was cooling systems that used water to cool the air. We now know, of course, that *Legionella* bacteria are widely disseminated in the water environment. However, the special conditions in cooling towers allow the multiplication of these organisms to the great numbers necessary to cause disease. Although not all cases of Legionnaires' disease are acquired from cooling towers, it is interesting to speculate whether the effort would have been put into identifying this important pathogen without the large outbreaks associated with these towers.

## Ecological change

Ecological change can affect infectious disease in several different ways. Several authors have discussed the potential impact of climate change on waterborne disease (Rose *et al.*, 2001; Hunter, 2003). Perhaps one of the most important effects may be due to changes in rainfall. For example, extreme rainfall events have been shown to be a significant risk factor for outbreaks of waterborne disease (Curriero *et al.*, 2001). Some ecological changes are directly related to human intervention. An example of this is the local emergence of schistosomiasis following the construction of dams (N'Goran *et al.*, 1997; Sow *et al.*, 2002). The appearance of a large body of still water provides opportunities for vectors to breed and the parasite to increase in the local human population.

## Demographic change

Demographic change will also have a number of effects on the emergence and re-emergence of infectious disease. The world's population is increasing and this is putting increased stress on available water resources, even in several industrialized nations such as the USA (Hunter, 1997). As water becomes scarcer lower quality water sources will be used and this can provide opportunities for increased pathogen transmission. The rapid increase in the acquired immune deficiency syndrome (AIDS) has been one of the main drivers affecting the emergence of new diseases. Much of our early interest in cryptosporidiosis has been because of its severe impact in people living with AIDS (Hunter and Nichols, 2002).

## International travel and trade

Clearly people who travel abroad, especially to tropical countries, are at risk of disease that they would not expect to experience in the home country. Many of these travel-associated illnesses are potentially waterborne (Payment and Hunter, 2001). Another route for international dissemination of waterborne pathogens has been carriage in ships' ballast waters (McCarthy and Khambaty, 1994).

## Breakdown in public health systems

Finally, one of the lessons that was associated with the collapse of the Soviet Union has been the risk to public health that comes with severe economic collapse. Such economic collapse can in turn threaten public health systems. This impacts on waterborne disease when such economic collapse impairs the effectiveness of water treatment and distribution systems. When water systems decay diseases once controlled, such as cholera and typhoid, can re-emerge (Semenza *et al.*, 1998). It is very sobering to consider what may be the impact of such economic collapse in some of the world's largest urban centres.

# Conclusions

As has been discussed, there are many factors that can lead to the emergence or re-emergence of new pathogens. Although some of these factors are outside the control of humanity, many are directly the result of human activity or human pressure on the environment. Emerging infectious diseases pose a particular problem to the water utility. It is often impossible to be certain of the eventual impact of the disease and the contribution to disease burden of drinking-water transmission. How many people realized the full importance of cryptosporidiosis in the years after the pathogen was first described in humans? On the other hand, we know many of the human factors that impact on risk of emergence and we can at least try to manage these factors effectively.

# References

Curriero, F.C., Patz, J.A., Rose, J.B. *et al.* (2001). The association between extreme precipitation and waterborne disease outbreaks in the United States, 1948–1994. *Am J Public Hlth*, **91**: 1194–1199.

Fraser, D.W., Tsai, T.R., Orenstein, W. *et al.* (1977). Legionnaires disease: description of an epidemic of pneumonia. *New Engl J Med*, **297**: 1189–1197.

Hunter, P.R. (1997). *Waterborne Disease: Epidemiology and Ecology*. Chichester: Wiley.

Hunter, P.R. (2003). Climate change and waterborne and vector-borne disease. *J Appl Microbiol* (in press).

Hunter, P.R. and Nichols, G. (2002). The epidemiology and clinical features of *cryptosporidium* infection in immune-compromised patients. *Clin Microbiol Rev*, **15**: 145–154.

McCarthy, S.A. and Khambaty, F.M. (1994). International dissemination of epidemic *Vibrio cholerae* by cargo ship ballast and other non-potable waters. *Appl Environ Microbiol*, **60**: 2597–2601.

Meisel, J.L., Perera, D.R., Meligro, C. *et al.* (1976). Overwhelming watery diarrhoea associated with a *cryptosporidium* in an immunosuppressed patient. *Gastroenterology*, **70**: 1156–1160.

Morse, S.S. (1995). Factors in the emergence of infectious diseases. *Emerg Infect Dis*, **1**: 7–15.

Morse, S.S. and Schluederberg, A. (1990). Emerging viruses: the evolution of viruses and viral diseases. *J Infect Dis*, **162**: 1–7.

N'Goran, E.K., Diabate, S., Utzinger, J. *et al.* (1997). Changes in human schistosomiasis levels after the construction of two large hydroelectric dams in central Cote d'Ivoire. *Bull World Health Org*, **75**: 541–545.

Nime, F.A., Burek, J.D., Page, D.L. *et al.* (1976). Acute enterocolitis in a human being infected with the protozoan *Cryptosporidium*. *Gastroenterology*, **70**: 592–598.

Payment, P.R. and Hunter, P.R. (2001). Endemic and epidemic infectious intestinal disease and its relation to drinking water. In *Water Quality: Guidelines, Standards and Health. Risk Assessment and Management for Water-related Infectious Disease*, Fewtrell, L. and Bartram, J. (eds). London: IWA Publishing, pp. 61–88.

Rose, J.B., Epstein, P.R., Lipp, E.K. *et al.* (2001). Climate variability and change in the United States: Potential impacts on water- and foodborne diseases caused by microbiologic agents. *Environ Hlth Perspec*, **109**(Suppl. 2): 211–221.

Rubin, E.J., Waldor, M.K. and Mekalanos, J.J. (1998). Mobile genetic elements and the evolution of new epidemic strains of *Vibrio cholerae*. In *Emerging Infections*, Krause, R.M. (ed.). New York: Academic Press, pp. 147–461.

Semenza, J.C., Roberts, L., Henderson, A. *et al.* (1998). Water distribution system and diarrheal disease transmission: A case study in Uzbekistan. *Am J Trop Med Hyg*, **59**: 941–946.

Sow, S., de Vlas, S.J., Engels, D. *et al.* (2002). Water-related disease patterns before and after the construction of the Diama dam in northern Senegal. *Ann Trop Med Parasitol*, **96**: 575–586.

Whittam, T.S., McGraw, E.A. and Reid, S.D. (1998). Pathogenic *Escherichia coli* O157:H7: a model for emerging infectious diseases. In *Emerging Infections*, Krause, R.M. (ed.). New York: Academic Press, pp. 163–183.

# Index